Algebraic Geometry in East Asia — Hanoi 2005

ADVANCED STUDIES
IN PURE MATHEMATICS 50

Chief Editor: Toshitake Kohno (The University of Tokyo)

Algebraic Geometry in East Asia — Hanoi 2005

Edited by

Kazuhiro Konno (Osaka University)
Viet Nguyen-Khac (Vietnam Information Security Lab.)

Mathematical Society of Japan

This book was typeset by $\mathcal{A}_{\mathcal{M}}\mathcal{S}$-TEX and $\mathcal{A}_{\mathcal{M}}\mathcal{S}$-LaTEX, the TEX macro systems of the American Mathematical Society, together with the style files `aspm.sty` *and* `aspmfm.sty` *for $\mathcal{A}_{\mathcal{M}}\mathcal{S}$-TEX written by Dr. Chiaki Tsukamoto and* `aspmproc.sty` *for $\mathcal{A}_{\mathcal{M}}\mathcal{S}$-LaTEX written by Dr. Akihiro Munemasa and* `aspm.cls` *for $\mathcal{A}_{\mathcal{M}}\mathcal{S}$-LaTEX provided by Tokyo Shoseki Printing Co., Ltd.*

TEX is a trademark of the American Mathematical Society.

Edited by the Mathematical Society of Japan.

Published by the Mathematical Society of Japan.

Distrubuted by the Mathematical Society of Japan, the American Mathematical Society and the World Scientific Publishing Co. Ltd.
Distributed exclusively in North America by the American Mathematical Society.

ISBN 978-4-931469-45-7

2000 Mathematics Subject Classification.
Primary 14–06.
Secondary 14H50, 20F36, 14B0, 14G32, 53C15, 32M10, 14J45, 14J27, 14J28, 14J17, 14E07, 14H30, 52C35, 06C05, 05C22, 14B05, 14C20, 11G30, 11G10, 14H31.

Advanced Studies in Pure Mathematics 50

PRINTED IN JAPAN
by Tokyo Shoseki Printing Co., Ltd.

Preface

The present volume is the Proceedings of the International Conference "Algebraic Geometry in East Asia, II" which was held at the Conference Hall (Hanoi, Vietnam), during October 10th – October 14th, 2005. There were about 50 participants who came from six countries: Vietnam, P. R. China, Korea, Japan, Russia and France, and 24 invited lectures were presented during the period of the conference. It was supported by Hanoi Institute of Mathematics, Grants-in Aid for Scientific Research by JSPS, ICTP, ARC-VNC, Spect Project.

The first conference under the same name was held in International Institute for Advanced Studies (Kyoto, Japan) in 2001 and this is the second meeting. One of our primary goals was to facilitate a chance to meet and exchange ideas among east Asian mathematical communities of algebraic geometry. Thanks to the active cooperations of all participants, we are quite confident that our aim was successfully achieved. We hope that this conference will continue to have a positive impact on the future relations among algebraic geometers in east Asia.

As the reader can see from the list of articles appearing in this volume, quite a number of different topics were touched upon. Some of those are: Zariski pairs, rational homogeneous manifolds, Kummer surfaces, arc spaces, Cremona groups, algebraic curves and dual varieties.

The book consists of two parts. The first part is devoted to the survey articles each of which provides the summary on the current status of research and gives a scope to the future, pointing out problems which deserve to be studied. The second part consists of the research articles showing various aspects of researches, methods and techniques, as well as the activity of our leading algebraic geometers.

During the conference, staffs and students in Hanoi Institute of Mathematics helped us. On behalf of the organizers, we wish to express our hearty thanks to all invited speakers for their interesting talks, and to all participants for their stimulating discussions and the warm response to the conference.

Osaka, October 2007
Kazuhiro Konno

All papers in this volume have been refereed and are in final form. No version of any of them will be submitted for publication elsewhere.

International Symposium "Algebraic Geometry in East Asia. II"

Place: Conference Hall, 15 B Ho Xuan Huong, Hanoi (Vietnam)
Date: October 10th – October 14th, 2005

Speakers

T. Ashikaga (Tohoku Gakuin Univ.), J.-P. Brasselet (Luminy), L. S. Choe (KIAS), Do Duc Thai (Hanoi Univ. of Education), T. Harui (Osaka Univ.), J.-M. Hwang (KIAS), S. Ishii (Tokyo Inst. of Tech.), V.A. Iskovskikh (Steklov Math. Inst.), G.-F. Jiang (Beijing Univ. of Chem. Tech.), J.-H. Keum (KIAS), Y.-R. Kim (Konkuk Univ.), S. Kitagawa (Osaka Univ.), K. Konno (Osaka Univ.), Y.-N. Lee (Sogang Univ.), K.-Z. Li (Chinese Acad. of Sci.), M. Murakami (Osaka City Univ.), A. Ohbuchi (Tokushima Univ.), K. Saito (RIMS, Kyoto Univ.), I. Shimada (Hokkaido Univ.), T. Shioda (Rikkyo Univ.), H. Tokunaga (Tokyo Metropolitan Univ.), Lê Dung Tráng (ICTP), S. Usui (Osaka Univ.), Nguyen-Khac Viet (Hanoi Inst. of Math.)

Scientific Committee

Dao Trong Thi (Hanoi National Univ.)
Ha Huy Khoai (Hanoi Institute of Math.)
Kazuhiro Konno (Osaka Univ.)
Nguyen-Khac Viet (Hanoi Institute of Math.)
Mutsuo Oka (Tokyo Univ. of Sci.)
Tetsuji Shioda (Rikkyo Univ.)
Sampei Usui (Osaka Univ.)

Organizing Committee

Dao Trong Thi (Hanoi National Univ.)
Ha Huy Khoai (Hanoi Institute of Math.)
Nguyen-Khac Viet (Hanoi Institute of Math.)
Vu The Khoi (Hanoi Institute of Math.)
Do Duc Thai (Hanoi Univ. of Education)

by Professor Tetsuji Shioda

CONTENTS

I. Survey Articles

II. Research Articles

Advanced Studies in Pure Mathematics 50, 2008
Algebraic Geometry in East Asia — Hanoi 2005
pp. 1–100

A survey on Zariski pairs

Enrique Artal Bartolo, José Ignacio Cogolludo
and Hiro-o Tokunaga

Dedicated to Professor Kenji Ueno on his sixtieth birthday.

§ Introduction

In 1929, O. Zariski published a paper entitled *"On the Problem of Existence of Algebraic Functions of Two Variables Possessing a Given Branch Curve"* [130] where the following question was considered:

Does an algebraic function z of x and y exist, possessing a preassigned curve f as branch curve?

As Zariski pointed out in the *Introduction* of [130], this question was first considered by Enriques and the problem is reduced to finding the fundamental group of the complement of the given curve (the word *complement* is understood and often omitted for short). Zariski considered some explicit cases and *proved* important results. Here we detail some of the most relevant:

(Z1) If two curves lie in a connected family of equisingular curves, then they have isomorphic fundamental groups.

(Z2) If a continuous family $\{\mathcal{C}_t\}_{t\in[0,1]}$ is equisingular for $t \in (0,1]$ and $\mathcal{C}_0$ is reduced, then there is a natural epimorphism $\pi_1(\mathbb{P}^2 \setminus \mathcal{C}_0, p_0) \twoheadrightarrow \pi_1(\mathbb{P}^2 \setminus \mathcal{C}_t, p_t)$, where the base point p_t ($t \in [0,1]$) depends on t continuously.

(Z3) The fundamental group of an irreducible curve of order n, possessing ordinary double points only, is cyclic of order n ([130, Theorem 7]), see Remark 1.

Received May 23, 2006.
Revised May 14, 2007.
2000 *Mathematics Subject Classification.* 14H50, 20F36, 14B0, 14G32.
Key words and phrases. Zariski pair, fundamental group, Alexander invariants, Galois cover.
The two first authors are partially supported by MTM2007-67908-C02-01.

(Z4) Consider the projection from the general cubic surface in $\mathbb{P}^3$ onto $\mathbb{P}^2$, centered at a general point outside the surface. Its branch locus is a sextic $\mathcal{C}_6$ with six cusps whose fundamental group is isomorphic to $\mathbb{Z}/2\mathbb{Z} * \mathbb{Z}/3\mathbb{Z}$.

(Z5) He noted that the six cusps of any sextic described in (Z4) satisfy the extra condition of lying on a conic – without decreasing the dimension of their family. Moreover, if $\mathcal{C}_6$ is a sextic with six cusps and its fundamental group has a representation onto the symmetric group of three letters, then $\mathcal{C}_6$ is the branch curve of a cubic surface and its six cusps lie on a conic. In particular if a sextic $\mathcal{C}'_6$ with six cusps not on a conic exists, then $\pi_1(\mathbb{P}^2 \setminus \mathcal{C}_6, p_o) \not\cong \pi_1(\mathbb{P}^2 \setminus \mathcal{C}'_6, p'_o)$.

(Z6) If Q is a three-cuspidal quartic, then $\pi_1(\mathbb{P}^2 \setminus Q, p_o)$ is isomorphic to the binary 3-dihedral group, which is a non-abelian finite group of order 12 presented by

$$\langle a, b \mid aba = bab, \quad a^2b^2 = 1\rangle. \tag{1}$$

Remark 1. Zariski's proof of (Z3) depended on the following claim of Severi: *The family of irreducible curves of degree n possessing a given number of ordinary double points is irreducible* [104]. Severi's proof was not correct, and the first rigorous proof of his claim was given in 1986 by J. Harris [54]. In the meantime (Z3) was known as the *Zariski conjecture* until the 70's, when it was proved by Deligne and Fulton in [40] and [48].

Zariski proved the commutativity of the fundamental group for certain smooth curves and then he used (Z1) for general smooth curves. He was also able to prove the commutativity for nodal arrangements of lines. He found degenerations of nodal curves into nodal arrangements of lines, thus proving the commutativity of the fundamental group for certain nodal curves. A combination of (Z1) with Severi's claim allowed him to complete a proof of (Z3).

In [131], Zariski proved another result regarding (Z5): in modern language, the Alexander polynomial of $\mathcal{C}_6$ equals $t^2 - t + 1$ and the Alexander polynomial of $\mathcal{C}'_6$ is 1 (provided $\mathcal{C}'_6$ exists); the key point for both claims is the position of the cusps. The story (almost) ends in [132] where Zariski shows the existence of curves $\mathcal{C}'_6$ using deformation arguments that allow him to prove that their fundamental group is abelian; explicit examples were found much later [2, 90]. He also claims that there are only two irreducible families of sextics with six cusps. It is not hard to prove that the family of sextics with six cusps on a conic is irreducible, and an analogue for the other family is announced by Degtyarëv in [39, Theorem 5.3.2].

Another important result of [130] is (Z6); for a long time the three-cuspidal quartic (the only quartic with a non-abelian fundamental group) was the only example of a curve whose complement has a non-abelian finite fundamental group. In the early nineties, several such examples have been found by A. Degtyarëv [38], M. Oka [91] and I. Shimada [106].

Fundamental groups of curves up to degree five are well known (see [36, 38]), but for now little is known about their structure in the general case. In this sense, questions like the one raised by Zariski ([134, Chapter VIII, §1]) on the residual finiteness of such groups are still open.

These results and open questions motivated many mathematicians to study the topology of the complements of plane curves. One of the most surprising phenomena in this field was the one found by Zariski and stated in (Z5), where two irreducible curves with the same singularities have non-isomorphic fundamental groups. This leads us to the definition of *Zariski pairs* which are, roughly speaking, two curves that have the same local topology but do not have the same embedded topology. Let us give a more precise definition of a Zariski pair.

Definition 2 ([2]). A pair $(\mathcal{C}_1, \mathcal{C}_2)$ of reduced plane curves in $\mathbb{P}^2$ is called a *Zariski pair* if it satisfies the following conditions:

(1) There exist tubular neighborhoods $T(\mathcal{C}_i)$ $(i = 1, 2)$ and a homeomorphism $h : T(\mathcal{C}_1) \to T(\mathcal{C}_2)$ such that $h(\mathcal{C}_1) = \mathcal{C}_2$.

(2) There exists no homeomorphism $f : \mathbb{P}^2 \to \mathbb{P}^2$ with $f(\mathcal{C}_1) = \mathcal{C}_2$.

Analogously $(\mathcal{C}_1, \dots, \mathcal{C}_k)$ is a *Zariski k-plet* if $(\mathcal{C}_i, \mathcal{C}_j)$ is a Zariski pair for any $i \neq j$.

Remark 3. The first condition in Definition 2 is replaced by the one about the combinatorial data on $\mathcal{C}_i$ $(i = 1, 2)$. More precisely, the *combinatorial type* of a curve $\mathcal{C}$ is given by a 7-tuple

$$(\mathrm{Irr}(\mathcal{C}), \deg, \mathrm{Sing}(\mathcal{C}), \Sigma_{\mathrm{top}}(\mathcal{C}), \sigma_{\mathrm{top}}, \{\mathcal{C}(P)\}_{P\in\mathrm{Sing}(\mathcal{C})}, \{\beta_P\}_{P\in\mathrm{Sing}(\mathcal{C})}),$$

where:

- $\mathrm{Irr}(\mathcal{C})$ is the set of irreducible components of $\mathcal{C}$ and $\deg : \mathrm{Irr}(\mathcal{C}) \to \mathbb{Z}$ assigns to each irreducible component its degree.
- $\mathrm{Sing}(\mathcal{C})$ is the set of singular points of $\mathcal{C}$, $\Sigma_{\mathrm{top}}(\mathcal{C})$ is the set of topological types of $\mathrm{Sing}(\mathcal{C})$, and $\sigma_{\mathrm{top}} : \mathrm{Sing}(\mathcal{C}) \to \Sigma_{\mathrm{top}}(\mathcal{C})$ assigns to each singular point its topological type.
- $\mathcal{C}(P)$ is the set of local branches of $\mathcal{C}$ at $P \in \mathrm{Sing}(\mathcal{C})$, (a local branch can be seen as an arrow in the dual graph of the minimal resolution of $\mathcal{C}$ at P, see [42, Chapter II.8] for details) and $\beta_P : \mathcal{C}(P) \to \mathrm{Irr}(\mathcal{C})$ assigns to each local branch the global irreducible component containing it.

We say that two curves $\mathcal{C}_1$ and $\mathcal{C}_2$ have the *same combinatorial data* (or simply the *same combinatorics*) if their combinatorial data are equivalent, that is, if $\Sigma_{\text{top}}(\mathcal{C}_1) = \Sigma_{\text{top}}(\mathcal{C}_2)$, and there exist bijections $\varphi_{\text{Sing}} : \text{Sing}(\mathcal{C}_1) \to \text{Sing}(\mathcal{C}_2)$, $\varphi_P : \mathcal{C}_1(P) \to \mathcal{C}_2(\varphi_{\text{Sing}}(P))$ (restriction of a bijection of dual graphs) for each $P \in \text{Sing}(\mathcal{C}_1)$, and $\varphi_{\text{Irr}} : \text{Irr}(\mathcal{C}_1) \to \text{Irr}(\mathcal{C}_2)$ such that $\deg_2 \circ \varphi_{\text{Irr}} = \deg_1$, $\sigma_{\text{top}_2} \circ \varphi_{\text{Sing}} = \sigma_{\text{top}_1}$, and $\beta_{2,\varphi_{\text{Sing}}(P)} \circ \varphi_P = \varphi_{\text{Irr}} \circ \beta_{1,P}$.

In the irreducible case, two curves have the same local topology if they have the same degree and the same topological types for local singularities. On the other extreme, for line arrangements, combinatorial type is just the set of incidence relations.

The fact that two curves have the same combinatorial data if and only if they satisfy Definition 2(1) is a consequence of Waldhausen graph manifold theory [125, 126]. The dual graph of the minimal resolution of the singularities of $\mathcal{C}$ is a plumbing graph. Waldhausen theory was developed in terms of plumbing graphs by Neumann [89]. His main result states that minimal normalized graphs are determined by the manifold. It is not hard to see that the topological type of $(\mathbb{P}^2, \mathcal{C})$ determines the combinatorial data. Since the graph coming from the minimal resolution may be not minimal, it is possible to find curves $\mathcal{C}_1, \mathcal{C}_2$ whose complements $\mathbb{P}^2 \setminus \mathcal{C}_i$, $i = 1, 2$ are homeomorphic but such that they do not have the same combinatorics using, for example, Cremona transformations. Jiang and Yau [60] proved that the homeomorphism type of the complement of a line arrangement determines its combinatorics. The connection between homeomorphism type of complements to curves and combinatorics was studied by Di Pasquale in [100], but not much is known about it.

Also, curves with the same combinatorics form a quasi-projective variety in a certain projective space $\mathbb{P}_d$ of dimension $\frac{d(d+3)}{2}$, where d is the total degree of the curves. We will refer to such a variety as the *combinatorial stratum of curves*. A *rigid isotopy* between curves is a smooth path in a combinatorial stratum.

A connected family of equisingular curves is contained in the connected component of the combinatorial stratum of curves determined by any curve of that family. Therefore, the topology of the pair $(\mathbb{P}^2, \mathcal{C})$ is an invariant of such a component. In particular $\pi_1(\mathbb{P}^2 \setminus \mathcal{C}, p_o)$ is also an invariant and hence (Z5) provides the first example of a Zariski pair. In the early nineties, some new examples were found ([2, 37]). Since then, the variety of such examples has been very broad and subtle.

In what follows we will give an insight on the different nature of each of these phenomena, the techniques used, and some open questions on the general study. The study of Zariski pairs may consist of two parts:

(I) To locate curves $\mathcal{C}_1$, $\mathcal{C}_2$ in different irreducible components of a combinatorial stratum, that is, two non rigidly-isotopic curves.
(II) To find an effective invariant $\mathcal{P}$ of the topology of the embedding, so that, if $\mathcal{P}(\mathcal{C}_1) \neq \mathcal{P}(\mathcal{C}_2)$ then $(\mathbb{P}^2, \mathcal{C}_1) \not\approx (\mathbb{P}^2, \mathcal{C}_2)$, that is, there is no homeomorphism $f : \mathbb{P}^2 \to \mathbb{P}^2$ with $f(\mathcal{C}_1) = \mathcal{C}_2$.

As for (I), different strategies have been used, for instance:

(I1) Certain geometrical properties such as position of singularities (e.g. sextics whose six cusps belong to a conic), or in more generality the existence of non-zero global sections for an ideal sheaf $\mathcal{I}$ on $\mathbb{P}^2$ twisted by a certain $\mathcal{O}(d)$. The expected dimension of such vector space of global sections is 0 (like the six cusps on a sextic, which in principle are not supposed to belong to a conic). This method is often used in combination with birational transformations in order to lower the degrees of the curves. In that case, the geometrical properties have to be *translated* into new properties on the transformation ([12, 91]). Also geometrical properties of flex points play an important role in finding irreducible components of the combinatorial strata ([95, 96]).
(I2) Arithmetical properties of the components. Such is the typical case with line arrangements, when built up by pasting together *smaller* arrangements whose combinatorial strata are disconnected ([102, 11, 10]). Also, in the special case of sextics, many of these arithmetical properties come from the existence of a double covering ramified along the curve. Classification of $K3$-surfaces with a given Picard group and some computer work finish the task in the case of simple singularities ([127]). Birational transformations of the covering and appropriate blowdowns can be used to produce equations ([6]).
(I3) Direct computation of strata. This can be used either in the negative sense (proving the irreducibility of a stratum) which tells us where there are no Zariski pairs ([61, 109]), or actually finding equations of the strata. This method often requires big Milnor numbers, so that the dimension of the strata is small and the problem becomes computationally feasible ([9, 98]).

As for (II), several methods have been developed:

(II1) Zariski-van Kampen Theorem. This is the classical method to find the fundamental group of the complement to a given curve from its braid monodromy ([130, 59, 28]). This technique will be treated in detail in §1. Though a very rich invariant, the fundamental group of a curve contains the topological information of the embedding in an intricate manner. The undecidability of the isomorphism problem in group theory justifies the need to construct new invariants that are effectively computable, simple to compare, and fine enough to keep the essential data.

(II2) Alexander invariants such as the Alexander module, Alexander polynomials, and characteristic varieties. This type of invariant shows the connection between algebraic curves and knot theory, since many of these invariants have been adopted from knot theory. Conversely, much of the original interest and techniques of knot theory had the study of algebraic curves at their root. This technique will be developed in §2.

(II3) Braid monodromy equivalence, also referred to as braid monodromy factorization. It is a much stronger invariant of the topology of the embedding than the monodromy group. Braid monodromy factorization was only recently proved to be an invariant of the (not-necessarily-rigid) isotopy class ([70] for cuspidal curves and [25] for any plane curve). This technique has been proved to be specially useful to study conjugated curves [9, 10], branch curves of surfaces of general type and the Chisini conjecture ([85, 70, 69]), as well as symplectic isotopies of curves and the realization problem ([99, 62, 64]). Definition and more details will be given in §1.

(II4) Branched Galois coverings. Based on geometric versions of the inverse Galois problem for certain elementary non-abelian groups. This technique will be treated in detail in §3.

(II5) Nikulin theory of integral lattices. Recently developed by Degtyarëv [39] and Shimada [107, 108]. Let $\mathcal{C}$ be a sextic curve with at worst simple singularities (see [20] for simple singularities). Let X be the double covering of $\mathbb{P}^2$ ramified along $\mathcal{C}$, and $\bar{X}$ its minimal resolution. Degtyarëv obtains a quadruple $Q := (L, h, \sigma, \omega)$ where $L = H_2(\bar{X})$, which is isomorphic to the integral lattice of the singularities of $\mathcal{C}$, $h \subset L$ is the pull-back of the hyperplane section class $[\mathbb{P}^1] \in H_2(\mathbb{P}^2)$, $\sigma \subset L$ is the set of classes of exceptional divisors appearing in the resolution $\bar{X} \to X$, and $\omega \subset L \otimes \mathbb{R}$ denotes the oriented 2-subspace spanned by the real and imaginary parts of the class of a holomorphic 2-form on $\bar{X}$. He proves that $Q_{\mathcal{C}_1} \cong Q_{\mathcal{C}_2}$ if and only if $\mathcal{C}_1$ and $\mathcal{C}_2$ are rigidly isotopy and the pairs $(\mathbb{P}^2, \mathcal{C}_1)$, $(\mathbb{P}^2, \mathcal{C}_2)$ are regularly diffeomorphic, that is, there is a diffeomorphism between $(\mathbb{P}^2, \mathcal{C}_1)$ and $(\mathbb{P}^2, \mathcal{C}_2)$ that can be extended to a homeomorphism between the $K3$-surfaces $\bar{X}_1$ and $\bar{X}_2$. Shimada proves that the isomorphism class of L is an invariant of the Γ-equivalence of the pair $(\mathbb{P}^2, \mathcal{C})$. This implies that L is an invariant of the homeomorphism class of $(\mathbb{P}^2, \mathcal{C})$ (for a discussion on Γ-equivalence see below).

In light of the previous list of strategies, one can also describe more precisely different examples of Zariski pairs according to which invariants are equal and which are different for each curve.

- **Alexander polynomial.** It is associated with a group and a homomorphism onto $\mathbb{Z}$ and with cyclic coverings ramified along each component with the same ramification index. A Zariski pair that can be distinguished using this invariant is a *classical Zariski pair*, otherwise it will be called an *Alexander-equivalent Zariski pair.* Examples of classical Zariski pairs are abundant in the literature ([131, 37, 2, 95] among many others). Alexander-equivalent Zariski pairs can be found in ([5, 92, 93]).
- **Characteristic varieties and Oka polynomials**. The first ones (introduced by A. Libgober in [75] for curves) are associated with a group and its abelianization morphism whereas the second ones (introduced by Oka in [94]) are associated with a group and any homomorphism onto $\mathbb{Z}$. They are both associated with Alexander modules and abelian coverings ramified (or not) along each component with any ramification index and basically provide the same information. The existence of certain irreducible components of characteristic varieties can be described in terms of algebraic conditions of the singular locus. Analogously we have *Libgober-Oka-equivalent Zariski pairs.*
- **Non-abelian coverings and twisted Alexander polynomials**. The first one is given by the existence or not of certain non-abelian coverings ramified along components. Algebraic conditions can be given for the existence of such coverings. Twisted Alexander polynomials are associated with a group and a representation. Zariski pairs whose algebraic fundamental groups are isomorphic are called *algebraically-equivalent Zariski pairs.* The main source of examples of algebraically-equivalent Zariski pairs is found among conjugated curves ([9, 10, 107, 108]) in a number field; we will call them *arithmetic Zariski pairs.* There are still open questions whether or not some pairs of conjugate, non rigidly-isotopic curves are Zariski pairs ([44]). Also, a famous example of a Zariski pair of line arrangements was produced by G. Rybnikov [102]. A proof was published in [11] using arguments of homological rigidity

(see Definition 2.30), and a final argument of Alexander modules. Such Zariski pair is distinguished by the fundamental group of the complement, and it is an Alexander-equivalent, but it is not known whether it is a Libgober-Oka-equivalent or algebraically-equivalent Zariski pair.

- **Fundamental group of the complement** $\pi_1(\mathbb{P}^2 \setminus \mathcal{C}, p_o)$. A presentation of it can be obtained via the Zariski-van Kampen Theorem from the action of generic (and sometimes even non-generic) braid monodromy groups of the curve $\mathcal{C}$. Sometimes groups can be compared directly ([45, 92, 93]), but oftentimes this is too hard of a task. Zariski pairs whose fundamental groups are isomorphic will be called *π_1-equivalent Zariski pairs.* Note that fundamental groups of algebraic-equivalent Zariski pairs have the same profinite completion. Finitely presented groups of infinite order with the same profinite completion are hard to distinguish. In fact, it is not known whether or not any of the algebraic-equivalent Zariski pairs is also a π_1-equivalent Zariski pair. Examples of π_1-equivalent Zariski pairs can be obtained from the list of arithmetic Zariski pairs given by Shimada [108]. For example, sextics with singularities $\mathbb{A}_{18} + \mathbb{A}_1$, $\mathbb{A}_{16} + \mathbb{A}_3$, and $\mathbb{A}_{16} + \mathbb{A}_2 + \mathbb{A}_1$ have abelian fundamental groups isomorphic to $\mathbb{Z}/6\mathbb{Z}$ (see [6, Remark 5.9]).
- **The homotopy type of the complement** $\mathbb{P}^2 \setminus \mathcal{C}$. It can be described as the homotopy type of the CW-complex associated with a presentation of $\pi_1(\mathbb{P}^2 \setminus \mathcal{C}, p_o)$ obtained from a very particular braid monodromy of $\mathcal{C}$ that will be referred to as the *Puiseux-braid monodromy of* $\mathcal{C}$ (see [73]). Analogously we have *homotopy type-equivalent Zariski pairs.* Known examples of *homotopy type-equivalent Zariski pairs* are related to Cremona transformations and conjugated curves. In fact, it would be very interesting to see whether or not any of the non-regular-diffeomorphic examples mentioned above are homotopy type-equivalent.
- **The topology of the complement** $\mathbb{P}^2 \setminus \mathcal{C}$. Any example of this sort will be called *complement-equivalent Zariski pairs.* The main information lost between the embedding and the complement is the *peripheral information*, that is the information on the location of meridians of the irreducible components. A complement-equivalent Zariski pair can be obtained from the problem proposed by Eyral-Oka in [44]. Also, in Example 1.39 we show in detail a complement-equivalent Zariski pair.

- **Γ-equivalence**. This equivalence relation, introduced by Shimada [107], has to do with the *peripheral information* mentioned above. Two curves $\mathcal{C}_1$ and $\mathcal{C}_2$ with the same combinatorics and homeomorphic complements are called Γ-equivalent, if the homeomorphism induces an isomorphism of fundamental groups preserving meridians. Examples of this sort will be called *Γ-equivalent Zariski pairs.*
- **The topology of** $(\mathbb{P}^2, \mathcal{C})$. Since it is determined by any generic braid monodromy factorization of $\mathcal{C}$ (as mentioned above), the ultimate tool to check for a Zariski pair is its generic braid monodromy. These techniques have been used in ([9, 10]).

The main purpose of this article is to review these different methods and to explain how they are used in the study of Zariski pairs.

Contents

§1. Fundamental group and braid monodromy

As indicated in the Introduction, the main goal of Zariski's foundational paper [130] is to study the fundamental group of the complement of a projective plane curve. A method for its computation is outlined in [130]. In [59], E.R. van Kampen gave a more rigorous presentation of this method which is now known as the *Zariski-van Kampen method.* Roughly speaking, Zariski showed that such a group is generated by meridians of a generic line and then described some relations by *moving* this generic line around non-generic lines in a pencil; van Kampen stated and proved his well-known theorem on fundamental groups (now known as the Seifert-van Kampen Theorem) and used it to prove that Zariski's relations define a system of relations for the fundamental group. D. Chéniot gave a modern approach to this method in [28]. In [29], O. Chisini realized that this method contains a finer invariant of the curve if one interprets the *motions* of the generic line in terms of a representation of a free group in a braid group. Much later, B. Moishezon [84] called this invariant *braid monodromy* and used it to study projective complex surfaces via ramification curves of projections.

1.1. Preliminaries

Before explaining the Zariski-van Kampen method and the braid monodromy invariant, let us introduce some settings and notations.

Let G be a group. Let $a, b \in G$. For simplicity, we introduce the following notations:

$$\begin{aligned} [a,b] &= a^{-1}b^{-1}ab \\ a^b &= b^{-1}ab \\ b * a &= bab^{-1}. \end{aligned}$$

We denote the free group with n generators $x_1, \ldots, x_n$ by $\mathbb{F}_n$ and the braid group on n strings by $\mathbb{B}_n$ given by the following presentation:

$$\mathbb{B}_n := \left\langle \sigma_1, \ldots, \sigma_{n-1} \;\middle|\; \underset{1<i+1<j<n}{[\sigma_i, \sigma_j] = 1}, \; \underset{1 \leq i \leq n-2}{\sigma_i\sigma_{i+1}\sigma_i = \sigma_{i+1}\sigma_i\sigma_{i+1}} \right\rangle \tag{2}$$

$\mathbb{B}_n$ naturally acts on $\mathbb{F}_n$ on the right as follows:

$$\Phi(x_i, \sigma_j) = x_i^{\sigma_j} := \begin{cases} x_{i+1} & \text{if } i = j \\ x_i * x_{i-1} & \text{if } i = j+1 \\ x_i & \text{if } i \neq j+1. \end{cases} \tag{3}$$

for $i = 1, \ldots, n-1$, and $j = 1, \ldots, n$.

We call the above right action of $\mathbb{B}_n$ on $\mathbb{F}_n$ the *Hurwitz action.* In particular, the actions of the generators σ_i $(i = 1, \ldots, n-1)$ on $x_1, \ldots, x_n$ are called *Hurwitz moves.* Clearly Φ induces an antihomomorphism $\mathbb{B}_n \to \mathrm{Aut}(\mathbb{F}_n)$, $\sigma \in \mathbb{B}_n \mapsto \Phi(\bullet, \sigma)$. This homomorphism is injective and its image, identified with $\mathbb{B}_n$, is characterized by the following result:

Proposition 1.1 ([22, Theorem 1.9.]). *Let $\tau \in \mathrm{Aut}(\mathbb{F}_n)$. Then $\tau \in \mathbb{B}_n$ if and only if*

$$\tau(x_n \cdot \ldots \cdot x_1) = x_n \cdot \ldots \cdot x_1$$

and there exists a permutation σ of n-letters such that

$$\tau(x_i) = y_i x_{\sigma(i)} y_i^{-1}, \quad y_i \in \mathbb{F}_n, \quad i = 1, \ldots, n.$$

We will now present a geometric interpretation of $\mathbb{B}_n$ and its action $\mathbb{F}_n$. For details in the following, we refer to [7] and [22].

Let us fix a subset, $\mathbf{Y} = \{t_1, \ldots, t_n\}$, of $\mathbb{C}$ consisting of n distinct elements. Let $\Delta \subset \mathbb{C}$ be a sufficiently big closed disk, *i.e.*, $\{z \in \mathbb{C} \mid |z| \leq R\}$ $R \gg 0$, such that $\mathbf{Y}$ is contained in its interior. Choose a point $\star$ on $\partial\Delta = \{z \in \Delta \mid |z| = R\}$.

Definition 1.2. We define some special elements of the fundamental group $\pi_1(\mathbb{C} \setminus \mathbf{Y}, \star)$, called *meridians.* Meridians are obtained as follows:

- Take a small disk S centered at $t \in \mathbf{Y}$ containing no other elements of $\mathbf{Y}$ and choose a point $\hat{\star} \in \partial S$.
- Consider a path α in $\mathbb{C} \setminus \mathbf{Y}$ joining $\hat{\star}$ and $\star$, and denote by $\eta_{\hat{\star},S}$ the closed path based at $\hat{\star}$ that runs counterclockwise along ∂S.
- The homotopy class of the loop $\alpha^{-1} \cdot \eta_{\hat{\star},S} \cdot \alpha$ is called a meridian of t in $\pi_1(\mathbb{C} \setminus \mathbf{Y}, \star)$. If the base point is understood, then we will simply speak of a meridian of t in $\mathbb{C} \setminus \mathbf{Y}$.
- It is easily checked that the set of meridians of $t \in \mathbf{Y}$ coincides with a conjugacy class in $\pi_1(\mathbb{C} \setminus \mathbf{Y}, \star)$ completely determined by t.
- It is also well known that suitable collections of n meridians in $\mathbb{C} \setminus \mathbf{Y}$ (one for each point of $\mathbf{Y}$) define bases of $\pi_1(\mathbb{C} \setminus \mathbf{Y}, \star)$.
- This construction of meridians also applies to the fundamental group of the complement of a divisor in a surface, see Figure 1.

Definition 1.3. Let Δ and $\star$ be as above. A *geometric basis* of $\pi_1(\mathbb{C} \setminus \mathbf{Y}, \star)$ is an ordered basis $(\gamma_1, \ldots, \gamma_n)$ of $\pi_1(\mathbb{C} \setminus \mathbf{Y}, \star)$ consisting of meridians such that $\gamma_n \cdot \ldots \cdot \gamma_1$ is homotopic to the loop $\gamma_\star$, the closed path based at $\hat{\star}$ that runs counterclockwise along $\partial\Delta$.

Fig. 1. A meridian

Definition 1.4. A *pseudo-geometric basis* of $\pi_1(\mathbb{C} \setminus \mathbf{Y}, \star)$ is an ordered basis $(\gamma_1, \ldots, \gamma_n)$ of meridians such that $\gamma_n \cdot \ldots \cdot \gamma_1$ is homotopic to the inverse of a meridian of $\{\infty\}$ in $\mathbb{C} \setminus \Delta$. The product $\gamma_n \cdot \ldots \cdot \gamma_1$ is called the *pseudo-Coxeter* element of the basis.

Note that a geometric basis $\Gamma := \{\gamma_1, \ldots, \gamma_n\}$ is a free basis of $\pi_1(\mathbb{C} \setminus \mathbf{Y}, \star) \cong \mathbb{F}_n$. Given any element $\sigma \in \mathbb{B}_n$, the Hurwitz action of σ on Γ produces another geometric basis. By Proposition 1.1, one has the following:

Proposition 1.5 (Artin). *The Hurwitz action of the group $\mathbb{B}_n$ on the set of all geometric bases of $\pi_1(\mathbb{C} \setminus \mathbf{Y}, \star)$ is free and transitive.*

Definition 1.6. A $\mathbf{Y}$-*special homeomorphism* is an orientation-preserving homeomorphism $f : \mathbb{C} \to \mathbb{C}$ such that

(i) $\mathbf{Y}$ is fixed as a set, not necessarily pointwise, and
(ii) f is the identity on $\overline{\mathbb{C} \setminus \Delta}$.

A $\mathbf{Y}$-*special isotopy* is an isotopy $H : \mathbb{C} \times [0,1] \to \mathbb{C}$ such that $H(\bullet, t)$ is a $\mathbf{Y}$-special homeomorphism for all $t \in [0,1]$. We denote the set of classes of $\mathbf{Y}$-special homeomorphisms up to $\mathbf{Y}$-special isotopy by $\mathbf{B}_{\mathbf{Y}}$.

Let f_1 and f_2 be $\mathbf{Y}$-special homeomorphisms. We denote their classes in $\mathbf{B}_{\mathbf{Y}}$ by $[f_i]$ $(i = 1, 2)$, respectively. The product $[f_1][f_2] := [f_1 \circ f_2]$ endows $\mathbf{B}_{\mathbf{Y}}$ with a group structure that acts on $\pi_1(\mathbb{C} \setminus \mathbf{Y}, \star)$ on the left.

Let $M_n :=$ the set of ordered n-distinct points in $\mathbb{C}^n$. The symmetric group of n letters acts on M_n freely via permutation of coordinates. Let $S_n(M)$ be the quotient space of M_n with respect to this action. One can regard $S_n(M)$ as the set of unordered n-distinct points; and $\mathbf{Y} \in S_n(M)$. Consider $B_{\mathbf{Y}} := \pi_1(S_n(M), \mathbf{Y})$, and note that any $\gamma \in B_{\mathbf{Y}}$ is the homotopy class, relative to $\{0, 1\}$, of a set of n-paths $\{\gamma_1(t), \ldots, \gamma_n(t)\}$, $\gamma_i : [0,1] \to \mathbb{C}$ such that $\mathbf{Y} = \{\gamma_1(0), \ldots, \gamma_n(0)\} = \{\gamma_1(1), \ldots, \gamma_n(1)\}$, $\gamma_1(t), \ldots, \gamma_n(t)$ are all distinct for $t \in [0,1]$. An element of $B_{\mathbf{Y}}$ is called a *braid based at* $\mathbf{Y}$. It is known that $B_{\mathbf{Y}} \cong \mathbb{B}_n$ (see [22] for details). For technical purposes, it is oftentimes convenient to consider *open braids starting at* $\mathbf{Y}_1$ *and ending at* $\mathbf{Y}_2$, where $\mathbf{Y}_1, \mathbf{Y}_2 \in S_n(M)$ as the homotopy class, relative to $\{0,1\}$, as before where $\mathbf{Y}_1 = \{\gamma_1(0), \ldots, \gamma_n(0)\}$, and $\mathbf{Y}_2 = \{\gamma_1(1), \ldots, \gamma_n(1)\}$. The obvious composition operations endow the class of open braids with a groupoid structure (see [7] for details).

Our purpose next is to describe natural identifications between $\mathbf{B}_{\mathbf{Y}}$ and $B_{\mathbf{Y}}$.

Lemma 1.7. $\mathbf{B}_{\mathbf{Y}}$ *and* $B_{\mathbf{Y}}$ *are naturally anti-isomorphic.*

Proof. Note that any orientation-preserving homeomorphism of $\mathbb{C}$ is isotopic to the identity (for a proof, see [53, Lemma 5.6.]). Also, let $[h] \in \mathbf{B}_{\mathbf{Y}}$ be an arbitrary element of $\mathbf{B}_{\mathbf{Y}}$ and let $H : \mathbb{C} \times [0,1] \to \mathbb{C}$ be an isotopy such that $H(\bullet, 0) = h$ and $H(\bullet, 1) = id_{\mathbb{C}}$. Then, $H(t_1, t), \ldots, H(t_n, t)$, $t \in [0,1]$, are paths such that:

(*i*) $\{H(t_1, 0), \ldots, H(t_n, 0)\} = \{H(t_1, 1), \ldots, H(t_n, 1)\} = \mathbf{Y}$, and

(*ii*) $\{H(t_1, t), \ldots, H(t_n, t)\}$ ($t \in [0,1]$) is a set of n distinct points for all $t \in [0,1]$, i.e., $\gamma_h := \{H(t_1, t), \ldots, H(t_n, t)\}$ gives an element of $B_{\mathbf{Y}}$.

This correspondence $[h] \mapsto \gamma_h$ gives the announced anti-isomorphism (for more details, see [53, Theorem 5.4.]). Q.E.D.

Lemma 1.7 assures the existence of a right action of $\mathbf{B}_{\mathbf{Y}}$ on $\pi_1(\mathbb{C} \setminus \mathbf{Y}, \star)$, which coincides with the Hurwitz action for a proper choice of the base point and the geometric basis (see for instance [7, Example 1.8]). In particular, any homomorphism from a group G to $\mathbb{B}_n$ induces a right action of G on $\pi_1(\mathbb{C} \setminus \mathbf{Y}, \star)$.

1.2. Zariski-van Kampen method and braid monodromy

Let $\mathcal{C} \subset \mathbb{P}^2$ be a reduced projective plane curve of degree d. Choose a line $L \pitchfork \mathcal{C}$ and a point $P \in L \setminus \mathcal{C}$, and consider the projection centered at P. Let us take homogeneous coordinates $[X : Y : Z]$ in $\mathbb{P}^2$ such that $L : Z = 0$ and $P = [0 : 1 : 0]$. In what follows we will focus on $\mathbb{C}^2 := \mathbb{P}^2 \setminus L$ with affine coordinates $x := \frac{X}{Z}$, $y := \frac{Y}{Z}$, and whose projection may be written as $\Pi : \mathbb{C}^2 \to \mathbb{C}$, $\Pi(x,y) := x$. The affine curve $\mathcal{C}^{\text{aff}} := \mathcal{C} \cap \mathbb{C}^2 = \mathcal{C} \setminus L$ can be defined by a reduced monic polynomial $f(x,y)$ in the variable y such that $\deg_y f = \deg f = d$.

Let $\mathcal{D}_f := \{x \in \mathbb{C} \mid \operatorname{disc}_y f(x) = 0\}$ and let $\mathbf{L}$ be the union of the lines $L_t : x = t$, $t \in \mathcal{D}_f$. The main point of the Zariski-van Kampen method is the following:

Lemma 1.8. *The restriction* $\pi := \Pi_{|\mathbb{C}^2\setminus(\mathcal{C}^{\mathrm{aff}}\cup\mathbf{L})} : \mathbb{C}^2 \setminus (\mathcal{C}^{\mathrm{aff}} \cup \mathbf{L}) \to \mathbb{C}\setminus\mathcal{D}_f$ *is a locally trivial fibration whose fiber is isomorphic to* $\mathbb{C}$ *with* d *punctures.*

This follows from Ehresmann's Fibration Theorem (see [41, p.15]), since the vertical lines having less than d distinct intersection points with $\mathcal{C}^{\mathrm{aff}}$ have been removed.

Since π is a locally trivial fibration whose fiber is diffeomorphic to $\mathbb{C}$ with d punctures, the polynomial f induces an algebraic mapping $\tilde{f} : \mathbb{C}\setminus\mathcal{D}_f \to S_n(M)$ given by $x_0 \mapsto \{y \in \mathbb{C} \mid f(x_0, y) = 0\} \in S_n(M)$. In order to define braid monodromy we need to consider $* \in \mathbb{C}\setminus\mathcal{D}_f$ a regular value on the boundary of a disk Δ_f containing $\mathcal{D}_f$ in its interior and denote by $\mathbf{Y}^*$ the set of roots of the polynomial $f(*, y) = 0$.

Definition 1.9.

(i) The homomorphism $\nabla_* : \pi_1(\mathbb{C}\setminus\mathcal{D}_f, *) \to B_{\mathbf{Y}} := \pi_1(S_n(M), \mathbf{Y}^*)$ induced by $\tilde{f}$ is called a *braid monodromy.*

(ii) Fix an isomorphism $\iota_{\mathbf{Y}^*} : B_{\mathbf{Y}^*} \cong \mathbb{B}_d$ (Note that the isomorphism depends on the choice of $\mathbf{Y}^*$, i.e. $*$) and define $\nabla_{*,\mathbf{Y}^*} := \iota_{\mathbf{Y}^*} \circ \nabla_*$. Given any geometric basis $\gamma_1, \ldots, \gamma_r$ of $\pi_1(\mathbb{C}\setminus\mathcal{D}_f, *)$, where $r := \#(\mathcal{D}_f)$, the r-tuple $(\nabla_{*,\mathbf{Y}^*}(\gamma_1), \ldots, \nabla_{*,\mathbf{Y}^*}(\gamma_r)) \in (\mathbb{B}_d)^r$ is called a *factorization of the braid monodromy of* $(\mathcal{C}, L, P)$ or simply a *braid monodromy factorization of* $(\mathcal{C}, L, P)$.

(iii) The image by $\nabla_{*,\mathbf{Y}^*}$ of a pseudo-Coxeter element (Definition 1.4) will be referred to as a *pseudo-Coxeter braid.*

Note that a braid monodromy factorization depends on the choice of the geometric basis of $\pi_1(\mathbb{C}\setminus\mathcal{D}_f, *)$ and $\iota_{\mathbf{Y}}$. By Proposition 1.5, any change of geometric basis is given by a Hurwitz move. We will expound upon this in section §1.3.

Now we are in the position to state the results obtained by Zariski-van Kampen in [130, 59] with the purpose of describing a presentation of the fundamental group of an affine curve. In order to do so consider Δ_f as above and Δ_y a closed disc such that $\mathcal{C}^{\mathrm{aff}} \cap \Pi^{-1}(\Delta_f) \subset \Delta_f \times \Delta_y$, and $\mathcal{C}^{\mathrm{aff}} \cap (\partial\Delta_f) \times (\partial\Delta_y) = \emptyset$. Let $\Gamma := \{\gamma_1, \ldots, \gamma_r\}$ be a geometric basis of $\pi_1(\mathbb{C}\setminus\mathcal{D}_f, *)$ and choose a base point $(*, \hat{*})$ ($\hat{*} \in \partial\Delta_y$). Since $\pi^{-1}(\Delta_f) \cap (\Delta_f \times \Delta_y) \to \Delta_f$ has a section, there exists a lifting α_t for each γ_t. Note that $\pi_1(\mathbb{C}\setminus\mathcal{D}_f, *) \cong \pi_1(\Delta_f \setminus \mathcal{D}_f, *)$ and α_t is a meridian of L_t. Under these conditions, and using the long exact sequence of homotopy, one has the following:

Proposition 1.10. *Let* $(\rho_1, \ldots, \rho_r) \in (\mathbb{B}_d)^r$ *be a braid monodromy factorization of* $(\mathcal{C}, L, P)$ *and let* $\{g_1, \ldots, g_d\}$ *be a geometric basis of*

$\pi_1(\mathbb{C}\setminus \mathbf{Y}^*, \hat{*})$. *Then* $\pi_1(\mathbb{C}^2\setminus(\mathcal{C}^{\text{aff}}\cup \mathbf{L}), (*, \hat{*}))$ *has the following presentation:*

$$(4)\quad \left\langle g_1,\dots,g_d,\alpha_1,\dots,\alpha_r \mid g_j^{\rho_i}=g_j^{\alpha_i},\quad i=1,\dots,r,\quad j=1,\dots,d\right\rangle.$$

As a corollary to Proposition 1.10, one obtains the celebrated Zariski-van Kampen Theorem:

Corollary 1.11 (Zariski-van Kampen Theorem). *Under the previous hypotheses,* $\pi_1(\mathbb{C}^2\setminus\mathcal{C}^{\text{aff}})$ *has the following presentation:*

$$(5)\qquad \left\langle g_1,\dots,g_d \mid g_j^{\rho_i}=g_j,\quad i=1,\dots,r,\quad j=1,\dots,d\right\rangle.$$

A presentation of $\pi_1(\mathbb{P}^2\setminus\mathcal{C})$ *is given by*

$$(6)\quad \left\langle g_1,\dots,g_d \mid g_d\cdot\ldots\cdot g_1=1,\ g_j^{\rho_i}=g_j,\ i=1,\dots,r,\ j=1,\dots,d\right\rangle.$$

The main tool of the proof of Corollary 1.11 is in [47, Lemma 4.18].

Lemma 1.12 ([47]). *Let* $A\subset X$ *be a divisor in a smooth quasi-projective variety* X*, and let* $B\subset X$ *be an irreducible divisor not contained in* A*. Then, the inclusion induces an epimorphism* $\pi_1(X\setminus(A\cup B))\twoheadrightarrow\pi_1(X\setminus A)$ *whose kernel is generated by the meridians of* B*.*

Proof of Corollary 1.11. It is an iterated application of Lemma 1.12. For (5), we use that the loops $\alpha_1,\dots,\alpha_r$ in (4) are meridians of the lines L_t, $t\in\Delta$. For (6), note that $(g_d\cdot\ldots\cdot g_1)^{-1}$ is a meridian of the projection point P in $\overline{L}_\star=L_\star\cup\{P\}\subset\mathbb{P}^2$, i.e. a meridian of the line at infinity L in $\mathbb{P}^2$. Q.E.D.

Since we are mostly interested in the isomorphism class of the fundamental group, and since the spaces studied are connected, unless necessary for technical reasons, the base point of the fundamental group of a curve complement will be omitted.

Remark 1.13. We can decrease the number of relations in (4), (5) and (6).

(a) Since the Hurwitz action fixes the product $g_d\cdot\ldots\cdot g_1$, it is enough to consider $j=1,\dots,d-1$.

(b) Since L is transversal to $\mathcal{C}$ it is well known that $\rho_r\cdot\ldots\cdot\rho_1=\Delta_d^2$, that is the full twist, where $\Delta_d^2=(\sigma_{d-1}\cdot\ldots\cdot\sigma_1)^d$, which is a generator of the center of $\mathbb{B}_d$. Thus its Hurwitz action coincides with the conjugation by $g_d\cdot\ldots\cdot g_1$. Therefore, in (5) it is enough to consider $i=1,\dots,r-1$, $j=1,\dots,d-1$, and add relations to make $g_d\cdot\ldots\cdot g_1$ into a central element. Analogously, in (6) it is enough to consider $i=1,\dots,r-1$, $j=1,\dots,d-1$, and $g_d\cdot\ldots\cdot g_1=1$.

(c) There is another possible reduction in the number of relations which was already indicated in [130]. Let us explain it in modern terms.

One can identify $\pi_1(S_d(M), \mathbf{d})$ (where $\mathbf{d} := \{-1, \ldots, -d\}$) and $\mathbb{B}_d$ as follows: each generator σ_i ($i = 1, \ldots, d-1$) represents a positive half-twist interchanging $-i$ and $-(i+1)$. Note that any open braid τ starting at $\mathbf{Y}^*$ and ending at $\mathbf{d}$ defines an isomorphism from $\pi_1(S_d(M), \mathbf{Y}^*)$ onto $\mathbb{B}_d$ by conjugation.

Any meridian γ of $t \in \Delta$ can be decomposed as $\alpha^{-1} \cdot \eta_{\hat{\star},S} \cdot \alpha$ as in Definition 1.2. On the other hand, any open braid τ' starting at the set of roots of $f(\hat{\star}, y)$ and ending at $\mathbf{d}$, can be decomposed as $\nabla(\gamma) = \rho^{-1} \cdot \beta \cdot \rho$, where $\rho, \beta \in \mathbb{B}_d$ satisfying:

- ρ is obtained by the juxtaposition of τ'^{-1}, $\tilde{f}_*(\alpha)$ and τ;
- β is obtained by the juxtaposition of τ'^{-1}, $\tilde{f}_*(\eta_{\hat{\star},S})$ and τ'.

The braid β reflects the local structure of the singularities of the projection Π with respect to $\mathcal{C}$ at t. Let us use the following notation:

- $L_t \cap \mathcal{C} := \{p_1, \ldots, p_s\}$,
- $(L_t \cdot \mathcal{C})_{p_i} = \ell_i + 1$.

Note that $s + \sum_{i=1}^{s} \ell_i = d$. Moreover, for generic projections one can assume that $\ell_i > 0 \Leftrightarrow i = 1$, also if p_1 is smooth then $\ell_1 = 1$, and finally if p_1 is singular then $\ell_1 + 1$ is the multiplicity of $\mathcal{C}$ at p_1.

The braid β is obtained as the unlinked union of s braids $\beta_i \in \mathbb{B}_{\ell_i+1}$, $i = 1, \ldots, s$. Note that each β_i is the local algebraic braid obtained via the Puiseux expansion of the branches of $\mathcal{C}$ at p_i with respect to the variable x. Also note that each β_i is a positive braid (i.e., represented as a word in positive powers of σ_i's). A braid β obtained as an unlinked union of local algebraic braids is called a *Puiseux braid.* In this scenario, the following relations suffice:

$$g_{n_i+k}^{\beta \cdot \rho} = g_{n_i+k}^{\rho}, \quad k = 1, \ldots, \ell_i, \quad i = 1, \ldots, s, \tag{7}$$

where $n_i := \sum_{j=1}^{i-1} (\ell_j + 1)$. Each point p_i produces ℓ_i relations which are *transported* via ρ. Regular points for the projection give no relations.

Example 1.14. Examples of braids β for several arrangements of $\{p_1, \ldots, p_s\}$ are presented below:

(a) $(\mathcal{C}, p_1)$ has a local equation $y^2 - x = 0$ and $\mathcal{C} \pitchfork_{p_i} L_t$, $i = 2, \ldots, d-1$. This means that L_t is a simple tangent to $\mathcal{C}$. Thus $\beta = \sigma_1$.

(b) $(\mathcal{C}, p_i)$ has local equations $y^2 - x = 0$, $i = 1, 2$, and $\mathcal{C} \pitchfork_{p_i} L_t$, $i = 3, \ldots, d-2$. This means that L_t is a simple bitangent to $\mathcal{C}$. Thus $\beta = \sigma_1\sigma_3$.

(c) $(\mathcal{C}, p_1)$ has a local equation $y^3 - x = 0$ and $\mathcal{C} \pitchfork_{p_i} L_t$, $i = 2, \ldots, d-2$. This means that L_t is a tangent at an ordinary inflection point of $\mathcal{C}$. Thus $\beta = \sigma_2\sigma_1$.

(d) $(\mathcal{C}, p_1)$ has local a equation $y^k - x = 0$ and $\mathcal{C} \pitchfork_{p_i} L_t$, $i = 2, \ldots, d-k+1$. This means that L_t is a tangent at a higher order flex of $\mathcal{C}$. Thus $\beta = \sigma_{k-1} \cdot \ldots \cdot \sigma_1$.

(e) $(\mathcal{C}, p_1)$ has a local equation $y^2 - x^{k+1} = 0$ and $\mathcal{C} \pitchfork_{p_i} L_t$, $i = 2, \ldots, d-1$. This means that L_t intersects $\mathcal{C}$ transversally at an $\mathbb{A}_k$-point. Thus $\beta = \sigma_1^{k+1}$.

(f) $(\mathcal{C}, p_1)$ has a local equation $y^3 - x^2 = 0$ and $\mathcal{C} \pitchfork_{p_i} L_t$, $i = 2, \ldots, d-2$. This means that L_t is tangent to $\mathcal{C}$ at an ordinary cusp. Thus $\beta = (\sigma_2\sigma_1)^2$.

(g) $(\mathcal{C}, p_1)$ has a local equation $y^3 - x^4 = 0$ and $\mathcal{C} \pitchfork_{p_i} L_t$, $i = 2, \ldots, d-2$. This means that L_t is transversal to $\mathcal{C}$ at an $\mathbb{E}_6$-point. Thus $\beta = (\sigma_2\sigma_1)^4$. One can check that local equation $y^3 + x^4 = 0$ provides a different braid, namely, $\beta = (\sigma_1\sigma_2)^4$.

(h) $(\mathcal{C}, p_1)$ has a local equation $y^k - x^k = 0$ and $\mathcal{C} \pitchfork_{p_i} L_t$, $i = 2, \ldots, d-k+1$. This means that L_t intersects $\mathcal{C}$ transversally at a k-fold ordinary point. Thus $\beta = (\sigma_{k-1} \cdot \ldots \cdot \sigma_1)^k = \Delta_k^2$.

A general algorithm to obtain a positive braid from Puiseux factorizations has been developed by S. Martínez Juste.

In general, the computation of a braid monodromy factorization is a hard numerical task. Computer-based algorithms have been produced by D. Bessis and J. Michel [21], and J. Carmona [25]. There are some particular cases where a more or less direct computation is possible:

(C1) Arrangements of lines by Arvola [18] via wiring diagrams.

(C2) Strongly real curves [3, 4, 5, 123, 12, 9, 88], i.e., curves with a real equation such that the real picture and the topological type of the singular points contain all the topological information of the embedding of the curve.

(C3) A combination of (C1) and (C2) was considered by M. Salvetti in [103]. Explicit constructions can be found in [10].

The computation of the fundamental group allows us to compute the first homology group.

Proposition 1.15. *Let $d_1, \ldots, d_r$ be the degrees of the irreducible components of $\mathcal{C}$, $d := \deg \mathcal{C} = \sum_{i=1}^r d_i$, and $d_0 := \gcd(d_1, \ldots, d_r)$. Then*

(8) $$H_1(\mathbb{P}^2 \setminus \mathcal{C}, \mathbb{Z}) = \langle x_1, \ldots, x_r \mid d_1x_1 + \cdots + d_rx_r = 0\rangle \cong \mathbb{Z}^{r-1} \times \mathbb{Z}/d_0\mathbb{Z}.$$

There is a natural mapping $\mathrm{Mer}:\pi_1(\mathbb{P}^2\setminus\mathcal{C})\to\mathbb{Z}/d\mathbb{Z}$ *sending any meridian to* 1 mod d.

Proof. It is enough to abelianize (6) and recall that two meridians are in the same conjugacy class if and only if they are meridians of the same irreducible component (Definition 1.2). Q.E.D.

In some sense $\pi_1(\mathbb{P}^2\setminus\mathcal{C})$ and $\pi_1(\mathbb{C}^2\setminus\mathcal{C}^{\text{aff}})$ determine each other (it is important to recall that we assume in this subsection that $L \pitchfork \mathcal{C}$).

Proposition 1.16. *The fundamental group* $\pi_1(\mathbb{C}^2\setminus\mathcal{C}^{\text{aff}})$ *is the pull-back of*

$$\begin{array}{ccc} & & \mathbb{Z} \\ & & \downarrow \\ \pi_1(\mathbb{P}^2\setminus\mathcal{C}) & \to & \mathbb{Z}/d\mathbb{Z} \end{array} \tag{9}$$

where the horizontal arrow is the homomorphism Mer *in Proposition* 1.15 *and the vertical one is the standard projection. In particular, if a presentation of* $\pi_1(\mathbb{P}^2\setminus\mathcal{C})$ *is given whose generators* $g_1,\dots,g_\delta$ *are meridians of* $\mathcal{C}$, *whose relations are of type* $g_{\ell_i}^{w_i(\bar g)}=g_{m_i}$, *and* $\prod_{j=1}^d g_{n_j}^{z_j(\bar g)}=1$, *where* $\ell_i,m_i,n_j\in\{1,\dots,\delta\}$, *and* $w_i(\bar g), z_j(\bar g)$ *are words in* $g_1,\dots,g_\delta$, *then*

$$\pi_1(\mathbb{C}^2\setminus\mathcal{C}^{\text{aff}})=\left\langle h_1,\dots,h_\delta \;\middle|\; h_{\ell_i}^{w_i(\bar h)}=h_{m_i}, \left[\prod_{j=1}^d h_{n_j}^{z_j(\bar h)},h_k\right]=1, 1\le k\le\delta\right\rangle, \tag{10}$$

where $w_i(\bar h)$ *and* $z_j(\bar h)$ *are words in* $h_1,\dots,h_\delta$ *obtained from* $w_i(\bar g), z_j(\bar g)$ *replacing* g_i*'s by* h_i*'s, respectively.*

Proof. The pull-back of (9) is given by

$$G:=\{(g,n)\in\pi_1(\mathbb{P}^2\setminus\mathcal{C})\times\mathbb{Z} \mid \mathrm{Mer}(g)\equiv n \bmod d\}.$$

Note that $\pi_1(\mathbb{C}^2\setminus\mathcal{C}^{\text{aff}})$ induces a commutative diagram in (9) using as a vertical arrow the mapping coming from inclusion and as a horizontal arrow the mapping that sends any meridian to $1\in\mathbb{Z}$. Hence there is a natural mapping $\pi_1(\mathbb{C}^2\setminus\mathcal{C}^{\text{aff}})\xrightarrow{\psi}G$. Using Reidemeister-Schreier, it is easily seen that (10) is a presentation of G.

For the first part, one may assume that the presentations of $\pi_1(\mathbb{P}^2\setminus\mathcal{C})$ and $\pi_1(\mathbb{C}^2\setminus\mathcal{C}^{\text{aff}})$ are those of (6) and (5), respectively. Since L is transversal to $\mathcal{C}$, Remark 1.13(b) implies that presentation (5) can also be obtained from (6). For convenience, the generators of $\pi_1(\mathbb{P}^2\setminus\mathcal{C})$ will be denoted by $\tilde g_i$. Note that $\psi(g_i)=(\tilde g_i,1)$ and $\psi(g_d\cdot\ldots\cdot g_1)=(1,d)$. Since G and $\pi_1(\mathbb{C}^2\setminus\mathcal{C}^{\text{aff}})$ have the same presentations, ψ is an isomorphism. Q.E.D.

Presentations of fundamental groups of affine plane curves $\pi_1(\mathbb{C}^2 \setminus \mathcal{C}^{\text{aff}})$ have special properties. For instance, $\pi_1(\mathbb{C}^2 \setminus \mathcal{C}^{\text{aff}})$ can be generated by meridians and any two meridians of the same irreducible component are in the same conjugacy class (Definition 1.2). A useful type of presentations of $\pi_1(\mathbb{C}^2 \setminus \mathcal{C}^{\text{aff}})$ is the following:

Let $\mathcal{C}_1, \ldots, \mathcal{C}_r$ be the irreducible components of $\mathcal{C}^{\text{aff}}$ and let us fix a meridian x_i of $\mathcal{C}_i$. Note that, as mentioned above, any other meridian $\tilde{x}_i$ of $\mathcal{C}_i$ is in the conjugacy class of x_i, in particular, it can be written as $x_i \cdot y$, $y \in \pi_1(\mathbb{C}^2 \setminus \mathcal{C}^{\text{aff}})'$, $\pi_1(\mathbb{C}^2 \setminus \mathcal{C}^{\text{aff}})'$ being the commutator subgroup of $\pi_1(\mathbb{C}^2 \setminus \mathcal{C}^{\text{aff}})$. We call a presentation of a group satisfying these properties a Zariski presentation, whose precise definition is as follows:

Definition 1.17. Let G be group. We denote its commutator group by G'. A presentation $\langle x_1, \ldots, x_r, y_1, \ldots, y_u \mid w_i(\bar{x}, \bar{y}) = 1 \rangle$, $\bar{x} := (x_1, \ldots, x_r)$, $\bar{y} := (y_1, \ldots, y_u)$, of a group G is called a *Zariski presentation* if

(ZP$_1$) The Abelian group $G/G' \cong \mathbb{Z}^q \oplus \mathbb{Z}/p_1\mathbb{Z} \oplus \cdots \oplus \mathbb{Z}/p_s\mathbb{Z}$ $(q+s=r)$ is generated by the classes of $x_1, \ldots, x_r$, where $x_{q+i}^{p_i} \in G'$ for $i = 1, \ldots, s$.

(ZP$_2$) The classes of $y_1, \ldots, y_u$ in G/G' vanish.

(ZP$_3$) The words $w_i(\bar{x}, \bar{1})$ are products of commutators in $\bar{x}$ and $x_{q+i}^{p_i}$.

To be more precise, we have just proved the following.

Proposition 1.18. *Under the conditions above, $\pi_1(\mathbb{C}^2 \setminus \mathcal{C}^{\text{aff}})$ admits a Zariski presentation, where $s \leq d - r$, and $d := \deg \mathcal{C}^{\text{aff}}$.*

1.3. Braid monodromy and topology

In this section we want to examine in more detail the type of objects presented in Definition 1.9 in the previous subsection. They have been used to obtain presentations of fundamental groups, but they actually contain much more information.

Consider $(\tau_1, \ldots, \tau_r) \in (\mathbb{B}_d)^r$ a braid monodromy factorization of the triple $(\mathcal{C}, L, P)$ obtained from the geometric basis $\gamma_1, \ldots, \gamma_r$. As mentioned after Definition 1.9, any change of geometric basis produces a new braid monodromy factorization, which is given by a Hurwitz move. Since the family of possible geometric bases is parametrized by $\mathbb{B}_r$ (Proposition 1.5), the braid group induces a *Hurwitz action* on $(\tau_1, \ldots, \tau_r)$. Also, a change of the base point $\iota_{\mathbf{Y}^*}$ produces a new braid monodromy factorization, which is given by conjugation $(\tau_1^\beta, \ldots, \tau_r^\beta)$, where β is a certain braid in $\mathbb{B}_d$.

The different factorizations derived from all the different choices can be described as natural $\mathbb{B}_r$ and $\mathbb{B}_d$ actions on $(\mathbb{B}_r)^r$ on the right as follows:

$\mathbb{B}_r$: Let $(\tau_1, \ldots, \tau_r) \in (\mathbb{B}_d)^r$ and let σ_i $(i = 1, \ldots, r-1)$ be canonical generators. Then the action is given by

$$\tau_i^{\sigma_j} := \begin{cases} \tau_{i+1} & \text{if } i = j \\ \tau_i * \tau_{i-1} & \text{if } i = j+1 \\ \tau_i & \text{if } i \neq j+1. \end{cases} \tag{11}$$

for $i = 1, \ldots, n-1, j = 1, \ldots, n$.

$\mathbb{B}_d$: Let $(\tau_1, \ldots, \tau_r) \in (\mathbb{B}_d)^r$ and let $\beta \in \mathbb{B}_d$. Then the action is given by

$$(\tau_1, \ldots, , \tau_r)^\beta := (\tau_1^\beta, \ldots, \tau_r^\beta)$$

These actions of $\mathbb{B}_r$ and $\mathbb{B}_d$ commute; and hence they define a $\mathbb{B}_r \times \mathbb{B}_d$ action on $(\mathbb{B}_d)^r$.

For more details, see [7]. Summing up, we have the following:

Proposition 1.19. *Let $(\tau_1, \ldots, \tau_r)$ be a braid monodromy factorization of $(\mathcal{C}, L, P)$. Then $(\tilde{\tau}_1, \ldots, \tilde{\tau}_r) \in (\mathbb{B}_d)^r$ is another braid monodromy factorization of $(\mathcal{C}, L, P)$ if and only if $(\tilde{\tau}_1, \ldots, \tilde{\tau}_r)$ is in the orbit of $(\tau_1, \ldots, \tau_r)$ with respect to the $\mathbb{B}_r \times \mathbb{B}_d$ action as above.*

Remarks 1.20.

(1) The definitions and results in §3.1 and §1.2 where *geometric bases* are required can be substituted by *pseudo-geometric bases*. In particular, the concept of braid monodromy factorization can also be obtained from a pseudo-geometric basis and Proposition 1.10, and Corollary 1.11 remain true.

(2) Note that a continuous change of the generic line L and the projection P produces a situation similar to the change of the base point in the following sense. A continuous change of the generic line L and fixing P (that is, in the pencil of lines through P) only produces a new pseudo-geometric basis, whose associated factorization is the same as the original one. On the other hand, a continuous change of P on L also defines a new pseudo-geometric basis, whose associated factorization is conjugated of the original one by the braid defined by the motion of P on the generic fiber. Combinations of these two motions allow one to move from $(\mathcal{C}, L, P)$ to any other triple $(\mathcal{C}, L', P')$.

This motivates the following:

Definition 1.21. Two triplets $(\mathcal{C}, L, P)$ and $(\mathcal{C}, L', P')$ as above are said to have *equivalent braid monodromies* if their braid monodromy factorizations are in the same orbit. We will refer to the class of equivalent braid monodromies as the *braid monodromy of the curve* $\mathcal{C}$.

A special – and very useful – type of braid monodromy factorization can be obtained as follows. Let us fix a pseudo-geometric basis $(\gamma_1, \ldots, \gamma_r)$ of $\pi_1(\mathbb{C} \setminus \Delta; \star)$ and decompose each γ_i in the form of a conjugation as in Definition 1.2. Following the ideas in Remark 1.13(c), one obtains a braid monodromy factorization of the form $(\rho_1 \cdot \beta_1 \cdot \rho_1^{-1}, \ldots, \rho_r \cdot \beta_r \cdot \rho_r^{-1})$, where $\beta_1, \ldots, \beta_r$ are Puiseux braids. These are not difficult to obtain from a Puiseux expansion of each local singularity. The difficult numerical part comes from the computation of the braids ρ_i.

Definition 1.22. We say that $((\rho_1, \beta_1), \ldots, (\rho_r, \beta_r))$ is a *Puiseux-braid monodromy factorization* of $\mathcal{C}$ if $(\rho_1^{-1} \cdot \beta_1 \cdot \rho_1, \ldots, \rho_r^{-1} \cdot \beta_r \cdot \rho_r)$ is a braid monodromy factorization of $\mathcal{C}$ and the braids ρ_i, β_i are obtained from the decomposition of a meridian (in a pseudo-geometric basis) as in Remark 1.13(c).

Remark 1.23. Note that if $((\rho_1, \beta_1), \ldots, (\rho_r, \beta_r))$ is a Puiseux-braid monodromy factorization of $\mathcal{C}$, then in particular $\beta_1, \ldots, \beta_r$ are Puiseux braids. Also, if $(\mathcal{C}_1, \mathcal{C}_2)$ is a Zariski pair one can obtain Puiseux-braid monodromy factorizations $((\rho_i^1, \beta_i))_{i=1}^r$ and $((\rho_i^2, \beta_i))_{i=1}^r$ such that the Puiseux braids coincide.

Remark 1.24. Note that Definition 1.22 is very restrictive. Let $((\rho_1, \beta_1), \ldots, (\rho_r, \beta_r))$ be a r-tuple of pairs of braids such that $(\rho_1^{-1} \cdot \beta_1 \cdot \rho_1, \ldots, \rho_r^{-1} \cdot \beta_r \cdot \rho_r)$ is a braid monodromy factorization of $\mathcal{C}$ and $\beta_1, \ldots, \beta_r$ are Puiseux braids. This is a necessary condition for $((\rho_1, \beta_1), \ldots, (\rho_r, \beta_r))$ to be a Puiseux-braid monodromy. Since Definition 1.22 imposes that the factorization $\rho_i^{-1} \cdot \beta_r \cdot \rho_i$ must come from a very particular geometrical decomposition of a meridian, one cannot ensure that $((\rho_1, \beta_1), \ldots, (\rho_r, \beta_r))$ is a Puiseux-braid monodromy factorization of $\mathcal{C}$.

The construction leading to the Zariski-van Kampen method is closer to Puiseux-braid than to general braid monodromy, but it is not easy to define good equivalence relations between Puiseux-braid monodromy factorizations. From Corollary 1.11 one deduces that braid monodromy factorizations determine $\pi_1(\mathbb{P}^2 \setminus \mathcal{C})$ and $\pi_1(\mathbb{C}^2 \setminus \mathcal{C}^{\text{aff}})$.

In [73], Libgober showed the relationship between Puiseux-braid monodromy factorizations and homotopy type as follows.

Theorem 1.25 (Libgober [73]). *The CW-complex associated with the finite presentation of* $\pi_1(\mathbb{C}^2 \setminus \mathcal{C}^{\text{aff}})$ *obtained from* (5), *with the reduction in the number of relations described in Remark* 1.13(c), *has the* *homotopy type of* $\mathbb{C}^2 \setminus \mathcal{C}^{\text{aff}}$. *In particular, a Puiseux-braid monodromy factorization of* $\mathcal{C}$ *determines the homotopy type of* $\mathbb{C}^2 \setminus \mathcal{C}^{\text{aff}}$.

The following results show the strength of Theorem 1.25 and are evidence of the importance of braid monodromy.

Theorem 1.26 (Kulikov-Teicher [70]). *If* $\mathcal{C}$ *has only ordinary nodes and cusps, then a braid monodromy factorization of* $\mathcal{C}$ *determines the diffeomorphism type of* $(\mathbb{P}^2, \mathcal{C})$.

Theorem 1.27 (Carmona [25]). *A braid monodromy factorization of* $\mathcal{C}$ *determines the oriented homeomorphism types of the pairs* $(\mathbb{C}^2, \mathcal{C}^{\text{aff}})$ *and* $(\mathbb{P}^2, \mathcal{C})$.

Carmona uses the local results of M. Namba and M. Takai [87] to prove that one can produce a topological model of the pairs $(\mathbb{C}^2, \mathcal{C}^{\text{aff}})$ and $(\mathbb{P}^2, \mathcal{C})$ from any decomposition as in Remark 1.24. With this model one can apply Libgober's techniques in [73] to prove that $\mathbb{C}^2 \setminus \mathcal{C}^{\text{aff}}$ has the homotopy type of the CW-complex associated with a presentation obtained from a Puiseux-braid factorization, after the reduction explained in Remark 1.13(c).

1.4. Generic and non-generic braid monodromies

Up to now, all the statements in §1 assume a generic projection, i.e., $P \notin \mathcal{C}$ and $L \pitchfork \mathcal{C}$. We also assume that the lines L_t, $t \in \Delta$, satisfy the following:

- Either L_t passes through a singular point of $\mathcal{C}$ and L_t does not belong to its tangent cone, or
- L_t is an ordinary tangent to a smooth point of $\mathcal{C}$ (i.e., not a flex).

Also, all other intersections of L_t are transversal, i.e., in the notation of Remark 1.13(c), one has $\ell_i = 0$ if $i > 1$. The braid monodromy obtained under these hypotheses does not depend on the particular choice of L and P, $P \in L$ (Remark 1.20(2)). Moreover, it is an invariant of the connected components of the combinatorial strata of curves.

Oftentimes, non-generic braid monodromies arise in a natural way, and in general they are very useful. Let us explain different situations where a braid monodromy is non-generic.

(NG1) We can choose a non-generic line L and P generic in L; in particular, $P \notin \mathcal{C}$ and $L \not\pitchfork \mathcal{C}$. In this case Proposition 1.10 and Corollary 1.11 are still true, because Libgober's proof of

Theorem 1.25 also applies here and so does Carmona's Theorem 1.27. One can eliminate relations for the fundamental group using Remark 1.13(c), but not (b).

(NG2) The line L is generic but P is non-generic and $P \notin \mathcal{C}$, i.e., for some $t \in \Delta$ either we have several non-transversal intersections or L_t is tangent to $\mathcal{C}$ at a flex. All the results of §1.2 and §1.3 are true word-for-word. In general, this braid monodromy is not an invariant of connected components of combinatorial strata of curves.

(NG3) P is a smooth (non-flex) point of $\mathcal{C}$ and L is the tangent line. In this case the affine parts of §1.2 and §1.3 remain true. This is due to the fact that $\mathcal{C}^{\text{aff}}$ admits an equation $f(x, y) = 0$, where f is monic in y. In general we cannot eliminate the relations as in Remark 1.13(b). This braid monodromy defines an invariant of connected components of combinatorial strata of curves.

(NG4) The point $P \in \mathcal{C}$ is either a flex or a singularity whose tangent cone is irreducible. In either case we choose L to be the tangent line. The braid monodromy in this case behaves as in (NG3) (because $\mathcal{C}$ admits a monic equation in y) though in general it will not be an invariant of the connected component of combinatorial strata of curves. In this case, as well as in (NG3), one can also compute $\pi_1(\mathbb{P}^2 \setminus \mathcal{C})$, but some additional information about the behavior of the strings of the braid at infinity is needed, see [5, 9, 12] for examples.

(NG5) Choose P, L in order to have vertical asymptotes, i.e., $P \in \mathcal{C}$ and at least one tangent line to $\mathcal{C}$ at P is not L. This case has been deeply studied in [25]. Braid monodromy and some additional data allow one to apply a modified version of the Zariski-van Kampen Theorem and to codify the embedding of $\mathcal{C}$ in $\mathbb{P}^2$.

Why are non-generic braid monodromies interesting? There are at least two reasons. The first one comes from effectiveness of computation. For a curve $\mathcal{C}$, generic braid monodromy factorizations are orbits in $(\mathbb{B}_d)^r$, where r is the sum of the degree of the dual curve and the number of singular points of $\mathcal{C}$. This number could be reduced significantly by considering non-generic projections. Also, under certain circumstances, a generic braid monodromy factorization can be recovered from a non-generic one. The second reason has to do with a partial converse of Theorem 1.27 and will be developed in page 26.

Proposition 1.28. *Let us assume that L_t, $t \in \Delta$, is as in* (NG2). *Let τ be the braid associated with L_t. If L_t, $t \in \Delta$, has h non-transversal intersections, then τ is the product of h pairwise commuting braids.*

Let us assume now that any L_t has only one non-transversal intersection $p_1 \in L_t \cap \mathcal{C}$. Then:

(1) *If L_t is tangent to $\mathcal{C}$ at a smooth point p_1 with $(\mathcal{C} \cdot L_t)_{p_1} = n$, then r decreases by $n-2$. Moreover, τ decomposes into $n-1$ braids conjugated to σ_1 when P is slightly moved in L.*

(2) *If L_t, $t \in \Delta$, passes through $p_1 \in \operatorname{Sing}(\mathcal{C})$ of multiplicity m and $(\mathcal{C} \cdot L_t)_{p_1} = n$, then r decreases by $n-m$ and τ decomposes into $n-m+1$ braids, $n-m$ of them conjugated to σ_1 when P is slightly moved in L.*

Proof. The situation becomes generic by slightly moving the point P in L, which implies a change in the projection direction. In this case, if L_t intersects non-transversally at h points, then after changing the projection direction, L_t splits into h non-transversal lines in a small neighborhood of L_t, which correspond to disjoint and unlinked strings. Thus the first statement follows. For instance, in the case of a bitangent (Figure 2) as in Example 1.14(b), $\tau = \rho \cdot (\sigma_3 \cdot \sigma_1) \cdot \rho^{-1}$ decomposes into the commuting braids $\rho \cdot \sigma_3 \cdot \rho^{-1}$ and $\rho \cdot \sigma_1 \cdot \rho^{-1}$.

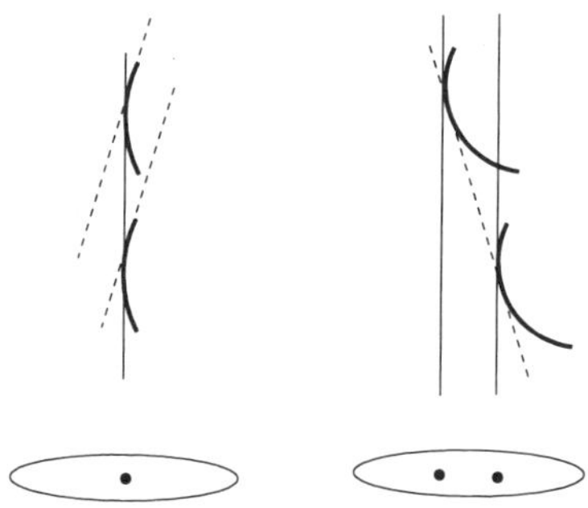

Fig. 2. Bitangent

Let us prove (1). Note that, in this situation $\tau = \rho\cdot(\sigma_{n-1}\cdot\ldots\cdot\sigma_1)\cdot\rho^{-1}$ as in Example 1.14(d), and hence τ decomposes into $n-1$ braids $\rho\cdot\sigma_i\cdot\rho^{-1}$, $i = 1, \ldots, n-1$ – see Figure 3.

If p_1 is a singular point of multiplicity m and $(\mathcal{C} \cdot L_t)_{p_1} = n$, then a perturbation of the projection produces $m-n+1$ non-transversal lines close to L_t: one of them passing through p_1 (not in the tangent cone) and the other ones ordinary tangents. This gives the statement of (2). Let us see what happens if p_1 is a cusp as in Example 1.14(f), where

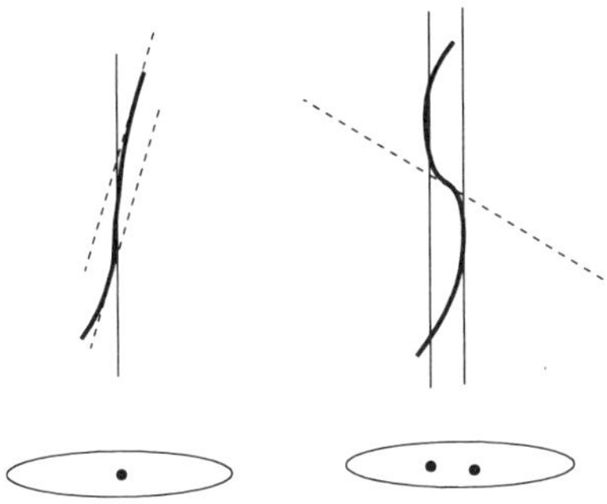

Fig. 3. Flex

$\tau = \rho \cdot (\sigma_2 \cdot \sigma_1)^2 \cdot \rho^{-1}$. The decomposition is given by $\rho \cdot \sigma_2^3 \cdot \rho^{-1}$ and $\rho \cdot \sigma_2^{-1} \cdot \sigma_1 \cdot \sigma_2 \cdot \rho^{-1}$. Q.E.D.

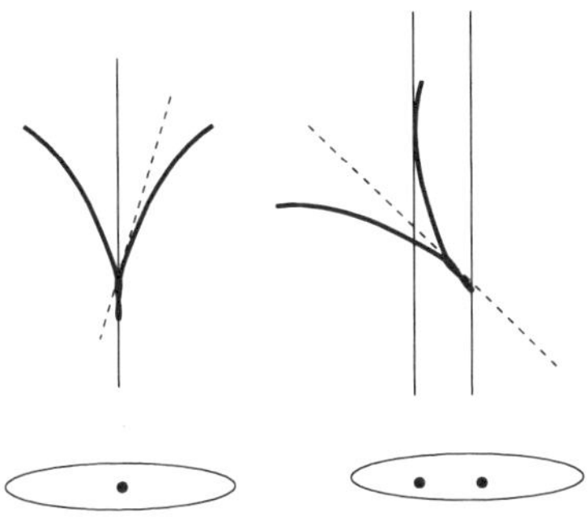

Fig. 4. Cusp

Also, for projections as in (NG1) one can prove the following.

Proposition 1.29. *For each non-transversal point* $p \in \mathcal{C} \cap L$, *with* $\operatorname{mult}_p \mathcal{C} = m$ *and* $(\mathcal{C} \cdot L)_p = n$, *the decreasing of* r *is given by:*

(1) $n - 1$ *if* L *is tangent to* $\mathcal{C}$ *at a smooth point* p.

(2) $n - m + 1$ *if* L *passes through* $p \in \operatorname{Sing}(\mathcal{C})$.

In this case the line at infinity can be deformed into a generic line through P. *The new braid factorization is of type* (NG2) *and has an extra term, which is obtained as the product of* Δ_d^2 *by the inverse of the product of the original braid factorization.*

Therefore a combination of Propositions 1.28 and 1.29 allows one to obtain a generic braid monodromy factorization.

Remark 1.30. If $P \in \mathcal{C}$ and $\operatorname{mult}_P \mathcal{C} = m$, then not only r decreases but also its braid monodromy takes values in $\mathbb{B}_{d-m}$. As for the fundamental group, one needs additional data in order to recover the generic braid monodromy but the complete description of these additional data and the recovering process has not been developed yet.

The second reason stands on this partial converse of Theorem 1.27. Let us introduce some notation. Let $L \not\subseteq \mathcal{C}$ be a line and $P \in L$. Let us also assume that the tangent cone of $\mathcal{C}$ at P is contained in L (the tangent cone is empty if $P \notin \mathcal{C}$). In other words, keeping the notations of §1.2, we are assuming that f is monic in y. For $t \in \Delta$ let $\bar{L}_t$ be the projective closure of L_t (recall that $P \in \bar{L}_t$).

Definition 1.31. A triple $(\mathcal{C}, L, P)$ as above is called a *horizontal triple.* The braid monodromy obtained choosing P as projection point and L as line at infinity is called the *braid monodromy of the horizontal triple.* The *fibered curve* $\mathcal{C}^\varphi$ associated with the horizontal triple is $\mathcal{C} \cup L \cup \bigcup_{t\in\Delta} \bar{L}_t$.

We recall the following results by the first two authors and Carmona.

Theorem 1.32 ([7]). *Let* $(\mathcal{C}_1, L, P)$ *and* $(\mathcal{C}_2, L, P)$ *be two horizontal triples. Let* $F : \mathbb{P}^2 \to \mathbb{P}^2$ *be an orientation-preserving homeomorphism such that:*

(i) $F(P) = P$, *i.e,* F *respects the base point of the fibration.*
(ii) $F(L) = L$ *preserving orientations.*
(iii) $F(\mathcal{C}_1^\varphi) = \mathcal{C}_2^\varphi$ *preserving orientations.*

Then $(\mathcal{C}_1, L, P)$ *and* $(\mathcal{C}_2, L, P)$ *have the same braid monodromy.*

For the special case of line arrangements this result has an *ordered* version.

Definition 1.33. An *ordered arrangement of lines* $\mathcal{L}$ is an ordered list of lines in $\mathbb{P}^2$. A triple $(\mathcal{L}, L, P)$ is called a *horizontal triple arrangement* if it defines a horizontal triple and the lines $\{\bar{L}_t\}_{t\in\Delta}$ are ordered. The *fibered arrangement* $\mathcal{L}^\varphi$ associated with the horizontal triple arrangement is the ordered arrangement $\mathcal{L} + (L) + (\bar{L}_t)_{t\in\Delta}$.

Let us consider a braid monodromy factorization of $(\mathcal{L}, L, P)$. Since $\mathcal{L}$ is an arrangement of lines, this representative belongs to $(\mathbb{PB}_d)^r$, where $\mathbb{PB}_d$ is the pure braid group. Moreover, by the choice of an order in $\mathcal{L}$ only conjugations by elements of $\mathbb{PB}_d$ are allowed (otherwise meridians of different components are interchanged). Moreover, since $\{\bar{L}_t\}_{t\in\Delta}$ is also ordered by the incidence relations, only *pure* Hurwitz moves are allowed.

Definition 1.34. A *pure braid monodromy* is an orbit of $(\mathbb{PB}_d)^r$ by the action of $\mathbb{PB}_d \times \mathbb{PB}_r$. A *horizontal triple arrangement* $(\mathcal{L}, L, P)$ has associated with it a pure braid monodromy.

Theorem 1.35 ([10]). *Let $(\mathcal{L}_1, L, P)$ and $(\mathcal{L}_2, L, P)$ be two horizontal triple arrangements. Let $F : \mathbb{P}^2 \to \mathbb{P}^2$ be an orientation-preserving homeomorphism such that:*

(i) $F(P) = P$.
(ii) $F(L) = L$ *preserving orientations.*
(iii) $F(\mathcal{L}_1^\varphi) = \mathcal{L}_2^\varphi$ *preserving orientations and orders.*

Then $(\mathcal{L}_1, L, P)$ and $(\mathcal{L}_2, L, P)$ have the same pure braid monodromy.

1.5. Fundamental group, braid monodromy and Zariski pairs

Fundamental groups are a primary tool in the problem of finding Zariski pairs. In the example mentioned in the Introduction (Z5), Zariski proved that the combinatorially-equivalent curves had different fundamental groups ($\mathbb{Z}/2\mathbb{Z} * \mathbb{Z}/3\mathbb{Z}$ and $\mathbb{Z}/2\mathbb{Z} \times \mathbb{Z}/3\mathbb{Z}$). In fact, before completing all the computations, Zariski had found other weaker invariants that also served to distinguish the members of the pair. The first one is the study of branched Galois coverings ramified along the curve: in only one case do D_6-coverings exist – a proof in modern terms can be found in §3.3. In §3, we present how the algebraic study of Galois coverings is a powerful tool to study Zariski pairs, obtaining information about fundamental groups without having to compute them. The second invariant is the Alexander polynomial which, along with some generalizations, is studied in §2.

These two kinds of invariants are useful for several reasons. As we have seen in this section, computation of fundamental groups can be a very tricky task. On one hand, some algebraic properties of the curves can give rise to invariants of the fundamental group, and thus a difference in such invariants means a difference in fundamental groups. On the other hand, even if the fundamental group is computed, what one obtains is a finite presentation of it. The undecidability of the isomorphism problem makes this task feasible only in the simplest examples.

As explained in the Introduction note that, even though $\pi_1(\mathbb{P}^2 \setminus \mathcal{C})$ is an invariant of the embedded topology of a curve $\mathcal{C}$, any homeomorphism of pairs $(\mathbb{P}^2, \mathcal{C})$ should send meridians to meridians. Therefore, $\pi_1(\mathbb{P}^2 \setminus \mathcal{C})$ with a peripheral structure given by the conjugacy class of meridians of the irreducible components of $\mathcal{C}$ is sometimes a more useful invariant. For example, if we are counting the number of irreducible representations of $\pi_1(\mathbb{P}^2 \setminus \mathcal{C})$ onto a given finite group, the peripheral structure introduces some restrictions on the images of the meridians. Analogously, if we are considering Betti numbers of abelian coverings, the peripheral structure allows us to describe such coverings canonically and thus point out possible differences. In general they will not be invariants of the sole fundamental group, but they will be useful for detecting Zariski pairs. These techniques fail when we look for *arithmetic Zariski pairs*, since in this case most invariants of finite coverings are of an arithmetic nature.

We end up this section with referring examples for various kinds of Zariski pairs. For details, see the cited references

Example 1.36. Once one has an example of a Zariski pair which is distinguished by the fundamental group, it is possible to give infinite families of Zariski pairs using Cremona transformations and covering maps. These techniques have been used by Oka [91], Shimada [105, 106] and A.M. Uludağ [123].

Example 1.37 ([62]). In their paper Kharlamov-Kulikov use braid monodromy factorizations to find a special kind of Zariski pair, called *oriented Zariski pair.* An oriented Zariski pair is characterized by the non-existence of orientation-preserving homeomorphisms. Note that complex conjugation preserves the orientation of $\mathbb{P}^2$, but reverses the orientations of the curves. In [63], they also find examples of complex-conjugated surfaces such that the complex conjugation does not preserve canonical divisors, thus they do not admit orientation-preserving homeomorphisms. If we apply Chisini's conjecture, the branch curves of generic coverings give oriented Zariski pairs. For each m the produced Zariski pair involves curves of degree $333m^2$.

Example 1.38 ([9]). Let us consider the combinatorial stratum of curves with the following combinatorics: sextics with two irreducible components of degrees 5 and 1. The quintic curve has three singular points of types $\mathbb{E}_6$, $\mathbb{A}_3$ and $\mathbb{A}_2$ and the line intersects the quintic at two smooth points with intersection multiplicities 4 and 1.

It is not hard to prove that this space has two connected components (each one is isomorphic to $\mathrm{PGL}(3;\mathbb{C})$). For one component there is a representative $\mathcal{C}_+$ with equation in $\mathbb{Q}(\sqrt{2})[X,Y,Z]$. Its conjugate $\mathcal{C}_-$ belongs to the other connected component. Let us denote by $(\mathcal{C}_\pm, L_\pm, P_\pm)$ the horizontal triples, where $P_\pm$ is the $\mathbb{E}_6$ point of $\mathcal{C}_\pm$ and $L_\pm$ is the tangent line.

Using techniques for strongly real curves one can compute a non-generic braid monodromy factorization of type (NG4) of $(\mathcal{C}_+, L_+, P_+)$:

$$(\sigma_2^8, \sigma_2^4\sigma_1^2\sigma_2^{-4}, \sigma_2^3\sigma_1^3\sigma_2^{-3}, \sigma_2\sigma_1^4\sigma_2^{-1}, \sigma_1^{-3}\sigma_2\sigma_1^3),$$

and one of $(\mathcal{C}_-, L_-, P_-)$:

$$(\sigma_2^3, (\sigma_2\sigma_1^{-1}\sigma_2)\sigma_1(\sigma_2\sigma_1^{-1}\sigma_2)^{-1}, \sigma_2\sigma_1^8\sigma_2^{-1}, \sigma_1^{-2}\sigma_2^4\sigma_1^2, \sigma_1^{-3}\sigma_2^2\sigma_1^3).$$

In fact, these are Puiseux-braid monodromy factorizations. Since the additional data at infinity are easy to obtain, one can compute $\pi_1(\mathbb{P}^2 \setminus \mathcal{C}_\pm)$. It turns out that both groups are isomorphic to $\mathbb{Z} \times \mathrm{GL}(2; \mathbb{F}_7)$.

The main point is that these two braid monodromy factorizations are non-equivalent. Taking representations of the braid group onto $\mathrm{GL}(2, \mathbb{Z}/32\mathbb{Z})$, the image of the braid monodromies becomes a finite set. One can simply check using GAP [49] that both images are disjoint. By Theorem 1.32, $(\mathcal{C}_+^\varphi, \mathcal{C}_-^\varphi)$ is an arithmetic Zariski pair. Similar examples in [9] also provide examples of oriented Zariski pairs.

Example 1.39. Let us consider the combinatorial stratum of sextic curves with 4 singular points of types $\mathbb{E}_7$, $\mathbb{E}_6$, $\mathbb{A}_4$, and $\mathbb{A}_2$. As in the previous example, this space consists of two irreducible components, each one isomorphic to $\mathrm{PGL}(3; \mathbb{C})$. Representatives can be taken in each component with equations in $\mathbb{Q}(\sqrt{5})[x, y, z]$ as follows:

$$\begin{aligned}
f_s(x, y, z) := &- (200 + 90\,s)\,x^6 - (1575 + 705\,s)\,x^5y - (552 + 254\,s)\,zx^5\\
&- (3963 + 1779\,s)\,zx^4y - (456 + 222\,s)\,z^2x^4\\
&- (63 + 27\,s)\,zx^3y^2 - (2817 + 1251\,s)\,z^2x^3y\\
&- (56 + 21\,s)\,z^3x^3 + (666 + 324\,s)\,z^2x^2y^2\\
&+ (-45 + 15\,s)\,z^3x^2y + (48 + 16\,s)\,z^4x^2 + (1737 + 783\,s)\,z^3xy^2\\
&+ (384 + 192\,s)\,z^4xy + 54z^3y^3 + (1008 + 432\,s)\,z^4y^2,
\end{aligned}$$

where $s^2 = 5$. Let us consider the triples associated with the $\mathbb{E}_7$-point $P = [0 : 1 : 0]$ and the tangent line $L = \{z = 0\}$ at P. First we will compute their braid monodromy factorization based on the real picture, since both curves are strongly real (see (C2)).

Figure 5 shows the real picture of $\mathcal{C}_{\sqrt{5}} := \{f_{\sqrt{5}} = 0\}$, the choice of the generic line $L_\star$, and the choice of the generators of the braid group based on $L_\star \setminus \mathcal{C}_{\sqrt{5}}$. We recall that the way the σ_i are chosen in general corresponds to the lexicographic order in $\mathbb{C}$ where $a_1 + b_1\sqrt{-1} < a_2 + b_2\sqrt{-1}$ if and only if $a_1 < a_2$, or $a_1 = a_2$ and $b_1 < b_2$.

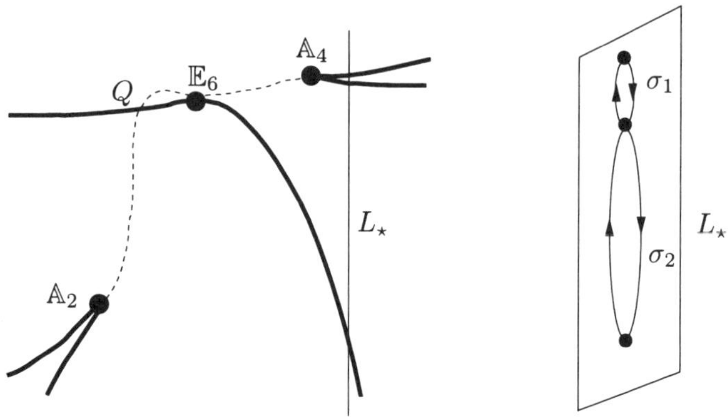

Fig. 5. Real graph of $\mathcal{C}_{\sqrt{5}}$.

The only singular fibers occur at $x = -\frac{32}{5}$ (where the $\mathbb{A}_2$-point $(-\frac{32}{5}, -\frac{1584}{5} - 144\sqrt{5})$ lies), at $x = -1$ (where the $\mathbb{E}_6$-point $(-1, 0)$ lies), and at $x = 0$ (where the $\mathbb{A}_4$-point $(0, 0)$ lies). This can be checked by factorizing the discriminant of $f_{\sqrt{5}}$ with respect to y.

The dotted curve in Figure 5 represents the real parts of the complex conjugated branches. When at most two branches are complex conjugated per fiber of the projection (as is our case) this picture plus the local braids contain all the necessary information to compute the braid monodromy factorization.

In our case, the local braids around the $\mathbb{A}_4$ and the $\mathbb{A}_2$ are obvious because the branches involved are real. Therefore the first factor of the factorization should be σ_1^5 and the last one should be a conjugated of σ_2^3 (Example 1.14(e)). In this case note that half a turn around the $\mathbb{A}_4$ point corresponds to σ_1^2. Therefore the factorization this far looks like $(\sigma_1^5, \sigma_1^2 * \beta, (\sigma_1^2\beta_1\alpha) * \sigma_2^3)$, where β is the local braid around the $\mathbb{E}_6$-point, β_1 is half the braid around the $\mathbb{E}_6$-point, and α is the braid from $\mathbb{E}_6$ to $\mathbb{A}_2$.

To obtain α it is enough to note that a local crossing of type Q as in Figure 5 corresponds to $\sigma_2^{-1}\sigma_1$ (always according to our lexicographic order in $\mathbb{C}$) as shown in Figure 6. Since there are no more crossings between $\mathbb{E}_6$ and $\mathbb{A}_2$ one has that $\alpha = \sigma_2^{-1}\sigma_1$.

For the $\mathbb{E}_6$-point one has to work a little bit more. One first considers a parametrization for the local branches of f_s at this point: something of the form $y = \omega_1 x + \omega_2 \xi^k x^{\frac{4}{3}}$, $k = 0, 1, 2$, where $\xi^2 + \xi + 1 = 1$. Basically the sign of the real part of ω_2 determines the local braid as

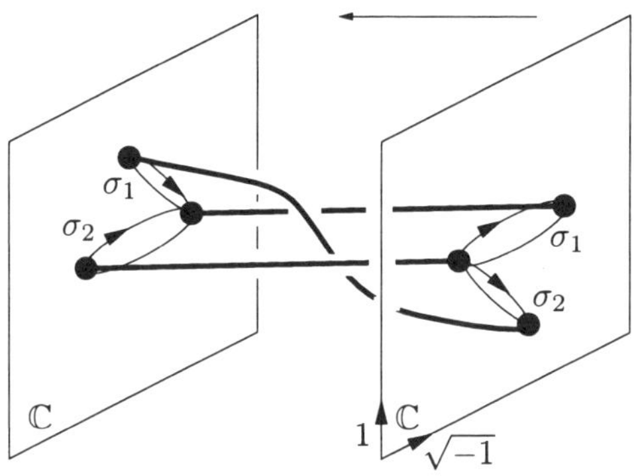

Fig. 6. Description of the crossing at Q.

shown in Example 1.14(g). In our case one obtains $\beta = (\sigma_1\sigma_2)^4$ and hence $\beta_1 = (\sigma_1\sigma_2)^2$.

Therefore the braid monodromy factorization in this example is

$$(\sigma_1^5, \sigma_1^2 * (\sigma_1\sigma_2)^4, (\sigma_1^2(\sigma_1\sigma_2)^2(\sigma_2^{-1}\sigma_1)) * \sigma_2^3).$$

Using the relation $\sigma_1 * \sigma_2^k = \sigma_2^{-1} * \sigma_1^k$ and the obvious $\sigma_i^r * \sigma_i^k = \sigma_i^k$, the last term can be reduced to $\sigma_1 * \sigma_2^3$, obtaining Table 1.

σ_1^5	$\sigma_1^2 * (\sigma_1\sigma_2)^4$	$\sigma_1 * \sigma_2^3$

Table 1.

Claim 1.40. *The (non-generic) braid monodromy factorizations of* $(\mathcal{C}_{\sqrt{5}}, L, P)$ *and* $(\mathcal{C}_{-\sqrt{5}}, L, P)$ *coincide.*

Proof. For $\mathcal{C}_{-\sqrt{5}}$ one has an analogous situation as shown in Figure 7, which has the same local and global information of the strongly real picture. Therefore, their braid monodromy factorizations coincide. Q.E.D.

Moreover, note that, even though the combinatorial stratum consists of two irreducible components, the associated affine curves are isomorphic. In particular

$$\tilde{f}_{\sqrt{5}}(\omega(x,y)) = \left(51841 + 23184\sqrt{5}\right) \tilde{f}_{-\sqrt{5}}(\ell(x,y)),$$

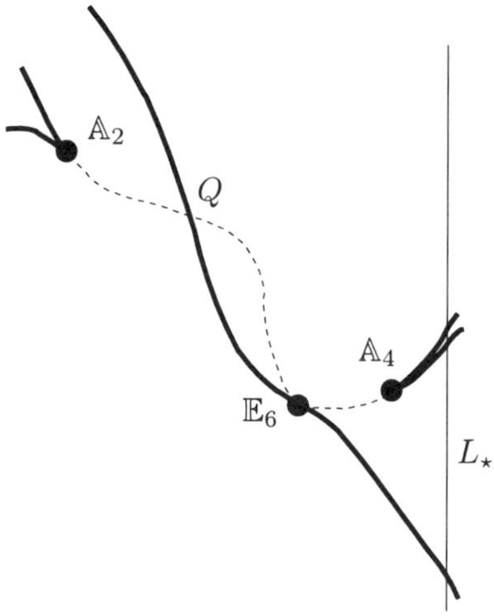

Fig. 7. Real graph of $\mathcal{C}_{-\sqrt{5}}$.

where $\tilde{f}_s$, denote the affine equation of f_s,

$$\omega(x,y) := \left(x, -y + \left(1620 + 648\sqrt{5}\right)x - \left(\frac{25}{2} + \frac{35\sqrt{5}}{6}\right)x^2\right)$$

is a Jung automorphism, and

$$\ell(x,y) := \left(x, \left(\frac{8075}{2} - \frac{10657\sqrt{5}}{6}\right)x + \left(-\frac{47}{2} + \frac{21\sqrt{5}}{2}\right)y\right)$$

is a linear automorphism of $\mathbb{C}^2$. Note that this does not give another proof of Claim 1.40 via Theorem 1.27 since the latter can only be applied in principle to generic projections. But it does tell us that the generic braid monodromies of $\mathcal{C}_{\sqrt{5}} \cup L$ and $\mathcal{C}_{-\sqrt{5}} \cup L$ are equivalent.

A geometrical interpretation of the Jung automorphism can be given as follows. In Figure 8, we show the dual graph of the total transform of L after an embedded resolution of the $\mathbb{E}_7$-point. The successive exceptional divisors are denoted by E_i, $i = 1, 2, 3$. The branch B_1 denotes the strict transform of the cusp of $\mathbb{E}_7$, the branch B_2 denotes the strict transform of the smooth branch of the $\mathbb{E}_7$ and the branch B_3 denotes the branch at the smooth point of $\mathcal{C}$ on $L \cap \mathcal{C}_s$.

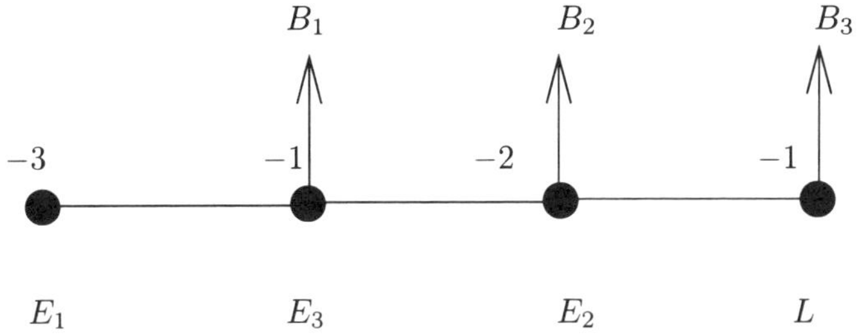

Fig. 8.

Note that, contracting the strict transform of L one achieves a combinatorially-symmetric situation where B_2 and B_3 cannot be distinguished. Note that, by the Jung automorphism, B_2 and B_3 are interchanged.

We can still recuperate valuable information to add to equivalent braid monodromies that can distinguish the different behavior at infinity. The idea is to *color* the different branches at infinity. This idea will be developed in what follows.

Remark 1.41. Let $(\mathcal{C}, L, P)$ be a horizontal triple of degree d and let us choose L_*, a generic member of $\mathcal{H}_P$, as the base line of the pencil. Let us also fix a continuous uniparametric family of lines L_t ($t \in [0,1]$) in the pencil such that $L_0 = L_*$ and $L_1 = L$. The continuity of L_t allows us to associate a branch of $\mathcal{C}$ at L to each point of $\mathcal{C} \cap L_*$. The combinatorics of $\mathcal{C}$ at L defines a partition on the set of such branches and hence induces a partition $\mathscr{P}^*$ on $\mathcal{C} \cap L_*$ (which turns out to be independent of the chosen path L_t). Ordering the points of $\mathcal{C} \cap L_*$ induces a partition $\mathscr{P}$ in $\{-1, \ldots, -d\}$. Let $\Sigma_{\mathscr{P}}$ be the subgroup of Σ_d preserving the partition and let $\mathbb{B}_{\mathscr{P}}$ be the preimage of $\Sigma_{\mathscr{P}}$ in $\mathbb{B}_d$. By restriction to $\mathbb{B}_{\mathscr{P}}$ one can define the $\mathscr{P}$-braid monodromy of $(\mathcal{C}, L, P)$. The same proof of Theorem 1.32 can be applied to this particular scenario to obtain the following.

Theorem 1.42. *The statement of Theorem* 1.32 *also holds if we replace braid monodromy by $\mathscr{P}$-braid monodromy, where $\mathscr{P}$ has the same combinatorial meaning at infinity for both triples.*

Now we can show that $\left(\mathbb{P}^2, \mathcal{C}^{\varphi}_{\sqrt{5}} \cup L\right)$ and $\left(\mathbb{P}^2, \mathcal{C}^{\varphi}_{-\sqrt{5}} \cup L\right)$ form an arithmetic-Zariski pair.

Proposition 1.43. *Let us consider the curves $\mathcal{C}_{\sqrt{5}}$ and $\mathcal{C}_{-\sqrt{5}}$ in Example* 1.39. *There is no homeomorphism between the pairs* $(\mathbb{P}^2, \mathcal{C}^{\varphi}_{\sqrt{5}} \cup L)$ *and* $(\mathbb{P}^2, \mathcal{C}^{\varphi}_{-\sqrt{5}} \cup L)$*, but their complements are homeomorphic, that is* $\mathbb{P}^2 \setminus \left(\mathcal{C}^{\varphi}_{\sqrt{5}} \cup L\right) \cong \mathbb{P}^2 \setminus \left(\mathcal{C}^{\varphi}_{-\sqrt{5}} \cup L\right)$.

Proof. Let us prove the last statement first. Note that the homeomorphism between $\mathbb{P}^2 \setminus (\mathcal{C}_{\sqrt{5}} \cup L)$ and $\mathbb{P}^2 \setminus (\mathcal{C}_{-\sqrt{5}} \cup L)$ preserves the pencil of lines through P, therefore, it induces a homeomorphism between $\mathbb{P}^2 \setminus \left(\mathcal{C}^{\varphi}_{\sqrt{5}} \cup L\right)$ and $\mathbb{P}^2 \setminus \left(\mathcal{C}^{\varphi}_{-\sqrt{5}} \cup L\right)$.

Let us prove now that the pairs are not homeomorphic. In order to do so we will consider the partitions of Remark 1.41. We order the points of $\mathcal{C} \cap L_*$ as in Figure 6. One has the following situation:

for $\mathcal{C}_{\sqrt{5}}$: $(1,2,3) \leftrightarrow (B_1, B_2, B_3)$.
for $\mathcal{C}_{-\sqrt{5}}$: $(1,2,3) \leftrightarrow (B_1, B_3, B_2)$.

Hence, the group $\mathbb{B}_{\mathscr{P}}$ is simply the pure braid group. Let us denote by $n_{\sqrt{5}}$ the braid monodromy in Table 1. It defines a $\mathscr{P}$-braid monodromy for $\mathcal{C}_{\sqrt{5}}$. In order to have a representative of the $\mathscr{P}$-braid monodromy for $\mathcal{C}_{-\sqrt{5}}$ we have to permute the second and third strings, for instance $n_{-\sqrt{5}} := \sigma_2 * n_{\sqrt{5}}$.

Let H_s, $s^2 = 5$, be the monodromy groups in $\mathbb{B}_3$. If these curves have the same $\mathscr{P}$-braid monodromy, then $H_{\sqrt{5}}$ and $H_{-\sqrt{5}}$ are conjugated by an element in $\mathbb{B}_{\mathscr{P}}$. Using GAP4 [49], it is easily seen that this is the case and that $H_{\sqrt{5}} \neq H_{-\sqrt{5}}$.

Let c_s be the pseudo-Coxeter braid of n_s, $s^2 = 5$, (see Definition 1.9*(iii)*). Note that

$$c := c_{\sqrt{5}} = c_{-\sqrt{5}} = (\sigma_1 * \sigma_2^3)(\sigma_1^2 * (\sigma_1\sigma_2)^4)\sigma_1^5 = \sigma_1\sigma_2^3\sigma_1\Delta_3^2\sigma_1\sigma_2\sigma_1^3 =$$

$$\Delta_3^2\sigma_1\sigma_2^3\sigma_1^2\sigma_2\sigma_1^3 = \Delta_3^2\sigma_1\sigma_2^2\sigma_1^{-1}\Delta_3^2\sigma_1^2 = \Delta_3^4\sigma_1\sigma_2^2\sigma_1,$$

where $\Delta_3^2 = (\sigma_1\sigma_2)^3 = (\sigma_1\sigma_2\sigma_1)^2$ is the generator of the center of $\mathbb{B}_3$ (see Remark 1.13(b)).

If $n_{-\sqrt{5}}$ and $n_{\sqrt{5}}$ are $\mathscr{P}$-equivalent, there exists a pure braid τ such that $\tau * H_{\sqrt{5}} = H_{-\sqrt{5}}$ and $[c, \tau] = 1$. It can be easily computed (for innstance, via the standard representation in the special linear group $\mathrm{SL}(2, \mathbb{Z})$) that the intersection of the pure braid group and the commutator of c is the subgroup generated by c and Δ_3^2. This group is contained in the normalizer of $H_{\sqrt{5}}$ and hence, such a τ cannot exist. Q.E.D.

Example 1.44 ([10]). This is the first example of arithmetic Zariski pairs of lines. It consists of two arrangements $\mathscr{M}_{\pm}$ of eleven lines having conjugate equations in with coefficients in $\mathbb{Q}(\sqrt{2})$. In particular, their fundamental groups have isomorphic profinite completions.

The real pictures of $\mathcal{M}_\pm$ are shown in Figure 9 (lines at infinity included). In order to prove that they provide an arithmetic Zariski pair, one can proceed by contradiction as follows.

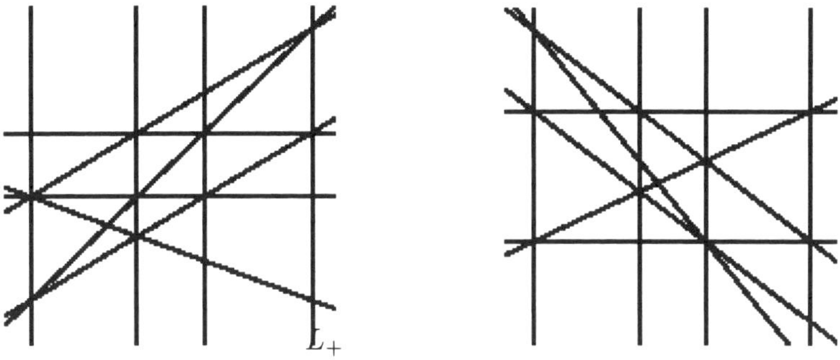

Fig. 9. Zariski pair of real line arrangements

Let us assume that a homeomorphism $\psi : (\mathbb{P}^2, \mathcal{M}_+) \to (\mathbb{P}^2, \mathcal{M}_-)$ exists. This homeomorphism must preserve the orientation of $\mathbb{P}^2$. Using standard intersection theory, one can assume that it either preserves the orientations of all the lines in $\mathcal{M}_\pm$ or it reverses them. Taking complex conjugation into account, one can assume that ψ preserves the orientations of the lines in $\mathcal{M}_\pm$. For combinatorial reasons $\psi(L_+) = L_-$ in Figure 9. Let us consider the arrangements $\mathcal{L}_\pm$ obtained by removing both the vertical lines and $L_\pm$ from $\mathcal{M}_\pm$. Since $\mathcal{M}_\pm$ has a unique point of multiplicity five, it is easy to see that $\psi(\mathcal{L}_+) = \mathcal{L}_-$. Thus, one can order these arrangements in such a way that the ith line of $\mathcal{L}_+$ and $\mathcal{L}_-$ are conjugate in $\mathbb{Q}(\sqrt{2})$. The choice of the lines $L_\pm$ implies that ψ preserves the order. Moreover, the vertical lines can be ordered so as to fulfill the same property.

Let P be the point of intersection of the vertical lines and let L_∞ be the line at infinity. Then $(\mathcal{L}_\pm, L_\infty, P)$ are horizontal triple arrangements such that $\mathcal{L}_\pm^\varphi = \mathcal{M}_\pm \setminus \{L_\pm\}$. By Theorem 1.35, $(\mathcal{L}_\pm, L_\infty, P)$ have the same pure braid monodromy, but this contradicts [10, Theorem 4.19].

Note that no *ordered* homeomorphism exists from $\mathcal{L}_+^\varphi$ to $\mathcal{L}_-^\varphi$, but it is not hard to prove that there exists a projective transformation in $\mathrm{PGL}(3, \mathbb{C})$ sending $\mathcal{L}_+^\varphi$ to $\mathcal{L}_-^\varphi$ (thus not respecting orders).

Whether or not this is an example of a π_1-equivalent Zariski pair (that is, if the groups are actually isomorphic) or a complement-equivalent

Zariski pair (that is, if the complements are homeomorphic) remains an open problem.

Example 1.45. [88] Example 1.44 and Rybnikov's example are particularly interesting cases of Zariski pairs, since they come from line arrangements. What happens with conic arrangements? A nice example of a Zariski pair of conic arrangements has been provided by Namba and Tsuchihashi [88]. An elementary and exhaustive approach to it occupies §4.

§2. Alexander invariants

2.1. Alexander polynomials

Alexander polynomials have been largely used for knots and links in connection with cyclic branched coverings of their complement (see [50] for a survey on the matter). The first application of cyclic coverings to complements of plane curves was already proposed by Zariski (as mentioned in the Introduction), and later formalized by Libgober [71]. Since then the bibliography on the subject has become extensive. In what follows, we will give the basic definitions and present the main results on this invariant.

Consider $X_{\mathcal{C}} := \mathbb{P}^2 \setminus \mathcal{C}$, where $\mathcal{C} = \mathcal{C}_0 \cup \mathcal{C}_1 \cup \cdots \cup \mathcal{C}_r$, $\mathcal{C}_i$ is an irreducible curve of degree d_i with equation $\{C_i = 0\}$, and $d_0 = 1$ (this last condition is purely technical to simplify notation). Note that, under these conditions,

$$H_1(X_{\mathcal{C}}; \mathbb{Z}) = \frac{\bigoplus_{i=0}^{r} \gamma_i \mathbb{Z}}{\langle \gamma_0 + d_1\gamma_1 + \cdots + d_r\gamma_r \rangle} \approx \mathbb{Z}^r,$$

where γ_i is the homology class of a meridian of $\mathcal{C}_i$. Let $\varepsilon : H_1(X_{\mathcal{C}}; \mathbb{Z}) \to \mathbb{Z}$ be an epimorphism. This epimorphism is defined by $(\varepsilon_1, \ldots, \varepsilon_r) \in \mathbb{Z}^r$, where $\varepsilon_i := \varepsilon(\gamma_i)$.

The kernel K_ε of the composition $G := \pi_1(X_{\mathcal{C}}) \overset{\text{ab}}{\to} H_1(X_{\mathcal{C}}; \mathbb{Z}) \overset{\varepsilon}{\to} \mathbb{Z}$ defines a covering of $X_{\mathcal{C}}$, say $\pi_\varepsilon : X_{\mathcal{C},\varepsilon} \to X_{\mathcal{C}}$, whose group of deck transformations is $G/K_\varepsilon = \mathbb{Z}$.

Remark 2.1. Given $n \in \mathbb{N}$ the composition of $\varepsilon \circ \text{ab}$ with the natural quotient $\mathbb{Z} \to \mathbb{Z}/n\mathbb{Z}$ produces an n-fold cyclic finite covering $\pi_{\varepsilon,n} : X^n_{\mathcal{C},\varepsilon} \to X_{\mathcal{C}}$ whose group of deck transformations is $\mathbb{Z}/n\mathbb{Z}$. Note that if n divides ε_i then $\pi_{\varepsilon,n}$ could be extended above $\mathcal{C}_i \setminus \bigcup_{j\neq i} \mathcal{C}_j$ as an unramified covering.

The group $G/K_\varepsilon = \mathbb{Z}$ acts on $K_\varepsilon/K'_\varepsilon = H_1(X_{\mathcal{C},\varepsilon};\mathbb{Z})$ by conjugation as follows

$$\begin{array}{rccc} * : & G/K_\varepsilon \times K_\varepsilon/K'_\varepsilon & \to & K_\varepsilon/K'_\varepsilon \\ & (\varepsilon(g),\bar{k}) & \mapsto & \varepsilon(g) * \bar{k} := \overline{g\cdot k\cdot g^{-1}}. \end{array}$$

Note that if $g' = gh_1$ ($h_1 \in K_\varepsilon$) and $k' = kh_2$ ($h_2 \in K'_\varepsilon$), then

$$\begin{aligned} \left(g'\cdot k'\cdot g'^{-1}\right)\left(g\cdot k^{-1}\cdot g^{-1}\right) &= \left((gh_1)\cdot(kh_2)\cdot(h_1^{-1}g^{-1})\right)\left(g\cdot k^{-1}\cdot g^{-1}\right) \\ &= g\cdot\left(h_1kh_2h_1^{-1}k^{-1}\right)\cdot g^{-1} \\ &= g\cdot\left((h_1k)\cdot h_2\cdot(h_1k)^{-1}\ h_1kh_1^{-1}k^{-1}\right)g^{-1} \\ &\in K'_\varepsilon. \end{aligned}$$

Hence "$*$" does not depend on the choice of g mod K_ε or k mod K'_ε.

This action endows $M^{\mathbb{Z}}_{\mathcal{C},\varepsilon} := H_1(X_{\mathcal{C},\varepsilon};\mathbb{Z})$ with a Λ_ε-module structure, where $\Lambda_\varepsilon := \mathbb{Z}[G/K_\varepsilon] \approx \mathbb{Z}[t^{\pm 1}]$. One can tensor such a module by a field $\mathbb{K} = \mathbb{Q}, \mathbb{C}, \mathbb{F}_p, \ldots$ to obtain a module $M^{\mathbb{K}}_{\mathcal{C},\varepsilon}$ over $\Lambda^{\mathbb{K}}_\varepsilon = \mathbb{K}[t^{\pm 1}]$. Since G is finitely presented, $M^{\mathbb{K}}_{\mathcal{C},\varepsilon}$ is finitely generated as a $\Lambda^{\mathbb{K}}_\varepsilon$-module (by as many 1-cells as generators of G). The rings $\Lambda^{\mathbb{K}}_\varepsilon$ are principal ideal domains and hence one can define $\Delta^{\mathbb{K}}_{\mathcal{C},\varepsilon}(t)$ as the *order* of $M^{\mathbb{K}}_{\mathcal{C},\varepsilon}$. We recall that, if R is a principal ideal domain, the *order* of an R-module M, is defined as

$$\Delta := \begin{cases} 0 & \text{if } M \text{ has a free summand,} \\ 1 & \text{if } M = 0, \\ \prod_{i=1}^m \lambda_i & \text{if } M \approx \frac{R}{(\lambda_1)} \oplus \cdots \oplus \frac{R}{(\lambda_m)}. \end{cases} \tag{12}$$

Such a polynomial can be assumed to be unique by adding the extra condition $\lambda_i(0) = 1$. This is known as the *Alexander polynomial* of $\mathcal{C}$ associated with ε. In general, if $\mathbb{K} = \mathbb{Q}$ or $\mathbb{C}$, then the reference to the field will be omitted.

The *classical Alexander polynomial* (denoted $\Delta_{\mathcal{C}}(t)$) corresponds to the special case when $\mathbb{K} = \mathbb{Q}$, $\mathcal{C}_0$ is transversal to $\mathcal{C}_i$ for any $i = 1, \ldots, r$, and ε is the epimorphism that sends any meridian γ_i around $\mathcal{C}_i$ to 1, except for $i = 0$, where $\varepsilon(\gamma_0) = -d$, where $d := \sum_{i=1}^r d_i$. We will refer to this morphism as the *trivial morphism*. If $\varepsilon(\gamma_i) \neq \pm 1$ for any $i = 0, 1, \ldots, r$ we will call ε a *non-coordinate* epimorphism. The *Oka polynomials* (denoted $\Delta_{\mathcal{C},\varepsilon}(t)$) correspond to $\mathbb{K} = \mathbb{Q}$, and a transversal $\mathcal{C}_0$ ([94]).

Remark 2.2.

(1) Note that $M^{\mathbb{Z}}_{\mathcal{C},\varepsilon}$ is not necessarily a torsion module in general. For example, if $\mathcal{C}$ is a union of $r+1$ lines passing through a common point and ε is the trivial morphism, then $\pi_1(X_{\mathcal{C}}) = \mathbb{Z} * \cdots * \mathbb{Z}$ a free product of rank r, and it is easy to see that $M^{\mathbb{K}}_{\mathcal{C},\varepsilon} = (\Lambda^{\mathbb{K}}_{\varepsilon})^{r-1}$.

(2) Also note that $M^{\mathbb{Z}}_{\mathcal{C},\varepsilon}$ depends only on $G = \pi_1(X_{\mathcal{C}})$ and ε. Hence one can associate an Alexander polynomial $\Delta_{G,\varepsilon}(t)$ to any finitely presented group G and epimorphism $\varepsilon : G/G' \to \mathbb{Z}$. In fact, such a polynomial corresponds to the Alexander polynomial of the CW-complex X_G associated with any finite presentation of G, and $\varepsilon : H_1(X_G;\mathbb{Z}) \to \mathbb{Z}$.

(3) Assume that

$$(\Lambda^{\mathbb{K}}_{\varepsilon})^m \xrightarrow{A} (\Lambda^{\mathbb{K}}_{\varepsilon})^n \to M^{\mathbb{K}}_{G,\varepsilon}$$

is a free resolution of $M^{\mathbb{K}}_{G,\varepsilon}$, where A is an $n \times m$ matrix with coefficients in $\Lambda^{\mathbb{K}}_{\varepsilon}$. Then $\Delta_{G,\varepsilon}(t)$ can also be defined as 0 if $m < n$, or as the greatest common divisor of all the minors of maximal order of A if $n \leq m$. From (2) above, n can be considered as the number of generators in a presentation of G.

A very useful remark on Alexander polynomials is the following:

Lemma 2.3. [71, Proposition 2.1] *Let $G \overset{\psi}{\twoheadrightarrow} H$ be an epimorphism of finitely presented groups and consider $\varepsilon_H : H/H' \twoheadrightarrow \mathbb{Z}$ another epimorphism. Then $\Delta^{\mathbb{K}}_{H,\varepsilon_H}$ divides $\Delta^{\mathbb{K}}_{G,\varepsilon_G}$, where $\varepsilon_G = \varepsilon_H \circ \psi_1$ and $\psi_1 : G/G' \to H/H'$ is induced by ψ.*

Proof. A presentation of H can be given from one of G just by adding a finite number of relations. Therefore from Remark 2.2(3), a presentation matrix for $M^{\mathbb{K}}_{H,\varepsilon_H}$ is the result of adding a finite number of columns to the presentation matrix of $M^{\mathbb{K}}_{G,\varepsilon_G}$. Therefore the ideal generated by the minors of maximal order of $M^{\mathbb{K}}_{G,\varepsilon_G}$ is contained in the one of $M^{\mathbb{K}}_{H,\varepsilon_H}$. Q.E.D.

This situation appears in a natural way when an equisingular family of curves $\{\mathcal{C}_t\}_{t\in(0,\delta]}$ degenerates into a reduced curve $\mathcal{C}_0$.

Proposition 2.4. *Under the above conditions there is an epimorphism of fundamental groups*

$$\pi_1(X_{\mathcal{C}_0}) \overset{j}{\twoheadrightarrow} \pi_1(X_{\mathcal{C}_\delta}).$$

Hence $\Delta^{\mathbb{K}}_{\mathcal{C}_\delta,\varepsilon_1}$ divides $\Delta^{\mathbb{K}}_{\mathcal{C}_0,\varepsilon_2}$, where $\varepsilon_2 = \varepsilon_1 \circ j_1$ as in Lemma 2.3.

Proof. A proof of the first part can be found in [41, Corollary §3 (3.2)]. The second part is an immediate consequence of Lemma 2.3. Q.E.D.

Example 2.5.

(1) Consider a family of $r+1$ lines $\mathcal{C}_t := \ell_{t,0} \cup \cdots \cup \ell_{t,r}$, $t \in (0,1]$ in general position degenerating into $r+1$ lines $\mathcal{C}_0 := \ell_0 \cup \cdots \cup \ell_r$ passing through a common point. If ε is the trivial morphism $\varepsilon(\gamma_i) = 1$, $i = 1, \ldots, r$ and $\varepsilon(\gamma_0) = -r$, then one has the following

$$
\begin{array}{lll}
M^{\mathbb{K}}_{\mathcal{C}_1,\varepsilon} = \left(\frac{\Lambda^{\mathbb{K}}_{\varepsilon}}{(t-1)}\right)^{r-1} & \Rightarrow & \Delta^{\mathbb{K}}_{\mathcal{C}_1,\varepsilon}(t) = (t-1)^{r-1} \\
M^{\mathbb{K}}_{\mathcal{C}_0,\varepsilon} = \left(\Lambda^{\mathbb{K}}_{\varepsilon}\right)^{r-1} & \Rightarrow & \Delta^{\mathbb{K}}_{\mathcal{C}_0,\varepsilon}(t) = 0.
\end{array} \tag{13}
$$

(2) Consider the three-cuspidal quartic $\mathcal{C}_1$ presented in (Z6) and a generic line $\mathcal{C}_0$. In order to give a presentation for the fundamental group of $\mathcal{C} := \mathcal{C}_0 \cup \mathcal{C}_1$ one can simply apply Proposition 1.16 to the presentation (1) and obtain

$$
\langle a, b \mid aba = bab, \quad [a, a^2b^2] = [b, a^2b^2] = 1 \rangle. \tag{14}
$$

Note that there is basically only one possible morphism ε, the abelianization morphism, which we will omit in the notation. An easy computation produces

$$
M^{\mathbb{K}}_{\mathcal{C}} = \frac{\mathbb{K}[t^{\pm 1}]}{(3, t+1)}
$$

and hence

$$
\Delta^{\mathbb{K}}_{\mathcal{C}}(t) = \begin{cases} t+1 & \text{if } \operatorname{char}(\mathbb{K}) = 3 \\ 1 & \text{otherwise.} \end{cases}
$$

Since the three-cuspidal quartic is dual to a nodal cubic, we know it has a bitangent, say ℓ_0. The fundamental group of $\mathcal{C}' := \ell_0 \cup \mathcal{C}_1$ has the following presentation (see [97, Example 4.5(3)])

$$
\langle a_1, a_2, a_3, a_4 \mid a_i a_{i+1} a_i = a_{i+1} a_i a_{i+1} (i = 1, .., 3), \quad a_2 a_4 = a_1 a_2 \rangle, \tag{15}
$$

which produces

$$
M^{\mathbb{K}}_{\mathcal{C}'} = \frac{\mathbb{K}[t^{\pm 1}]}{(t^2 - t + 1)} \oplus \frac{\mathbb{K}[t^{\pm 1}]}{(t^2 - t + 1)}
$$

and hence

$$\Delta_{\mathcal{C}'}^{\mathbb{K}}(t) = (t^2 - t + 1)^2.$$

Note that if $\operatorname{char}(\mathbb{K}) = 3$, then $(t^2 - t + 1) = (t+1)^2$, and hence $\Delta_{\mathcal{C}'}^{\mathbb{K}}(t) = (t+1)^4$.

All computations in Example 2.5 above have been performed directly from a presentation of the fundamental group. We refer the reader to Section §2.5 for more details on this.

Remark 2.6. Very often the Alexander polynomial is defined as the torsion of $H_1(X_{\mathcal{C},\varepsilon}, \pi_\varepsilon^{-1}(p); \mathbb{Q})$ for some $p \in X_{\mathcal{C}}$. This definition comes handy for computational purposes, since it can be obtained as the determinantal variety of corank 1 of the Fox derivative matrix associated with the group G, as in the case of knots ([35]).

The geometrical interpretation of the classical Alexander polynomial is given as follows (see [101]). The polynomial $C_1 \cdot \ldots \cdot C_r$ defines a non-isolated singularity at the origin of $\mathbb{C}^3$. The monodromy of the Milnor fiber defines an automorphism on the H_1 and the classical Alexander polynomial is the characteristic polynomial of the monodromy of the Milnor fiber.

Theorem 2.7. [94, Theorem 43] *The Alexander polynomial of $\mathcal{C}$ with respect to the epimorphism $\varepsilon : H_1(X_{\mathcal{C}}) \twoheadrightarrow \mathbb{Z}$ $(\varepsilon_i \geq 0,\ i = 1, \ldots, r)$ is equal to the characteristic polynomial of the monodromy $h_* : H_1(F) \to H_1(F)$ where F is the Milnor fiber of the polynomial $C_1^{\varepsilon_1} \cdot \ldots \cdot C_r^{\varepsilon_r}$.*

Since the monodromy has a finite order, this implies the following.

Corollary 2.8. *All the zeroes of the Alexander polynomial $\Delta_{\mathcal{C},\varepsilon}(t)$ of a curve $\mathcal{C}$ with respect to an epimorphism ε are roots of unity.*

Alexander polynomials depend on the local type of singularities of $\mathcal{C}$. To describe this dependency we will consider $L_1, \ldots, L_s$ the local links of the singularities of the affine part $\mathcal{C}^{\text{aff}} := \mathcal{C}_1 \cup \cdots \cup \mathcal{C}_r$ and L_∞ the link at infinity, that is, the intersection of $\mathcal{C}^{\text{aff}}$ with the boundary of a tubular neighborhood of the line at infinity $\mathcal{C}_0$. The inclusion $\mathbb{S}^3 \setminus L_k \hookrightarrow X_{\mathcal{C}}$ induces a map $\pi_1(\mathbb{S}^3 \setminus L_k) \to \pi_1(X_{\mathcal{C}})$. Therefore ε also induces epimorphisms $\pi_1(\mathbb{S}^3 \setminus L_k) \twoheadrightarrow \mathbb{Z}$. The Alexander polynomials associated with such maps will be called local Alexander polynomials and denoted by $\Delta_{L_k,\varepsilon}$ for simplicity.

This dependency can be described for classical Alexander polynomials.

Theorem 2.9 ([71]). *The classical Alexander polynomial of $\mathcal{C}$ divides both the product of the local Alexander polynomials $\prod_{k=1}^{s} \Delta_{L_k}(t)$ and $\Delta_{L_\infty}(t)$.*

This dependency also has an expression for general Alexander polynomials.

Theorem 2.10. *Under the above conditions*

$$\left(\prod_{i=1}^{r}(1-t^{\varepsilon_i})^{s_i-\chi(\mathcal{C}_i)}\right)\prod_{k=1,\ldots,s,\infty}\Delta_{L_k,\varepsilon}(t)=\Delta^2_{\mathcal{C},\varepsilon}(t)\cdot\det\varphi^t(\mathcal{C}),$$

where $s_i := \#\operatorname{Sing}(\mathcal{C}^{\text{aff}})\cap\mathcal{C}_i$, χ *is the Euler characteristic and* $\varphi^t(\mathcal{C})$ *is an intersection form on* $H_2(X_{\mathcal{C},\varepsilon},\mathbb{Q}[t^{\pm 1}])$ *with twisted coefficients.*

Proof. It is an immediate consequence of [31, Theorem 5.6] and the fact that $\Delta^{\mathbb{Q}}_{\mathcal{C},\varepsilon}(t^{-1})=\Delta^{\mathbb{Q}}_{\mathcal{C},\varepsilon}(t)$ (by Corollary 2.8). Q.E.D.

The fact that Alexander polynomials are not combinatorial invariants was already known (with a different language) by Zariski as mentioned in the Introduction with the first example of a classical Zariski pair.

A topological interpretation of classical Alexander polynomials can be given as follows: an n-th root of unity ($n>1$) is a root of the classical Alexander polynomial of a curve $\mathcal{C}$ if and only if the cyclic covering of the complement $X_{\mathcal{C}}$ ramified along each irreducible component of $\mathcal{C}^{\text{aff}}$ with order n has a bigger first Betti number than $h_1(X_{\mathcal{C}};\mathbb{C})$ ([71, Corollary 3.2]). Moreover the difference between the two Betti numbers is exactly the sum of the multiplicities of such roots. The reason for this is that the Alexander invariant $M_{\mathcal{C}}$ is *semisimple* in this case, that is, it is a direct sum of modules with no proper submodules (also called *simple modules*).

Analogously, for general Alexander polynomials one has the following:

Theorem 2.11. *Let* $\mathcal{C}$ *be a curve and* $\varepsilon: H_1(X_{\mathcal{C}};\mathbb{Z})\to\mathbb{Z}$ *an epimorphism. If an* n*-th primitive root of unity* ($n>1$) *is a root of the Alexander polynomial* $\Delta_{\mathcal{C},\varepsilon}(t)$ *then the covering* $X^n_{\mathcal{C},\varepsilon}$ *has a bigger first Betti number than* $h_1(X_{\mathcal{C}};\mathbb{C})$.

Moreover,

$$h_1(X^n_{\mathcal{C},\varepsilon};\mathbb{C})=h_1(X_{\mathcal{C}};\mathbb{C})+\sum_{i=1}^{m}\alpha_i^n, \tag{16}$$

where α_i^n *is the number of common roots between* $\frac{t^n-1}{t-1}$ *and* $\lambda_i(t)$ *from* (12).

Remark 2.12.

(1) A similar formula for homology with coefficients in other fields exists [83, Theorem 4.6]. The field needs to contain all the n-th roots of unity.

(2) Note that in general one cannot just count multiplicities in order to compute the first Betti number of cyclic (or abelian) coverings since the Alexander invariant need not be semisimple. For instance, consider the Example 2.5(2) of the three cuspidal quartic and the bitangent line $\mathcal{C}'$. According to Matei-Suciu [83] if ε is the trivial morphism and $n = 2$, then the formula (16) is still valid

$$h_1(X^2_{\mathcal{C}',\varepsilon}; \mathbb{F}_3) = h_1(X_{\mathcal{C}'}; \mathbb{F}_3) + \sum_{i=1}^{2} \alpha_i^2,$$

but in this case $h_1(X^2_{\mathcal{C}',\varepsilon}; \mathbb{F}_3) - h_1(X_{\mathcal{C}'}; \mathbb{F}_3) = 2$ even though $t = -1$ has multiplicity 4 in $\Delta^{\mathbb{F}_3}_{\mathcal{C}'}(t)$.

2.2. Characteristic varieties

Characteristic varieties were introduced by Hillman [55] for links, then systematically studied by Arapura [1] for Kähler manifolds, and first applied to algebraic curves by Libgober [74]. They can be defined analogously to Alexander polynomials as follows.

Let $\mathcal{C} := \mathcal{C}_1 \cup \cdots \cup \mathcal{C}_r$ similarly as at the beginning of this section, except that we are not asking any component to be a line. Let $\tau := \gcd(d_1, \ldots, d_r)$. Then

$$H_1(X_{\mathcal{C}}; \mathbb{Z}) = \frac{\bigoplus_{i=1}^{r} \gamma_i \mathbb{Z}}{\langle d_1\gamma_1 + \cdots + d_r\gamma_r \rangle} \approx \mathbb{Z}^{r-1} \oplus \frac{\mathbb{Z}}{\tau\mathbb{Z}}, \tag{17}$$

where γ_i is the homology class of a meridian of $\mathcal{C}_i$.

We can study the projective plane curve $\mathcal{C}$ as follows; let $\mathrm{ab} : G := \pi_1(X_{\mathcal{C}}) \to H_1(X_{\mathcal{C}}; \mathbb{Z})$ be the abelianization epimorphism. The kernel G' of ab defines the universal abelian covering of $X_{\mathcal{C}}$, say $X_{\mathcal{C},\mathrm{ab}} \to X_{\mathcal{C}}$, whose group of deck transformations is $G/G' = H_1(X_{\mathcal{C}}; \mathbb{Z})$. Such a group acts on $G'/G'' = H_1(X_{\mathcal{C},\mathrm{ab}}; \mathbb{Z})$ by conjugation as before endowing $M^{\mathbb{Z}}_{\mathcal{C},\mathrm{ab}} := H_1(X_{\mathcal{C},\mathrm{ab}}; \mathbb{Z})$ with a $\Lambda_{\mathcal{C}}$-module structure, where $\Lambda_{\mathcal{C}} := \mathbb{Z}[G/G'] \approx \mathbb{Z}[t_1^{\pm 1}, \ldots, t_r^{\pm 1}]/(t_1^{d_1} \cdot \ldots \cdot t_r^{d_r} - 1)$.

One can tensor $M^{\mathbb{Z}}_{\mathcal{C},\mathrm{ab}}$ by a field $\mathbb{K} = \mathbb{Q}, \mathbb{C}, \mathbb{F}_p, \ldots$ to obtain a module $M^{\mathbb{K}}_{\mathcal{C},\mathrm{ab}}$ over $\Lambda^{\mathbb{K}}_{\mathcal{C}} = \mathbb{K}[t_1^{\pm 1}, \ldots, t_r^{\pm 1}]/(t_1^{d_1} \cdot \ldots \cdot t_r^{d_r} - 1)$ (in general we only ask $\Lambda^{\mathbb{K}}_{\mathcal{C}}$ to be integrally closed and Noetherian). Since G is finitely presented,

$M^{\mathbb{K}}_{\mathcal{C},\mathrm{ab}}$ is again a finitely generated $\Lambda^{\mathbb{K}}_{\mathcal{C}}$-module (by as many 1-cells as generators of G). If $r \geq 2$, then $\Lambda^{\mathbb{K}}_{\mathcal{C}}$ is not a Principal Ideal Domain and hence one has to study the module invariants of $M^{\mathbb{K}}_{\mathcal{C},\mathrm{ab}}$, that is, the Fitting ideals of $M^{\mathbb{K}}_{\mathcal{C},\mathrm{ab}}$.

Let us briefly recall the notion of Fitting ideals. Let R be a commutative Noetherian ring with unity and M a finitely generated R-module. One has a finite free presentation for M, say $\phi : R^m \to R^n$, where $M = \operatorname{coker} \phi$. The homomorphism ϕ has an associated $(n \times m)$ matrix A_ϕ with coefficients in R such that $\phi(x) = A_\phi x^t$.

Definition 2.13. The k-th *Fitting ideal* $F_k(M)$ of M is defined as the ideal generated by

$$\begin{cases} 0 & \text{if } k \leq \max\{0, n-m\} \\ 1 & \text{if } k > n \\ \text{minors of } A_\phi \text{ of order } (n-k+1) & \text{otherwise.} \end{cases}$$

It will be denoted F_k if no ambiguity seems likely to arise.

Definition 2.14. [74] Under the above conditions the *k-th characteristic variety* of M can be defined as

$$\operatorname{Char}_k(M) := \operatorname{Supp}_R\left(R/F_k(M)\right) \subset \operatorname{Spec} R.$$

The subindex k is also known as the *depth* of a characteristic variety.

Similarly, the *k-th projective characteristic variety* $\operatorname{Char}^{\mathbb{K}}_{k,\mathbb{P}}(\mathcal{C})$ of a curve $\mathcal{C}$ is the k-th characteristic variety of $M^{\mathbb{K}}_{\mathcal{C},\mathrm{ab}}$ as a $\Lambda^{\mathbb{K}}_{\mathcal{C}}$-module.

If L is a line not contained in $\mathcal{C}$ then $\Lambda^{\mathbb{K}}_{L\cup\mathcal{C}}$ is naturally isomorphic to $\mathbb{K}[t_1^{\pm 1}, \ldots, t_r^{\pm 1}]$. Moreover, if $L \pitchfork \mathcal{C}$ then the $\Lambda^{\mathbb{K}}_{L\cup\mathcal{C}}$-module $M^{\mathbb{K}}_{L\cup\mathcal{C},\mathrm{ab}}$ does not depend on L by Proposition 1.16 and $\operatorname{Char}^{\mathbb{K}}_{k,\mathbb{P}}(L \cup \mathcal{C})$ is called the *k-th affine characteristic variety* and denoted simply by $\operatorname{Char}^{\mathbb{K}}_k(\mathcal{C})$.

One can also define the k-th characteristic variety $\operatorname{Char}^{\mathbb{K}}_k(G)$ of a finitely presented group G as the k-th characteristic variety of the module $M^{\mathbb{K}}_G$ obtained by considering the CW-complex associated with a given presentation (of course, such invariant is independent of the finite presentation of G).

In the particular case when $\mathbb{K} = \mathbb{C}$ and $M = M^{\mathbb{C}}_{\mathcal{C},\mathrm{ab}}$ one has:

- $\operatorname{Spec} \Lambda_{L\cup\mathcal{C}} = \mathbb{T}^r = (\mathbb{C}^*)^r$, for the affine case, and
- $\operatorname{Spec} \Lambda_{\mathcal{C}} = \mathbb{T}_{\mathcal{C}} = \{\omega^i\}_{i=0}^{\tau-1} \times (\mathbb{C}^*)^{r-1} = V(t_1^{d_1} \cdot \ldots \cdot t_r^{d_r} - 1) \subset \mathbb{T}^r$, where ω is a τ-th primitive root of unity for the projective case.

In the case of a finitely presented group G where $G/G' = \mathbb{Z}^r \oplus \mathbb{Z}/\tau_1\mathbb{Z} \oplus \cdots \oplus \mathbb{Z}/\tau_s\mathbb{Z}$ we obtain

$$\operatorname{Spec} \Lambda_G = \mathbb{T}_G = \{(\omega_1^{i_1}, \ldots, \omega_s^{i_s}) \mid i_k = 0, \ldots, \tau_k - 1,\ k = 1, \ldots, s\} \times (\mathbb{C}^*)^r,$$

where $\Lambda_G = \mathbb{C}[G/G']$ and ω_i is a τ_i-th primitive root of unity.

One might want to keep the non-reduced structure of the Fitting ideal. In that case we define the *projective (resp. affine) k-th Fitting ideal* of the curve $\mathcal{C}$ over the field $\mathbb{K}$ and denote it as $F_{k,\mathbb{P}}^{\mathbb{K}}(\mathcal{C})$ (resp. $F_k^{\mathbb{K}}(\mathcal{C})$).

Remarks 2.15.

(1) Note that any isomorphism between two finitely presented groups G_1 and G_2 produces an automorphism of the ambient torus $\mathbb{T}_{G_1} = \mathbb{T}_{G_2}$ such that $\operatorname{Char}_k(G_1)$ and $\operatorname{Char}_k(G_2)$ are isomorphic.

Note however, that in some particular cases, like fundamental groups of link complements (which will not be considered here) or curve complements, the ambient torus has a natural system of coordinates. For instance in the latter case, a natural system of coordinates for $\mathbb{T}_{\mathcal{C}}$ is given by the preferred basis of $H_1(X_{\mathcal{C}}; \mathbb{Z})$ described in (17). In that respect, the characteristic varieties of $\mathcal{C}$ as subspaces of the torus $\mathbb{T}_{\mathcal{C}}$ are not directly an invariant of the group G, rather they are invariant of the embedding of $\mathcal{C}$ in $\mathbb{P}^2$, which is the group G with some peripheral information about the homology classes of the meridians of the irreducible components of $\mathcal{C}$.

(2) In the particular case of plane algebraic curves, due to the Hodge decomposition of $H_1(X_{\varepsilon,\mathcal{C}}; \mathbb{C})$ for appropriate branched coverings, the ring homomorphism $\Lambda_{\mathcal{C}}^{\mathbb{Z}} \to \Lambda_{\mathcal{C}}^{\mathbb{Z}}$ given by $t_i \to t_i^{-1}$ induces an automorphism of the components of $\operatorname{Char}_k(\mathcal{C})$ containing ε (see [75, Theorem 3.1.c]).

Proposition 2.16. *If $(\mathbb{P}^2, \mathcal{C})$ and $(\mathbb{P}^2, \mathcal{D})$ are homeomorphic, then $\Lambda^{\mathbb{C}} := \Lambda_{\mathcal{C}}^{\mathbb{C}} = \Lambda_{\mathcal{D}}^{\mathbb{C}}$ in a natural way, and $M_{\mathcal{C},\mathrm{ab}}^{\mathbb{C}} \approx M_{\mathcal{D},\mathrm{ab}}^{\mathbb{C}}$ are isomorphic as $\Lambda^{\mathbb{C}}$-modules. In particular $\operatorname{Char}_k^{\mathbb{C}}(\mathcal{C}) = \operatorname{Char}_k^{\mathbb{C}}(\mathcal{D})$.*

Proof. Let us denote by $f : (\mathbb{P}^2, \mathcal{C}) \to (\mathbb{P}^2, \mathcal{D})$ the homeomorphism of pairs. Note that the image by f of any disk transversal to a component $\mathcal{C}_i$ of $\mathcal{C}$ will be sent to a disk transversal to a component, say $\mathcal{D}_i$, of $\mathcal{D}$. Since irreducible components intersect pairwise and always positively, it is possible to prove that f must either respect or reverse orientations on all the irreducible components of the curves. Therefore meridians will be sent to meridians (up to sign) and the induced homomorphism of groups $f_* : H_1(X_{\mathcal{C}};\mathbb{Z}) \to H_1(X_{\mathcal{D}};\mathbb{Z})$ has the expected property $f_*(\gamma_{\mathcal{C}_i}) = \delta\gamma_{\mathcal{D}_i}$, $i = 1,\dots,r$ ($\delta = \pm 1$). By Remark 2.22 below, one can assume that $\delta = 1$. Finally note that $d_i := \deg \mathcal{C}_i = \deg \mathcal{D}_i$ is preserved, since it is a topological invariant and hence the first part follows and $\Lambda^{\mathbb{C}} = \mathbb{C}[t_1^{\pm 1},\dots,t_r^{\pm 1}]/(t_1^{d_1}\cdot\ldots\cdot t_r^{d_r}-1)$. The fundamental groups of $X_{\mathcal{C}}$ and $X_{\mathcal{D}}$ are isomorphic and the action of each t_i on both $M^{\mathbb{C}}_{\mathcal{C},\mathrm{ab}} = \Lambda^{\mathbb{C}} \otimes_\Lambda \pi_1(X_{\mathcal{C}})'/\pi_1(X_{\mathcal{C}})''$ and $M^{\mathbb{C}}_{\mathcal{D},\mathrm{ab}} = \Lambda^{\mathbb{C}} \otimes_\Lambda \pi_1(X_{\mathcal{D}})'/\pi_1(X_{\mathcal{D}})''$ is preserved by f. Therefore the second part follows. Q.E.D.

Remark 2.17.

(1) Note that the isomorphism of $\Lambda^{\mathbb{K}}$-modules exhibited in Proposition 2.16 is of a very special kind, since it comes from an isomorphism of fundamental groups inducing the identity on the abelianization. Since the Alexander invariant G/G'' of a group G can be seen as an extension of G/G' by G'/G'', this type of isomorphisms of $\Lambda^{\mathbb{K}}$-modules will be called *extension isomorphisms.*

(2) Closely related to Remark 2.15.(1), if one wants to say something about whether or not the fundamental groups of two curves are isomorphic, verifying that $M^{\mathbb{K}}_{\mathcal{C},\mathrm{ab}} \not\cong M^{\mathbb{K}}_{\mathcal{D},\mathrm{ab}}$ as modules over an abstract Λ or $\mathrm{Char}_k(\mathcal{C}) \neq \mathrm{Char}_k(\mathcal{D})$ is not enough. Instead, invariants of the isomorphism class of $\mathrm{Char}_k(\mathcal{C})$ and $\mathrm{Char}_k(\mathcal{D})$ should be used such as their total number of irreducible components of a certain dimension (see Section §4), or their combinatorial structure (see §2.3).

Example 2.18.

(1) Let $G = \mathbb{Z}^q * \mathbb{Z}/p_1\mathbb{Z} * \cdots * \mathbb{Z}/p_s\mathbb{Z}$. According to Proposition 2.39

$$M^{\mathbb{K}}_G = \frac{\bigoplus_{1\leq i<j\leq r} \Lambda^{\mathbb{K}}_G x_{i,j}}{\mathcal{T}+\mathcal{J}},$$

where

- $\Lambda^{\mathbb{K}}_G = \mathbb{K}[t_1^{\pm 1},\dots,t_{q+s}^{\pm 1}]/(t_{q+1}^{p_1}-1,\dots,t_{q+s}^{p_s}-1)$,
- $\mathcal{T}$ is the submodule generated by $\frac{t_{q+i}^{p_i}-1}{t_{q+i}-1}$, $i = 1,\dots,s$, and
- $\mathcal{J}$ is the Jacobian submodule.

In this situation it is easy to see that

$$F^{\mathbb{K}}_1(G) = \left(\frac{t_{q+1}^{p_1}-1}{t_{q+1}-1},\dots,\frac{t_{q+s}^{p_s}-1}{t_{q+s}-1}\right).$$

And hence,

$$\operatorname{Char}_1(G) = (\mathbb{C}^*)^q \times \{(\omega_1, \dots, \omega_s) \mid \omega_i^{p_i} = 1, \omega_i \neq 1, i = 1, \dots, s\}.$$

(2) Consider the Hopf link of d components. For convenience we denote its components as $L_{i,j(i)}$, $j(i) = 1, \dots, d_i$, $i = 1, \dots, r$, where $\sum_{i=1}^r d_i = d$. A natural presentation of its fundamental group G is given by meridians $\gamma_{i,j(i)}$ for each component and an extra generator σ, and whose relations are $[\sigma, \gamma_{i,j(i)}] = 1$ and

$$\sigma \prod_{i=1}^{r} \prod_{j(i)=1}^{d_i} \gamma_{i,j(i)} = 1,$$

i.e, G is the direct product of $\mathbb{Z}$ and the free group in $d-1$ generators. Then, $\Lambda_G^{\mathbb{K}}$ is a ring of Laurent polynomials in the variables s and $t_{i,j(i)}$, $j(i) = 1, \dots, d_i$, $i = 1, \dots, r$, where the product of all variables equals 1. Applying Proposition 2.39 one obtains

$$M_G^{\mathbb{K}} = \frac{\bigoplus_{1 \leq i < j < d} \Lambda_G^{\mathbb{K}} x_{i,j}}{\mathcal{W}_d + \mathcal{J}},$$

where $\mathcal{W}_d$ is the submodule generated by $s-1$, and $\mathcal{J}$ is the Jacobian submodule. Hence

$$F_1^{\mathbb{K}}(G) = (s-1) = (\prod_{i=1}^{r} \prod_{j(i)=1}^{d_i} t_{i,j(i)} - 1).$$

Remark 2.19. Note that if $0 \to M' \to M \to M'' \to 0$ is an exact sequence of R-modules, then $\operatorname{Char}_1^{\mathbb{K}}(M) = \operatorname{Char}_1^{\mathbb{K}}(M') \cup \operatorname{Char}_1^{\mathbb{K}}(M'')$. Let L be a line transversal to $\mathcal{C}$, and consider $T(L)$ a tubular neighborhood of L. In this situation there is a surjection

$$G_\infty := \pi_1(\partial T(L) \setminus \mathcal{C}) \twoheadrightarrow G_L := \pi_1(\mathbb{P}^2 \setminus (L \cup \mathcal{C})).$$

Following the notations of Example 2.18(2), we assume that the image of $\gamma_{i,j(i)}$ by this surjection is a meridian of $\mathcal{C}_i$.

Since $\Lambda_{G_L}^{\mathbb{K}} = \mathbb{K}[t_1^{\pm 1}, \dots, t_r^{\pm 1}]$, the above morphism induces a ring morphism $\Lambda_{G_\infty}^{\mathbb{K}} \twoheadrightarrow \Lambda_{G_L}^{\mathbb{K}}$ given by $t_{i,j(i)} \mapsto t_i$ and $s \mapsto (t_1^{d_1} \cdot \ldots \cdot t_r^{d_r})^{-1}$. Therefore, the surjection $M_{G_\infty, G_L} := M_{G_\infty} \otimes \Lambda_{G_L}^{\mathbb{K}} \to M_{L \cup \mathcal{C}, \mathrm{ab}}$ induces an inclusion

$$\operatorname{Char}_1^{\mathbb{K}}(\mathcal{C}) = \operatorname{Char}_1^{\mathbb{K}}(G_L) \subset \operatorname{Char}_1^{\mathbb{K}}(M_{G_\infty} \otimes \Lambda_{G_L}^{\mathbb{K}}) = V(t_1^{d_1} \cdot \ldots \cdot t_r^{d_r} - 1),$$

where the last equality can be computed from Example 2.18(2) above.

In other words, even though the affine $\mathrm{Char}_1(\mathcal{C})$ seems to sit in the bigger torus $\mathbb{T}^r$ than the projective $\mathrm{Char}_{1,\mathbb{P}}(\mathcal{C})$, the fact is that both are contained in $\mathbb{T}_\mathcal{C}$. Moreover, they coincide as subtori of $\mathbb{T}_\mathcal{C}$ ([75, Proposition 1.2.3]). In what follows we will use either one indistinctly.

D. Arapura, in [1, Theorem 1.6] gives the following description of the structure of first characteristic varieties for certain Kähler varieties. We adapt the original statement for our particular case of curve complements. In order to do that we need the following concept.

Definition 2.20. A (fixed component free) pencil of curves in $\mathbb{P}^2$ is said to *completely contain a curve* $\mathcal{C}$, if the induced morphism $\hat{f}$: $\widehat{\mathbb{P}}^2 \to \mathbb{P}^1$ (after blowing-up the corresponding base points) satisfies that $\widehat{\mathcal{C}} \subset \hat{f}^{-1}(P)$, where P is a finite subset of $\mathbb{P}^1$, and $\widehat{\mathcal{C}}$ is the strict transform of $\mathcal{C}$ in $\widehat{\mathbb{P}}^2$. Note that, if a pencil completely contains $\mathcal{C}$, then f restricts to a well-defined holomorphic map $f : \mathbb{P}^2 \setminus \mathcal{C} = X_\mathcal{C} \to \mathbb{P}^1 \setminus P$. A pencil is said to *contain a curve* $\mathcal{C}$, if it completely contains at least one irreducible component of $\mathcal{C}$. In addition, a pencil is called *primitive* if the fibers of $\hat{f}$ are connected.

Theorem 2.21 ([1]). *There exist a finite number of torsion points $\varepsilon_i \in \mathbb{T}_\mathcal{C}$, unitary points $\tilde{\varepsilon}_j \in \mathbb{T}_\mathcal{C}$, and primitive pencils containing $\mathcal{C}$, $f_i : X_\mathcal{C} \to \mathcal{D}_i = \mathbb{P}^1 \setminus P_i$ such that*

$$\mathrm{Char}_{1,\mathbb{P}}(\mathcal{C}) := \bigcup \varepsilon_i f_i^* H^1(\mathcal{D}_i; \mathbb{C}^*) \cup \bigcup \tilde{\varepsilon}_j.$$

An analogous result follows for the affine case $\mathrm{Char}_1(\mathcal{C})$.

Note that any element of $H^1(\mathcal{D}_i; \mathbb{C}^*) = \mathrm{Hom}(H_1(\mathcal{D}_i; \mathbb{C}), \mathbb{C}^*)$, i.e. any character on $H_1(\mathcal{D}_i; \mathbb{C})$ can be seen as a point of $\mathrm{Spec}(\Lambda^\mathbb{C}_{\mathcal{D}_i}) = \mathrm{Spec}(\mathbb{C}[H_1(\mathcal{D}_i; \mathbb{C})]) = \mathbb{T}^{h_1(\mathcal{D}_i;\mathbb{C})}$ and vice versa. Therefore $f_i^* H^1(\mathcal{D}_i; \mathbb{C}^*)$ is a subset of $\mathbb{T}_\mathcal{C}$. Also note that $\varepsilon_i f_i^* H^1(\mathcal{D}_i; \mathbb{C}^*)$ refers to coordinatewise multiplication in $\mathbb{T}_\mathcal{C}$. Finally, ε_i is a torsion point if $\varepsilon_i^n = (1, \ldots, 1) =: \mathbb{1}_r$ for some $n \in \mathbb{Z}$ and $\tilde{\varepsilon}_j$ is unitary if $\tilde{\varepsilon}_j \in (\mathbb{S}^1)^r \subset \mathbb{T}_\mathcal{C}$.

Remark 2.22. According to a recent work by Libgober [78] unitary non-torsion isolated points cannot exist in $\mathrm{Char}_{k,\mathbb{P}}(\mathcal{C})$. Therefore, according to Remark 2.15(2), the ring automorphism $\Lambda^\mathbb{Z}_\mathcal{C} \to \Lambda^\mathbb{Z}_\mathcal{C}$ given by $t_i \to t_i^{-1}$ induces a skew automorphism of the corresponding modules $M^\mathbb{C}_{\mathcal{C},\mathrm{ab}}$.

Certain components of $\mathrm{Char}_{k,\mathbb{P}}(\mathcal{C})$ can be inherited from subarrangements of $\mathcal{C}$. More specifically, let us assume that $V \subset \mathrm{Char}_{k,\mathbb{P}}(\mathcal{C}_{(i)})$

is a non-empty component of the k-th characteristic variety of $\mathcal{C}_{(i)} = \mathcal{C}_1 \cup \cdots \cup \mathcal{C}_{i-1} \cup \mathcal{C}_{i+1} \cup \cdots \cup \mathcal{C}_r$. Then the natural surjection

$$\begin{array}{ccc} \Lambda_{\mathcal{C}} & \twoheadrightarrow & \Lambda_{\mathcal{C}_{(i)}} \\ t_j & \mapsto & \begin{cases} t_j & \text{if } j \neq i \\ 1 & \text{if } i = j \end{cases} \end{array}$$

induces an inclusion

$$\operatorname{Spec}(\Lambda_{\mathcal{C}_{(i)}}) = \mathbb{T}_{\mathcal{C}_{(i)}} \subset \mathbb{T}^i_{\mathcal{C}}$$

where $\mathbb{T}^i_{\mathcal{C}} = \{(\varepsilon_1, \ldots, \varepsilon_r) \in \mathbb{T}_{\mathcal{C}} \mid \varepsilon_i = 1\}$.

This allows us to see both $M_{\mathcal{C},\mathrm{ab}}$ and $M_{\mathcal{C}_{(i)},\mathrm{ab}}$ as $\Lambda_{\mathcal{C}}$-modules and consider the natural surjection $M_{\mathcal{C},\mathrm{ab}} \twoheadrightarrow M_{\mathcal{C}_{(i)},\mathrm{ab}}$, as a surjection of modules. Remark 2.19 and the previous discussion imply that $\operatorname{Char}_{k,\mathbb{P}}(\mathcal{C}_{(i)}) \subset \operatorname{Char}_{k,\mathbb{P}}(\mathcal{C}) \cap \mathbb{T}^i_{\mathcal{C}}$.

Definition 2.23. We call a component V of $\operatorname{Char}_{k,\mathbb{P}}(\mathcal{C})$ *essential* if it is not contained in any $\operatorname{Char}_{k,\mathbb{P}}(\mathcal{C}_{(i)})$. Otherwise we say V is *non-essential.* We call V a *coordinate component* if it is contained in a coordinate torus $\mathbb{T}^i_{\mathcal{C}}$. Otherwise we say V is *non-coordinate.*

Note that non-coordinate components are necessarily essential by the discussion above.

Libgober proved in [75] that any *positive dimensional* coordinate component is necessarily non-essential. In [75] he also introduced ideal sheaves called *ideals of quasi-adjunction* denoted by $\mathcal{A}^{\bar{X}}_{\mathcal{C}}$, where $\bar{X} \in (0,1)^r$ and $\bar{X}$ is determined by the configuration of singularities of $\mathcal{C}$, and showed that points in $\operatorname{Char}^{\mathrm{nc}}_{1,\mathbb{P}}(\mathcal{C})$ (the non-coordinate components of $\operatorname{Char}_{1,\mathbb{P}}(\mathcal{C})$) can be detected by studying the superabundance (see below for its definition) of a finite family of ideal sheaves $\mathcal{A}^{\bar{X}}_{\mathcal{C}}(d-3-\ell_{\bar{X}})$, where $\ell_{\bar{X}} := \sum_{i=1}^r d_i X_i \in \mathbb{N}$ (see [75] for details).

It is known that, for a given $\mathcal{C}$, $\mathcal{A}^{\bar{X}}_{\mathcal{C}}$ satisfies the following properties.

- The number of points $\bar{X} \in (0,1)^r$ determined by the configuration of singularities of $\mathcal{C}$ is finite.
- $\mathcal{O}_{\mathbb{P}^2}/\mathcal{A}^{\bar{X}}_{\mathcal{C}}$ is supported on the singularities of $\mathcal{C}$.
- Let $\bar{\xi} := (\xi_1, \ldots, \xi_r) \in \mathbb{T}_{\mathcal{C}}$ be a torsion point such that $\xi_1^{d_1} \cdot \ldots \cdot \xi_r^{d_r} = 1$, and $\xi_i \neq 1$. We define $X_i := \frac{\log \xi_i}{2\pi\sqrt{-1}} \in (0,1)$. Under these notations, $\bar{\xi} \in \operatorname{Char}_{1,\mathbb{P}}(\mathcal{C})$ if the homomorphism $\sigma_{\bar{X}}$ is *not* surjective.

$$\begin{aligned} &0 \to H^0(\mathbb{P}^2, \mathcal{A}^{\bar{X}}_{\mathcal{C}}(d-3-\ell_{\bar{X}})) \to \\ &\to H^0(\mathbb{P}^2, \mathcal{O}(d-3-\ell_{\bar{X}})) \stackrel{\sigma_{\bar{X}}}{\to} \bigoplus_{P \in \operatorname{Sing} \mathcal{C}} \mathcal{O}_{\mathbb{P}^2,P}/(\mathcal{A}^{\bar{X}}_{\mathcal{C}})_P, \end{aligned} \tag{18}$$

We say that there is a *superabundance* of dimension $\dim \operatorname{coker} \sigma_{\bar{X}}$. In Example 2.24 below, we describe $(\mathcal{A}_{\mathcal{C}}^{\bar{X}})_P$ in the case when $P \in \operatorname{Sing}(\mathcal{C})$ is a double point.

Moreover, in that case the exponential of the irreducible system of equations given by the local and global conditions provides an irreducible component of $\operatorname{Char}_{1,\mathbb{P}}(\mathcal{C})$ (by an *irreducible system of equations* with integer coefficients we mean an equivalent system where the integer coefficients of the variables are relatively prime).

Example 2.24.

(1) One of the simplest examples of positive dimensional non-coordinate characteristic varieties is the case of two conics $\mathcal{C} = \mathcal{C}_1 \cup \mathcal{C}_2$ intersecting at two tacnodes. It is not hard to see that the pencil generated by C_1 and C_2 induces an epimorphism $G := \pi_1(X_{\mathcal{C}}) \twoheadrightarrow \mathbb{Z} * \mathbb{Z}/2\mathbb{Z} =: G_2$, where a meridian of $\mathcal{C}_1$ is sent to $(1,0)$ and a meridian of $\mathcal{C}_2$ is sent to $(-1,1)$ (see [14, §4] for more details on this). By Example 2.18(1) $\operatorname{Char}_1^{\mathbb{K}}(\mathcal{C}) \subset \operatorname{Char}_1^{\mathbb{K}}(G_2) = V(\bar{t}_2 + 1)$, where the embedding comes from the following morphism of rings:

$$\begin{array}{ccc} \Lambda_{\mathcal{C}}^{\mathbb{K}} := \mathbb{K}[t_1^{\pm 1}, t_2^{\pm 1}]/(t_1^2 t_2^2 - 1) & \to & \Lambda_{G_2}^{\mathbb{K}} := \mathbb{K}[\bar{t}_1^{\pm 1}, \bar{t}_2^{\pm 1}]/(\bar{t}_2^2 - 1) \\ t_1 & \mapsto & \bar{t}_1 \\ t_2 & \mapsto & \bar{t}_1^{-1}\bar{t}_2 \end{array}$$

Since the classical Alexander polynomial of $\mathcal{C}$ is trivial, $\operatorname{Char}_1(\mathcal{C}) \neq \mathbb{T}_{\mathcal{C}}$, and hence $\{(t, -t^{-1}) \mid t \in \mathbb{C}^*\} \subset \mathbb{T}_{\mathcal{C}}$ is an irreducible component of $\operatorname{Char}_1(\mathcal{C})$.

(2) Let us describe the local quasiadjunction ideals of the $\mathbb{A}_k$-singularities (locally described as $y^2 - x^{k+1}$).

(a) If $k = 2s$, then there is only one local branch. In this case, $r = 1$. Given $x_1 \in (0,1)$ we associate

$$(\mathcal{A}_{\mathcal{C}}^{x_1})_P := \begin{cases} (y, x^\delta) & \begin{array}{l} \text{if there exists } \delta \in \mathbb{N} \text{ such that} \\ 2s - 2\delta - 1 < 2(2s+1)x_1 \leq 2s - 2\delta + 1 \end{array} \\ \mathcal{O} & \text{if } 2s - 1 < 2(2s+1)x_1, \end{cases}$$

(b) If $k = 2s - 1$, then there are two local branches. In this case, $r = 2$. Given $(x_1, x_2) \in (0,1)^2$ we associate

$$(\mathcal{A}_{\mathcal{C}}^{(x_1,x_2)})_P := \begin{cases} (y, x^\delta) & \begin{array}{l} \text{if there exists } \delta \in \mathbb{N} \text{ such that} \\ s - \delta - 1 < s(x_1 + x_2) \leq s - \delta \end{array} \\ \mathcal{O} & \text{if } s - 1 < s(x_1 + x_2), \end{cases}$$

Finally note that the description of ideals given here is local, analytical and not global, algebraic. In other words, if $\mathcal{C}$ has an $\mathbb{A}_k$ singularity at P, then the equation D of a curve $\mathcal{D}$ belongs to the local ideal $\mathfrak{m}_\delta := (y, x^\delta)$ at P if the multiplicity of intersection of $\mathcal{D}$ with each branch of $\mathcal{C}$ at the singular point P is at least δ.

The role of essential coordinate components is oftentimes important. For instance, in [8] an example of an Alexander equivalent Zariski pair was exhibited by computing the characteristic varieties. It turns out that they only differ in essential coordinate components. The problem with such components is that up to now, no algebro-geometrical condition has been found for their existence, so one needs to compute them via a presentation of the fundamental group.

As we mentioned above, non-coordinate components can be detected by considering the singularities of $\mathcal{C}$ both locally and globally. To be more specific, all singularities seem to play a role, except for nodes, as the following result claims.

Proposition 2.25. [31, Proposition 6.1] *Let $\mathcal{C}_\lambda$, $\lambda \in (0,1]$ be an equisingular continuous family of curves degenerating into a curve $\mathcal{C}_0$ with the same non-nodal singularities as $\mathcal{C}_\lambda$. Consider also a continuous family of non-coordinate epimorphisms $\varepsilon_\lambda : H_1(X_{\mathcal{C}_\lambda};\mathbb{Z}) \to \mathbb{Z}$ degenerating into $\varepsilon_0 : H_1(X_{\mathcal{C}_0};\mathbb{Z}) \to \mathbb{Z}$, then*

$$\Delta_{\mathcal{C}_\lambda,\varepsilon_\lambda}(t) = \Delta_{\mathcal{C}_0,\varepsilon_0}(t).$$

Moreover, if $\mathcal{C}_0$ has the same number of irreducible components as $\mathcal{C}_\lambda$, then one also has

$$\operatorname{Char}_1^{\rm nc}(\mathcal{C}_\lambda) = \operatorname{Char}_1^{\rm nc}(\mathcal{C}_0) \subset \mathbb{T}_{\mathcal{C}_0},$$

where $\operatorname{Char}_1^{\rm nc}(\mathcal{C})$ *denotes the union of the non-coordinate components of* $\operatorname{Char}_1(\mathcal{C})$.

Note that the definition of non-coordinate epimorphisms is given in the paragraph preceding Remark 2.2.

Finally we want to compare Fitting ideals, characteristic varieties, and Oka-polynomials.

Theorem 2.26. *Let $\mathcal{C} = \mathcal{C}_0 \cup \mathcal{C}_1 \cup \cdots \cup \mathcal{C}_r$ be a curve where $\mathcal{C}_0$ is a transversal line, denote by G its fundamental group, consider $\varepsilon = (\varepsilon_1, ..., \varepsilon_r) \in \operatorname{Hom}(G,\mathbb{Z})$ an epimorphism, and the evaluation morphism*

$$\begin{array}{rccc}\varphi_\varepsilon : & \Lambda_{\mathcal{C}}^{\mathbb{K}} = \mathbb{K}[t_1^{\pm1},\ldots,t_r^{\pm1}] & \to & \Lambda_\varepsilon^{\mathbb{K}} = \mathbb{K}[t^{\pm1}] \\ & t_i & \mapsto & t^{\varepsilon_i}.\end{array}$$

Then $(t-1)^q\varphi_\varepsilon(F_1^{\mathbb{K}}(\mathcal{C}))$ *is a principal ideal generated by the Oka-polynomial* $\Delta_{\mathcal{C},\varepsilon}^{\mathbb{K}}(t)$.

In particular, $\operatorname{Char}_1^{\mathbb{K}} \cap \{(t^{\varepsilon_1},\dots,t^{\varepsilon_r}) \mid t\in\mathbb{K}^*\} = \operatorname{Supp}\left(\Lambda_\varepsilon^{\mathbb{K}}/\Delta_{\mathcal{C},\varepsilon}^{\mathbb{K}}(t)\right)$.

Proof. Let $M_{\mathcal{C},\mathrm{ab}}^{\mathbb{K}}$ be the Alexander invariant. Let

$$\tilde{\Lambda}^{\mathbb{K}} = \frac{\Lambda_{\mathcal{C}}^{\mathbb{K}}[t^{\pm1}]}{(t_1-\varphi_\varepsilon(t_1),\dots,t_r-\varphi_\varepsilon(t_r))}.$$

Using for example the identity

$$M_{\mathcal{C},\mathrm{ab}}^{\mathbb{K}} = \frac{\ker(\partial_1 : C_1(X_{\mathcal{C},\mathrm{ab}}) \to C_0(X_{\mathcal{C},\mathrm{ab}}))}{\operatorname{Im}(\partial_2 : C_2(X_{\mathcal{C},\mathrm{ab}}) \to C_1(X_{\mathcal{C},\mathrm{ab}}))},$$

and the fact that $\gcd(\varepsilon_1,\dots,\varepsilon_r)=1$ it is easy to check that

$$\begin{aligned}&\ker(\partial_1 : C_1(X_{\mathcal{C},\mathrm{ab}}) \to C_0(X_{\mathcal{C},\mathrm{ab}})) \otimes \tilde{\Lambda}^{\mathbb{K}}\\ &= \ker(\partial_1 : C_1(X_{\mathcal{C},\varepsilon}) \to C_0(X_{\mathcal{C},\varepsilon})) \otimes (t-1)\Lambda_\varepsilon^{\mathbb{K}}\end{aligned}$$

and $\operatorname{Im}(\partial_2 : C_2(X_{\mathcal{C},\mathrm{ab}}) \to C_1(X_{\mathcal{C},\mathrm{ab}})) \otimes \tilde{\Lambda}^{\mathbb{K}} = \operatorname{Im}(\partial_2 : C_2(X_{\mathcal{C},\varepsilon}) \to C_1(X_{\mathcal{C},\varepsilon}))$. Hence

$$M_{\mathcal{C},\mathrm{ab}}^{\mathbb{K}} \otimes \tilde{\Lambda}^{\mathbb{K}} = M_{\mathcal{C},\varepsilon}^{\mathbb{K}} \otimes (t-1)\Lambda_\varepsilon^{\mathbb{K}}$$

as $\Lambda_\varepsilon^{\mathbb{K}}$-modules. The results follows from the exact sequence

$$0 \to M_{\mathcal{C},\varepsilon}^{\mathbb{K}} \otimes (t-1)\Lambda_\varepsilon^{\mathbb{K}} \to M_{\mathcal{C},\varepsilon}^{\mathbb{K}} \to \frac{M_{\mathcal{C},\varepsilon}^{\mathbb{K}}}{M_{\mathcal{C},\varepsilon}^{\mathbb{K}} \otimes (t-1)\Lambda_\varepsilon^{\mathbb{K}}} = \left(\frac{\Lambda_\varepsilon^{\mathbb{K}}}{(t-1)}\right)^q \to 0$$

and from [55, Lemma III.6]. Q.E.D.

In other words, varying the epimorphisms $\varepsilon \in \operatorname{Hom}(G,\mathbb{Z}^r)$ and computing their corresponding Oka-polynomials, one is able to recuperate $\operatorname{Char}_1(\mathcal{C})$.

2.3. The special case of line arrangements

Characteristic varieties and Alexander polynomials of line arrangements have been largely studied in the recent years by Cohen-Orlik [32], Cohen-Suciu [33, 34], M. Falk [46], E. Hironaka [56, 57], Libgober [75], Libgober-Yuzvinsky [76, 77], M. Marco [81], and S. Yuzvinsky [129] among others. It turns out that the set of positive dimensional components passing through the origin $\mathbb{1}$ of the characteristic variety of a line arrangement is combinatorially determined (this is also the case for rational arrangements [30]). Components not passing through the origin sometimes exist [110, Example 10.6], but it is not known whether or not they are combinatorially determined.

The example we want to describe in more detail was proposed by G.Rybnikov in the mid 90's ([102]). He presented a Zariski pair of line arrangements (in particular, both arrangements had the same combinatorics). His final purpose was to prove that both arrangements had non-isomorphic fundamental groups. An alternative proof was proposed in [11] using the Alexander invariants plus an extra property that made the Alexander invariant an invariant of the fundamental group (this is usually not the case as mentioned in Remark 2.17(2)).

We will briefly recall the concept of combinatorial type or (abstract) line combinatorics:

Definition 2.27. A *combinatorial type* (or simply a *(line) combinatorics*) is a couple $\mathscr{C} := (\mathcal{L}, \mathcal{P})$, where $\mathcal{L}$ is a finite set and $\mathcal{P}$ is a family of subsets of $\mathcal{L}$, satisfying that:

(1) For all $P \in \mathcal{P}$, $\#P \geq 2$;
(2) For any $\ell_1, \ell_2 \in \mathcal{L}$, $\ell_1 \neq \ell_2$, there exists a unique $P \in \mathcal{P}$ such that $\ell_1, \ell_2 \in P$.

An *ordered combinatorial type* $\mathscr{C}^{\text{ord}}$ is a combinatorial type where $\mathcal{L}$ is an ordered set.

In what follows we will present a typical technique that allows one to find Zariski pairs of line arrangements. It is directly related to the Alexander invariant and extension isomorphisms (Remark 2.17(1)). Other possible techniques related to Massey products have also been explored [82]. First we will describe some combinatorial types.

Example 2.28 (Rybnikov's combinatorics)**.** For details on this example see [11]. Let us consider $V := \mathbb{F}_3^2 \setminus \{(0,0)\}$, where $\mathbb{F}_3^2$ is the 2-dimensional affine space on the field $\mathbb{F}_3$ of three elements. We define $\mathscr{C}_{\text{ML}} := (\mathcal{L}_{\mathscr{C}_{\text{ML}}}, \mathcal{P}_{\mathscr{C}_{\text{ML}}})$, where $\mathcal{L}_{\mathscr{C}_{\text{ML}}}$ is the set of points in V and $\mathcal{P}_{\mathscr{C}_{\text{ML}}}$ is the set of affine lines in V. Note that any affine line in V contains either two or three points of V (which implies property (1) in Definition 2.27). Also note that any two points in V define exactly one line in V (which implies property (2) in Definition 2.27). Thus $\mathscr{C}_{\text{ML}}$ is a combinatorial type that will be referred to as *MacLane's combinatorial type.*

Recall that a combinatorics is called *real* if it admits a realization in $\mathbb{CP}^2$ whose global equation has real coefficients, whereas it is called *strongly real* if each line admits a real equation. Note that strongly real combinatorics admit strongly real curves as equations in the sense of (C2).

It is well known that the MacLane combinatorics is real but not strongly real and that its combinatorial stratum is connected, however its ordered combinatorial stratum has two connected components whose representatives have eight complex conjugated lines. Moreover, five of them are real and the remaining three have coefficients in $\mathbb{Q}[\omega]$, where $\omega := \exp\left(\frac{2\pi\sqrt{-1}}{3}\right)$.

We will refer to such ordered realizations as

$$L_\omega := \{\ell_0, \ell_1, \ell_2, \ell_3, \ell_4, \ell_5, \ell_6, \ell_7\} \text{ and } L_{\bar\omega} := \{\ell_0, \ell_1, \ell_2, \ell_3, \ell_4, \bar\ell_5, \bar\ell_6, \bar\ell_7\}.$$

Let us decompose $L_\omega = L_0 \cup L_+$ and $L_{\bar\omega} = L_0 \cup L_-$, where $L_0 := \{\ell_0, \ell_1, \ell_2\}$ and consider a projective transformation ρ fixing the initial ordered set L_0 (that is, $\rho(\ell_i) = \ell_i$ $i = 0, 1, 2$) and such that ρL_+ and ρL_- intersect both L_+ and L_- only in double points. Note that ρ can be chosen with real coefficients. Let us consider the following ordered arrangements of thirteen lines: $R_{\alpha,\beta} = L_0 \cup L_\alpha \cup \rho L_\beta$, where $\alpha, \beta \in \{+, -\}$. They produce a combinatorics $\mathscr{C}_{\mathrm{RYB}}$ with 13 lines, 33 double points, and 15 triple points.

Using complex conjugation one can see that $(\mathbb{P}^2, R_{+,+}) \approx (\mathbb{P}^2, R_{-,-})$ and that $(\mathbb{P}^2, R_{+,-}) \approx (\mathbb{P}^2, R_{-,+})$. Hence, we will only deal with the Alexander invariants M_+ (resp. M_-) of $R_{+,+}$ (resp. $R_{+,-}$).

One can prove ([11, Theorem 3.8]) that there is no extension isomorphism (see Remark 2.17(1)) from M_+ to M_- as Λ-modules, where $\Lambda = \mathbb{Z}[t_1, \ldots, t_{12}]$ and each t_i represents a meridian around each affine line of $R_{\pm,\pm}$ in $\mathbb{P}^2 \setminus \ell_0$.

Let $\mathcal{C} = \mathcal{C}_0 \cup \mathcal{C}_1 \cup \cdots \cup \mathcal{C}_r$ be a line arrangement. We will denote by $\mathcal{I}^{\mathbb{Z}}$ the submodule of $\Lambda^{\mathbb{Z}}_{\mathcal{C},\mathrm{ab}}$ generated by $(t_1 - 1, \ldots, t_r - 1)$ and it will be referred to as the *augmentation* ideal.

Proposition 2.29. *The truncation $M^{\mathbb{Z}}_{\mathcal{C},\mathrm{ab}} \otimes \Lambda^{\mathbb{Z}}_{\mathcal{C}}/\mathcal{I}^{\mathbb{Z}}$ of the Alexander invariant of $\mathcal{C}$ is completely determined by the combinatorics of $\mathcal{C}$.*

Proof. A Zariski presentation of $G := \pi_1(X_{\mathcal{C}})$ can be given as in Definition 1.17, where the set of relations of the presentation are combinatorial up to conjugation. For instance, at each ordinary multiple point P of multiplicity k one obtains the following relations $[x_{i_j}^{\alpha_j}, X] = 1$ where $X := \prod_{j=1}^k x_{i_j}^{\alpha_j}$, $a^b = b^{-1} \cdot a \cdot b$, x_i is a meridian of the line ℓ_i, and $i_1, \ldots, i_k$ are subindices of the k lines intersecting at P. The result is an immediate consequence of [11, Proposition 2.15] which assures that the class of $[x_{i_j}^{\alpha_j}, X]$ in $M^{\mathbb{Z}}_{\mathcal{C},\mathrm{ab}} \otimes \Lambda^{\mathbb{Z}}_{\mathcal{C}}/\mathcal{I}^{\mathbb{Z}}$ only depends on the abelian class of each $x_{i_j}^{\alpha_j}$, that is x_{i_j}, and hence it is combinatorial. Q.E.D.

Let $\mathcal{L}$ denote the combinatorial type of a line arrangement $\mathcal{C}$. Since $H_1(X_\mathcal{C})$ only depends on $\mathcal{L}$, it will be denoted it by $H_\mathcal{L}$. Consider $\operatorname{Aut}(H_\mathcal{L})$ the set of automorphisms of $H_\mathcal{L}$. Any $h \in \operatorname{Aut}(H_\mathcal{L})$ induces a transformation of $M^\mathbb{Z}_{\mathcal{C},\text{ab}} \otimes \Lambda^\mathbb{Z}_\mathcal{C}/\mathcal{I}^\mathbb{Z}$ as follows $h([x_i, x_j]) = [h(x_i), h(x_j)]$ (since the elements $[x_i, x_j]$ generate G' this defines a transformation, but not necessarily an automorphism). The set of those elements in $\operatorname{Aut}(H_\mathcal{L})$ that induce automorphisms of $M^\mathbb{Z}_{\mathcal{C},\text{ab}} \otimes \Lambda^\mathbb{Z}_\mathcal{C}/\mathcal{I}^\mathbb{Z}$ (as a group) will be denoted by $\operatorname{Aut}^1(H_\mathcal{L})$. Let $\Gamma(\mathcal{L}) \subset \operatorname{Aut}(H_\mathcal{L})$ denote the set of automorphisms that preserve the combinatorics. Note that if $\varphi \in \Gamma(\mathcal{L})$, then $\pm\varphi \in \operatorname{Aut}^1(H_\mathcal{L})$. Thus $\{\pm 1\} \times \Gamma(\mathcal{L}) \subset \operatorname{Aut}^1(H_\mathcal{L})$.

Definition 2.30. A line combinatorics $\mathcal{L}$ is called *homologically rigid* if $\operatorname{Aut}^1(H_\mathcal{L}) = \{\pm 1\} \times \Gamma(\mathcal{L})$.

Proposition 2.31. *If $\mathcal{C}$ is a line arrangement whose combinatorics, say $\mathcal{L}$, is homologically rigid, then the Alexander invariant $M^\mathbb{Z}_{\mathcal{C},\text{ab}}$ as a $\Lambda^\mathbb{Z}_\mathcal{C}$-module is an invariant of the fundamental group $\pi_1(X_\mathcal{C})$ (up to* extension isomorphisms*).*

Example 2.32. Rybnikov's combinatorics $\mathscr{C}_{\text{RYB}}$ is homologically rigid ([11, Proposition 4.22]). We have mentioned above that no extension isomorphism exists from M_+ to M_-. Therefore one concludes that $\pi_1(X_{R_{+,+}}) \not\cong \pi_1(X_{R_{+,-}})$.

2.4. Twisted Alexander polynomials

Twisted Alexander polynomials have been developed and extensively studied in the mid 90's for knots. In many instances where abelian invariants were not able to identify a certain property of knots, non-abelian invariants such as these, were able to do it – see [67, 79, 124]. Later P. Kirk and C. Livingston [65, 66] were able to give partial answers to questions of mutation and concordance for general CW-complexes. Our purpose here is to define and briefly describe twisted Alexander polynomials for curves and some of their recent applications following [31].

Let us consider the general setting of §2.1, that is, a curve $\mathcal{C}$, its complement $X_\mathcal{C}$, the epimorphism $\varepsilon : G := \pi_1(X_\mathcal{C}) \to \mathbb{Z}$, $K_\varepsilon := \ker \varepsilon$, and the infinite cyclic covering $X_{\mathcal{C},\varepsilon}$. In addition let us consider a $\mathbb{K}$-vector space V of finite dimension and an (anti)representation

$$\rho : G \longrightarrow \operatorname{GL}(V).$$

Note that V inherits a right $\mathbb{K}[G]$-module structure denoted by V_ρ. Let $X_{\mathcal{C},\text{ab}} \to X_\mathcal{C}$ denote the universal abelian covering of $X_\mathcal{C}$. Analogously as mentioned in Remark 2.1 and the subsequent discussion, the cellular chain complex $C_*(X_{\mathcal{C},\mathrm{ab}};\mathbb{K})$ also becomes a finitely generated (left) $\mathbb{K}[G]$-module generated by the lifts of the cells of $X_{\mathcal{C}}$. Hence, one defines

$$C_*^{\varepsilon,\rho}(X_{\mathcal{C}};\mathbb{K}[t^{\pm 1}]) := V_\rho \otimes_{\mathbb{K}[K_\varepsilon]} C_*(X_{\mathcal{C},\mathrm{ab}})$$

as a $\mathbb{K}[t^{\pm 1}]$-module, where $\mathbb{K}[t^{\pm 1}]$ is a trivial $\mathbb{K}[G]$-module, as follows:

$$t^n \cdot (v \otimes c) = v\gamma^{-n} \otimes \gamma^n c$$

where $\gamma \in G$ verifies $\varepsilon(\gamma) = t$.

Definition 2.33. The homology of $(X_{\mathcal{C}}, \varepsilon, \rho)$ is defined as the $\mathbb{K}[t^{\pm 1}]$-module

$$H_*^{\varepsilon,\rho}(X_{\mathcal{C}};\mathbb{K}[t^{\pm 1}]) = H_*(C^{\varepsilon,\rho}(X_{\mathcal{C}},\mathbb{K}[t^{\pm 1}])).$$

Definition 2.34. The k-th *twisted Alexander polynomial* $\Delta^k_{\mathcal{C},\varepsilon,\rho}(t)$ of $(X_{\mathcal{C}}, \varepsilon, \rho)$ is the order of $H_k^{\varepsilon,\rho}(X_{\mathcal{C}};\mathbb{K}[t^{\pm 1}])$. For short, we denote by $\Delta_{\mathcal{C},\varepsilon,\rho}(t) = \frac{\Delta^1_{\mathcal{C},\varepsilon,\rho}(t)}{\Delta^0_{\mathcal{C},\varepsilon,\rho}(t)}$ the element of $\mathbb{K}(t)$.

See (12) for a definition of order of a module over a principal ideal domain.

Remark 2.35.

(1) Note that even if $\varepsilon : G \to \mathbb{Z}/m\mathbb{Z}$ was an epimorphism onto the finite cyclic group $\mathbb{Z}/m\mathbb{Z}$, all the definitions can be modified accordingly to suit this case.

(2) Note that $\Delta_{\mathcal{C},\varepsilon,\rho}(t)$ does not have to be a polynomial. For example, we can consider the projective three-cuspidal quartic $\mathcal{Q}$, whose fundamental group G is shown in (1). Since G is finite we can consider the regular representation $\rho : G \to \mathrm{GL}(12,\mathbb{K})$ and the trivial morphism $\varepsilon : G \to \mathbb{Z}/4\mathbb{Z}$. In this situation

$$\Delta_{\mathcal{Q},\varepsilon,\rho}(t) = \begin{cases} \frac{1}{t-1} & \text{if } \operatorname{char}\mathbb{K} = 3 \\ \frac{1}{t^2-1} & \text{otherwise.} \end{cases}$$

Under certain very general conditions, however, $\Delta_{\mathcal{C},\varepsilon,\rho}(t)$ is a polynomial [31, Proposition 5.4].

(3) An alternative definition of $\Delta_{\mathcal{C},\varepsilon,\rho}(t)$ can be given by means of Fox calculus – see [124].

Twisted Alexander polynomials can also be seen ([120, 121]) as the Reidemeister torsion of the complex of vector spaces obtained by tensoring the usual CW-complex C_* describing the homotopy type of $X_{\mathcal{C}}$

by V_ρ and by $\mathbb{K}(t)$. Since the Reidemeister torsion behaves well with respect to surgery of complexes, a division formula for twisted Alexander polynomials that generalizes Theorem 2.10 can be obtained. In order to do so, one needs the following construction:

Suppose that we are given a curve $\mathcal{C}$, an epimorphism $\varepsilon : H_1(X_\mathcal{C}) \to \mathbb{Z}$, and a representation $\rho : \pi_1(X_\mathcal{C}) \to \mathrm{GL}(V)$. Let $\mathbb{S}^3_1, \ldots, \mathbb{S}^3_s$ be sufficiently small 3-spheres around the singular points $\{P_1, ..., P_s\}$ of $\mathcal{C}$. Denote by $L_k = \mathcal{C} \cap \mathbb{S}^3_k$ the link of the singularity at P_k. Also choose a base point $Q_k \in \mathbb{S}^3_k \setminus L_k$ and denote by $\pi^k = \pi_1(\mathbb{S}^3_k \setminus L_k; Q_k)$ the local fundamental groups at P_k. The inclusion maps $i_k : \pi^k \to \pi_1(X_\mathcal{C})$ and (ε, ρ) induce morphisms

$$\varepsilon_k : \pi^k \longrightarrow \mathbb{Z}$$

$$\text{and } \rho_k : \pi^k \longrightarrow \mathrm{GL}(V),$$

for any $k = 1, ..., s$. Analogously, one can consider $\mathbb{S}_\infty$ a sufficiently large 3-sphere, $L_\infty = \mathcal{C} \cap \mathbb{S}_\infty$ the link at infinity and define accordingly π^∞, ε_∞, and ρ_∞.

Theorem 2.36. [31, Theorem 5.6] *Let $\mathcal{C}$ be a curve, ε an epimorphism, and ρ a unitary representation. Suppose also that the induced triples $(X_\mathcal{C} \cap \mathbb{S}^3_k, \varepsilon_k, \rho_k)$, $k = 1, ..., s, \infty$ are acyclic. Then*

$$\begin{aligned}\Big(\prod_{\ell=1}^{r} \det(\mathrm{Id} - \rho(\nu_\ell) t^{q_\ell})^{s_\ell - \chi(\mathcal{C}_\ell)}\Big) \cdot \prod_{k=1,\ldots,s,\infty} \Delta_{L_k, \rho_k} \\ = \Delta_{\mathcal{C},\varepsilon,\rho} \cdot \overline{\Delta}_{\mathcal{C},\varepsilon,\rho} \cdot \det \varphi^{\varepsilon,\rho}(\mathcal{C}),\end{aligned}$$

where ν_ℓ is the homology class of a meridian of the irreducible component $\mathcal{C}_\ell$, $s_\ell = \#\,\mathrm{Sing}(\mathcal{C}) \cap \mathcal{C}_\ell$, and $\varphi^{\varepsilon,\rho}(\mathcal{C})$ is an intersection form on $H_2^{\varepsilon,\rho}(X_\mathcal{C}, \mathbb{Q}[t^{\pm 1}])$ with twisted coefficients.

Remark 2.37. The condition of acyclicity is purely technical and can be expressed as follows. A triple (X, ε, ρ) is *acyclic* if the chain complex $C_*^{\varepsilon,\rho}(X; \mathbb{K}(t))$ is acyclic over $\mathbb{K}(t)$.

In the irreducible case, something is known about the roots of the twisted Alexander polynomial of unitary representations.

Theorem 2.38. [78, Theorem 5.3.] *Let $\mathcal{C}$ be an irreducible curve and L is a line at infinity. Let ρ be a unitary representation of the fundamental group and let $\mathbb{K}$ be the extension of $\mathbb{Q}$ generated by the eigenvalues of $\rho(\gamma)$ where γ is a meridian of $\mathcal{C}$. Then the roots of $\Delta_\rho(C)$ belong to a cyclotomic extension of $\mathbb{K}$.*

Note that in this case the morphism ε is uniquely determined up to orientation.

Finally, let us point out that twisted Alexander polynomials are sensitive to nodal degenerations, that is, Proposition 2.25 is no longer true for twisted Alexander polynomials as the following example illustrates.

We say a plane projective curve $\mathcal{D}$ of degree, say d, is a type-I curve if $\mathcal{D}$ is irreducible and has an ordinary $(d-2)$-ple point at some point, say P. Consider $\mathcal{D}$ a type-I curve and let L_1 and L_2 be lines through P such that either L_i is tangent to a smooth point $P_i \in \mathcal{D}$ or L_i passes through a double point $P_i \neq P$ of type $\mathbb{A}_{2r}$. Let us denote $\mathcal{C} = L_1 + L_2 + \mathcal{D}$. Assume that $\mathcal{D}$ has only nodes as singular points apart from P. We recall the following properties of such curves:

- There exist nodal degenerations $\mathcal{D}_\lambda \to \mathcal{D}_0$ of non-rational type-I curves $\mathcal{D}_\lambda$ $(\lambda > 0)$ into a rational type-I curve $\mathcal{D}_0$ ([13, Corollary 3]).
- Let $\mathcal{D}_\lambda \to \mathcal{D}_0$ be a nodal degeneration as above. This produces a degeneration $\mathcal{C}_\lambda \to \mathcal{C}_0$, where $\mathcal{C}_\lambda = L_1 + L_2 + \mathcal{D}_\lambda$, $\lambda \geq 0$. If G_λ denotes $\pi_1(X_{\mathcal{D}_\lambda})$, then G_λ, $\lambda > 0$ is Abelian, whereas G_0 is not ([13, Proposition 6.1]). Moreover, a presentation of G_0 can be given as follows:

$$G_0 = \langle\ \ell, x_1, x_2 \mid [x_1, x_2] = 1, \ell^{-1}x_1\ell = x_2, \ell^{-1}x_2\ell = x_1\ \rangle,$$

  where ℓ is a meridian around a line and x_1, x_2 are meridians around $\mathcal{D}_0$.

Consider $\mathcal{C}_\lambda \to \mathcal{C}_0$ a degeneration as above. Let us denote by ν_1, ν_2 and ν_ℓ the homology classes of the corresponding generators of G_0. Since x_1 and x_2 are meridians of the same irreducible component, one has that $\nu = \nu_1 = \nu_2$. Our purpose is to find a suitable representation that produces a sensitive twisted Alexander polynomial. Let us consider ε the usual morphism $\varepsilon(\nu_\ell) = \varepsilon(\nu) = 1$, and the rank 2 representation

$$\rho(\ell) = \begin{pmatrix} 1 & 0 \\ 0 & -1 \end{pmatrix}, \quad \rho(x_1) = \begin{pmatrix} -1 & 0 \\ 1 & -1 \end{pmatrix}, \quad \rho(x_2) = \begin{pmatrix} -1 & 0 \\ -1 & -1 \end{pmatrix}.$$

Using ε and ρ one obtains

$$\Delta_{\mathcal{C}_0,\varepsilon,\rho}(t) = (t^2 - 1). \tag{19}$$

Note that $\rho(G_0) \cong \mathbb{Z}/2\mathbb{Z} * \mathbb{Z}/2\mathbb{Z}$.

Finally note that, by Proposition 2.25, the classical Alexander polynomial $\Delta_{\mathcal{C}_\lambda}(t)$ and the torsion non-coordinate characteristic varieties are invariant for $\lambda \in [0,1]$. Since G_λ is abelian, this implies that $\Delta_{\mathcal{C}_1,\varepsilon} = \Delta_{\mathcal{C}_0,\varepsilon} = (t-1)$ and $\operatorname{Char}_1^*(\mathcal{C}_1) = \operatorname{Char}_1^*(\mathcal{C}_0) = \emptyset$.

On the other hand, formula (19) shows that $\mathcal{C}_0$ has a non-trivial twisted Alexander polynomial, whereas any twisted Alexander polynomial of $\mathcal{C}_\lambda$, $\lambda \in (0,1]$ is trivial since G_λ is abelian.

2.5. Computational methods

We have used the following result which is a straightforward generalization of [11, Proposition 2.8].

Proposition 2.39. *Let $\langle \bar{x}, \bar{y}; \bar{W}\rangle$ be a presentation of G such that*

(1) *$\bar{x}$ be the free group generated the words $x_1, ..., x_r$ whose abelian classes generate $G/G' = \mathbb{Z}^q \oplus \mathbb{Z}/p_1\mathbb{Z} \oplus \cdots \oplus \mathbb{Z}/p_s\mathbb{Z}$, $q+s=r$,*
(2) *$\bar{y} = \langle y_1, \ldots, y_u\rangle \subset G'$,*
(3) *$\forall w(\bar{x}, \bar{y}) \in \bar{W}$, one has that $w(\bar{x}, 1)$ is a product of commutators in $\bar{x}$ and $x_{q+i}^{p_i}$.*

Then the module $M_G^{\mathbb{Z}}$ admits a presentation $\tilde{\Gamma}/(\mathcal{T}+\mathcal{J}+\mathcal{W})$, where

$$\tilde{\Gamma} := \left(\bigoplus_{1\leq i<j\leq r} x_{ij}\Lambda^{\mathbb{Z}}\right) \oplus \left(\bigoplus_{1\leq k\leq u} y_k\Lambda^{\mathbb{Z}}\right),$$

(1) *$\Lambda^{\mathbb{Z}} = \mathbb{Z}[t_1^{\pm 1}, \ldots, t_r^{\pm 1}]/(t_{q+1}^{p_1}-1, \ldots, t_r^{p_s}-1)$,*
(2) *$\mathcal{T}$ is the submodule of $\tilde{\Gamma}$ generated by the torsion relations $\dfrac{t_{i+q}^{p_i}-1}{t_{i+q}-1}$, $i=1,\ldots,s$,*
(3) *$\mathcal{J}$ is the Jacobian submodule of $\tilde{\Gamma}$ generated by the relations*

$$J(i,j,k) := (t_i-1)x_{jk}-(t_j-1)x_{ik}+(t_k-1)x_{ij}, \quad 1\leq i<j<k\leq r, \text{ and}$$

(4) *$\mathcal{W}$ is the submodule of $\tilde{\Gamma}$ generated by subset of $\tilde{\Gamma}$ obtained by rewriting the relations $\bar{W}$ in terms of $\tilde{\Gamma}$.*

Proof. The same proof used in [11, Proposition 2.8] can be applied using the Reidemeister-Schreier method to obtain a presentation of G'/G'' (which is not finitely presented in general) and then apply the module structure to give the finite presentation as a module. Q.E.D.

By Proposition 1.18, the fundamental group of the complement of any plane curve admits a presentation as in Proposition 2.39. In order to obtain the submodule $\mathcal{W}$ the following properties are very useful.

Proposition 2.40. *The following equalities hold in $M_G^{\mathbb{Z}}$:*

(1) $[x,x]=0$,
(2) $[x,y]=-[y,x]$,
(3) $[x^{-1},y]=-t_x^{-1}[x,y]$,

(4) $[x,p] = (t_x - 1)p \quad \forall p \in G'$,
(5) $[xy, z] = [x, z] + t_x[y, z]$,
(6) $[x * y, z] = [y, z] + (t_z - 1)[y, x]$, *where* $x * y = xyx^{-1}$,
(7) $[x_1 \cdot \ldots \cdot x_n, y_1 \cdot \ldots \cdot y_m] = \sum_{i=1}^{n}\sum_{j=1}^{m} T_{ij}[x_i, y_j]$, *where* $T_{ij} = \prod_{k=1}^{i-1} t_{x_k} \cdot \prod_{\ell=1}^{j-1} t_{y_\ell}$.
(8) Jacobi relations:

$$J(x, y, z) := (t_x - 1)[y, z] + (t_y - 1)[z, x] + (t_z - 1)[x, y] = 0,$$

where t_x *represents the image of* $x \in G$ *in* $\Lambda_G^{\mathbb{Z}}$.

Proof. Property (1) is obvious (it is even true in G). For (2) note that $xyx^{-1}y^{-1} = \left(yxy^{-1}x^{-1}\right)^{-1}$. Property (3) follows from $x^{-1}yxy^{-1} = x^{-1}\left(xyx^{-1}y^{-1}\right)^{-1}x$. To prove (4) note that $xpx^{-1}p^{-1} = t_x p - p$. Property (5) follows from $[xy, z] = x\left(yzy^{-1}z^{-1}\right)x^{-1}\left(xzx^{-1}z^{-1}\right)$. For (6),

$$\begin{aligned}[x * y, z] &= [[x, y]y, z] \overset{(5)}{=} [[x, y], z] + [y, z] \overset{(4)}{=} [y, z] - (t_z - 1)[x, y] \\ &\overset{(2)}{=} [y, z] + (t_z - 1)[y, x].\end{aligned}$$

Property (7) follows by induction and using property (2). Finally, for the Jacobi relations, note that on the one hand, by (5)

$$[xy, z] = [x, z] + t_x[y, z]. \tag{20}$$

On the other hand, $[xy, z] = [(x * y)x, z]$, then by properties (5) and (6), one has

$$[xy, z] = [(x*y)x, z] \overset{(5)}{=} [(x*y), z] + t_y[x, z] = [y, z] - (t_z - 1)[x, y] + t_y[x, z]. \tag{21}$$

The difference between (20) and (21) equals zero and the result follows. Q.E.D.

§3. Non-abelian branched coverings and Zariski pairs

In §2 we have mainly dealt with invariants associated with different sorts of abelian coverings. In this section we will give an approach to invariants related to non-abelian coverings. A more group-theoretical approach is given by the *Hall invariants* studied by Matei-Suciu [83] in relation with the Alexander invariant. The Hall invariant $\delta_\Gamma(G)$ of a group G associated with a finite group Γ is defined as the number of epimorphisms from G to Γ up to automorphisms of Γ. Matei-Suciu [83] prove that in the case of metabelian groups $\Gamma = (\mathbb{Z}/q\mathbb{Z})^s \rtimes \mathbb{Z}/p\mathbb{Z}$, where p and q are distinct primes and s is the order of $q \bmod p$ in $(\mathbb{Z}/p\mathbb{Z})^\times$, the Hall invariant $\delta_\Gamma(G)$ can be computed in terms of the characteristic varieties $\operatorname{Char}^{\mathbb{F}_{q^s}}(G)$.

Our approach here is more algebraic, in the sense that we ask ourselves whether or not there are any algebraic conditions on the singular points of a curve $\mathcal{C}$ that can characterize the existence of certain metabelian coverings (in this case dihedral coverings). *A posteriori* torsion points in $\operatorname{Char}^{\mathbb{F}_{q^s}}(\mathcal{C})$ have an algebraic interpretation in terms of position of singularities.

3.1. Preliminaries

Let X and Y be normal varieties. We call X a *(branched) covering of* Y if there exists a finite surjective morphism $\pi : X \to Y$. When needed, the covering morphism will be specified as *a covering* $\pi : X \to Y$.

Let $\pi : X \to Y$ be a covering. The corresponding rational function fields will be denoted by $\mathbb{C}(X)$ and $\mathbb{C}(Y)$, respectively. Note that $\mathbb{C}(X)$ is an algebraic extension of $\mathbb{C}(Y)$ and $\deg \pi = [\mathbb{C}(X) : \mathbb{C}(Y)]$ (see e.g. [86, p. 46, Proposition 3.17]).

Definition 3.1. Let X, Y and $\pi : X \to Y$ be as above.

(*i*) We call X a *Galois covering* of Y if the field extension is Galois.

(*ii*) Let G be a finite group. We call X a *G-covering* if X is a Galois covering of Y with $\operatorname{Gal}(\mathbb{C}(X)/\mathbb{C}(Y)) \cong G$.

We say that $x \in X$ is a *ramification point* of π if $\pi^*\mathfrak{m}_{Y,f(x)}\mathcal{O}_{X,x} \neq \mathfrak{m}_{X,x}$, where $\mathfrak{m}_{X,x}$ and $\mathfrak{m}_{Y,f(x)}$ are the maximal ideals of $\mathcal{O}_{X,x}$ and $\mathcal{O}_{Y,f(x)}$, respectively. Geometrically, this means that π is not a local isomorphism around x. The set of all ramification points will be denoted by R_π. Its image $\pi(R_\pi)$ is the branch locus of π and will be denoted by Δ_π or $\Delta(X/Y)$. By the purity of the branch locus [133], if Y is smooth then Δ_π is an algebraic subset of pure codimension 1.

When we apply the algebraic theory of branched coverings to the study of Zariski pairs, we consider their associated analytic spaces. Here we summarize some results from algebraic geometry and analytic geometry which will be needed later.

Let Y be a normal algebraic variety over $\mathbb{C}$. We denote by Y^{an} its associated analytic space. The following statements are key in relating branched coverings of Y with those of Y^{an}.

Theorem 3.2. *Let Y be a proper normal variety over $\mathbb{C}$. Let $\mathcal{X}$ be a normal complex analytic space and let $f : \mathcal{X} \to Y^{\text{an}}$ be a proper* *morphism with finite fiber. Then there exists a unique normal variety X (up to isomorphism over Y) and a finite morphism $\pi : X \to Y$ such $X^{\mathrm{an}} \cong \mathcal{X}$ and $\pi^{\mathrm{an}} = f$ (up to isomorphism between X^{an} and $\mathcal{X}$).*

For a proof, see [52, EXPOSÉ XII, Corollaire 4.6]. The following theorem can also be found in [52, EXPOSÉ XII] or [51, Theorem 5.4].

Theorem of Grauert-Remmert 3.3. *Let $\mathcal{Y}$ be a normal analytic space and let B be a closed analytic subset of codimension 1. Let $\pi_o : U \to \mathcal{Y} \setminus B$ be an étale finite covering of $\mathcal{Y} \setminus B$. Then there exist a normal analytic space $\mathcal{X}$ containing U and a finite surjective morphism $\pi : \mathcal{X} \to \mathcal{Y}$ such that $\pi^{-1}(\mathcal{Y} \setminus B) = U$ and $\pi|_U = \pi_o$. Moreover $\mathcal{X}$ is unique up to isomorphism over $\mathcal{Y}$.*

Notation 3.4. Let $\pi : X \to Y$ be a G-covering of a smooth projective variety Y. Let B be a reduced divisor on Y whose irreducible decomposition will be denoted by $B = B_1 + \cdots + B_r$. Given a morphism $\sigma : X \to Y$ between smooth projective varieties and a divisor D in Y, $\sigma^*(D)$ (resp. $\sigma^{-1}(D)$) will denote its inverse image as a divisor (resp. as a set). Its strict transform will be denoted by $\sigma_q^{-1}(D)$.

Definition 3.5. A covering π is said to be *branched at* $e_1B_1 + \cdots + e_rB_r$ ($e_i \geq 2$) if

- $\Delta_\pi = B$ and
- the ramification index along the smooth part of B_i^{an} is e_i. Namely, for any smooth point $y \in B_i^{\mathrm{an}}$ and $x \in (\pi^{\mathrm{an}})^{-1}(y)$, there exist neighborhoods U_x and V_y, respectively, such that π^{an} is locally given by

$$(z_1, \ldots, z_n) \mapsto (w_1, \ldots, w_n) = (z_1^{e_i}, \ldots, z_n),$$

where $(z_1, \ldots, z_n)$ and $(w_1, \ldots, w_n)$ denote local coordinates on U_x and U_y, respectively, such that $x = (0, \ldots, 0), y = (0, \ldots, 0)$ and $B_i^{\mathrm{an}} \cap V_y$ is given by $w_1 = 0$.

Let γ_i be a meridian around B_i as in Figure 1, and $[\gamma_i]$ denote its class in $\pi_1(Y^{\mathrm{an}} \setminus B^{\mathrm{an}}, p_o)$.

Proposition 3.6. *Let Y be a smooth projective variety and let $B = B_1 + \cdots + B_r$ be the decomposition into irreducible components of a reduced divisor B on Y. If there exists a G-covering $\pi : X \to Y$ branched at $e_1B_1 + \cdots + e_rB_r$, then there exists a normal subgroup H_π of $\pi_1(Y^{\mathrm{an}} \setminus B^{\mathrm{an}}, p_o)$ such that:*

(*i*) $[\gamma_i]^{e_i} \in H_\pi, [\gamma_i]^k \notin H_\pi, (1 \leq k \leq e_i - 1)$, *and*
(*ii*) $\pi_1(Y^{\mathrm{an}} \setminus B^{\mathrm{an}}, p_o)/H_\pi \cong G$.

Conversely, if there exists a normal subgroup H of $\pi_1(Y^{\mathrm{an}} \setminus B^{\mathrm{an}}, p_o)$ satisfying the above two conditions for H_π, then there exists a G-covering $\pi_H : X_H \to Y$ branched at $e_1B_1 + \cdots + e_rB_r$.

Proof. Since G acts on X such that $Y = X/G$ ([114]), G also acts on X^{an} ([52, §3]) transitively on each fiber $(\pi^{\mathrm{an}})^{-1}(y)$, $y \in Y$. Hence $X^{\mathrm{an}}/G = Y^{\mathrm{an}}$. Since $X^{\mathrm{an}} \setminus (\pi^{\mathrm{an}})^{-1}(B^{\mathrm{an}}) \to Y^{\mathrm{an}} \setminus B^{\mathrm{an}}$ is étale, our statement easily follows from the standard theory of covering spaces.

Conversely, let H be the normal subgroup in the statement. Let $\mathcal{X}'_H$ be an étale covering of $Y^{\mathrm{an}} \setminus B^{\mathrm{an}}$ corresponding to H. By Theorem 3.3, there exists a normal analytic space $\mathcal{X}_H$ and a finite morphism $\pi_H^{\mathrm{an}} : \mathcal{X}_H \to Y^{\mathrm{an}}$ extending the covering morphism $\mathcal{X}'_H \to Y^{an} \setminus B^{an}$. Since $G \cong \pi_1(Y^{\mathrm{an}} \setminus B^{\mathrm{an}}, p_o)/H$ acts on $\mathcal{X}'_H$ so that $\mathcal{X}'_H/G = Y^{\mathrm{an}} \setminus B^{\mathrm{an}}$, G also acts on $\mathcal{X}_H$ ([52, Proposition 5.3]). Hence one has a morphism $\mathcal{X}_H/G \to Y$, which is finite and an isomorphism on $Y^{\mathrm{an}} \setminus B^{\mathrm{an}}$. By Zariski's main theorem (e.g., see [122, Theorem 1.11]), $\mathcal{X}_H/G \cong Y$.

By Theorem 3.2, there exists a normal variety X_H and a finite morphism $\pi_H : X_H \to Y$. Since G acts on $\mathcal{X}_H$, it also acts on X_H over Y. This implies that $G \subset \mathrm{Aut}_{\mathbb{C}(Y)}(\mathbb{C}(X))$. Since $\deg \pi_H = \#G$, X_H is a G-covering of Y and thus the statement on the ramification index follows from how we extend $\mathcal{X}'_H$ to $\mathcal{X}_H$ along B_i. Q.E.D.

Remark 3.7. We recall some facts on Galois theory of Galois coverings. Let Y be a normal algebraic variety. Let K be a finite extension of $\mathbb{C}(Y)$ and let X_K be the normalization of Y in K called the "*K-normalization of Y*". There exists a canonical finite surjective morphism $\pi_K : X_K \to Y$. Hence X_K is a covering of Y with $\mathbb{C}(X_K) = K$. If K is a Galois extension, then $\pi_K : X_K \to Y$ is a Galois covering. Conversely, note that any covering $\pi : X \to Y$ defines a finite field extension $\mathbb{C}(X)$ of $\mathbb{C}(Y)$.

Let G be a finite group and let H be a normal subgroup. Consider a Galois extension K of $\mathbb{C}(Y)$ with $\mathrm{Gal}(K/\mathbb{C}(Y)) \cong G$.

Let $\pi : X \to Y$ be a G-covering corresponding the extension $K/\mathbb{C}(Y)$. Let K^H be the fixed field by H. The field K^H is also a Galois extension of $\mathbb{C}(Y)$ with $\mathrm{Gal}(K^H/\mathbb{C}(Y)) \cong G/H$. Let $D_H(X/Y)$ be the K^H-normalization of $\mathbb{C}(Y)$. Since K/K^H is an H-extension and $K^H/\mathbb{C}(Y)$ is a G/H-extension, X_K is an H-covering of $D_H(X/Y)$ and $D_H(X/Y)$ is a G/H-covering of Y. The corresponding covering morphisms will be denoted by

$$\beta_{1,H}(\pi) : D_H(X/Y) \to Y, \quad \text{and} \quad \beta_{2,H}(\pi) : X \to D_H(X/Y). \tag{22}$$

Note that $\pi = \beta_{1,H}(\pi) \circ \beta_{2,H}(\pi)$.

3.2. Dihedral coverings

Let D_{2n} denote the dihedral group of order $2n$. The following presentation of D_{2n} will be extensively used throughout this section $\langle \sigma, \tau \mid \sigma^2 = \tau^n = (\sigma\tau)^2 = 1 \rangle$.

Remark 3.8. Since we consider non-abelian branched coverings, we will always assume that $\# D_{2n} \geq 6$.

Following the notation introduced in §3.1, we will study the case $G = D_{2n}$ and $H = \langle \tau \rangle$. Since there is no ambiguity for H, we will use notations $D(X/Y)$, $\beta_1(\pi)$ and $\beta_2(\pi)$ for simplicity. The notion of *generic* and *non-generic* D_{2n}-coverings will be key in our arguments.

Definition 3.9. A D_{2n}-covering $\pi : X \to Y$ of a smooth variety Y is said to be *generic* if $\Delta_\pi = \Delta_{\beta_1(\pi)}$, otherwise π is said to be *non-generic.*

Note that, if $\pi : X \to Y$ is a D_{2n}-covering, then $D(X/Y)$ is a double covering of Y and X is an n-cyclic covering of $D(X/Y)$, whose morphisms will be denoted by $\beta_1(\pi)$ and $\beta_2(\pi)$ respectively, as in (22).

Remark 3.10. In what follows, Y will be assumed to be smooth and simply connected. Also note that $\sim$ will denote linear equivalence of divisors.

Let us start with a sufficient condition for the existence of D_{2n}-coverings.

Proposition 3.11. *Let Z be a smooth double covering of Y with covering morphism $f : Z \to Y$ and D be an effective divisor on Z such that:*

- (i) *D and $\sigma^* D$ have no common component, where σ denotes the covering transformation,*
- (ii) *if $D = \sum_{i=1}^h a_i D_i$ is the irreducible decomposition, then for all $i = 1, \ldots, h$, $0 < a_i$ and $\gcd\{a_1, \ldots, a_h, n\} = 1$, and*
- (iii) *there exists a line bundle $\mathcal{L}$ on Z such that $D - \sigma^* D \sim n\mathcal{L}$.*

Then there exists a D_{2n}-covering X of Y such that

- (a) *$D(X/Y) = Z$,*
- (b) *the branch locus $\Delta_{\beta_2(\pi)}$ of $\beta_2(\pi)$ is contained in* $\operatorname{Supp}(D + \sigma^* D)$, *i.e., $\Delta(X/Y) \subset \Delta_{\beta_1(\pi)} \cup f(\operatorname{Supp}(D))$ and*
- (c) *if $D_i \subset \Delta_{\beta_2(\pi)}$, then the ramification index along D_i is* $\dfrac{n}{\gcd(n, a_i)}$.

Proof. For n odd, our statement is a special case of [117, Proposition 1.1], except for part (c), which follows from the proof of [117,

Proposition 0.4]. For n even, a similar proof to that of [117, Proposition 1.1] also works by [117, Remark 3.1, Proposition 0.6]. Q.E.D.

As for a necessary condition for the existence of D_{2n}-coverings, one has the following:

Proposition 3.12 ([112, §2, §3]). *Let $\pi : X \to Y$ be a D_{2n}-covering such that $D(X/Y)$ is smooth. Let us denote by σ the covering transformation of $\beta_1(\pi)$. Then there exist (possibly empty) effective divisors, D_1 and D_2, and a line bundle $\mathcal{L}$ on $D(X/Y)$ satisfying the following conditions:*

(*i*) *D_1 and $\sigma^* D_1$ have no common components. Moreover, if we denote its irreducible decomposition by $\sum_j a_j D_{1,j}$, then $0 \leq a_j < n$.*

(*ii*) *If $D_2 \neq \emptyset$, then n is even and D_2 is a reduced divisor such that there exists a divisor B_2 on Y satisfying $D_2 = f^* B_2$.*

(*iii*) *$D_1 + \frac{n}{2} D_2 - \sigma^* D_1 \sim n\mathcal{L}$.*

(*iv*) *$\Delta(X/D(X/Y)) = \mathrm{Supp}(D_1 + \sigma^* D_1 + D_2)$. The ramification index along $D_{1,j}$ (resp. an irreducible component of D_2) is $\dfrac{n}{\gcd(a_i, n)}$ (resp. 2).*

Corollary 3.13. *Let D be an irreducible component of $\beta_1(\pi)(\Delta_{\beta_2(\pi)})$. If the ramification index of $\beta_2(\pi)$ along $\beta_1(\pi)^{-1}(D)$ is > 2, then the divisor $\beta_1(\pi)^* D$ is of the form $D' + \sigma^* D'$ for some irreducible divisor D' on $D(X/Y)$ such that $D' \neq \sigma^* D'$.*

3.3. Zariski's example and D_6-coverings

Let us review Zariski's example of sextics with six cusps using D_6-coverings as in Zariski's original proof. Our purpose is to give a detailed proof in modern language of (Z5). Let us start with the following Lemma.

Lemma 3.14. *Let B be a sextic with 6 cusps such that a D_6-covering $\pi : S \to \mathbb{P}^2$ with $\Delta(S/\mathbb{P}^2) = B$ exists. Then the following statements hold:*

(*i*) *$\beta_1(\pi) : D(S/\mathbb{P}^2) \to \mathbb{P}^2$ is a double covering branched at $2B$.*

(*ii*) *The branch locus of $\beta_2(\pi) : S \to D(S/\mathbb{P}^2)$ is contained in $\mathrm{Sing}(D(S/\mathbb{P}^2))$ and S is smooth.*

Proof. For (*i*), since $\mathbb{P}^2$ is simply connected one has $\Delta_{\beta_1(\pi)} \neq \emptyset$. This means $\Delta_{\beta_1(\pi)} = B$. For (*ii*), we first show that $\Delta_{\beta_2(\pi)} \subset \mathrm{Sing}(D(S/\mathbb{P}^2))$ and $\Delta_{\beta_2(\pi)} \neq \emptyset$. Since D_6 has no element of order 6,

$\beta_2(\pi)$ cannot be branched along $\beta_1(\pi)^{-1}(B)$. This means that $\Delta_{\beta_2(\pi)} \subset \mathrm{Sing}(D(S/\mathbb{P}^2))$. Let $\mu : Z \to D(S/\mathbb{P}^2)$ be the minimal resolution. Since $K_Z \sim 0$ and the irregularity $q = 0$ by a general theory of double coverings (see [58, §2]) one has that Z is a $K3$-surface. In particular, Z is simply connected. If $\beta_2(\pi) : S \to D(S/\mathbb{P}^2)$ is unramified, then $Z \times_{D(S/\mathbb{P}^2)} S$ gives an étale cyclic triple covering of Z, but this is impossible. Hence $\Delta_{\beta_2(\pi)} \neq \emptyset$. By [118, Lemma 8.8], $\Delta_{\beta_2(\pi)} = \mathrm{Sing}(D(S/\mathbb{P}^2))$. Finally, the local structure of π around a cusp of B (described in [111, §2, Example 3]) forces S to be smooth. Q.E.D.

Lemma 3.15. *Let B be a sextic with 6 cusps. If a D_6-covering $\pi : S \to \mathbb{P}^2$ branched at $2B$ exists, then*

- (i) *the quotient surface $X := S/\langle\sigma\rangle$ is smooth for any element of $\sigma \in D_6$ of order 2,*
- (ii) *$K_X \sim -\overline{\pi}^* l$, where $\overline{\pi} : X \to \mathbb{P}^2$ denotes the induced non-Galois triple covering and l denotes a line of $\mathbb{P}^2$, and*
- (iii) *X is a del-Pezzo surface of degree 3.*

Proof. For (i), due to the local structure of $\pi : S \to \mathbb{P}^2$ around each cusp of B ([112, Example 3, §2]), π^*B is of the form $2(R_1 + R_2 + R_3)$, where R_i is a smooth divisor such that τ acts on the set $\{R_1, R_2, R_3\}$ transitively. One may assume that the fixed locus of σ is R_1, and this implies that X is smooth.
For (ii) and (iii), note that S is a $K3$-surface. Let us assume that the ramification locus of the quotient morphism $\alpha : S \to X$ is R_1. Since $R_1^2 = 6$ and R_1 is smooth, R_1 is numerically effective by [19, Proposition VIII 13]. Also note that $0 \sim K_S \sim \alpha^* K_X + R_1$, and hence $\alpha^* K_X \sim -R_1$. Thus $-K_X$ is numerically effective and $K_X^2 = 3$. This implies that X is a rational surface.
Now choose a general point x of $\mathbb{P}^2$. Let $\rho : \hat{X} \to X$ be the composition of the blowing-ups at the three points of $\overline{\pi}^{-1}(x)$. The pencil of lines through x on $\mathbb{P}^2$ induces an elliptic fibration $\varphi_x : \hat{X} \to \mathbb{P}^1$ and the three exceptional curves of ρ give sections of φ_x.

In order to complete the proof, we need the following result.

Claim 3.16. *$\varphi_x : \hat{X} \to \mathbb{P}^1$ is relatively minimal.*

Proof of Claim. Since $K_{\hat{X}}^2 = 0$ and $\hat{X}$ is a rational surface, the topological Euler number of $\hat{X}$ is 12. This implies that φ_x is relatively minimal. Q.E.D.

By this Claim, $K_{\hat{X}} \sim -F$, F being a fiber of φ_x. By our construction of $\hat{X}$, $\rho^*(\overline{\pi}^* l) \sim F + E_1 + E_2 + E_3$, where E_i ($i = 1, 2, 3$) denote the exceptional curves of ρ. Hence

$$\rho^*(\overline{\pi}^* l) \sim F + E_1 + E_2 + E_3 \sim -K_{\hat{X}} + E_1 + E_2 + E_3 \sim -\rho^*(K_X).$$

Therefore $\overline{\pi}^* l \sim -K_X$. In particular, $-K_X$ is ample and thus X is a del-Pezzo surface of degree 3. Q.E.D.

We are now in a position to prove the following:

Proposition 3.17. *Let B be a sextic with 6 cusps. Then there exists a D_6-covering branched at $2B$ if and only if B is given by an equation of the form $F^3 + G^2 = 0$, where $F(X_0, X_1, X_2)$ and $G(X_0, X_1, X_2)$ are homogeneous polynomials of degree 2 and 3, respectively.*

Proof. Suppose that B is given by the equation $F^3 + G^2 = 0$ as above. Consider the cubic surface X in $\mathbb{P}^3$ given by

$$X : X_3^3 + 3F(X_0, X_1, X_2)X_3 + 2G(X_0, X_1, X_2) = 0,$$

where $[X_0 : X_1 : X_2 : X_3]$ denotes a homogeneous coordinate system of $\mathbb{P}^3$. Let $P = [0 : 0 : 0 : 1]$ and let $pr_P : \mathbb{P}^3 \dashrightarrow \mathbb{P}^2$ be the projection centered at P. The restriction pr_P to X gives a non-Galois triple covering $pr_{P|X} : X \to \mathbb{P}^2$. By its defining equation, $\Delta(X/\mathbb{P}^2) = B$. The Galois closure K of $\mathbb{C}(X)$ is a D_6-extension of $\mathbb{C}(\mathbb{P}^2)$ and the K-normalization S of $\mathbb{P}^2$ is a D_6-covering $\pi : S \to \mathbb{P}^2$. By [111, Lemma 1.4], $\Delta(S/\mathbb{P}^2) = \Delta(X/\mathbb{P}^2) = B$ and by Lemma 3.14, π is branched at $2B$. These arguments follow Zariski's original idea.

The converse is the less detailed part in [130]. Suppose that there exists a D_6-covering $\pi : S \to \mathbb{P}^2$ branched at $2B$. Let $\overline{\pi} : X \to \mathbb{P}^2$ be a non-Galois triple covering as in Lemma 3.15. By [111, Lemma 1.4], $\Delta(S/\mathbb{P}^2) = \Delta(X/\mathbb{P}^2) = B$ and X is a del-Pezzo surface of degree 3 according to Lemma 3.15. Hence X is embedded as a cubic surface in $\mathbb{P}^3$ and its embedding is given by $\phi_{|-K_X|}$. Moreover, since $K_X \sim -\overline{\pi}^* l$ by Lemma 3.15, one has the following commutative diagram:

$$\begin{array}{ccc} X & \xrightarrow{\phi_{|-K_X|}} & \mathbb{P}^3 \setminus \{P_0\} \\ {\scriptstyle\overline{\pi}}\downarrow & & \downarrow{\scriptstyle pr} \\ \mathbb{P}^2 & = = = & \mathbb{P}^2, \end{array}$$

where pr denotes the projection centered at a suitable point $P_0 \in \mathbb{P}^3 \setminus \phi_{|-K_X|}(X)$. By choosing homogeneous coordinates $[X_0 : X_1 : X_2 : X_3]$ appropriately, one may assume that $P_0 = [0 : 0 : 0 : 1]$. This implies that pr is given by

$$[X_0 : X_1 : X_2 : X_3] \mapsto [X_0 : X_1 : X_2],$$

and that $\phi_{|-K_X|}(X)$ is given by the equation

$$X_3^3 + g_1(X_0, X_1, X_2)X_3^2 + g_2(X_0, X_1, X_2)X_3 + g_3(X_0, X_1, X_2) = 0,$$

where $g_i(X_0, X_1, X_2)$ are homogeneous polynomials of degree i. Now the defining equation of B is given by the discriminant of the above cubic equation which is

$$-4\left(g_2 - \frac{1}{3}g_1^2\right)^3 - 27\left(-\frac{1}{3}g_1g_2 + g_3 + \frac{2}{27}g_1^3\right)^2.$$

Q.E.D.

Remark 3.18. Any sextic given by an equation of the form $F^3 + G^2 = 0$, where F and G are homogeneous polynomials of degrees 2 and 3, respectively, is called a $(2,3)$-*torus sextic.* Zariski pairs of sextics given by $(2,3)$-torus and non-torus sextics are extensively studied by Oka [95, 96]. He uses Alexander polynomials to distinguish the topology of the complements. It may be interesting to revisit his proofs using the geometry of cubic surfaces.

3.4. Generic D_{2n}-coverings and Zariski pairs

In this section, an application of generic D_{2n}-coverings for the study of Zariski pairs will be shown. This method was used in [12, 15, 113, 115, 116]. Let Σ be a smooth projective surface and let B be a reduced divisor on Σ.

Remark 3.19. Thoughout this section, Σ is assumed to be simply connected.

Our purpose is to answer the following question:

Question 3.20. *Are there necessary and sufficient algebraic conditions on B for the existence of generic D_{2n}-coverings with $\Delta_\pi = B$?*

Suppose that Question 3.20 has a positive answer and let (P) be such a condition. The existence of a pair (B_1, B_2) of reduced divisors on Σ such that B_1 satisfies (P), while B_2 does not, implies that $(\Sigma, B_1) \not\approx (\Sigma, B_2)$. Hence if $\Sigma = \mathbb{P}^2$ and the combinatorial data of B_1 and B_2 are the same, (B_1, B_2) is a Zariski pair.

Now let us consider Question 3.20 in the case of n odd. The existence of a double covering $f' : Z' \to \Sigma$ with $\Delta_{f'} = B$ will always be assumed.

Let

$$\begin{array}{ccc} Z' & \xleftarrow[\mu]{} & Z \\ {\scriptstyle f'}\downarrow & & \downarrow{\scriptstyle f} \\ \Sigma & \xleftarrow[\rho]{} & W. \end{array}$$

denote the canonical resolution of Z' (see [58] for a definition).

Lemma 3.21. *Let $\pi : S \to \Sigma$ be a generic D_{2n}-covering with $\Delta_\pi = B$ (n is not necessarily odd in this lemma). Then*

(*i*) $D(S/\Sigma) \cong Z'$ *over* Σ *and*

(*ii*) $\Delta_{\beta_2(\pi)} \subset \mathrm{Sing}(D(S/\Sigma))$.

Proof. By hypothesis, the branch locus of $\beta_1(\pi) : D(S/\Sigma) \to \Sigma$ is B. Since Σ is simply connected, any double covering of Σ is determined by its branch locus up to isomorphism over Σ. This implies (*i*). The statement (*ii*) is immediate by hypothesis. Q.E.D.

Suppose that a generic D_{2n}-covering $\pi : S \to \Sigma$ with $\Delta_\pi = B$ exists. Let $\tilde{S}$ be the $\mathbb{C}(S)$-normalization of W and let $\tilde{\mu} : \tilde{S} \to S$ be an induced morphism. The induced covering morphism from $\tilde{S}$ to W will be denoted by $\tilde{\pi}$. Note that $\tilde{S}$ is again a D_{2n}-covering and one may assume that $D(\tilde{S}/W) = Z$, since $\mathbb{C}(Z') = \mathbb{C}(Z)$ and the $\mathbb{C}(S)$-normalization of Z is also $\tilde{S}$. Thus one has the following commutative diagram:

$$\begin{array}{ccc} S & \xleftarrow[\tilde{\mu}]{} & \tilde{S} \\ {\scriptstyle \beta_2(\pi)}\downarrow & & \downarrow{\scriptstyle \beta_2(\tilde{\pi})} \\ Z' & \xleftarrow[\mu]{} & Z \\ {\scriptstyle f'=\beta_1(\pi)}\downarrow & & \downarrow{\scriptstyle f=\beta_1(\tilde{\pi})} \\ \Sigma & \xleftarrow[\rho]{} & W. \end{array}$$

By Lemma 3.21(*ii*), $\beta_2(\tilde{\pi}) : \tilde{S} \to Z$ is an n-cyclic covering whose branch locus is contained in the exceptional locus of μ.

Conversely, suppose that there exists an n-cyclic covering $g : X \to Z$ such that:

(D1) Δ_g is contained in the exceptional locus of μ and

(D2) the composition $f \circ g : X \to W$ gives rise to a D_{2n}-covering.

The Stein factorization of $\rho \circ f \circ g : X \to \Sigma$ gives a generic D_{2n}-covering of Σ with $\Delta(X/\Sigma) = B$. Thus Question 3.20 is reduced to the following:

Question 3.22. *Find a sufficient and necessary condition for the existence of an n-cyclic covering $g : X \to Z$ satisfying* (D1) *and* (D2) *above.*

As usual, let us denote by σ the covering transformation of the double covering f. Propositions 3.11 and 3.12 provide a partial answer to Question 3.22 as follows:

Proposition 3.23. *Assume that Z is simply connected. A generic D_{2p}-covering (p odd prime) $\pi : S \to \Sigma$ with $\Delta_\pi = B$ exists if and only if there exist a non-empty effective divisor D and a line bundle $\mathcal{L}$ on Z satisfying the following conditions:*

(*i*) *D and σ^*D have no common components. Moreover, if $D = \sum_i a_i D_i$ denotes its irreducible decomposition, then* $\gcd(a_i, p) = 1$.

(*ii*) $\mathrm{Supp}(D + \sigma^*D)$ *is contained in the exceptional set of μ.*

(*iii*) $D - \sigma^*D \sim p\mathcal{L}$.

Proof. As noted above, a generic D_{2n}-covering $\pi : S \to \Sigma$ with $\Delta_\pi = B$ exists if and only if there exists another D_{2n}-covering $\tilde{\pi} : \tilde{S} \to W$ satisfying conditions (D1)-(D2) above. Applying Propositions 3.11 and 3.12 to $\beta_2(\tilde{\pi}) = f : Z \to W$ the statement follows, since the branch locus of $\beta_2(\tilde{\pi})$ is non-empty and is contained in the exceptional set of μ. Q.E.D.

From now on, we assume that singularities of B are at most simple singularities (see [20] for simple singularities). In this case, another version of Proposition 3.23 can be stated. In order to explain it, we need some preparation.

Let us keep the assumption in Proposition 3.23. Since Z is simply connected, $H^2(Z, \mathbb{Z})$ is a unimodular lattice with respect to the intersection pairing. Let $\mathrm{NS}(Z)$ be the Néron-Severi group of Z. It is a sublattice of $H^2(Z, \mathbb{Z})$ such that $H^2(Z, \mathbb{Z})/\mathrm{NS}(Z)$ is torsion-free. Since Z is simply connected, the Picard group $\mathrm{Pic}(Z)$ coincides with $\mathrm{NS}(Z)$. For $x \in \mathrm{Sing}(Z')$, R_x denotes the subgroup of $\mathrm{NS}(Z)$ generated by the irreducible components of the exceptional set arising from x. The lattice R_x is a negative definite sublattice of $\mathrm{NS}(Z)$. One can define the following sublattice of $\mathrm{NS}(Z)$:

$$T := \bigoplus_{x \in \mathrm{Sing}(Z')} R_x.$$

Given a lattice L, its dual lattice will be denoted by $L^\vee$ and its quotient modulo L by $G_L := L^\vee/L$. Associated with T one has $G_T \cong$

The type of x	G_{R_x}
$\mathbb{A}_n$	$\mathbb{Z}/(n+1)\mathbb{Z}$
$\mathbb{D}_n$ $n \equiv 1 \bmod 2$	$\mathbb{Z}/4\mathbb{Z}$
$\mathbb{D}_n$ $n \equiv 0 \bmod 2$	$(\mathbb{Z}/2\mathbb{Z})^{\oplus 2}$
$\mathbb{E}_6$	$\mathbb{Z}/3\mathbb{Z}$
$\mathbb{E}_7$	$\mathbb{Z}/2\mathbb{Z}$
$\mathbb{E}_8$	$\{0\}$

Table 2.

$\bigoplus_{x\in \mathrm{Sing}(Z')} G_{R_x}$. For a rational double point x, the results of Table 2 are well known.

One can consider both R_x and $R_x^\vee$ as subgroups of $R_x \otimes \mathbb{Q}$ and give a $\mathbb{Q}$-divisor which produces a generator of G_{R_x} in the cases of $x = \mathbb{E}_6$ and $\mathbb{A}_n$. This will come in handy for later use. For this purpose, let us label the irreducible components of the exceptional divisors for singularities of type $\mathbb{A}_n$ and $\mathbb{E}_6$ as in Figure 10. Note that $\sigma^*\Theta_k = \Theta_{n+1-k}$ if x is of type $\mathbb{A}_n$, and $\sigma^*\Theta_1 = \Theta_6$, $\sigma^*\Theta_2 = \Theta_5$ if x is of type $\mathbb{E}_6$.

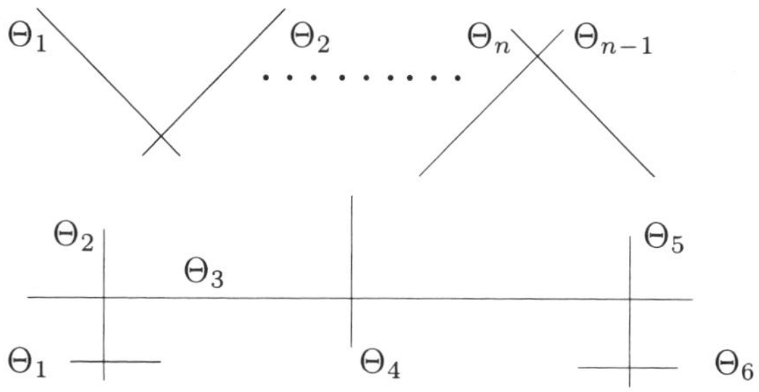

Fig. 10.

Lemma 3.24. *G_{R_x} is generated by the class of $\mathbb{Q}$-divisors $\dfrac{D_x}{n+1}$ (resp. $\dfrac{D_x}{3}$) for x of type $\mathbb{A}_n$ (resp. x of type $\mathbb{E}_6$), where*

$$D_x = \begin{cases} \displaystyle\sum_{k=1}^{\frac{n}{2}} (n+1-k)(\Theta_k - \Theta_{n+1-k}) & \text{if } x \text{ is of type } \mathbb{A}_n,\ n \text{ even} \\ \sum_{k=1}^{\frac{n-1}{2}} (n+1-k)(\Theta_k - \Theta_{n+1-k}) + \frac{n+1}{2}\Theta_{\frac{n+1}{2}} & \text{if } x \text{ is of type } \mathbb{A}_n,\ n \text{ odd,} \end{cases}$$

and

$$D_x = (\Theta_1 - \Theta_6) + 2(\Theta_2 - \Theta_5) \quad \textit{if } x \textit{ is of type } \mathbb{E}_6.$$

Proof. Our statement easily follows by considering the inverse of the intersection matrix of R_x. Q.E.D.

Let us now concentrate on the torsion part $(\mathrm{NS}(Z)/T)_{\mathrm{tor}}$ of $\mathrm{NS}(Z)/T$.

Lemma 3.25.

$$(\mathrm{NS}(Z)/T)_{\mathrm{tor}} \cong T^{\perp\perp}/T,$$

where $\bullet^{\perp}$ *denotes the orthogonal complement in* $H^2(Z, \mathbb{Z})$.

Proof. Since $H^2(Z,\mathbb{Z})/\mathrm{NS}(Z)$ is torsion free, one has $\mathrm{NS}(Z)^{\perp\perp} = \mathrm{NS}(Z)$. This implies that $T^{\perp\perp} \subset \mathrm{NS}(Z)$. Since $(\mathrm{NS}(Z)/T))_{\mathrm{tor}} \subset T^{\perp\perp}/T$, the result follows. Q.E.D.

Let ν be the homomorphism $T^{\perp\perp} \to T^{\vee} \to G_T$. Let L be an element in $T^{\perp\perp}$ such that its image in G_T has order p (p odd prime). Since $G_T \cong \bigoplus_{x \in \mathrm{Sing}(Z')} G_{R_x}$, one has

$$\nu(L) = (\gamma_x)_{x \in \mathrm{Sing}(Z')} \in \bigoplus_{x \in \mathrm{Sing}(Z')} G_{R_x}.$$

Note that $\gamma_x = 0$ unless x is a singular point of type $\mathbb{A}_n$ ($n+1 \equiv 0 \bmod p$) or type $\mathbb{E}_6$ (the latter case happens only when $p = 3$).

Lemma 3.26. *Assume that* $\gamma_x \neq 0$. *If* x *is a singular point of type* $\mathbb{A}_n$ $(n+1 \equiv 0 \bmod p)$, *then*

$$\gamma_x = \textit{ the class of } \tfrac{k_x}{p} D_x \qquad 0 < k_x \leq p-1,$$

and if x *is a singular point of type* $\mathbb{E}_6$, *then* $p = 3$ *and*

$$\gamma_x = \textit{ the class of } \tfrac{k_x}{3} D_x \qquad k_x = 1, 2,$$

where D_x *denotes the divisor in Lemma* 3.24.

Proof. Since R_x is a cyclic group generated by the class described in Lemma 3.24, the result follows. Q.E.D.

Now we are in a position to state another version of Proposition 3.23.

Theorem 3.27. *A generic* D_{2p}*-covering* (p *odd prime*) *of* Σ *with* $\Delta_\pi = B$ *exists if and only if* $\mathrm{NS}(Z)/T$ *has a* p*-torsion element.*

Proof. Suppose that there exists a generic D_{2p}-covering with branch locus B. Then by Proposition 3.23, there exists a non-empty divisor D and a line bundle $\mathcal{L}$ satisfying the three conditions. One can show that $\mathcal{L}$ gives rise to a p-torsion element of $\mathrm{NS}(Z)/T$. By the condition (*iii*) of Proposition 3.23, $p\mathcal{L} \in T$. Hence it is enough to show that $\mathcal{L} \notin T$. Suppose that $\mathcal{L} \in T$. Hence $\mathcal{L} \sim \sum_{x\in\mathsf{Sing}(Z')} \sum_i m_{i,x}\Theta_{i,x}$, where $\Theta_{i,x}$ denotes the irreducible components of the exceptional set arising from x. By plugging this relation into the one given in condition (*iii*), one obtains a non-trivial linear relation among $\Theta_{i,x}$'s. This leads to contradiction.

For the converse, let us suppose that $\mathrm{NS}(Z)/T$ has a p-torsion element. By Lemma 3.25, there exists an element L_1 in $T^{\perp\perp}$ such that whose class in $T^{\perp\perp}/T$ has torsion p. By Lemma 3.26,

$$L_1 \sim_{\mathbb{Q}} \sum_{x\in\mathsf{Sing}(Z')} \frac{k_x}{p} D_x \bmod T,$$

where $\sim_{\mathbb{Q}}$ denotes $\mathbb{Q}$-linear equivalence of divisors. This implies that there exists an element L_2 in T such that

$$L_1 + L_2 \sim_{\mathbb{Q}} \sum_{x\in\mathsf{Sing}(Z'),\gamma_x\neq 0} \frac{k_x}{p} D_x.$$

Let us define a divisor D on Z as follows:

$$D = \sum_{x\in\mathsf{Sing}(Z')} D_x^+,$$

where D_x^+ is defined as:

- $D_x^+ = 0$, if $\gamma_x = 0$.
- $D_x^+ = k_x\left(\sum_{i=1}^{\frac{n}{2}}(n+1-i)\Theta_i\right)$, if $\gamma_x \neq 0$ and x is of type $\mathbb{A}_n$ (n even).
- $D_x^+ = k_x\left(\sum_{i=1}^{\frac{n-1}{2}}(n+1-i)\Theta_i\right)$, if $\gamma_x \neq 0$ and x is of type $\mathbb{A}_n$ (n odd).
- $D_x^+ = k_x(\Theta_1 + 2\Theta_2)$, if $\gamma_x \neq 0$ and x is of type $\mathbb{E}_6$.

By the definition of D,

$$D - \sigma^* D \sim p(L_1 + L_2 - L_3),$$

where

$$L_3 = \sum_{x=\mathbb{A}_{n_x}(n_x\equiv 1 \bmod 2),\gamma_x\neq 0} \frac{k_x(n_x+1)}{2p}\Theta_{\frac{n_x+1}{2}}.$$

Moreover, the greatest common divisor of the coefficients of the irreducible components of D and p is 1. Thus the pair $(D, \mathcal{L})$, $\mathcal{L} = L_1 + L_2 - L_3$, satisfies the conditions in Proposition 3.23. Q.E.D.

Example 3.28. Let us consider Zariski's example from the viewpoint of Theorem 3.27, where $\Sigma = \mathbb{P}^2$ and B is a sextic with 6 cusps. In this case, the double covering $f' : Z' \to \mathbb{P}^2$ with $\Delta_{f'} = B$ has 6 $\mathbb{A}_2$-singularities x_i $(i = 1, \dots, 6)$ and its canonical resolution Z is a $K3$-surface. Hence Z is simply connected. Let us denote by $\Theta_{i,j}$ the exceptional curves arising at x_i $(j = 1, 2)$. By Proposition 3.17, a generic D_6-covering $\pi : S \to \mathbb{P}^2$ of $\mathbb{P}^2$ with $\Delta_\pi = B$ exists if and only if B is a $(2,3)$-torus curve.

If B is a $(2,3)$-torus curve, there exists a conic $\mathcal{Q}$ passing through the 6 cusps. One can show that $\mathcal{Q}$ gives rise to a 3-torsion element in $\mathrm{NS}(Z)/T$. Let $\rho_q^{-1}\mathcal{Q}$ be the proper transform of $\mathcal{Q}$ in W. Then $f^*(\rho_q^{-1}\mathcal{Q})$ is of the form $\mathcal{Q}^+ + \mathcal{Q}^-$. After relabeling $\Theta_{i,j}$ if necessary, we may assume that $\mathcal{Q}^+ \cdot \Theta_{i,1} = 1$, $\mathcal{Q}^- \cdot \Theta_{i,2} = 0$ $(i = 1, \dots, 6)$.

Claim 3.29. *$3\mathcal{Q}^+ \sim 3\tilde{f}^*l - \sum_{i=1}^{6}(2\Theta_{i,1} + \Theta_{i,2})$, where l denotes a generic line in $\mathbb{P}^2$ and $\tilde{f} = \rho \circ f$.*

Proof of Claim. Let us consider

$$D := 3\mathcal{Q}^+ - 3\tilde{f}^*l + \sum_{i=1}^{6}(2\Theta_{i,1} + \Theta_{i,2}).$$

One can see that $(\tilde{f}^*l)\cdot D = 0$ and $D^2 = 0$. By the Hodge index theorem, one has that $D \approx 0$. Since Z is simply connected, $D \sim 0$. Q.E.D.

By the Claim, note that the class of $\mathcal{Q}^+ - \tilde{f}^*l$ in $\mathrm{NS}(Z)/T$ gives a 3-torsion element. On the other hand, if B is not a $(2,3)$-torus curve, no D_6-covering branched at $2B$ exists. Hence, $\mathrm{NS}(Z)/T$ has no 3-torsion.

Remark 3.30.

(*i*) When using Theorem 3.27, we often replace T by $M \oplus T$, where M is a sublattice of $\mathrm{NS}(Z)$ orthogonal to T and such that $p \not| \operatorname{disc} M$, $\operatorname{disc} \bullet$ being the discriminant of a lattice $\bullet$. For example, $M = \tilde{f}^*\,\mathrm{NS}(\Sigma)$ in case $p \not| \operatorname{disc} \mathrm{NS}(\Sigma)$.

(*ii*) By Theorem 3.27, the problem of the existence of generic D_{2p}-coverings of Σ with $\Delta_\pi = B$ is reduced to that of primitive and non-primitive embeddings of T into $\mathrm{NS}(Z)$. In the case when $\Sigma = \mathbb{P}^2$ and B is a sextic with at most simple singularities, Z is a $K3$-surface. In this case, using Nikulin's lattice theory and the surjectivity of the period map, more detailed results than the existence of generic D_{2p}-coverings are obtained in [39].

3.5. Non-generic D_{2n}-coverings and Zariski k-plets

In this section, we will consider non-generic D_{2n}-coverings and their application to Zariski k-plets.

Let $\mathbf{B} = \mathbf{B}_1 + \mathbf{B}_2$ be a reduced divisor such that:

(i) there exists a double covering $f' : Z' \to \Sigma$ with $\Delta_{f'} = \mathbf{B}_1$, and

(ii) $\mathbf{B}_2$ is irreducible.

Let

$$\begin{array}{ccc} Z' & \xleftarrow[\mu]{} & Z \\ {\scriptstyle f'}\downarrow & & \downarrow{\scriptstyle f} \\ \Sigma & \xleftarrow[\rho]{} & W \end{array}$$

be the canonical resolution of Z'.

Proposition 3.31. *Suppose that Σ is simply connected and the divisor $f^*(\rho_q^{-1}\mathbf{B}_2)$ consists of two distinct irreducible components $\mathbf{B}_2^+$ and $\mathbf{B}_2^-$. Assume also that there exist both an effective divisor D and a line bundle $\mathcal{L}$ on Z satisfying:*

(i) $D = \mathbf{B}_2^+ + D'$. *D' and $\sigma^* D'$ have no common components,*

(ii) $\mathrm{Supp}(D' + \sigma^* D')$ *is contained in the exceptional set of μ, and*

(iii) $D - \sigma^* D \sim n\mathcal{L}$.

Then there exists a D_{2n}-covering $\pi : S \to \Sigma$ branched at $2\mathbf{B}_1 + n\mathbf{B}_2$ such that $\Delta_{\beta_1(\pi)} = \mathbf{B}_1$.

Proof. By Proposition 3.11, there exists a D_{2n}-covering $\tilde{\pi} : \tilde{S} \to W$ such that $\beta_1(\tilde{\pi}) = f$, $D(\tilde{S}/W) = Z$, $\mathbf{B}_2^+ \cup \mathbf{B}_2^- \subset \Delta_{\beta_2(\pi)} \subset \mathrm{Supp}(D + \sigma^* D)$ and whose ramification index along $\mathbf{B}_2^{\pm}$ is n. Since the irreducible components of D' are in the exceptional set of μ, the Stein factorization of $\rho \circ \tilde{\pi}$ gives the desired D_{2n}-covering. Q.E.D.

Proposition 3.32. *Under the notation above, if a D_{2n}-covering branched at $2\mathbf{B}_1 + n\mathbf{B}_2$ with $\Delta_{\beta_1(\pi)} = \mathbf{B}_1$ exists, then the following holds:*

(i) *$f^*(\rho_q^{-1}\mathbf{B}_2)$ consists of two irreducible components, $\mathbf{B}_2^{\pm}$,*

(ii) *there exist effective divisors D_1 and D_2, and a line bundle $\mathcal{L}$ on Z such that*

- $\mathrm{Supp}(D_1 + \sigma^* D_1 + D_2)$ *is contained in the exceptional set of μ,*

- *D_1 and $\sigma^* D_1$ have no common components,*
- *if $D_2 \neq \emptyset$, then n is even, D_2 is reduced, and $D' = \sigma^* D'$ for each irreducible component D' of D_2, and*
- $(\mathbf{B}_2^+ + D_1 + \frac{n}{2} D_2) - (\mathbf{B}_2^- + \sigma^* D_1) \sim n\mathcal{L}$.

Proof. Let us denote by $\pi : S \to \Sigma$ the D_{2n}-covering given by hypothesis. Let $\tilde{S}$ be the $\mathbb{C}(S)$-normalization of W. The induced morphism $\tilde{\pi} : \tilde{S} \to W$ is a D_{2n}-covering with $D(\tilde{S}/W) = Z$ and $\Delta_{\beta_2(\tilde{\pi})}$ is contained in the union of $\mathrm{Supp}(f^*(\rho_q^{-1}\mathbf{B}_2))$ with the exceptional subset of μ. Since $f^*(\rho_q^{-1}\mathbf{B}_2)$ is a part of $\Delta_{\beta_2(\tilde{\pi})}$, by Corollary 3.13, $f^*(\rho_q^{-1}\mathbf{B}_2)$ is of the form $\mathbf{B}_2^+ + \mathbf{B}_2^-$, which implies part (*i*). For part (*ii*) let $\tilde{D}_1$, $\tilde{D}_2$, and $\tilde{\mathcal{L}}$ be the two effective divisors and the line bundle on Z respectively, given by Proposition 3.12 applied to $\tilde{\pi} : \tilde{S} \to W$. Then by hypothesis, $\tilde{D}_1$ is of the form $a\mathbf{B}_2^+ + D_1'$. Moreover, $\mathrm{Supp}(D_1' + \sigma^* D_1' + \tilde{D}_2)$ is contained in the exceptional set of μ. By the assumption on the ramification index along $\mathbf{B}_2$, one has that $\gcd(a, n) = 1$ and there exists an integer a' $(0 < a < n)$ such that $aa' \equiv 1 \bmod n$. Note that a' is odd if n is even, therefore

$$a'(\tilde{D}_1 + \frac{n}{2}\tilde{D}_2) = \mathbf{B}_2^+ + a'D_1' + \frac{n}{2}\tilde{D}_2 + nM,$$

$$a'\sigma^*\tilde{D}_1 = \mathbf{B}_2^- + a'\sigma^* D_1' + nM'$$

for some effective divisors M and M'. Hence,

$$(\mathbf{B}_2^+ + a'D_1' + \frac{n}{2}\tilde{D}_2) - (\mathbf{B}_2^- + a'\sigma^* D_1') \sim n(a'\tilde{\mathcal{L}} + M' - M).$$

The result follows by considering $D_1 := a'D_1'$, $D_2 := \tilde{D}_2$ and $\mathcal{L} := a'\tilde{\mathcal{L}} + M' - M$. Q.E.D.

As an application of Propositions 3.31 and 3.32 one has the following:

Theorem 3.33. *Let $B_1 + B_{2,j}$ $(j = 1, \ldots, k)$ be reduced divisors on Σ satisfying:*

- *B_1 is smooth,*
- *$B_{2,j}$ $(j = 1, \ldots, k)$ are irreducible and not homeomorphic to B_1,*
- *$B_1 + B_{2,j}$ $(j = 1, \ldots, k)$ have the same combinatorial data,*
- *there exists a double covering $f : Z \to \Sigma$ with $\Delta_f = B_1$ such that*
  - *Z is simply connected and*
  - *$f^* B_{2,j}$ is of the form $B_{2,j}^+ + B_{2,j}^-$, $B_{2,j}^+ \neq B_{2,j}^-$,*

- *there exist distinct positive integers* n_j $(j = 1, \ldots, k)$ *and non-trivial line bundles* $\mathcal{L}_1, \ldots, \mathcal{L}_k$ *such that*
  - $B^+_{2,j} - B^-_{2,j} \sim n_j\mathcal{L}_j$, *and*
  - *no line bundle* M_j *satisfies* $L_j \sim d_jM_j$ *for any* j *and* $d_j \geq 2$.

Then $(\Sigma, B_1 + B_{2,i}) \not\approx (\Sigma, B_1 + B_{2,j})$.

Proof. By Proposition 3.31, there exists a non-generic D_{2n_j}-covering $\pi_j : S_j \to \Sigma$ branched at $2B_1 + n_jB_{2,j}$ with $\Delta_{\beta_1(\pi_j)} = B_1$ for each j. Since $B_{2,j}$ $(j = 1, \ldots, k)$ are not homeomorphic to B_1, there does not exist any homeomorphism $f : \mathbb{P}^2 \to \mathbb{P}^2$ such that $f(B_1) = B_{2,j}$ and $f(B_{2,j'}) = B_1$ for any j, j'. Hence, in order to prove this statement, it is enough to show that there exists no D_{2n_l}-covering $\pi'_l : S'_l \to \Sigma$ branched at $2B_1 + n_lB_{2,j}$ with $\Delta_{\beta_1(\pi'_l)} = B_1$ if $l \neq j$. If such a covering existed, then by Proposition 3.32, there should exist a line bundle $\mathcal{L}'$ such that $B^+_2 - B^-_2 \sim n_l\mathcal{L}'$. On the other hand, $B^+_2 - B^-_2 \sim n_j\mathcal{L}_j$. Let $d = \gcd(n_l, n_j)$ and set $n_j = n'_jd$, $n_l = n'_ld$. Thus $n'_l\mathcal{L}' \sim n'_j\mathcal{L}_j$, as Z is simply connected. Choose an integer b so that $n'_jb = mn'_l + 1$. Thus $\mathcal{L}_j \sim n'_l(b\mathcal{L}' - m\mathcal{L}_j)$ and $n'_l > 1$, which contradicts the hypothesis. Q.E.D.

Theorem 3.33 serves as the main tool to find the Zariski k-plet given in [16], where Zariski k-plets are explicitly obtained for any k. We recall that Zariski k-plets for any k were also obtained by V.S Kulikov in [68] in a more theoretical way. He proves the Chisini conjecture in many cases, i.e., if a curve $\mathcal{C}$ is the branch locus of a generic projection, then $\mathcal{C}$ determines the monodromy of the associated covering. In that case, $\mathcal{C}$ is an irreducible curve having only ordinary nodes and cusps as singularities, and its numerical invariants are determined by the numerical invariants of the surface. F. Catanese [26, 27] had shown that there exist moduli spaces of surfaces with given numerical invariants but different topologies and that is how the theoretical existence of Zariski k-plets was proved.

In what follows we will sketch an explicit construction of a Zariski k-plet.

Example 3.34. Let C_0 be a smooth conic on $\mathbb{P}^2$ and let $f : Z \to \mathbb{P}^2$ be a double covering with $\Delta_f = C_0$. It is well known that $Z = \mathbb{P}^1 \times \mathbb{P}^1$, the covering transformation σ exchanges the two rulings on Z, and $\mathrm{Pic}(Z) \cong \mathbb{Z} \oplus \mathbb{Z}$. Hence a class in $\mathrm{Pic}(Z)$ can be described by a pair of integers. Note that $\sigma^*(a, b) = (b, a)$ and $D_0 := (f^*C_0)_{\mathrm{red}} \sim (1, 1)$.

Let $g_n(t) \in \mathbb{C}(t)$ be a rational function of degree n. Given any $(a,b) \in \mathrm{Pic}(Z)$ one can define a morphism $\eta_{a,b}$ from $\mathbb{P}^1$ to $\mathbb{P}^1 \times \mathbb{P}^1$ by

$$t \mapsto (g_a(t), g_b(t)),$$

where t denotes a non-homogeneous coordinate of $\mathbb{P}^1$. If $g_a(t)$ and $g_b(t)$ are generic, then the image $D_{a,b} := \eta_{a,b}(\mathbb{P}^1)$ satisfies the following properties:

- $D_{a,b}(\sim (a,b))$ is a rational curve with $ab - (a+b) + 1$ distinct nodes.
- $D_{a,b}$ and $\sigma^* D_{a,b}$ meet at $a^2 + b^2$ distinct points, $a+b$ of which are on D_0.

These two properties imply that $f(D_{a,b})$ is a rational curve of degree $a+b$ tangent to C_0 at $a+b$ distinct points and with $\binom{a+b-1}{2}$ distinct nodes.

Let us fix an integer $m \geq 4$. Take $\lfloor \frac{m}{2} \rfloor$ distinct pairs of integers: $(m-j, j)$, $j = 1, \ldots, \lfloor \frac{m}{2} \rfloor$, where $\lfloor \bullet \rfloor$ denotes the greatest integer not exceeding $\bullet$, and consider a nodal rational curve $D_{m-j,j}$ as above. Consider $B_1 = C_0$, $B_{2,j} = f(D_{m-j,j})$ $(j = 1, \ldots, \lfloor \frac{m}{2} \rfloor)$. Since $D_{m-j,j} - \sigma^* D_{m-j,j} \sim (m-2j, 2j-m) = (m-2j)(1,-1)$ and $m-2j$ $(j = 1, \ldots, \lfloor \frac{m}{2} \rfloor)$ are all different, $(B_1 + B_{2,1}, \ldots, B_1 + B_{2,\lfloor \frac{m}{2} \rfloor})$ is a Zariski $\lfloor \frac{m}{2} \rfloor$-plet by Theorem 3.33.

In Example 3.34, when m is odd, one has a stronger statement. In this case, the fundamental groups themselves (disregarding the peripheral information) distinguish the Zariski pair.

Proposition 3.35. *Let $(B_1 + B_{2,1}, \ldots, B_1 + B_{2,\lfloor \frac{m}{2} \rfloor})$ be as in Example 3.34. If m is odd, then*

$$\pi_1(\mathbb{P}^2 \setminus (B_1 + B_{2,i}), p_o) \not\cong \pi_1(\mathbb{P}^2 \setminus (B_1 + B_{2,j}), p_o),$$

for any $i < j$.

Let us start by proving the following lemma.

Lemma 3.36. *Let $B_1 + B_{2,j}$ be as in Proposition 3.35. If a D_{2n}-covering (n odd) $\pi : S \to \mathbb{P}^2$ with $\Delta_\pi \subset B_1 + B_{2,j}$ exists, then*

(i) *$\Delta_\pi = B_1 + B_{2,j}$, $D(S/\mathbb{P}^2) = Z$ and $\beta_1(\pi) = f$, and*
(ii) *π is branched at $2B_1 + nB_{2,j}$.*

Proof. Since $\mathbb{P}^2$ is simply connected, $\Delta_{\beta_1(\pi)} \neq \emptyset$. Also note that the branch locus of a double covering is a reduced curve of even degree, $\Delta_{\beta_1(\pi)} = B_1$. This implies that $D(S/\mathbb{P}^2) = Z$ and $\beta_1(\pi) = f$. Since Z

is also simply connected, $\Delta_{\beta_2(\pi)} \neq \emptyset$ and thus $\Delta_{\beta_2(\pi)} = f^*(B_{2,j})$, which proves (i).

In order to prove (ii) it is enough to show that the ramification index along $B_{2,j}$ is n. Since n is odd, by [119], there exists a rational function $\theta \in \mathbb{C}(S)$ such that:

- $\mathbb{C}(S) = \mathbb{C}(\mathbb{P}^2)(\theta)$ and
- the action of $D_{2n} = \langle \sigma, \tau \mid \sigma^2 = \tau^n = (\sigma\tau)^2 = 1 \rangle$ on θ is given by

$$\theta^\sigma = \frac{1}{\theta}, \qquad \theta^\tau = \zeta_n \theta, \quad \zeta_n = \exp\left(\frac{2\pi\sqrt{-1}}{n}\right).$$

Considering $\varphi := \theta^n$, one has that $\varphi \in \mathbb{C}(Z)$. Since $\mathbb{C}(S) = \mathbb{C}(Z)(\sqrt[n]{\varphi})$, one may assume that the divisor (φ) of φ is of the form

$$(\varphi) = (aD_{m-j,j} + nD') - (a\sigma^* D_{m-j,j} + nD'')$$

for some effective divisors D' and D''. We claim that $\gcd(a, n) = 1$. If $\gcd(a, n) = d > 1$, then one has

$$\left(\frac{a}{d} D_{m-j,j} + \frac{n}{d} D'\right) - \left(\frac{a}{d}\sigma^* D_{m-j,j} + \frac{n}{d} D''\right) \sim 0,$$

for Z is simply connected. Hence there exists $\psi \in \mathbb{C}(Z)$ such that $\varphi = \psi^d$. This means that the polynomial $x^n - \varphi = x^n - \psi^d$ is reducible in $\mathbb{C}(Z)[x]$, which contradicts $\mathbb{C}(Z) = \mathbb{C}(Z)(\sqrt[n]{\varphi})$. Therefore $\gcd(a, n) = 1$ and hence $\beta_2(\pi)$ is branched at $n(D_{m-j,j} + \sigma^* D_{m-j,j})$. Q.E.D.

Proof of Proposition 3.35. Choose any $i < j$. Since a $D_{2(m-2i)}$-covering branched at $2B_1 + (m-2i)B_{2,i}$ exists, there is a surjective homomorphism $\pi_1(\mathbb{P}^2 \setminus (B_1 + B_{2,i}), p_o) \to D_{2(m-2i)}$. If $\pi_1(\mathbb{P}^2 \setminus (B_1 + B_{2,j}), p_o) \cong \pi_1(\mathbb{P}^2 \setminus (B_1 + B_{2,j}), p_o)$, then there also exists a surjective homomorphism $\pi_1(\mathbb{P}^2 \setminus (B_1 + B_{2,j}), p_o) \to D_{2(m-2i)}$. Therefore a $D_{2(m-2i)}$-covering $\pi : S \to \mathbb{P}^2$ with $\Delta_\pi \subset B_1 + B_{2,j}$ has to exist. By Lemma 3.36, $D(S/\mathbb{P}^2) = Z$, $\beta_1(\pi) = f$, and π is branched at $2B_1 + (m-2i)B_{2,j}$. Hence

$$D_{m-j,j} - \sigma^* D_{m-j,j} \sim (m - 2j, -m + 2j) \sim (m - 2i)\mathcal{L}$$

for some $\mathcal{L} \in \operatorname{Pic}(Z)$, which is not possible. Therefore

$$\pi_1(\mathbb{P}^2 \setminus (B_1 + B_{2,i}), p_o) \not\cong \pi_1(\mathbb{P}^2 \setminus (B_1 + B_{2,j}), p_o),$$

for any $i < j$. Q.E.D.

§4. The Zariski pair of Namba and Tsuchihashi

4.1. Description of the combinatorial stratum

In this section, an elementary construction of the first Zariski pair of an arrangement of conics ([88]) will be given. In their paper Namba and Tsuchihashi show the existence of two arrangements of smooth conics both having the following combinatorics: if $\mathcal{C}_1, \mathcal{C}_2, \mathcal{C}_3, \mathcal{C}_4$ denote four smooth conics, then $\mathcal{C}_1 \pitchfork \mathcal{C}_2$, $\mathcal{C}_3 \pitchfork \mathcal{C}_4$, and $\mathcal{C}_i, \mathcal{C}_j$ are bitangent if $i = 1,2$, $j = 3,4$. They computed the fundamental groups of two such arrangements and proved that they are not isomorphic.

From a more geometrical point of view, our interest is to describe the irreducible components of the combinatorial strata of such curves in terms of position of singularities. In order to do so, other combinatorial strata of curves will be defined along the way.

Let $\mathcal{M} \subset \mathbb{P}_8$ be the combinatorial stratum described above. The ordered version of $\mathcal{M}$ will be denoted by $\widetilde{\mathcal{M}} \subset (\mathbb{P}_2)^4$. Following the notations in [88], $\mathcal{A} \subset \mathbb{P}_6$ will be the family of all curves which decompose in three smooth conics $\mathcal{C}_1, \mathcal{C}_2, \mathcal{C}_3$ whose combinatorics results from $\mathcal{M} \subset \mathbb{P}_8$ by removing any conic. Finally, $\widetilde{\mathcal{A}} \subset (\mathbb{P}_2)^3$ will denote the ordered stratum associated with $\mathcal{A} \subset \mathbb{P}_6$, that is, triples $(\mathcal{C}_1, \mathcal{C}_2, \mathcal{C}_3)$ such that $\mathcal{C}_1 \pitchfork \mathcal{C}_2$ and $\mathcal{C}_i, \mathcal{C}_3$ are bitangent $(i = 1,2)$.

Notation 4.1. Given a projective space $\mathbb{P}$ and $A \subset \mathbb{P}$ we will denote $\Sigma(A)$ the smallest projective subspace of $\mathbb{P}$ containing A.

Lemma 4.2. *The families $\mathcal{A}$ and $\widetilde{\mathcal{A}}$ are irreducible as algebraic varieties (and thus, connected).*

Proof. By a natural mapping $\widetilde{\mathcal{A}} \twoheadrightarrow \mathcal{A}$ given by $(\mathcal{C}_1, \mathcal{C}_2, \mathcal{C}_3) \mapsto \mathcal{C}_1 + \mathcal{C}_2 + \mathcal{C}_3$, it is enough to prove the statement for $\widetilde{\mathcal{A}}$.

Let $\widetilde{\mathcal{A}}_1$ be the subset of $(\mathbb{P}_1)^2 \times \mathbb{P}_2$ such that $(L_1, L_2, \mathcal{C}_3) \in \widetilde{\mathcal{A}}_1$ if and only if $\mathcal{C}_3$ is smooth and $L_1 + L_2 + \mathcal{C}_3$ has only ordinary double points. Given $(\mathcal{C}_1, \mathcal{C}_2, \mathcal{C}_3) \in \widetilde{\mathcal{A}}$, one can consider the lines L_i joining the bitangent intersection points of $\mathcal{C}_i$ and $\mathcal{C}_3$, $i = 1,2$. This defines a natural mapping $\widetilde{\mathcal{A}} \to \widetilde{\mathcal{A}}_1$. It is straightforward to show that $\widetilde{\mathcal{A}}_1$ is irreducible. Therefore, it only remains to show that this mapping is surjective with irreducible fibers. Note that given $(L_1, L_2, \mathcal{C}_3) \in \widetilde{\mathcal{A}}_1$, its fiber is a Zariski open subset of $\Sigma(\mathcal{C}_3, 2L_1) \times \Sigma(\mathcal{C}_3, 2L_2) \times \{\mathcal{C}_3\}$. Q.E.D.

Definition 4.3. Let $S \subset \mathbb{P}_2$ be a pencil of conics. A point $P \in \mathbb{P}^2$ is said to be *associated with* S if P is a singular point of a member of S (recall that a multiple curve is singular).

Lemma 4.4. *Let $\mathcal{C}_1, \mathcal{C}_2, \mathcal{C}_3$ be conics (not in a pencil) such that $P \notin \mathcal{C}_i$, $i = 1, 2, 3$, is associated with both $\Sigma(\mathcal{C}_1, \mathcal{C}_2)$ and $\Sigma(\mathcal{C}_2, \mathcal{C}_3)$. Then, P is associated with $\Sigma(\mathcal{C}_1, \mathcal{C}_3)$. Moreover, if $\mathcal{D}_{12}$ and $\mathcal{D}_{23}$ are singular conics in $\Sigma(\mathcal{C}_1, \mathcal{C}_2)$ and $\Sigma(\mathcal{C}_2, \mathcal{C}_3)$, respectively, such that P is a double point of both, then there is a singular conic $\mathcal{D}_{13} \in \Sigma(\mathcal{C}_1, \mathcal{C}_3)$, containing P as a double point, such that $\mathcal{D}_{13} \in \Sigma(\mathcal{D}_{12}, \mathcal{D}_{23})$.*

Proof. If C_i (resp. D_{jk}) denotes an equation for $\mathcal{C}_i$ (resp. $\mathcal{D}_{jk}$), then there exist constants such that

$$\begin{aligned} \alpha_1 C_1 + \alpha_2 C_2 &= D_{12} \\ \beta_2 C_2 + \beta_3 C_3 &= D_{23}. \end{aligned}$$

Since $P \notin \mathcal{C}_i$ we have $\alpha_1\alpha_2\beta_1\beta_2 \neq 0$. Then $\beta_2\alpha_1 C_1 - \alpha_2\beta_3 C_3 = \beta_2 D_{12} - \alpha_2 D_{23}$. Since P is a double point of $\mathcal{D}_{12}$ and $\mathcal{D}_{23}$, the result follows. Q.E.D.

Let $\mathcal{B} \subset \mathbb{P}_4$ be the family of curves which decompose into two transversal smooth conics and $\widetilde{\mathcal{B}} \subset (\mathbb{P}_2)^2$ its ordered version. The pencil spanned by each element in $\mathcal{B}$ has exactly three associated points, which according to Lemma 4.4 are the double points of the three singular conics in the pencil generated by the two transversal smooth conics. Let us define the following combinatorial stratum

$$\mathcal{P} := \{(\mathcal{C}_1 + \mathcal{C}_2, P) \in \mathbb{P}_4 \times \mathbb{P}^2 \mid \mathcal{C}_1 + \mathcal{C}_2 \in \mathcal{B}, P \text{ associated with } \Sigma(\mathcal{C}_1, \mathcal{C}_2)\}$$

and denote by $\widetilde{\mathcal{P}} \subset (\mathbb{P}_2)^2 \times \mathbb{P}^2$ its ordered version.

Proposition 4.5. *Let $(\mathcal{C}_1, \mathcal{C}_2, \mathcal{C}_3) \in \widetilde{\mathcal{A}}$ and consider L_i the line joining the tangency points of $\mathcal{C}_i$ and $\mathcal{C}_3$, $i = 1, 2$. Then, $P := L_1 \cap L_2$ is associated with $\Sigma(\mathcal{C}_1, \mathcal{C}_2)$. Also, if $\mathcal{D}_{12}$ is the reducible conic of $\Sigma(\mathcal{C}_1, \mathcal{C}_2)$ containing P, then $\mathcal{D}_{12} \in \Sigma(2L_1, 2L_2)$.*

Moreover, given $(\mathcal{C}_1, \mathcal{C}_2, P) \in \widetilde{\mathcal{P}}$, there exists an irreducible quasiprojective subvariety U of $\mathbb{P}_2$ such that $\mathcal{C}_3 \in U$ if and only if $(\mathcal{C}_1, \mathcal{C}_2, \mathcal{C}_3) \in \widetilde{\mathcal{A}}$ and P is obtained as above.

Proof. Let us fix $(\mathcal{C}_1, \mathcal{C}_2, \mathcal{C}_3) \in \widetilde{\mathcal{A}}$ and let us consider L_1, L_2, P as in the statement. Note that P is associated with $\Sigma(\mathcal{C}_1, 2L_1) = \Sigma(\mathcal{C}_1, \mathcal{C}_3)$ and also with $\Sigma(\mathcal{C}_2, 2L_2) = \Sigma(\mathcal{C}_2, \mathcal{C}_3)$. Then, by Lemma 4.4, P is associated with $\Sigma(\mathcal{C}_1, \mathcal{C}_2)$. Note also that $2L_1$ and $2L_2$ are the reducible conics of the *Moreover* part of Lemma 4.4.

For the last statement, let $\mathcal{D}_{12}$ be the reducible conic of $\Sigma(\mathcal{C}_1, \mathcal{C}_2)$ containing P and fix a line L_1 through P transversal to $\mathcal{C}_1$. In the pencil $\Sigma(2L_1, D_{12})$ there is another double line $2L_2$. For *generic* L_1, L_2 is transversal to $\mathcal{C}_2$. The projective subspace

$$S := \Sigma(\mathcal{C}_1, \mathcal{C}_2, 2L_1, 2L_2) = \Sigma(\mathcal{C}_1, \mathcal{D}_{12}, 2L_1, 2L_2) \subset \mathbb{P}_2$$

is of dimension 2 and contains the pencils $\Sigma(2L_i, \mathcal{C}_i)$ which are lines in S and thus intersect at a conic $\mathcal{C}_3$. Q.E.D.

Definition 4.6. The point P in Proposition 4.5 is said to be *associated* with $(\mathcal{C}_1, \mathcal{C}_2, \mathcal{C}_3)$.

Corollary 4.7. *The spaces $\widetilde{\mathcal{P}}$ and $\mathcal{P}$ are irreducible.*

Note that the natural projection $\widetilde{\mathcal{M}} \twoheadrightarrow \mathcal{M}$ is 4:1. For simplicity if $\widetilde{\mathcal{C}} := (\mathcal{C}_1, \mathcal{C}_2, \mathcal{C}_3, \mathcal{C}_4) \in \widetilde{\mathcal{M}}$, then $\mathcal{C} = \mathcal{C}_1 \cup \mathcal{C}_2 \cup \mathcal{C}_3 \cup \mathcal{C}_4$ will denote its projection. Conversely, we will add a tilde for any element in the fiber of $\mathcal{C}$.

Corollary 4.8. *The spaces $\widetilde{\mathcal{M}}$ (resp. $\mathcal{M}$) have two irreducible (and connected) components $\widetilde{\mathcal{M}}_+, \widetilde{\mathcal{M}}_-$ (resp. $\mathcal{M}_+, \mathcal{M}_-$).*

Moreover, curves in $\mathcal{M}_+$ and $\mathcal{M}_-$ can be distinguished as follows: given $\mathcal{C} \in \mathcal{M}$, $\widetilde{\mathcal{C}} := (\mathcal{C}_1, \mathcal{C}_2, \mathcal{C}_3, \mathcal{C}_4) \in \widetilde{\mathcal{M}}$, and P_i the point associated with $(\mathcal{C}_1, \mathcal{C}_2, \mathcal{C}_i)$, $i = 3, 4$, then $\mathcal{C} \in \mathcal{M}_+$ (resp. $\mathcal{M}_-$) if and only if $P_3 = P_4$ (resp. $P_3 \neq P_4$).

There is another geometrical property which distinguishes the component $\mathcal{M}_+$ from $\mathcal{M}_-$.

Theorem 4.9. *Let $\mathcal{C} \in \mathcal{M}$. Then $\mathcal{C} \in \mathcal{M}_+$ if and only if there exists a conic passing through its eight tacnodes.*

Proof. Let us fix some notation. Given $i \in \{1, 2\}$, $j \in \{3, 4\}$, we will consider $\mathcal{C}_i \cap \mathcal{C}_j := \{P_{ij}, Q_{ij}\}$ and denote by L_{ij} the line joining P_{ij} and Q_{ij}. Let us also define $P_j := L_{1j} \cap L_{2j}$, $j = 3, 4$, which are both points associated with $\Sigma(\mathcal{C}_1, \mathcal{C}_2)$ according to Proposition 4.5. Let us denote by $\mathcal{D}_j$ the conic in $\Sigma(\mathcal{C}_1, \mathcal{C}_2)$ containing P_j as a double point. Note that $\mathcal{D}_j \in \Sigma(2L_{1j}, 2L_{2j})$ by Proposition 4.5.

Let us consider the pencils of conics $\Lambda_i := \Sigma(\mathcal{C}_i, L_{i3} + L_{i4})$, $i = 1, 2$, and let $S := \Sigma(\Lambda_1, \Lambda_2)$. The desired conic should belong to $\Lambda_1 \cap \Lambda_2$, and it exists (uniquely) if and only if $\dim S = 2$. Also note that $S = \Sigma(\mathcal{C}_1, \mathcal{D}_3, L_{13} + L_{14}, L_{23} + L_{24})$. Since $P := P_3 \notin \mathcal{C}_1$ and $P \in \mathcal{D}_3 \cap L_{13} \cap L_{23}$, one has:

$$S_P := \{\mathcal{C} \in S \mid P \in \mathcal{C}\} = \Sigma(\mathcal{D}_3, L_{13} + L_{14}, L_{23} + L_{24}) \subsetneq S.$$

Let us suppose that the conic exists, i.e., $\dim S = 2$ and $\dim S_P = 1$. Since P is a base point of the pencil S_P and a double point for one element in S_P, either it is a double point for *any* element of the pencil, or the tangent line at P of the *general* member is constant. The second possibility cannot happen since $L_{13} \neq L_{23}$. Since P is a double point, $P_3 \in L_{14} \cap L_{24}$, i.e. $P_3 = P_4$.

Let us assume now that $P = P_3 = P_4$, in particular $\mathcal{D}_3 = \mathcal{D}_4$. Let us choose coordinates such that $P := [0 : 0 : 1]$, $\mathcal{D}_3 : xy = 0$, and $L_{13} : x - y = 0$. Since $\mathcal{D}_3 \in \Sigma(2L_{13}, 2L_{23})$, it is easily seen that $L_{23} : x + y = 0$. Analogously, one can prove that there exists $\alpha \in \mathbb{C} \setminus \{0, \pm 1\}$ such that $L_{14} : x - \alpha y = 0$ and $L_{24} : x + \alpha y = 0$. An easy computation concludes that $\mathcal{D}_3 \in \Sigma(L_{13} + L_{14}, L_{23} + L_{24})$, i.e., $\dim S_P = 1$ and $\dim S = 2$. Q.E.D.

For the computation of Alexander polynomials and characteristic varieties it is useful to calculate the space of curves of a given degree passing through some points. Let $\mathcal{C} \in \mathcal{M}$ and let $P_1, \ldots, P_8$ be the eight tacnodes in $\mathcal{C}$. Let us consider

$$\begin{aligned} \mathcal{K}_{k,8} &:= \{\mathcal{D} \in \mathbb{P}_k \mid P_1, \ldots, P_8 \in \mathcal{D}\}, \\ \mathcal{K}_{k,6} &:= \{\mathcal{D} \in \mathbb{P}_k \mid P_1, \ldots, P_6 \in \mathcal{D}\}. \end{aligned}$$

In principle there are several choices for $\mathcal{K}_{k,6}$, but the kind of results we will obtain for $\mathcal{K}_{k,6}$ do not depend on the choice of such points. Let us consider the mappings

$$\begin{aligned} \sigma_{k,8} &: H^0(\mathbb{P}^2; \mathcal{O}(k)) \to \mathbb{C}^8 \\ \sigma_{k,6} &: H^0(\mathbb{P}^2; \mathcal{O}(k)) \to \mathbb{C}^6 \end{aligned}$$

defined as in the exact sequence (18) in §2.2. Then $\mathcal{K}_{k,8}$ (resp. $\mathcal{K}_{k,6}$) is the projective space of $\ker \sigma_{k,8}$ (resp. $\ker \sigma_{k,6}$).

Proposition 4.10. *If* $\mathcal{C} \in \mathcal{M}_+$ *then* $\dim \mathcal{K}_{3,8} = 2$, *whereas if* $\mathcal{C} \in \mathcal{M}_-$ *then* $\dim \mathcal{K}_{3,8} = 1$.

Proof. Assume first that $\mathcal{C} \in \mathcal{M}_+$. Since there is a conic $\mathcal{Q}$ passing through the eight points, $\mathcal{Q}$ is a base component of $\mathcal{K}_{3,8}$, i.e., any element of $\mathcal{K}_{3,8}$ is of the form $\mathcal{Q}+L$, $L \in \mathbb{P}_1$. This is trivial if $\mathcal{Q}$ is smooth. If it is singular, then $\mathcal{Q} = L_1 + L_2$. It is easily seen that no line can contain five of these points; then both L_1 and L_2 are also base components. On the other hand, $\mathcal{Q} + L \in \mathcal{K}_{3,8}$, for all $L \in \mathbb{P}_1$. Then, $\dim \mathcal{K}_{3,8} = \dim \mathbb{P}_1 = 2$.

By Theorem 4.9, it is enough to prove that $\dim \mathcal{K}_{3,8} \geq 2$ forces the eight points to be on a conic. Two cases will be considered:

Case. *No three points among* $P_1, \ldots, P_8$ *are aligned.*

Let us consider a conic $\mathcal{Q}$ passing through $P_1, \ldots, P_5$. The hypothesis added for this case implies that $\mathcal{Q}$ is irreducible. Take $P_9, P_{10} \in \mathcal{Q}$

different from the other eight points. Since $\dim \mathcal{K}_{3,8} \geq 2$, the elements of $\mathcal{K}_{3,8}$ passing through the two extra points P_9 and P_{10} form a subspace of dimension at least 0. Hence this space is not empty and $\mathcal{Q}$ is a base component. Since the other three points cannot be aligned, one may suppose that, say $P_6 \in \mathcal{Q}$. One can now repeat the above argument on the elements of $\mathcal{K}_{3,8}$ passing through one extra point $P_9 \in \mathcal{Q}$. The given subspace has dimension at least 1 and $\mathcal{Q}$ as a base component. Therefore, say $P_7 \in \mathcal{Q}$, and thus, $\mathcal{Q}$ is a base component of $\mathcal{K}_{3,8}$. Since the space of lines passing through P_8 is of dimension 1 and $\dim \mathcal{K}_{3,8} \geq 2$, $\mathcal{Q}$ passes through the eight points $P_1, \dots, P_8$.

Case. *P_1, P_2, P_3 are aligned.*

We will first prove that this case forces four points to be aligned. Let L be the line through P_1, P_2, P_3 and suppose that no other P_i is in L. Choosing a generic extra point in L as above, one can deduce that L is a base component of a subspace of $\mathcal{K}_{3,8}$ of dimension at least 1. Therefore there exists a pencil of conics passing through $P_4, \dots, P_8$, which can only happen if at least four of these points are aligned.

After reordering, one can assume that $P_1, \dots, P_4 \in L$. Then L is a base component of $\mathcal{K}_{3,8}$ and hence there is a linear family of conics through four points of dimension at least 2. Choosing an extra generic point on $\mathbb{P}^2$ one obtains as a subspace a pencil of conics through five points, hence four points are aligned. Note that the extra point can be chosen in an open Zariski set of $\mathbb{P}^2$, thus the four aligned points must be $P_5, \dots, P_8$. Therefore there is also a conic passing through $P_1, \dots, P_8$. Q.E.D.

Proposition 4.11. *If $\mathcal{C} \in \mathcal{M}$ then* $\dim \mathcal{K}_{3,6} = 3$.

Proof. It is enough to show that $\dim \mathcal{K}_{3,6} > 3$ leads to contradiction. As in Proposition 4.10 two cases will also be considered:

Case. *No three points among $P_1, \dots, P_6$ are aligned.*

Let us consider the irreducible conic $\mathcal{Q}$ passing through $P_1, ..., P_5$. Consider two extra points $P_9, P_{10} \in \mathcal{Q}$ and two extra points $P_{11}, P_{12} \notin \mathcal{Q}$. The subspace of cubics in $\mathcal{K}_{3,6}$ passing through $P_9, P_{10}, P_{11}, P_{12}$ is not empty and contains $\mathcal{Q}$ as a base component. Since P_{11}, P_{12} can be chosen in such a way that P_6, P_{11}, P_{12} are not on a line, this implies that $P_6 \in \mathcal{Q}$. Repeating the argument with $P_9 \in \mathcal{Q}$ and three extra points $P_{10}, P_{11}, P_{12} \notin \mathcal{Q}$ one obtains again a non-empty subspace of $\mathcal{K}_{3,6}$ containing $\mathcal{Q}$ as a base component. Since P_{10}, P_{11}, P_{12} need not be on a line, a contradiction results.

Case. *P_1, P_2, P_3 are aligned.*

Let L_1 be the line joining P_1, P_2, P_3 and assume that no other point P_i $(i = 4, 5, 6)$ belongs to L_1. Considering $P_9 \in L$ and $P_{10}, P_{11} \notin L$ one obtains a pencil in $\mathcal{K}_{3,6}$ having L_1 as a base component. Hence the five points $P_4, P_5, P_6, P_{10}, P_{11}$ belong to a pencil of conics. Since P_{10}, P_{11} can be chosen so that the line joining them does not contain any other P_i, one concludes that P_4, P_5, P_6 are also aligned. Let L_2 be such a line. Since we assumed that no four points are aligned, $P_9 \in L_1 \cap L_2$ is an extra point. The subspace of curves in $\mathcal{K}_{3,6}$ passing through $P_1, P_2, P_3, P_4, P_5, P_6, P_9$ has dimension at least 3 and has $L_1 \cup L_2$ as a base component. This leads to contradiction since $\dim \mathbb{P}_1 = 2$.

Therefore four points, say P_1, P_2, P_3, P_4, belong to a line L, which automatically becomes a base component of $\mathcal{K}_{3,6}$. Note that neither P_5 nor P_6 can belong to L since $\operatorname{mult}_{P_i}(L, \mathcal{C}) \geq 2$ and $\deg \mathcal{C} = 8$. One can now choose four extra points $P_9, P_{10}, P_{11}, P_{12} \notin L$ such that $P_5, P_6, P_9, P_{10}, P_{11}, P_{12}$ do not belong to a conic. Since $\dim \mathcal{K}_{3,6} > 3$, the subspace of curves in $\mathcal{K}_{3,6}$ passing through $P_1, ..., P_6, P_9, ..., P_{12}$ is not empty and has L as base component. This contradicts the choice of $P_9, P_{10}, P_{11}, P_{12}$. Q.E.D.

Proposition 4.12. *If* $\mathcal{C} \in \mathcal{M}$ *then* $\dim \mathcal{K}_{4,8} = 6$.

Proof. It is enough to prove that $\sigma_{4,8}$ is surjective. If $\mathcal{C} \in \mathcal{M}_-$, since $\sigma_{3,8}$ is surjective, so is $\sigma_{4,8}$. If $\mathcal{C} \in \mathcal{M}_+$, then let $\sum_{j=1}^{8} a_i x_i = 0$ be the equation of the image of $\sigma_{3,8}$ and let us suppose that $a_8 \neq 0$. It is enough to find a quartic curve passing through $P_1, \ldots, P_7$ and not through P_8; we can order this points such that P_8 is not in the line through P_5, P_6. In order to do so, one can choose a conic through $P_1, \ldots, P_4$, the line through P_5, P_6, and a generic line through P_7. Q.E.D.

4.2. Computation of characteristic varieties

We will first describe the geometrical method to compute some components of the characteristic varieties of a curve $\mathcal{C} \in \mathcal{M}$ as proposed by Libgober in [75] – see also a brief sketch of it on page 48. After that, a similar geometrical argument allows for a computation of its Alexander polynomial as proposed in [80, 43, 2].

According to Example 2.24(2), a tacnode (that is, an $\mathbb{A}_3$-singularity) has the sequence of ideals of quasiadjunction associated with it shown in Figure 11.

We follow the notation introduced on page 48. Let $\bar{\xi} := (\xi_1, \xi_2, \xi_3, \xi_4) \in (\mathbb{C}^*)^4$ be a torsion point such that $\ell_{\bar{X}} = 2(X_1 + X_2 + X_3 + X_4) \in \mathbb{N}$,

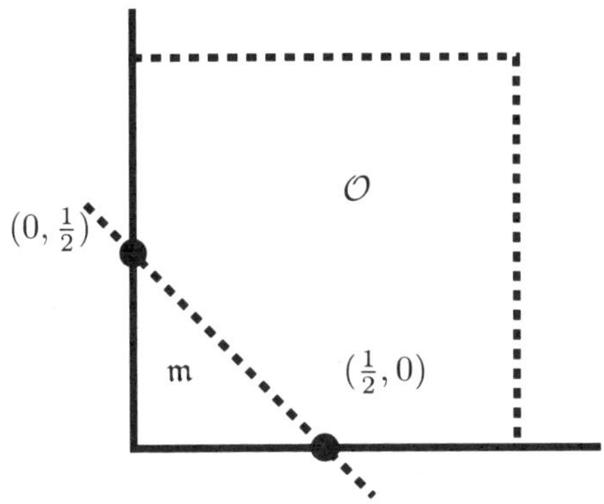

Fig. 11. Quasiadjunction ideals for tacnodes

where $X_i = \frac{\log \xi_i}{2\pi\sqrt{-1}} \in (0,1)$. Then $\bar{\xi} \in \operatorname{Char}_1^*(\mathcal{C})$ if and only if

$$\sigma_{\bar{X}} : H^0(\mathbb{P}^2, \mathcal{O}(5-\ell_{\bar{X}})) \to \bigoplus_{P \in \operatorname{Sing}\mathcal{C}} \mathcal{O}_{\mathbb{P}^2,P}/(\mathcal{A}_{\mathcal{C}}^{\bar{X}})_P =: V_{\bar{X}}$$

is not surjective. Let P be a tacnode of $\mathcal{C}_i + \mathcal{C}_j$. Following Example 2.24(2), $(\mathcal{A}_{\mathcal{C}}^{\bar{X}})_P \neq \mathcal{O}_{\mathbb{P}^2,P}$ if and only if $2(X_i + X_j) \leq 1$, in which case $(\mathcal{A}_{\mathcal{C}}^{\bar{X}})_P = \mathfrak{m}$ the maximal ideal.

Note that we can restrict ourselves to the case $\ell_{\bar{X}} \leq 5$. Also note that, using [75], at least one of these equations is satisfied:

$$2(X_i + X_j) = 1, \quad i = 1,2, \quad j = 3,4. \tag{23}$$

Without loss of generality, it can be assumed that $2(X_1 + X_3) = 1$. In that case $2(X_2 + X_4) = \ell_{\bar{X}} - 1$. The case $\ell_{\bar{X}} = 5$ is not possible since $X_2, X_4 < 1$. If $\ell_{\bar{X}} = 1$, then $X_2 = X_4 = 0$, which corresponds to coordinate components.

For $\ell_{\bar{X}} = 4$, one has $2(X_2 + X_4) = 3$ and $\dim H^0(\mathbb{P}^2, \mathcal{O}(1)) = 3$. Since only non-surjective $\sigma_{\bar{X}}$ matter, an extra equation $2(X_i + X_j) \leq 1$ must be satisfied for some appropriate indices (i,j). Without loss of generality, it can be assumed that $2(X_1 + X_4) \leq 1$, and $2(X_2 + X_3) \geq 3$, which has no solution in the open hypercube $(0,1)^4$.

The same arguments apply to $\ell_{\bar{X}} = 3$ but in this case one has $\dim H^0(\mathbb{P}^2, \mathcal{O}(2)) = 6$, $X_2 + X_4 = 1$, $2(X_1 + X_4) \leq 1$, and $X_2 + X_3 \geq 1$. There are solutions of the system in $(0,1)^4$. In this case $\dim V_{\bar{X}} = 4$ and it is easily seen that $\sigma_{\bar{X}}$ is surjective.

We finish with the case $\ell_{\bar{X}} = 2$, where $2(X_2 + X_4) = 1$. In this situation $2(X_1+X_4) = 1+2(X_4-X_3)$ and $2(X_2+X_3) = 1-2(X_4-X_3)$, hence, either $X_3 = X_4$, or say $2(X_1 + X_4) > 1$ and $2(X_2 + X_3) < 1$. In the latter case $\dim V_{\bar{X}} = 6$ and, by Proposition 4.11, $\sigma_{3,6} = \sigma_{\bar{X}}$ is surjective.

Therefore *all* of the equations in (23) are satisfied, in which case, $\sigma_{\bar{X}} = \sigma_{3,8}$ and its cokernel has dimension 1 (resp. 0) for curves in $\mathcal{M}_+$ (resp. $\mathcal{M}_-$) by Proposition 4.10.

Thus for $\mathcal{M}_+$ there is superabundance for solutions $\bar{X} \in (0,1)^4$ of $X_1 + X_2 + X_3 + X_4 = 1$ and (23) i.e. for

$$\{(X_1, X_1, 1/2 - X_1, 1/2 - X_1) \mid X_1 \in (0,1)\},$$

whose exponential is $\{(t, t, -t^{-1}, -t^{-1}) \mid t \in \mathbb{C}^*\}$. We have proved the following.

Proposition 4.13.

$$\operatorname{Char}_1^*(\mathcal{C}) = \begin{cases} \{(t,t,-t^{-1},-t^{-1}) \mid t \in \mathbb{C}^*\} & \text{if } \mathcal{C} \in \mathcal{M}_+ \\ \emptyset & \text{if } \mathcal{C} \in \mathcal{M}_-. \end{cases}$$

Remark 4.14. As mentioned in Remark 2.17(2), since the number of positive dimensional components of $\operatorname{Char}_1^*(\mathcal{C})$ is different in these cases, the fundamental groups are non-isomorphic.

Note that, according to Theorem 2.26 one can calculate the roots of the Alexander polynomial of $\mathcal{C}$ as follows

$$Z(\Delta_{\mathcal{C}}(t)) \setminus \{1\} = \begin{cases} \pm\sqrt{-1} & \text{if } \mathcal{C} \in \mathcal{M}_+ \\ \emptyset & \text{if } \mathcal{C} \in \mathcal{M}_- \end{cases}.$$

We will consider $\mathcal{A}_k := \mathcal{A}_{\mathcal{C}}^{(\frac{1-k}{d},\dots,\frac{1-k}{d})}$ (according to the definition of ideal sheaf of quasi-adjunction) and referred to this ideal as an *Alexander ideal sheaf* of $\mathcal{C}$. One can also obtain $\Delta_{\mathcal{C}}(t)$ geometrically using the exponents of the singularities as follows.

Theorem 4.15 ([72, 2]). *The Alexander polynomial of $\mathcal{C}$ can be written as the product*

$$\Delta_{\mathcal{C}}(t) = (t-1)^{r-1} \prod_{k=1}^{d-1} \Delta_k^{b_k}(t),$$

where $\Delta_k = \left(t - \exp(\frac{2\pi k\sqrt{-1}}{d})\right)\left(t - \exp(\frac{-2\pi k\sqrt{-1}}{d})\right)$ *and b_k is the defect of the Alexander ideal sheaf $\mathcal{A}_k(k-3)$.*

Each exponent $\frac{k}{8}$, $k = 4, 5, 6, 7$ has an Alexander ideal $\mathcal{A}_k(k-3)$ associated with it. The quotient sheaf $\mathcal{O}(k-3)/\mathcal{A}_k(k-3)$ is supported at the eight tacnodes of $\mathcal{C}$ as follows

$$(\mathcal{A}_k)_P = \begin{cases} \mathcal{O}_P & \text{if } k = 4, 5 \\ \mathfrak{m}_P & \text{if } k = 6, 7, \end{cases}$$

where $\mathfrak{m}_P$ is the maximal ideal at P –we refer to [43, 80, 2] for the details. Hence $h^1(\mathcal{A}_4(1)) = h^1(\mathcal{A}_5(2)) = 0$, and thus the only two interesting cases are the exponents $\frac{6}{8}$ and $\frac{7}{8}$. For the first exponent, by Proposition 4.10, $h^0(\mathcal{A}_6(3)) = \dim \ker \sigma_{3,8} = \begin{cases} 3 & \text{if } \mathcal{C} \in \mathcal{M}_+ \\ 2 & \text{if } \mathcal{C} \in \mathcal{M}_- \end{cases}$, hence

$$h^1(\mathcal{A}_6(3)) = h^0(\mathcal{A}_6(3)) - 2 = \begin{cases} 1 & \text{if } \mathcal{C} \in \mathcal{M}_+ \\ 0 & \text{if } \mathcal{C} \in \mathcal{M}_- \end{cases}.$$

For the second exponent, by Proposition 4.12, $h^0(\mathcal{A}_7(4)) = \dim \ker \sigma_{4,8} = 7$, hence $h^1(\mathcal{A}_7(4)) = h^0(\mathcal{A}_7(4)) - 7 = 0$. By Theorem 4.15 one has

$$\Delta_{\mathcal{C}}(t) = \begin{cases} (t-1)^3(t^2+1) & \text{if } \mathcal{C} \in \mathcal{M}_+ \\ (t-1)^3 & \text{if } \mathcal{C} \in \mathcal{M}_- \end{cases}.$$

4.3. A tower of D_{2n}-coverings

In this subsection, we will explain another way to study this example by using D_{2n}-coverings. Let us start by introducing the notion of a tower of dihedral coverings.

Definition 4.16. Let Y be a normal projective variety. A sequence $\{\pi_i : X_i \to Y\}_{i \in I}$ of Galois coverings is called a *tower of dihedral coverings* if it satisfies the following conditions:

(*i*) $\pi_i : X_i \to Y$ is a D_{2n_i}-coverings ($n_i \geq 3$) for each i.

(*ii*) If $n_i | n_j$, then there exists a morphism $\eta_{ji} : X_j \to X_i$ such that $\pi_j = \pi_i \circ \eta_{ji}$.

Here is an example of a tower of dihedral coverings, which we need later.

Example 4.17. Let $\varphi_n : \mathbb{P}^1 \to \mathbb{P}^1$ ($n \geq 3$) be the family of morphisms given by

$$t \mapsto s = \frac{1}{2}\left(t^n + \frac{1}{t^n}\right),$$

where s, t are non-homogeneous coordinates. It is easy to see that φ_n is a D_{2n}-covering branched at $2[1 : \pm 1] + n[0 : 1]$, where $[s_0 : s_1]$ are homogeneous coordinates with $s = \frac{s_1}{s_0}$. In fact, D_{2n} acts on the source $\mathbb{P}^1$ in such a way that $t^\sigma = t^{-1}$, $t^\tau = \zeta_n t$, $\zeta_n = \exp(\frac{2\pi\sqrt{-1}}{n})$. The set of dihedral coverings $\{\varphi_n : \mathbb{P}^1 \to \mathbb{P}^1\}_{n\geq 3}$ is a tower of dihedral coverings.

Proposition 4.18. *Let $\mathcal{C} = \mathcal{C}_1 \cup \mathcal{C}_2 \cup \mathcal{C}_3 \cup \mathcal{C}_4 \in \mathcal{M}$. Then the following two statements are equivalent:*

(1) *There exists a tower of dihedral coverings $\{\pi_n : X_n \to \mathbb{P}^2\}_{n\in\mathbb{N}}$ such that:*
   (a) $\mathrm{Gal}(X_n/\mathbb{P}^2) \cong D_{2p_n}$, *$p_n$ odd prime,*
   (b) *π_n is branched at $2(\mathcal{C}_1 + \mathcal{C}_2) + p_n(\mathcal{C}_3 + \mathcal{C}_4)$, and*
   (c) *$p_n \neq p_m$ for any $n \neq m$.*

(2) *There exists a conic through the 8 tacnodes of $\mathcal{C}$, i.e., $\mathcal{C} \in \mathcal{M}_+$.*

Note that Proposition 4.18 implies that a pair $(\mathcal{C}_+, \mathcal{C}_-)$ $(\mathcal{C}_+ \in \mathcal{M}_+$, $\mathcal{C}_- \in \mathcal{M}_-)$ is a Zariski pair. We need several steps to prove Proposition 4.18.

Lemma 4.19. *Let D_1 and D_2 be reduced plane curves of degree $2m$ such that all intersection points between D_1 and D_2 give rise to tacnodes in $D_1 + D_2$ (i.e., D_1 is tangent to D_2 at $2m^2$ distinct points). Let $\Lambda = \{\lambda_1 D_1 + \lambda_2 D_2\}_{[\lambda_1,\lambda_2]\in\mathbb{P}^1}$ be the pencil of curves spanned by D_1 and D_2. If there exists a unique reduced plane curve, E, of degree m passing through all the $2m^2$ intersection points of D_1 and D_2, then $2E \in \Lambda$.*

Proof. Since E meets D_1 at $2m^2$ distinct points, E is smooth at each intersection point and meets D_1 transversely. Choose a general point x on E. Let C_x be a member of Λ passing through x. Then C_x meets E at $2m^2 + 1$ points. Hence E is contained in C_x. Write $C_x = E + E'$. Then E' is a curve of degree m. Since the base points of Λ consists of $D_1 \cap D_2$ with multiplicity 2 at each base point, E' also passes through all the intersection points of $D_1 \cap D_2$. This implies $E' = E$.

Lemma 4.20. *Let $\mathcal{C} = \mathcal{C}_1 \cup \mathcal{C}_2 \cup \mathcal{C}_3 \cup \mathcal{C}_4 \in \mathcal{M}$. If there exists a conic passing through 8 tacnodes, then there exists a D_{2m}-covering of $\mathbb{P}^2$ branched at $2(\mathcal{C}_1 + \mathcal{C}_2) + m(\mathcal{C}_3 + \mathcal{C}_4)$ for any $m \geq 3$.*

Proof. Let $\Lambda := \Sigma(\mathcal{C}_1 + \mathcal{C}_2, \mathcal{C}_3 + \mathcal{C}_4)$ be the pencil of plane curves spanned by $\mathcal{C}_1 + \mathcal{C}_2$ and $\mathcal{C}_3 + \mathcal{C}_4$. Since the conic $\mathcal{Q}$ in the assumption passes through the eight (double) base points of Λ, we have $2\mathcal{Q} \in \Lambda$ by Lemma 4.19. Let C_i (resp. Q) be defining homogeneous polynomials for $\mathcal{C}_i$ (resp. $\mathcal{Q}$). Hence we may assume that $C_3C_4 = C_1C_2 - Q^2$. Let $\Phi : \mathbb{P}^2 \dashrightarrow \mathbb{P}^1$ be the rational function given by

$$[z_0 : z_1 : z_2] \mapsto [C_3C_4 : -(C_1C_2 + Q^2)].$$

Let $\varphi_m : \mathbb{P}^1 \to \mathbb{P}^1$ be the D_{2m}-covering described in Example 4.17. Let $\rho : \hat{\mathbb{P}}^2 \to \mathbb{P}^2$ be the resolution of indeterminacy of Φ and put $\hat{\Phi} = \rho \circ \Phi$. Let X' be the fibered product $\hat{\mathbb{P}}^2 \times_{\mathbb{P}^1} \mathbb{P}^1$ by $\hat{\Phi}$ and φ_m. Also let X_m be the $\mathbb{C}(X')$-normalization of $\mathbb{P}^2$. Then X_m is a D_{2m}-covering branched at $2(\mathcal{C}_1 + \mathcal{C}_2) + m(\mathcal{C}_3 + \mathcal{C}_4)$. Q.E.D.

Proof of Proposition 4.18 ($\Leftarrow$). Let $\pi_m : X_m \to \mathbb{P}^2$ be the D_{2m}-covering as in Lemma 4.20. Then $\{\pi_m : X_m \to \mathbb{P}^2\}_{m\geq 3}$ is a tower of dihedral coverings such that $\mathrm{Gal}(X_m/\mathbb{P}^2) \cong D_{2m}$. Also π_m is branched at $2(\mathcal{C}_1 + \mathcal{C}_2) + m(\mathcal{C}_3 + \mathcal{C}_4)$. In particular, one obtains the desired D_{2p_n}-coverings for odd primes p_n $(n = 1, 2, 3, \ldots)$. Q.E.D.

Finally, our purpose until the end of this section is to prove the converse. Let $f' : Z' \to \mathbb{P}^2$ be a double covering branched at $\mathcal{C}_1 + \mathcal{C}_2$, and let $\mu : Z \to Z'$ be the canonical resolution of Z'. Let

$$\begin{array}{ccc} Z' & \xleftarrow{\ \mu\ } & Z \\ {\scriptstyle f'}\downarrow & & \downarrow{\scriptstyle f} \\ \mathbb{P}^2 & \xleftarrow{\ \rho\ } & \hat{\mathbb{P}}^2 \end{array}$$

denote the diagram for the canonical resolution. In our case, the morphism $\rho : \hat{\mathbb{P}}^2 \to \mathbb{P}^2$ is a composition of 4 blowing-ups at the 4 nodes of $\mathcal{C}_1 + \mathcal{C}_2$. Let $f : Z \to \hat{\mathbb{P}}^2$ be the induced double covering. Note that Z is simply connected, as it is a rational surface. Suppose that there exists a D_{2p_n}-covering $\pi_n : X_n \to \mathbb{P}^2$ branched at $2(\mathcal{C}_1 + \mathcal{C}_2) + p_n(\mathcal{C}_3 + \mathcal{C}_4)$. Then $D(X_n/\mathbb{P}^2) = Z'$ and $\beta_1(\pi_n) = f'$. Let $\tilde{X}_n$ be the $\mathbb{C}(X_n)$-normalization of $\hat{\mathbb{P}}^2$. $\tilde{X}_n$ is a D_{2p_n}-covering of $\hat{\mathbb{P}}^2$ such that $D(\tilde{X}_n/\hat{\mathbb{P}}^2) = Z$ and $\beta_1(\tilde{\pi}_n) = f$, $\tilde{\pi}_n$ being the covering morphism. Summing up, one obtains the following commutative diagram:

$$\begin{array}{ccc} X_n & \longleftarrow & \tilde{X}_n \\ \downarrow & & \downarrow \\ Z' & \xleftarrow{\ \mu\ } & Z \\ \downarrow & & \downarrow \\ \mathbb{P}^2 & \xleftarrow{\ \rho\ } & \hat{\mathbb{P}}^2. \end{array}$$

Since the irreducible components of the exceptional divisor μ are invariant under the covering transform of f, they are not contained in $\Delta_{\beta_2(\tilde{\pi}_n)}$ (Corollary 3.13). Hence $\Delta_f = \rho_q^{-1}(\mathcal{C}_1+\mathcal{C}_2)$, and $\Delta_{\beta_2(\tilde{\pi}_n)} = (\rho\circ f)^*(\mathcal{C}_3+\mathcal{C}_4)$. Let us denote $(\rho\circ f)^*\mathcal{C}_3 = \mathcal{C}_3^+ + \mathcal{C}_3^-$ and $(\rho\circ f)^*\mathcal{C}_4 = \mathcal{C}_4^+ + \mathcal{C}_4^-$. One has the following.

Lemma 4.21. *There exists an integer k with $0 < k \leq \frac{p_n-1}{2}$ such that either $(\mathcal{C}_3^+ + k\mathcal{C}_4^+) - (\mathcal{C}_3^- + k\mathcal{C}_4^-)$ or $(\mathcal{C}_3^+ + k\mathcal{C}_4^-) - (\mathcal{C}_3^+ + k\mathcal{C}_4^+)$ is p_n-divisible in* $\mathrm{Pic}(Z)$.

Proof. By Proposition 3.12, either $(a_3\mathcal{C}_3^+ + a_4\mathcal{C}_4^+) - (a_3\mathcal{C}_3^- + a_4\mathcal{C}_4^-)$ or $(a_3\mathcal{C}_3^+ + a_4\mathcal{C}_4^-) - (a_3\mathcal{C}_3^- + a_4\mathcal{C}_4^+)$, where a_i $(i=3,4)$ are integers with $0 < a_i < p_n$ $(i=3,4)$, is p_n-divisible. The latter case is reduced to the first case by considering

$$\left(a_3\mathcal{C}_3^+ + (p_n-a_4)\mathcal{C}_4^+\right) - \left(a_3\mathcal{C}_3^- + (p_n-a_4)\mathcal{C}_4^-\right) + p_n(\mathcal{C}_4^- - \mathcal{C}_4^+).$$

So one may assume that

$$(a_3\mathcal{C}_3^+ + a_4\mathcal{C}_4^+) - (a_3\mathcal{C}_3^- + a_4\mathcal{C}_4^-) \sim p_n L$$

for some $L \in \mathrm{Pic}(Z)$. Choose an integer b $(0 < b < p_n)$ so that $a_3 b \equiv 1 \bmod p_n$. Let us define

$$\begin{aligned} D_1 &:= b(a_3\mathcal{C}_3^+ + a_4\mathcal{C}_4^+) - p_n\left(\left\lfloor\frac{a_3b}{p_n}\right\rfloor \mathcal{C}_3^+ + \left\lfloor\frac{a_4b}{p_n}\right\rfloor \mathcal{C}_4^+\right), \\ &= \mathcal{C}_3^+ + \left(a_4b - p_n\left\lfloor\frac{a_4b}{p_n}\right\rfloor \mathcal{C}_4^+\right). \end{aligned}$$

If $0 < a_4b - \left\lfloor\frac{a_4b}{p_n}\right\rfloor \leq \frac{p_n-1}{2}$, then define $k = a_4b - \left\lfloor\frac{a_4b}{p_n}\right\rfloor$ so that the divisor $(\mathcal{C}_3^+ + k\mathcal{C}_4^+) - (\mathcal{C}_3^- + k\mathcal{C}_4^-)$ is p_n-divisible.

If $a_4b - \left\lfloor\frac{a_4b}{p_n}\right\rfloor > \frac{p_n-1}{2}$, then define $k = p_n + \left\lfloor\frac{a_4b}{p_n}\right\rfloor - a_4b$ so that the divisor $(\mathcal{C}_3^+ + k\mathcal{C}_4^-) - (\mathcal{C}_3^- + k\mathcal{C}_4^+)$ is p_n-divisible. Q.E.D.

Lemma 4.22. *The self-intersection numbers of $(\mathcal{C}_3^+ + k\mathcal{C}_4^+) - (\mathcal{C}_3^- + k\mathcal{C}_4^-)$ and $(\mathcal{C}_3^+ + k\mathcal{C}_4^-) - (\mathcal{C}_3^- + k\mathcal{C}_4^+)$ are either $-8(k-1)^2$, $-8(k+1)^2$ or $-8(k^2+1)$.*

Proof. Since

$$8 = ((\rho\circ f)^*\mathcal{C}_3)^2 = (\mathcal{C}_3^+)^2 + (\mathcal{C}_3^-)^2 + 2\mathcal{C}_3^+\cdot\mathcal{C}_3^-,\ \mathcal{C}_3^+\cdot\mathcal{C}_3^- = 4,\ (\mathcal{C}_3^+)^2 = (\mathcal{C}_3^-)^2,$$

one has $(\mathcal{C}_3^+)^2 = (\mathcal{C}_3^-)^2 = 0$. Similarly $(\mathcal{C}_4^+)^2 = (\mathcal{C}_4^-)^2 = 0$. Hence $(\mathcal{C}_i^+ - \mathcal{C}_i^-)^2 = -8 \quad (i=3,4)$. Also, since

$$8 = ((\rho\circ f)^*\mathcal{C}_3)\cdot((\rho\circ f)^*\mathcal{C}_4) = \mathcal{C}_3^+\cdot\mathcal{C}_4^+ + \mathcal{C}_3^+\cdot\mathcal{C}_4^- + \mathcal{C}_3^-\cdot\mathcal{C}_4^+ + \mathcal{C}_3^-\cdot\mathcal{C}_4^-$$

and

$$\mathcal{C}_3^+ \cdot \mathcal{C}_4^+ = \mathcal{C}_3^- \cdot \mathcal{C}_4^-,\ \mathcal{C}_3^+ \cdot \mathcal{C}_4^- = \mathcal{C}_3^- \cdot \mathcal{C}_4^+,$$

one concludes that either

(a) $\mathcal{C}_3^+\mathcal{C}_4^+ = \mathcal{C}_3^-\mathcal{C}_4^- = 4$, $\mathcal{C}_3^+\mathcal{C}_4^- = \mathcal{C}_3^-\mathcal{C}_4^+ = 0$, or
(b) $\mathcal{C}_3^+\mathcal{C}_4^+ = \mathcal{C}_3^-\mathcal{C}_4^- = 4$, $\mathcal{C}_3^+\mathcal{C}_4^- = \mathcal{C}_3^-\mathcal{C}_4^+ = 0$, or
(c) $\mathcal{C}_3^+\mathcal{C}_4^+ = \mathcal{C}_3^-\mathcal{C}_4^- = 2$, $\mathcal{C}_3^+\mathcal{C}_4^- = \mathcal{C}_3^-\mathcal{C}_4^+ = 2$

holds. Thus

$$\left\{(\mathcal{C}_3^+ - \mathcal{C}_3^-) \pm k(\mathcal{C}_4^+ - \mathcal{C}_4^-)\right\}^2 = \begin{cases} -8(k \mp 1)^2 & \text{for the case (a),} \\ -8(k \pm 1)^2 & \text{for the case (b),} \\ -8(k^2+1) & \text{for the case (c).} \end{cases}$$

Q.E.D.

Lemma 4.23. *Let k be as above. Then $k = 1$.*

Proof. We will only consider the case when $(\mathcal{C}_3^+ - \mathcal{C}_3^-) + k(\mathcal{C}_4^+ - \mathcal{C}_4^-)$ is p_n-divisible (the other case is analogous). Suppose that

$$(\mathcal{C}_3^+ - \mathcal{C}_3^-) + k(\mathcal{C}_4^+ - \mathcal{C}_4^-) \sim p_n L,$$

for some $L \in \mathrm{Pic}(Z)$. After computing the self-intersection numbers of both sides, one deduces that either $-8(k-1)^2 = p_n^2 L^2$, or $-8(k+1)^2 = p_n^2 L^2$, or $-8(k^2+1) = p_n^2 L^2$ holds. Since p_n is odd, L^2 is an integer and $0 < k \leq \frac{p_n - 1}{2}$, which leads to $k = 1$ and $L^2 = 0$. Q.E.D.

Lemma 4.24. *Either*

$$(\mathcal{C}_3^+ + \mathcal{C}_4^+) - (\mathcal{C}_3^- + \mathcal{C}_4^-) \sim 0, \qquad or \qquad (\mathcal{C}_3^+ + \mathcal{C}_4^-) - (\mathcal{C}_3^- + \mathcal{C}_4^+) \sim 0$$

holds.

Proof. By Lemma 4.23, either $(\mathcal{C}_3^+ + \mathcal{C}_4^+) - (\mathcal{C}_3^- + \mathcal{C}_4^-)$, or $(\mathcal{C}_3^+ + \mathcal{C}_4^-) - (\mathcal{C}_3^- + \mathcal{C}_4^+)$ is p_n-divisible. By the assumption in Proposition 4.18, at least one of them is p_n-divisible for infinitely many odd prime numbers p_n. Since Z is simply connected, $\mathrm{Pic}(Z)$ is a finitely generated free $\mathbb{Z}$-module. Therefore, either $(\mathcal{C}_3^+ + \mathcal{C}_4^+) - (\mathcal{C}_3^- + \mathcal{C}_4^-) \sim 0$, or $(\mathcal{C}_3^+ + \mathcal{C}_4^-) - (\mathcal{C}_3^- + \mathcal{C}_4^+) \sim 0$ holds. Q.E.D.

Proof of Proposition 4.18 ($\Rightarrow$). In what follows, only the case $(\mathcal{C}_3^+ - \mathcal{C}_3^-) + (\mathcal{C}_4^+ - \mathcal{C}_4^-) \sim 0$ will be considered (the other case is analogous). Let φ be a rational function on Z such that

$$(\varphi) = (\mathcal{C}_3^+ + \mathcal{C}_4^+) - (\mathcal{C}_3^- + \mathcal{C}_4^-).$$

Note that one can choose φ in such a way that $\varphi^{\sigma_f} = 1/\varphi$, where σ_f denote the involution of the double covering $f : Z \to \hat{\mathbb{P}}^2$. Let $K_n := \mathbb{C}(Z)(\sqrt[p_n]{\varphi})$ and let S_n be the K_n-normalization of $\mathbb{P}^2$.

One can easily see that $\rho_n : S_n \to \mathbb{P}^2$ is a D_{2p_n}-covering such that $D(S_n/\mathbb{P}^2) = Z'$ and ρ_n is branched at $2(\mathcal{C}_1 + \mathcal{C}_2) + p_n(\mathcal{C}_3 + \mathcal{C}_4)$. In fact, S_n is isomorphic to X_n, but we do not need it here. Set

$$u := \frac{\varphi + \dfrac{1}{\varphi}}{2}.$$

Since u is a σ_f-invariant rational function, $u \in \mathbb{C}(\mathbb{P}^2)$. Let $\Phi_n : S_n \dashrightarrow \mathbb{P}^1$ and $\overline{\Phi}_n : \mathbb{P}^2 \dashrightarrow \mathbb{P}^1$ be the rational maps given by $\sqrt[p_n]{\varphi}$ and u. Note that Φ_n is D_{2p_n}-equivariant and the following diagram

$$\begin{array}{ccc} S_n & \overset{\Phi_n}{\dashrightarrow} & \mathbb{P}^1 \\ \rho_n \downarrow & & \downarrow \varphi_{p_n} \\ \mathbb{P}^2 & \underset{\overline{\Phi}_n}{\dashrightarrow} & \mathbb{P}^1, \end{array} \tag{24}$$

is commutative, where $\varphi_{p_n} : \mathbb{P}^1 \to \mathbb{P}^1$ is the D_{2p_n}-covering in Example 4.17. Note that S_n is birational to the fibered product $\mathbb{P}^2 \times_{\mathbb{P}^1} \mathbb{P}^1$ over $\mathbb{P}^2$.

Write $u = \frac{F_0}{F_\infty}$, where F_0 and F_∞ are homogeneous polynomials. The polar divisor of u is $\mathcal{C}_3^+ + \mathcal{C}_3^- + \mathcal{C}_4^+ + \mathcal{C}_4^-$. This implies that the plane curve given by $F_\infty = 0$ is $\mathcal{C}_3 + \mathcal{C}_4$ and $\deg F_\infty = 4$. Using the commutative diagram (24) and since φ_{p_n} is branched at $2[1 : \pm 1] + p_n[0 : 1]$, the curves $\mathcal{D}_1$ and $\mathcal{D}_2$ given by the equations $F_0 - F_\infty = 0$ and $F_0 + F_\infty = 0$, respectively, satisfy either (i) or (ii) below:

- (i) $\mathcal{D}_1 = \mathcal{C}_1 + \mathcal{C}_2$ and $\mathcal{D}_2$ is of the form $2\mathcal{Q}$ for some conic (or vice versa).
- (ii) $\mathcal{D}_1 = \mathcal{C}_1 + 2L_1$ for some line L_1 and $\mathcal{D}_2 = \mathcal{C}_2 + 2L_2$ for some line L_2.

If (ii) occurs, it implies that $\mathcal{C}_3 + \mathcal{C}_4 \in \Sigma(\mathcal{C}_1 + 2L_1, \mathcal{C}_2 + 2L_2)$. This means that $\mathcal{C}_3 + \mathcal{C}_4$ passes through $\mathcal{C}_1 \cap \mathcal{C}_2$, but this is impossible by the combinatorics of $\mathcal{C} \in \mathcal{M}$. Hence (i) must occur and $\mathcal{Q}$ is the unique conic passing through all the tacnodes of $\mathcal{C}$. Q.E.D.

Remark 4.25. The proof above shows that there exists a tower of dihedral coverings $\{\tilde{\pi}_n : \tilde{X}_n \to \mathbb{P}^2\}_{n \geq 1}$ such that $\mathrm{Gal}(\tilde{X}_n/\mathbb{P}^2) \cong D_{2p_n}$ (p_n odd prime), where $\tilde{\pi}_n$ is branched at $2(\mathcal{C}_1 + \mathcal{C}_2) + p_n(\mathcal{C}_3 + \mathcal{C}_4)$, if and only if there exists another tower of dihedral coverings $\{\pi_m : X_m \to$

$\mathbb{P}^2\}_{m\geq 3}$ such that $\mathrm{Gal}(X_m/\mathbb{P}^2) \cong D_{2m}$, where π_m is branched at $2(\mathcal{C}_1 + \mathcal{C}_2) + m(\mathcal{C}_3 + \mathcal{C}_4)$.

References

[1] D. Arapura, Geometry of cohomology support loci for local systems. I, J. Algebraic Geom., **6** (1997), 563–597.

[2] E. Artal, Sur les couples de Zariski, J. Algebraic Geom., **3** (1994), 223–247.

[3] ———, Fundamental group of a class of rational cuspidal curves, Manuscripta Math., **93** (1997), 273–281.

[4] ———, A curve of degree five with non-Abelian fundamental group, Topology Appl., **79** (1997), 13–29.

[5] E. Artal and J. Carmona, Zariski pairs, fundamental groups and Alexander polynomials, J. Math. Soc. Japan, **50** (1998), 521–543.

[6] E. Artal, J. Carmona and J. I. Cogolludo, On sextic curves with big Milnor number, In: Trends in Singularities (eds. A. Libgober and M. Tibăr), Trends Math. Birkhäuser, Basel, 2002, pp. 1–29.

[7] ———, Braid monodromy and topology of plane curves, Duke Math. J., **118** (2003), 261–278.

[8] ———, Essential coordinate components of characteristic varieties, Math. Proc. Cambridge Philos. Soc., **136** (2004), 287–299.

[9] ———, Effective invariants of braid monodromy, Trans. Amer. Math. Soc., **359** (2007), 165–183.

[10] E. Artal, J. Carmona, J. I. Cogolludo and M.Á. Marco, Topology and combinatorics of real line arrangements, Compos. Math., **141** (2005), 1578–1588.

[11] ———, Invariants of combinatorial line arrangements and Rybnikov's example, In: Singularity theory and its applications, Adv. Stud. Pure Math., **43**, Math. Soc. Japan, 2007, pp. 1–34.

[12] E. Artal, J. Carmona, J. I. Cogolludo and H. Tokunaga, Sextics with singular points in special position, J. Knot Theory Ramifications, **10** (2001), 547–578.

[13] E. Artal, J. I. Cogolludo, and H. Tokunaga, Nodal degenerations of plane curves and Galois covers, Geom. Dedicata, **121** (2006), 129–142.

[14] ———, Pencils and infinite dihedral covers of $\mathbb{P}^2$, Proc. Amer. Math. Soc., **136** (2008), 21–29.

[15] E. Artal and H. Tokunaga, Zariski pairs of index 19 and Mordell-Weil groups of $K3$ surfaces, Proc. London Math. Soc., **80** (2000), 127–144.

[16] ———, Zariski k-plets of rational curve arrangements and dihedral covers, Topology Appl., **142** (2004), 227–233.

[17] E. Artin, Theory of braids, Abh. Math. Sem. Hamburgischen Univ., **4** (1926), 47–72.

[18] W. A. Arvola, The fundamental group of the complement of an arrangement of complex hyperplanes, Topology, **31** (1992), 757–765.

[19] A. Beauville, Complex algebraic surfaces, London Math. Soc. Lecture Notes Ser., **68**.

[20] W. Barth, C. Peters and A. Van de Ven, Compact complex surfaces, Erg. der Math. und ihrer Grenz., A Series of Modern Surveys in Math., **3**, vol. 4, Springer-Verlag, Berlin, 1984.

[21] D. Bessis and J. Michel, Explicit presentations for exceptional braid groups, Experiment. Math., **13** (2004), 257–266.

[22] J. S. Birman, Braids, links, and mapping class groups, Princeton Univ. Press, Princeton, N. J., 1974, Ann. of Math. Stud., **82**.

[23] E. Brieskorn, Über die Auflösung gewisser Singularitäten von holomorphen Abbildungen, Math. Ann., **166** (1966), 76–102.

[24] ———, Die Auflösung der rationalen Singularitäten holomorpher Abbildungen, Math. Ann., **178** (1968), 255–270.

[25] J. Carmona, Monodromía de trenzas de curvas algebraicas planas, Ph. D. thesis, Universidad de Zaragoza, 2003.

[26] F. Catanese, On the moduli spaces of surfaces of general type, J. Differential Geom., **19** (1984), 483–515.

[27] ———, Connected components of moduli spaces, J. Differential Geom., **24** (1986), 395–399.

[28] D. Chéniot, Une démonstration du théorème de Zariski sur les sections hyperplanes d'une hypersurface projective et du théorème de Van Kampen sur le groupe foundamental du complémentaire d'une courbe projective plane, Compositio Math., **27** (1973), 141–158.

[29] O. Chisini, Una suggestiva rappresentazione reale per le curve algebriche piane, Ist. Lombardo, Rend., II. Ser., **66** (1933), 1141–1155.

[30] J. I. Cogolludo, Topological invariants of the complement to arrangements of rational plane curves, Mem. Amer. Math. Soc. **159** (2002), no. 756.

[31] J. I. Cogolludo and V. Florens, Twisted Alexander polynomials of plane algebraic curves, J. London Math. Soc., **76** (2007), 105–121.

[32] D. C. Cohen and P. Orlik, Some cyclic covers of complements of arrangements, Topology Appl., **118** (2002), 3–15, Arrangements in Boston: a Conference on Hyperplane Arrangements (1999).

[33] D. C. Cohen and A. I. Suciu, Alexander invariants of complex hyperplane arrangements, Trans. Amer. Math. Soc., **351** (1999), 4043–4067.

[34] ———, Characteristic varieties of arrangements, Math. Proc. Cambridge Philos. Soc., **127** (1999), 33–53.

[35] R. H. Crowell and R. H. Fox, Introduction to knot theory, Ginn and Co., Boston Mass., 1963.

[36] A. I. Degtyarëv, Isotopic classification of complex plane projective curves of degree 5, Leningrad Math. J., **1** (1990), 881–904.

[37] ———, Alexander polynomial of a curve of degree six, J. Knot Theory Ramifications, **3** (1994), 439–454.

[38] ———, Quintics in $\mathbb{CP}^2$ with non-Abelian fundamental group, Algebra i Analiz, **11** (1999), 130–151.

[39] ———, On deformations of singular plane sextics, J. Alg. Geom., **17** (2008), 101–135.

[40] P. Deligne, Le groupe fondamental du complément d'une courbe plane n'ayant que des points doubles ordinaires est abélien (d'après W. Fulton), Bourbaki Seminar, Vol. 1979/80, Lecture Notes in Math., **842**, Springer, Berlin, 1981, pp. 1–10.

[41] A. Dimca, Singularities and topology of hypersurfaces, Springer-Verlag, New York, 1992.

[42] D. Eisenbud and W. Neumann, Three-dimensional link theory and invariants of plane curve singularities, Ann. of Math. Stud., **110**, Princeton Univ. Press, Princeton, N. J., 1985.

[43] H. Esnault, Fibre de Milnor d'un cône sur une courbe plane singulière, Invent. Math., **68** (1982), 477–496.

[44] C. Eyral and M. Oka, Fundamental groups of the complements of certain plane non-tame torus sextics, Topology Appl., **153** (2006), 1705–1721.

[45] ———, On the fundamental groups of the complements of plane singular sextics, J. Math. Soc. Japan, **57** (2005), 37–54.

[46] M. Falk, Arrangements and cohomology, Ann. Comb., **1** (1997), 135–157.

[47] T. Fujita, On the topology of noncomplete algebraic surfaces, J. Fac. Sci. Univ. Tokyo Sect. IA Math., **29** (1982), 503–566.

[48] W. Fulton, On the fundamental group of the complement of a node curve, Ann. of Math. (2), **111** (1980), 407–409.

[49] The GAP Group, GAP – Groups, Algorithms, and Programming, Version 4.4, 2004, available at `http://www.gap-system.org`.

[50] C. McA. Gordon, Some aspects of classical knot theory, In: Knot theory, Proc. Sem., Plans-sur-Bex, 1977, Lecture Notes in Math., **685**, Springer-Verlag, Berlin, 1978, pp. 1–60.

[51] H. Grauert and R. Remmert, Komplexe Räume, Math. Ann., **136** (1958), 245–318.

[52] A. Grothendieck, Revêtements étales et groupe fondamental, Lecture Notes in Math., **224**, Springer-Verlag, Berlin, 1971.

[53] V. L. Hansen, Braids and Coverings, London Math. Soc. Student Texts, **18**, 1989.

[54] J. Harris, On the Severi problem, Invent. Math., **84** (1986), 445–461.

[55] J. A. Hillman, Alexander ideals of links, Lecture Notes in Math., **895**, Springer-Verlag, Berlin, 1981.

[56] E. Hironaka, Abelian coverings of the complex projective plane branched along configurations of real lines, Mem. Amer. Math. Soc., **105** (1993), no. 502, vi+85.

[57] ———, Alexander stratifications of character varieties, Ann. Inst. Fourier (Grenoble), **47** (1997), 555–583.

[58] E. Horikawa, On deformation of quintic surfaces, Invent. Math., **31** (1975), 43–85.

[59] E. R. van Kampen, On the fundamental group of an algebraic curve, Amer. J. Math., **55** (1933), 255–260.

[60] T. Jiang and S. S.-T. Yau, Intersection lattices and topological structures of complements of arrangements in $\mathbb{CP}^2$, Ann. Scuola Norm. Sup. Pisa Cl. Sci. (4), **26** (1998), 357–381.

[61] P.-L. Kang, On the variety of plane curves of degree d with δ nodes and κ cusps, Trans. Amer. Math. Soc., **316** (1989), 165–192.

[62] V. M. Kharlamov and V. S. Kulikov, Diffeomorphisms, isotopies, and braid monodromy factorizations of plane cuspidal curves, C. R. Acad. Sci. Paris Sér. I Math., **333** (2001), 855–859.

[63] ———, On real structures on rigid surfaces, Izv. Ross. Akad. Nauk Ser. Mat., **66** (2002), 133–152.

[64] ———, On braid monodromy factorizations, Izv. Ross. Akad. Nauk Ser. Mat., **67** (2003), 79–118.

[65] P. Kirk and C. Livingston, Twisted Alexander invariants, Reidemeister torsion, and Casson-Gordon invariants, Topology, **38** (1999), 635–661.

[66] ———, Twisted knot polynomials: inversion, mutation and concordance, Topology, **38** (1999), 663–671.

[67] T. Kitano, Twisted Alexander polynomial and Reidemeister torsion, Pacific J. Math., **174** (1996), 431–442.

[68] V. S. Kulikov, On Chisini's conjecture, Izv. Ross. Akad. Nauk Ser. Mat., **63** (1999), 83–116.

[69] ———, Generic coverings of the plane and braid monodromy invariants, The Fano Conference, Univ. Torino, Turin, 2004, pp. 533–558.

[70] V. S. Kulikov and M. Teicher, Braid monodromy factorizations and diffeomorphism types, Izv. Ross. Akad. Nauk Ser. Mat., **64** (2000), 89–120.

[71] A. Libgober, Alexander polynomial of plane algebraic curves and cyclic multiple planes, Duke Math. J., **49** (1982), 833–851.

[72] ———, Alexander invariants of plane algebraic curves, In: Singularities, Part 2, Arcata, Calif., 1981, Proc. Sympos. Pure Math., **40**, Amer. Math. Soc., Providence, RI, 1983, pp. 135–143.

[73] ———, On the homotopy type of the complement to plane algebraic curves, J. Reine Angew. Math., **367** (1986), 103–114.

[74] ———, On the homology of finite Abelian coverings, Topology Appl., **43** (1992), 157–166.

[75] ———, Characteristic varieties of algebraic curves, Applications of algebraic geometry to coding theory, physics and computation, Eilat, 2001, Kluwer Acad. Publ., Dordrecht, 2001, pp. 215–254.

[76] A. Libgober and S. Yuzvinsky, Cohomology of local systems, Arrangements—Tokyo, 1998, Kinokuniya, Tokyo, 2000, pp. 169–184.

[77] ———, Cohomology of the Orlik-Solomon algebras and local systems, Compositio Math., **121** (2000), 337–361.

[78] ———, Non vanishing loci of Hodge numbers of local systems, preprint available at `http://arXiv.org/math.AG/0701597`, 2007.

[79] X. S. Lin, Representations of knot groups and twisted Alexander polynomial, Acta Math. Sin., **17** (2001), 361–380.

[80] F. Loeser and M. Vaquié, Le polynôme d'Alexander d'une courbe plane projective, Topology, **29** (1990), 163–173.

[81] M. Á. Marco, Resonance varieties, admissible line combinatorics, and combinatorial pencils, preprint available at `http://arXiv.org/math.CO/0505435`, 2005.

[82] D. Matei, Massey products of complex hypersurface complements, In: Singularity theory and its applications, Adv. Stud. Pure Math., **43**, Math. Soc. Japan, Tokyo, 2006, pp. 205–219.

[83] D. Matei and A. I. Suciu, Hall invariants, homology of subgroups, and characteristic varieties, Int. Math. Res. Not., **2002**, no. 9, 465–503.

[84] B. Moishezon, Stable branch curves and braid monodromies, In: Algebraic geometry, Chicago, Ill., 1980, Lecture Notes in Math., **862**, Springer-Verlag, Berlin, 1981, pp. 107–192.

[85] ———, Braid group techniques in complex geometry. IV. Braid monodromy of the branch curve S_3 of $V_3 \to \mathbb{CP}^2$ and application to $\pi_1(\mathbb{CP}^2 - S_3, *)$, Classification of algebraic varieties, L'Aquila, 1992, Contemp. Math., **162**, Amer. Math. Soc., Providence, RI, 1994, pp. 333–358.

[86] D. Mumford, Algebraic geometry I, complex projective varieties, Grundlehren der mathematischen Wissenschaften, **221**, Springer-Verlag.

[87] M. Namba and M. Takai, Degenerating families of finite branched coverings, Osaka J. Math., **40** (2003), 139–170.

[88] M. Namba and H. Tsuchihashi, On the fundamental groups of Galois covering spaces of the projective plane, Geom. Dedicata, **104** (2004), 97–117.

[89] W. D. Neumann, A calculus for plumbing applied to the topology of complex surface singularities and degenerating complex curves, Trans. Amer. Math. Soc., **268** (1981), 299–344.

[90] M. Oka, Symmetric plane curves with nodes and cusps, J. Math. Soc. Japan, **44** (1992), 375–414.

[91] ———, Two transforms of plane curves and their fundamental groups, J. Math. Sci. Univ. Tokyo, **3** (1996), 399–443.

[92] ———, Flex curves and their applications, Geom. Dedicata, **75** (1999), 67–100.

[93] ———, A new Alexander-equivalent Zariski pair, Acta Math. Vietnam., **27** (2002), 349–357, Dedicated to the memory of Le Van Thiem (Hanoi, 1998).

[94] ———, A survey on Alexander polynomials of plane curves, Singularités Franco-Japonaises, Sémin. Congr., **10**, Soc. Math. France, Paris, 2005, pp. 209–232.

[95] ———, Zariski pairs on sextics. I, Vietnam J. Math., **33** (2005), Special Issue, 81–92.

[96] ———, Zariski pairs on sextics. II, In: Singularity Theory, Proceedings of the 2005 Marseille Singularity School and Conference, (eds. C. Murolo, A. Pichon, D. Chéniot, N. Dutertre and D. Trotman), 2007, pp. 837–863.

[97] ———, Tangential Alexander polynomials and non-reduced degeneration, Singularities in geometry and topology, World Sci. Publ., Hackensack, NJ, 2007, pp. 669–704.

[98] M. Oka and D. T. Pho, Classification of sextics of torus type, Tokyo J. Math., **25** (2002), 399–433.

[99] S. Y. Orevkov, Realizability of a braid monodromy by an algebraic function in a disk, C. R. Acad. Sci. Paris Sér. I Math., **326** (1998), 867–871.

[100] A. Di Pasquale, Links and complements of arrangements of complex projective plane algebraic curves, Ph. D. thesis, The Univ. of Melbourne, 1999.

[101] R. Randell, Milnor fibers and Alexander polynomials of plane curves, Singularities, Part 2, Arcata, Calif., 1981, Proc. Sympos. Pure Math., **41**, Amer. Math. Soc., Providence, RI, 1983, pp. 415–419.

[102] G. Rybnikov, On the fundamental group of the complement of a complex hyperplane arrangement, preprint available at `http://arXiv.org/math.AG/9805056`.

[103] M. Salvetti, Arrangements of lines and monodromy of plane curves, Compositio Math., **68** (1988), 103–122.

[104] F. Severi, Vorlesungen über algebraische Geometrie, Teubner, Leipzig, 1921.

[105] I. Shimada, A note on Zariski pairs, Compositio Math., **104** (1996), 125–133.

[106] ———, Fundamental groups of complements to singular plane curves, Amer. J. Math., **119** (1997), 127–157.

[107] ———, On arithmetic Zariski pairs in degree 6, preprint available at `http://arXiv.org/math.AG/0611596`.

[108] ———, Non-homeomorphic conjugate complex varieties, preprint available at `http://arXiv.org/math.AG/0701115`.

[109] E. Shustin, Smoothness and irreducibility of varieties of plane curves with nodes and cusps, Bull. Soc. Math. France, **122** (1994), 235–253.

[110] A. I. Suciu, Fundamental groups of line arrangements: enumerative aspects, Advances in algebraic geometry motivated by physics, Lowell, MA, 2000, Contemp. Math., **276**, Amer. Math. Soc., 2001, pp. 43–79.

[111] H. Tokunaga, Triple coverings of algebraic surfaces accoring to the Cardano formula, J. of Math. Kyoto Univ., **31** (1991), 359–375.

[112] ———, On dihedral Galois coverings, Canadian J. Math., **46** (1994), 1299–1317.

[113] ———, A remark on E. Artal-Bartolo's paper, Kodai Math. J., **19** (1996), 207–217.

[114] ———, Dihedral coverings branched along maximizing sextics, Math. Ann., **308** (1997), 633–648.

[115] ———, Some examples of Zariski pairs arising from certain elliptic $K3$ surfaces, Math. Z., **227** (1998), 465–477.

[116] ———, Some examples of Zariski pairs arising from certain elliptic $K3$ surfaces, II: Degtyarev's conjecture, Math. Z., **230** (1999), 389–400.

[117] ———, Dihedral coverings of algebraic surfaces and its application, Trans. Amer. Math. Soc., **352** (2000), 4007–4017.

[118] ———, Galois covers for $\mathfrak{S}_4$ and $\mathfrak{A}_4$ and their applications, Osaka J. Math., **39** (2002), 621–645.

[119] H. Tsuchihashi, Galois coverings of projective varietics for the dihedral groups and the symmetric groups, Kyushu J. of Math., **57** (2003), 411–427.

[120] V. G. Turaev, Reidemeister torsion in knot theory, Russian Math. Survey, **41** (1986).

[121] ———, Introduction to combinatorial torsions, Notes taken by F. Schlenk, Lectures in Math. ETH Zurich, Birkhauser, Bassel, 2001.

[122] K. Ueno, Classification theory of algebraic varieties and compact complex spaces, Lecture Note in Math., **439**, Springer-Verlag, 1975.

[123] A. M. Uludağ, More Zariski pairs and finite fundamental groups of curve complements, Manuscripta Math., **106** (2001), 271–277.

[124] M. Wada, Twisted Alexander polynomial for finitely presentable groups, Topology, **33** (1994), 241–256.

[125] F. Waldhausen, Eine Klasse von 3-dimensionalen Mannigfaltigkeiten I, Invent. Math., **3** (1967), 308–333.

[126] ———, Eine Klasse von 3-dimensionalen Mannigfaltigkeiten II, Invent. Math., **4** (1967), 87–117.

[127] J. G. Yang, Sextic curves with simple singularities, Tôhoku Math. J. (2), **48** (1996), 203–227.

[128] H. Yoshihara, On plane rational curves, Proc. Japan Acad. Ser. A Math. Sci., **55** (1979), 152–155.

[129] S. Yuzvinsky, Realization of finite Abelian groups by nets in $\mathbb{P}^2$, Compos. Math., **140** (2004), 1614–1624.

[130] O. Zariski, On the problem of existence of algebraic functions of two variables possessing a given branch curve, Amer. J. Math., **51** (1929), 305–328.

[131] ———, On the irregularity of cyclic multiple planes, Ann. Math., **32** (1931), 445–489.

[132] ———, The topological discriminant group of a Riemann surface of genus p, Amer. J. Math., **59** (1937), 335–358.

[133] ———, On the purity of the branch locus of algebraic functions, Proc. Nat. Acad. Sci. U. S. A., **44** (1958), 791–796.

[134] ———, Algebraic surfaces, Springer-Verlag, Heidelberg, 1971, 2nd. ed.

Enrique Artal Bartolo, José Ignacio Cogolludo
Departamento de Matemáticas, IUMA
Universidad de Zaragoza
CL. Pedro Cerbuna, 12
E-50009 Zaragoza SPAIN
E-mail address: artal@unizar.es,jicogo@unizar.es

Hiro-o Tokunaga
Department of Mathematics
Tokyo Metropolitan University
Minamiohsawa 1-1 Hachoji
Tokyo 192-0357 JAPAN
E-mail address: tokunaga@tmu.ac.jp

Advanced Studies in Pure Mathematics 50, 2008
Algebraic Geometry in East Asia — Hanoi 2005
pp. 101–175

Towards a categorical construction of Lie algebras

Kyoji Saito

To the memory of Nguyen Huu Duc
(13 August 1950 - 7 June 2007)

Preface

This is an introduction to the program which we call "towards a categorical construction of Lie Algebras". That is, from the data of a system of 4 integers $W := (a, b, c; h)$, called a *regular system of weights*, satisfying an arithmetic condition, we want to construct a certain generalization $\mathfrak{g}_W$ of a simple Lie algebra. Precisely, to a weight system, we first associate a surface with a singular point. Then, using the geometry of the singularity, a triangulated category is attached. Finally, we want to read Lie theoretic data from the category and to construct the algebra $\mathfrak{g}_W$.[1] The program is still in its early stages, and, in the present paper, we are mainly concerned with some categorical aspects of the program, and then ask questions on the possible constructions of Lie algebras.

The organization of the paper is as follows. In §1-9, we start by recalling the classical relations of simple or simply elliptic singularities with simple or elliptic Lie algebras, respectively, as the prototype of relations between singularities and Lie algebras. This part is rather

---

Received August 13, 2006.
Revised November 14, 2007.
2000 *Mathematics Subject Classification.* 18F99.
The author is grateful to H. Asashiba, B. Forbes, S. Iyama, H. Kajiura and A. Takahashi for their interest and help during the preparation of the present paper.

[1]This is a part of the long program "a categorical construction of primitive forms" (see [Mat][Od1][Sa7] and Footnote 11 for a definition of a primitive form, and consult the overview articles [Sa15]and [Sa19]). We expect that a good class of primitive forms are constructed from the Lie algebra $\mathfrak{g}_W$ associated with regular systems of weights W (see §4 and 12). In the present paper, we are concerned with the part of the program before the construction of the Lie algebra, and most parts are readable without a knowledge of a primitive form.

sketchy and we suggest the reader either look at the references or skip details. In §10-15, we start anew by introducing the concept of a regular system of weights and by associating a singularity to it. We discuss about two geometric (algebraic and topological) aspects of the singularity and about the possibly associated Lie algebra. We discuss also about the $*$-duality on the set of regular weight systems. This part may look somehow loose and involved without a clear focus. However, these considcrations seem to get converged to a clearer forcus by introducing a categorical approach in §16-18. In §16, we descripbe the triangulated category $\mathrm{HMF}^{gr}_{A_W}(f_W)$ associated with the singularity. Then we determine the generating structure of the category for two basic cases in §17 and 18, which are the goal of the present paper.

Let us explain the contents in more details. One key observation in §1-9 is that the Lie algebra side data: the Coxeter transformation $\mathbf{c}$ on the root lattice is identified with the singularity side data: the Milnor monodromy action $\mathbf{c}$ on the lattice of vanishing cycles (see §5). As in the classical Lie theory [Bou], we consider exponents $m_i \in \mathbb{Z}_{\geq 0}$ of eigenvalues of $\mathbf{c}$ (see §8), and then, inspired by the theory of primitive forms (see Footnotes 23, 24), we look at the *generating function of the exponents*:

$$\text{(A)} \qquad \chi(T) \;=\; T^{m_1} + T^{m_2} + \cdots + T^{m_\mu}.$$

Then, we observe that, for any of the simple or elliptic Lie algebras (corresponding to simple or simply elliptic singularities), $\chi(T)$ decomposes as:

$$\text{(B)} \qquad \chi(T) \;=\; T^{-h}\frac{(T^h - T^a)(T^h - T^b)(T^h - T^c)}{(T^a - 1)(T^b - 1)(T^c - 1)}$$

for some integers a, b, c and $h :=$ order of $\mathbf{c}$ with

$$\text{(C)} \qquad 0 < a, b, c < h \;\text{ and }\; \gcd(a, b, c) = 1.$$

In §10, we reverse our view point; we call a system of 4 integers $W = (a, b, c; h)$ satisfying (C) a *regular system of weights* (or, a *regular weight system*), if the rational function in the RHS of (B) becomes a Laurent polynomial. Then, we use the regular weight system as the starting point for all of the later constructions. Actually, the Laurent polynomial becomes a finite sum of monomials as in (A), where the *exponents* m_i of the monomials are allowed to be negative in general.

The regular weight systems are concisely classified by the *smallest exponent* $(= a + b + c - h)$, denoted by $\varepsilon_W \in \mathbb{Z}$. In fact, we see $\varepsilon_W \leq 1$ in general, and that regular weight systems with $\varepsilon_W = 1$ or 0 correspond to simple or simply elliptic singularities, respectively. As for the next class, $\varepsilon_W = -1$, we obtain 14+8+9 regular weight systems, which are the objects of our main interest in the present paper.

In §11-15, associated with a regular weight system W, we introduce and study a surface $X_{W,0}$ which has an isolated singular point at the origin 0. Namely, let f_W be a generic weighted homogeneous polynomial in coordinates x, y, z of weights a, b, c with the total degree h. Then, the regularity of W is equivalent to the equation $f_W = 0$ defining a hypersurface $X_{W,0}$ which has an isolated singular point at the origin 0. This is also equivalent to say that $C_W := (X_{W,0} \setminus \{0\})/\mathbb{G}_m$ being a smooth orbifold curve, where the orbifold data (i.e. signature, see §11, a)) is arithmetically determined from W. In other words, the curve C_W is equipped with a fractional $(=\varepsilon_W^{-1})$ power of the canonical bundle, and the blowing down of its zero-section is the surface $X_{W,0}$ with an isolated singular point which we want to study (see §11).

As described in §3-7, in order to get the Lie algebra $\mathfrak{g}_W$ from the simple or simply elliptic singularity, historically, there were two approaches: the algebraic one, using a resolution of the singularity, and the topological one, using the set of vanishing cycles (see §5) in a smoothing (Milnor fiber) of the singularity. Let us see below how these two approaches work for each of the cases ε_W=1 and 0.

Case ε_W=1 (the simple singularity): in the first approach, the *resolution diagram of the simple singularity* is identified with the Dynkin diagram of a simple Lie algebra (Du Val, see §3), and defines its Cartan matrix. Then, as is standard in Lie theory, by the use of Chevalley generators and Serre relations associated to the Cartan matrix, we obtain a simple Lie algebra $\mathfrak{g}_W$. On the other hand, in the second approach, *the set of vanishing cycles in the middle homology group of a smoothing (= Milnor fiber) of the singularity is identified with the set of roots of a finite root system in its root lattice of a simple Lie algebra* (see §7). Then, inside the lattice vertex algebra [Bo1] of the root lattice, we consider the Lie-algebra $\mathfrak{g}'_W$ generated by the vertex operators e^α of the roots α ([S-Y]§1). The Lie algebras $\mathfrak{g}_W$ and $\mathfrak{g}'_W$ constructed by these two approaches are canonically isomorphic, due to the fact that the vertices of the Dynkin diagram obtained by the first approach gives arise a simple basis of the root system obtained by the second approach, because of the existence of the simultaneous resolution of the simple singularity due to Brieskorn (§4 [Br1]). Further, Brieskorn's description of the universal family of the simple singularity enables us to describe a primitive form by the Kostant-Kirillov forms on co-adjoint orbits of a simple Lie group.

Case $\varepsilon_W = 0$ (the simply elliptic singularity): the first approach to use the exceptional set of the resolution of the singularity gives merely a single elliptic curve, and Lie theoretic data is not apparent (see Footnote 3). On the other hand, the data of the second approach, i.e. the set of vanishing cycles of a simply elliptic singularity, is characterized as the set of roots of an elliptic root system ([Sa 14] I, see §7 and Footnote 17). As in the case of $\varepsilon_W = 1$, we get the Lie algebra $\mathfrak{g}'_W$ generated by the vertex operators of elliptic roots inside the lattice vertex algebra of the elliptic root lattice. On the other hand, we construct arithmetically a certain root basis for the elliptic root system, called the elliptic diagram (Table 7). Then, as in the first approach for the case of $\varepsilon_W = 1$, we can construct a Lie algebra $\mathfrak{g}_W$ by generalizing the Serre relations associated to the Cartan matrix of the elliptic diagram. Actually, these two Lie algebras $\mathfrak{g}_W$ and $\mathfrak{g}'_W$ are shown to be naturally isomorphic; we call this algebra the *elliptic Lie algebra* (see §6 and [S-Y]).[2]

At this stage, we remark that there is a third approach for the construction of Lie algebras $\mathfrak{g}''_W$ by use of the representation theory of finite dimensional algebras, which is sometimes called the Ringel-Hall construction. Namely, Ringel [Ri 2,3,4] has determined the structure constant among the Chevalley basis of a simple Lie algebra by using the data of representations of a *hereditary algebra* (c.f. [Ga]). The idea was further extended to the representation theory of *tubular algebras* by Lin-Peng [L-P 1,2], and they obtained the elliptic Lie algebras of types $D_4^{(1,1)}$, $E_6^{(1,1)}$, $E_7^{(1,1)}$ and $E_8^{(1,1)}$ (which are exactly the cases when the elliptic Lie algebras are expected to admit primitive forms, [Sa14]II). In fact, those hereditary algebras and tubular algebras are obtained as the path algebras (see §16 **6.**(32)) of quivers associated to the classical Dynkin diagrams or to the elliptic diagrams, respectively. Since the Lie algebra depends only on the derived category of the abelian category of modules over the path algebra, some generalizations of the method in terms of triangulated category are in progress. The reader is referred to [P-X], [Toë], [D-X] and [X-X-Z] for details.

We examine, in the present paper, the "Lie theoretic data" of the above mentioned three approaches for the case $\varepsilon_W = -1$.

The singularities associated with the 14 weight systems with $\varepsilon_W = -1$ are called exceptional uni-modular singularities by Arnold [Ar3]. 1. Topological approach: certain distinguished bases of the lattices of vanishing cycles for them have been obtained by Gabrielov ([Gab2], see

[2]As in simple Lie algebra case, the symplectic structures on the co-adjoint orbits of the elliptic Lie group are expected to form a primitive form. See [Sa14] VI (Integrable Highest Weight Modules), VII (Elliptic Groups and their Invariants), in preparation, and [Ya3]).

Table 12), where the triplet (p,q,r) of lengths of the three branches of the diagram is called the Gabrielov number. 2. Algebraic approach: the exceptional set of the minimal resolution of the 14 singularities is given by a star-shape configuration of 4 rational curves (see Table 11), where the triplet (p,q,r) of the minus of the self-intersection numbers of the three branching curves is called the Dolgachev number. Then Arnold observed that there is an involutive one to one correspondence from the set of 14 exceptional uni-modular singularities to itself, which exchange the Gabrielov number and the Dolgachev number. The involution is called the *Strange duality* ([Ar3],§13). In the other words, the "Lie theoretic data" of the two approaches are exchanged by the strange duality.

The strange duality, which is nowadays understood as an appearance of *mirror symmetry* [3], admitted several interpretations and explanations. Among these, in §14, we introduce $*$*-duality on regular systems of weights*, which is an involution $*$ on a set of regular systems of weights characterized as follows: let us introduce the characteristic polynomial of the weight system W by $\varphi_W(\lambda) := \prod_{i=1}^{\mu}(\lambda - \exp(2\pi\sqrt{-1}\frac{m_i}{h})) \in \mathbb{Z}[\lambda]$. As a cyclotomic polynomial, we decompose it as $\varphi_W(\lambda) = \prod_{i|h}(\lambda^i - 1)^{e_W(i)}$. Then, another regular weight system W^* is the $*$-dual of W if and only if $h = h^*$ and $e_W(i) + e_{W^*}(h/i) = 0$ for all $i \in \mathbb{Z}_{>0}$ with minor additional conditions.[4] Then, we prove that any weight system with $\varepsilon_W = 1$ is selfdual; $W = W^*$, and that the $*$-duality induces the strange duality on the set of 14 weight systems with $\varepsilon_W = -1$. Therefore, we expect in general that the $*$-duality exchanges the algebraic approach for a weight system W with the topological approach for the dual system W^*. Then, instead of the naive study of resolution diagrams of the singularity $X_{W,0}$ in the algebraic side of W, what stands for the lattice and the basis of vanishing cycles of $X_{W^*,0}$ in the topological side of W^*?

Inspired by the recent studies of D-branes on mirror symmetry in mathematical physics ([K-L 1,2], [H-W], [Wal] and [Or1], see §15), we study *the homotopy category* $\mathrm{HMF}^{gr}_{A_W}(f_W)$ *of matrix factorizations of*

---

[3] The reader is referred to [Kon],[Yau] for mirror symmetry in general and to [K-Y][Ta1] for the Landau-Ginzburg orbifold case. Already in case of $\varepsilon_W = 0$, the algebraic data, i.e. the elliptic curve in the exceptional set in the resolution of the singularity, is not "mirror dual" to the elliptic root system of vanishing cycles obtained topologically. In order to get mirror symmetry here, one should think of the elliptic curve with a group action [Ta1]. A more comprehensive description is obtained by considering the pairs of a regular weight system and a group action. However, in the present paper, we do not get into such details.

[4] The $*$-dual of W may not exist for all W, but is unique if it exists and is denoted by W^* with $W^{**} = W$ [Sa17]. It seems interesting to extend the concept of regular systems of weights (by considering group actions (Footnote 3) and non-hypersurface singularities), which is closed under the $*$-duality.

the polynomial f_W as the algebraic approach.[5] We devote §16 to the descriptions of three different definitions of this category and its basic properties. We expect that the advantage of this approach is that this category carries a "universality" such that it can recover all the three approaches to the Lie algebra, which we have discussed above.[6]

In §17 and 18, we observe and explain this fact in the case of the category for simple singularities with $\varepsilon_W = 1$ and for the exceptional singularities with $\varepsilon_W = -1$.

We show that the category $\mathrm{HMF}^{gr}_{A_W}(f_W)$ for $\varepsilon_W=1$ is generated by a strongly exceptional collection $\mathcal{E}$ (see §16 **4.**), whose associated quiver is a Dynkin quiver $\vec{\Delta}$ of type W, and that the path-algebra $\mathbb{C}\vec{\Delta}$ (see §16 **6.**) is isomorphic to the algebra $\mathrm{End}(\mathcal{E})$ consisting of all morphisms among the objects of the exceptional collection. Therefore, we have the equivalence $\mathrm{HMF}^{gr}_{A_W}(f_W) \simeq D^b(\text{mod-}\mathbb{C}\vec{\Delta})$ due to a theorem of Bondal-Kaplanov (see §16 **4**). Hence, using the classical result by Gabriel [Ga], the K-group $K_0(\mathrm{HMF}^{gr}_{A_W}(f_W))$ and the image set in the K-group of indecomposable objects of the category are isomorphic to the root lattice and the set of roots of a finite root system, respectively. That is, $\mathrm{HMF}^{gr}_{A_W}(f_W)$ recovers all three data for the Lie algebra discussed above, inducing the natural isomorphisms $\mathfrak{g}_W \simeq \mathfrak{g}'_W \simeq \mathfrak{g}''_W$ among them.

In the case $\varepsilon_W=-1$, the category $\mathrm{HMF}^{gr}_{A_W}(f_W)$ is generated again by a strongly exceptional collection $\mathcal{E}$ whose associated quiver Δ_A is given in Table 14, where A is the signature set (13) of W (see Footnote 32). We show again an isomorphism $\mathrm{End}(\mathcal{E}) \simeq \mathbb{C}(\Delta_A, R)$ and an equivalence $\mathrm{HMF}^{gr}_{A_W}(f_W) \simeq D^b(\text{mod-}\mathbb{C}(\Delta_A, R))$ of the categories, where $\mathbb{C}(\Delta_A, R)$ is the quotient of the path-algebra $\mathbb{C}\Delta_A$ by the relations R (see (32)and §18 Theorem). Hence, in the 14 uni-modular exceptional cases, comparing Table 12 with 14 and in view of the strange duality, we conclude that the K-group $K_0(\mathrm{HMF}^{gr}_{A_W}(f_W))$ is isomorphic to the lattice of vanishing cycles for the $*$-dual weight system W^*; this is what we expected.

We conjecture that the image set in the K-group of exceptional indecomposable objects of the category coincides with the set of vanishing cycles for the singularity $X_{W^*,0}$, and, hence, the three approaches to the Lie algebra are available from the category $\mathrm{HMF}^{gr}_{A_W}(f_W)$. Whether the three Lie algebras $\mathfrak{g}_W$, $\mathfrak{g}'_W$ and $\mathfrak{g}''_W$ for them are isomorphic to each other or not is an interesting and important open problem.

[5]This was proposed by Takahashi [Ta2] (c.f. Orlov [Orl1]) answering a problem posed by the author [Sa15] (5.3) Problem. The sections §16, 17 and 18 are based on the joint works [K-S-T 1-2].

[6]It is also remarkable that the stability condition space [Br1][H-M-S] on this category seems to have a close relationship with the period domain for period maps of primitive forms [Sa22].

§1. Simple polynomials

There are a finite number of regular polyhedra, namely, the icosahedron, dodecahedron, octahedron, hexahedron and the tetrahedron, known at the time of Platon. The regular dihedron, which has only two faces of the n-gon ($n \geq 3$), is nowadays included in the list of regular polyhedra. The subgroup G of $SO(3)$ consisting of rotations of three dimensional Euclidean space, which moves a regular polyhedron (centered at the origin) to itself, is called the regular polyhedral group. The binary extension $\tilde{G}$ of the regular polyhedral group G is obtained by taking the inverse image of G through the surjective homomorphism $SU(2) \to SO(3)$. It is well-known that the binary regular polyhedral groups (including binary dihedral groups) and the cyclic subgroups $Z_n := \left\langle \begin{pmatrix} \exp \frac{2\pi\sqrt{-1}}{n} & 0 \\ 0 & \exp\left(-\frac{2\pi\sqrt{-1}}{n}\right) \end{pmatrix} \right\rangle$ for $n \in \mathbb{Z}_{>0}$ together form a complete list of finite subgroups of $SU(2)$ up to conjugacy. As an abstract group, all of the groups have a presentation:

$$\langle p, q, r \rangle := \langle\ x, y, z \ \mid\ x^p = y^q = z^r = xyz \rangle$$

for suitable integers $p, q, r \in \mathbb{Z}_{>0}$, given in the next Table 1 (here, x, y and z induces the rotation of the polyhedron centered at the barycentre of an edge, a face and a vertex).

$\langle 1, b, c \rangle$	$\simeq$	Z_n	$\simeq$	cyclic group of order $n = b + c$
$\langle 2, 2, n \rangle$	$\simeq$	$\mathcal{D}_{2n}$	$\simeq$	binary dihedral group of n-gon $n \geq 2$
$\langle 2, 3, 3 \rangle$	$\simeq$	$\mathcal{A}_4$	$\simeq$	binary regular tetrahedral group
$\langle 2, 3, 4 \rangle$	$\simeq$	S_4	$\simeq$	binary regular octahedral group
$\langle 2, 3, 5 \rangle$	$\simeq$	$\mathcal{A}_5$	$\simeq$	binary regular icosahedral group

Table 1.

In fact, these are the only cases when the group $\langle p, q, r \rangle$ is finite (see [C-M]). The group is sometimes called the Kleinean group because of the following result due to A. Schwarz [Sc] and F. Klein [Kl1].

Theorem. *Let $\tilde{G} \subset SU(2)$ be a Kleinean group. Let it act linearly on $\mathbb{C}^2$, and, hence, on the ring $\mathbb{C}[u, v]$ of polynomial functions on $\mathbb{C}^2$ (where u, v ares coordinates of $\mathbb{C}^2$). Then the subring $\mathbb{C}[u, v]^{\tilde{G}} := \{P \in \mathbb{C}[u, v] \mid gP = P\ {}^\forall g \in \tilde{G}\}$ of invariants is generated by 3-homogeneous elements, say x, y and z, which satisfy a single relation, say $f_{\tilde{G}} = f(x, y, z)$. That is:*

$$\mathbb{C}[u, v]^{\tilde{G}} \simeq \mathbb{C}[x, y, z]/(f_{\tilde{G}}).$$

The polynomial $f_{\tilde{G}}$ is called a *simple polynomial*, which is listed in the following table.

Type	$f_{\tilde{G}}$	Kleinean group
A_l	$x^{l+1} + yz$	Z_n
D_l	$x^2y + y^{1-1} + z^2$	$\langle 2, 2, n \rangle$
E_6	$x^4 + y^3 + z^2$	$\langle 2, 3, 3 \rangle$
E_7	$x^4 + xy^3 + z^2$	$\langle 2, 3, 4 \rangle$
E_8	$x^5 + y^3 + z^2$	$\langle 2, 3, 5 \rangle$

The Types in the left-side shall be explained in §3.

Table 2.

Note. From the polynomial $f_{\tilde{G}}$, one can recover $\tilde{G}$. See Appendix 3.

F. Klein, in the introduction to his lecture notes on the icosahedron [Kl1], described the time when he and Lie studied together in Berlin and Paris during the years 1869-70: "At that time we jointly conceived the scheme of investigating geometric or analytic forms susceptible of transformation by means of groups of changes. This purpose has been of directing influence in our subsequent labors, though these may have appeared to lie far asunder. Whilst I primary directed my attention to groups of discrete operations, and was thus led to the investigation of regular solids and their relations to the theory of equations, Professor Lie attacked the more recondite theory of continued groups of transformations, and therewith of differential equations".

§2. Simple Lie algebras and root systems

Let us explain another stream of mathematics started from Lie and Killing-Cartan.

The Lie algebras describe "the infinitesimal structure of continuous groups". The series of works [Ki] by Killing starting from the year 1888, determining the structure of *simple Lie algebras* (which was completed by E. Cartan [Ca]) has introduced a new mathematical structure (see [Ha]) which goes far beyond the class of simple Lie algebras, and is strongly influential on the present program.

Killing looked at the adjoint action of the maximal abelian (Cartan) subalgebra of a simple Lie algebra and decomposed the Lie algebra into a direct sum of equi-eigenspaces of the action. Since an equi-eigenvalue (as an element of the dual space of the Cartan subalgebra) is a root of the characteristic eigen-equation, he called it a *root* (*Wurzel*), and showed that the system of roots for a simple Lie algebra satisfies some properties, which are nowadays known as the axioms for a finite root system (see ([Bou]§6n°1)). The classification of simple Lie algebras is reduced to the classification of finite root systems. In fact, it is achieved by determining the matrix $(2I(\alpha,\beta)/I(\alpha,\alpha))_{\alpha,\beta\in\Gamma}$ (called the Cartan matrix), where I is Killing form on the root lattice and Γ is a simple basis of the root system[7].

§3. Du Val diagrams and Coxeter diagrams

Let us see how the two streams of mathematics, one starting with Klein and the other with Lie-Killing, meet again in the year 1934, when Du Val and Coxeter were together at Trinity college in Cambridge. At that time, the concept of the Weyl group, generated by reflections s_α for all roots α of the Lie algebra, was established in connection with the representation theory of simple Lie algebras (Weyl [We] (1925-6) and Cartan [Ca]). The classification of root systems is reduced to the classification of the Weyl group [Wae]. Then Coxeter, by use of the fundamental domain (=Weyl chamber) of the Weyl group, classified all finite reflection groups acting on Euclidean space. Namely, he gave an explicit presentation of the Weyl group in terms of generators and relations, known as the Coxeter relations [Co1].[8] For the classification, he introduced a diagram (tree) Γ, where the vertices correspond to the generators and an edge is drawn between two vertices which are non-commutative (see [Bou] for more details on reflection groups). In Table 3, the Coxeter's diagram for the Weyl groups of types A_l, D_l, or E_l are given by removing i) the vertex ρ_0 of the diagram and ii) the "tilde $\tilde{}$ "

[7]Recall [Bou](chap.6 §1 5.) that a simple basis of a (finite) root system is characterized as a system of linear forms on the Cartan algebra, whose zeros define the system of walls (oriented to the inside) of a Weyl chamber. It is admirable that, even at such an early stage (1888) of the study of simple Lie algebras, Killing (see [Ki]S12,13) began to study root basis Γ, the product $\prod_{\alpha\in\Gamma} s_\alpha$ of the reflections s_α associated to the basis (presently known as the Coxeter-Killing transformation) and its eigenvalues (which presently defines the exponents). However, for their geometric significance in terms of the Weyl group and chambers, one must wait until Weyl's work [We]. As we shall see, finding generalizations of the simple root basis, Coxeter- Killing transformations and the exponents are central problem in the present paper.

[8]The generators are given by the reflections attached to the walls of the chamber (which is bijective to the set Γ of simple basis of Killing) and the relations are given by the dihedral group relations for every pair of generators along 2-codimensional facets of the chamber. The higher codimensional facets of the chamber do not play a role in determining the group.

from the types in RHS of table (see Appendix for more details on the table).

Kleinean group	Diagram	Type
Z_n	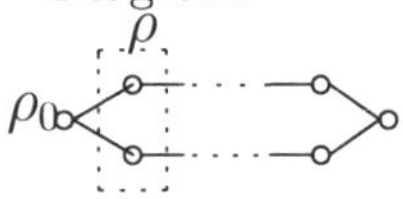	$\tilde{A}_{n-1}$
$\langle 2,2,n\rangle$	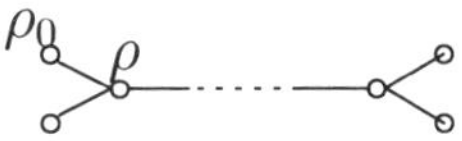	$\tilde{D}_{n+2}$
$\langle 2,3,3\rangle$	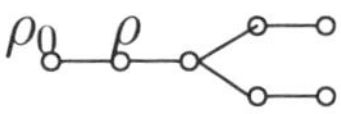	$\tilde{E}_6$
$\langle 2,3,4\rangle$	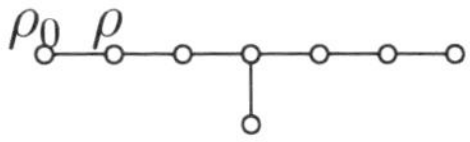	$\tilde{E}_7$
$\langle 2,3,5\rangle$	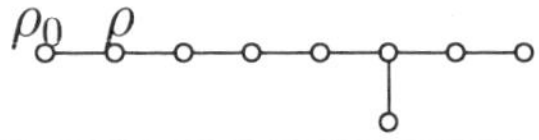	$\tilde{E}_8$

Table 3.

The complex hypersurface X_0 in $\mathbb{C}^3$ defined by the zero-loci of a simple polynomial in the list of Klein (Table 2) has an isolated singular point at the origin 0 (cf. §11 Fact4.), called a *simple singularity* [Dur]. In the year 1934, Du Val [Du] studied the (minimal) resolution $\pi : \tilde{X}_0 \to X_0$ of the simple singularity. He associated a diagram Γ to the resolution: decompose the exceptional set $E := \pi^{-1}(0)$ into irreducible components $\cup_{i=1}^{l} E_i$, then, vertices x_i of the diagram are in one to one correspondence with irreducible components E_i and an edge is drawn between x_i and x_j if and only if $E_i \cap E_j \neq \emptyset$. He observed that for each Kleinean group on the LHS of Table 3, the diagram he obtained is exactly the one given in the middle of the Table 3, deleting the vertex ρ_0. In the introduction of [Du], he wrote "It may be noted that the "trees" of curves which we have had to consider bear a strict formal resemblance to the spherical simplices whose submultiple of π, considered by Coxeter". In the same volume of the London Journal, Coxeter [Co1] listed diagrams for reflection groups, answering to a request of Du Val (for the definitions of diagrams for a basis of a lattice, see Footnote 41, and for a quiver, see §16, 6).

§4. Universal unfolding of simple singularities by Brieskorn

We observed in §3 that *there is a one to one correspondence between the diagrams of Du Val associated to simple polynomials and those of Coxeter in the classification of simple Lie algebras* (recall Table 3). However, at this stage, their relation remained a "strict resemblance", as Du Val wrote. A more direct and decisive relationship was found 40years later in the work of Brieskorn and Grothendieck. In ICM Nice 1970, Brieskorn [Br4] reported the following result.

Theorem. (Brieskorn [Br4]) *Let $X \to S$ be the universal unfolding* [9] *of a simple singularity, and let $\mathfrak{g}$ be the corresponding simple Lie algebra. Then, one has a commutative diagram:*

$$\begin{array}{ccccc} X & \subset & \mathfrak{g} & & \\ \downarrow & & \downarrow & & \\ S & \simeq & \mathfrak{g}/\!/Ad(\mathfrak{g}) & \simeq & \mathfrak{h}/\!/W \end{array}$$

where i) the vertical arrow in right side of the diagram is the adjoint quotient morphism due to Chevalley's theorem, and ii) $X \subset \mathfrak{g}$ is an embedding of X onto a transversal slice to the nilpotent subvariety of $\mathfrak{g}$ at a subregular element.

Brieskorn further described the simultaneous resolution (c.f. [Br1,2]) of the universal family.[10] He wrote "Maybe the two theories do not lie so far asunder".

Remark 1. The Brieskorn's description of the universal unfolding $X \to S$ of a simple singularity by use of a simple Lie algebra has the advantage in determining certain global differential geometric structures on the family $X \to S$, since, in the Lie algebra $\mathfrak{g}$, the integrability conditions are already built in. For instance, the primitive form of the family $X \to S$ [11], which is defined by an infinite system of non-linear equations, for the simple singularity is described by the Kostant-Kirirov symplectic form

---

[9]The concept of an unfolding of a singularity of a function f is due to R. Thom [Th]. We shall give in §5 and in Footnote 12. a brief description of them. From an algebraic geometric view point, it is essentially the same concept as a semi-universal deformation of the hypersurface defined by $f = 0$ near at the singular point (see [Sch] and [Tu]).

[10]This was reproven by a use of representation of quivers [Kr] (see the works by H. Nakajima for further studies on the relationship between Lie algebras and representations of quivers).

[11]For a primitive form, see [Mat][Od1][Sa7][Sa19]. It is a relative de-Rham cohomology class $\zeta \in \mathbf{H}_{DR}(X/S)$ which 1) generates all the other de-Rham

[Sa7] [Yah] [Ya1] [Yo]. The flat structure (Frobenius mfd structure) on the deformation parameter space S is described by the Coxeter-Killing transformation of the Weyl group [Sa16] [He] [Sab].

These facts motivated the author to convince the following: *for a further class of singularities, using suitable Lie algebras, construct primitive forms and flat structures globally.* However, the list of regular polyhedral groups and that of the simple Lie algebras have already been used up. Are these the only cases where singularity theory and Lie theory come happily together?

§5. Universal unfolding of a hypersurface singularity

Before we go further, we prepare some terminologies on vanishing cycles of a hypersurface isolated singular point studied by authors [Br3] [Le1] [Gab1] [Eb1].

Let $f(\underline{x})$ with $\underline{x} := (x_0, \cdots, x_n)$ $(n \geq 0)$ be a holomorphic function defined in a neighborhood $\mathcal{U}$ of the origin 0 of $\mathbb{C}^{n+1}$ with the coordinate $\underline{x}$. Assume that the hypersurface $X_0 := \{(\underline{x}) \in \mathcal{U} \mid f(\underline{x}) = 0\}$ has an isolated singular point at the origin $0 \in X_0$. This is equivalent to that $J_f := \mathbb{C}\{\underline{x}\}/\left(\frac{\partial f(x)}{\partial x_0}, \cdots, \frac{\partial f(x)}{\partial x_n}\right)$ is of finite rank over $\mathbb{C}$, where $\mathbb{C}\{\underline{x}\}$ is the local ring of all convergent series in $\underline{x}$.

Theorem. (Milnor [Mi]) *Consider a map* $f : X_{\delta,\varepsilon} \to D_\varepsilon$ *where* $X_{\delta,\varepsilon} := \{\underline{x} \in \mathcal{U} \mid |\underline{x}| < \delta\} \cap f^{-1}(D_\varepsilon)$ *and* $D_\varepsilon := \{t \in \mathbb{C} \mid |t| \leq \varepsilon\}$ *for positive real numbers* δ, ε *such that* $0 < \varepsilon \ll \delta \ll 1$. *Then,* $f|_{X \setminus f^{-1}(0)} : X \setminus f^{-1}(0) \to D_\varepsilon \setminus \{0\}$ *is a locally trivial topological fibration such that the general fiber is homotopic to a bouquet of* μ_f*-copies of* n*-sphere* S^n*, where* $\mu_f := \dim_{\mathbb{C}} J_f$ *is called the Milnor number.*

The fibration is called the Milnor fibration, whose general fiber over a base point $1 \in D_\varepsilon$, denoted by X_1, is called the *Milnor fiber. If* f *is globally defined weighted homogeneous polynomial of positive weights, then we may choose* $\delta = \varepsilon = \infty$.

As a consequence of this result, the (reduced) homology group of the Milnor fiber is non-trivial only in dimension n, and we have $\tilde{\mathrm{H}}_n(X_1, \mathbb{Z}) \simeq \mathbb{Z}^{\mu_W}$. Let us introduce particular elements of $\tilde{\mathrm{H}}_n(X_1, \mathbb{Z})$, called *vanishing*

cohomology classes as a $\mathcal{D}_S$-module, and 2) satisfies an infinite system of bilinear differential equation (by means of residue pairings). Its local existence on S is known by [Sai]. Global existence on S is known only for simple or simply elliptic singularities. It is believable that $\mathfrak{g}$ is the Cartan prolongation of X with respect to the primitive form. Such global construction of primitive forms by means of globally defined integrable systems (such as Lie algebras) is the basic motivation in the present paper. However, we shall not discuss the primitive form itself in any further detail.

cycles: let us consider a *universal unfolding* of f (Thom [Th]), which is a function $F(\underline{x},\underline{t})$ in $\underline{x}\in\mathbb{C}^{n+1}$ and $\underline{t}=(t_1,\cdots,t_{\mu_f})\in\mathbb{C}^{\mu_f}$ defined in a neighborhood of the origin $(\underline{0},\underline{0})\in\mathbb{C}^{n+1}\times\mathbb{C}^{\mu_f}$ satisfying i) $F(\underline{x},\underline{0})=f(\underline{x})$, and ii) $\frac{\partial F(\underline{x},0)}{\partial t_i}$ $(i=1,\cdots,\mu_f)$ span the $\mathbb{C}$-vector space J_f.

For a small value of $\underline{t}$, again by choosing δ and ε suitably for $f_{\underline{t}}(\underline{x})=F(\underline{x},\underline{t})$, we consider the map $f_{\underline{t}}:X_{\delta,\varepsilon}\to D_\varepsilon$ such that, excluding finite number of its fivers over the critical values, it gives a locally trivial fibration, whose general fiber is homeomorphic to the Milnor fiber. If $\underline{t}$ is general, then $f_{\underline{t}}|_X$ has exactly μ_f-number of non-degenerate critical points and the (critical) values are distinct (that is, $f_{\underline{t}}$ is a Morsification of f). We may choose the "base point" 1 whose fiber $f_{\underline{t}}^{-1}(1)$ is the Milnor fiber X_1 on the boundary of the disc D_ε. Let $g:[0,1]\to D_\varepsilon$ be any continuous path starting at the base point $1\in D_\varepsilon$ and ending at a critical value c, without passing any critical points on $[0,1)$. Then the pull-back $X_{[0,1]}$ of the fibration $X\to D_\epsilon$ over the interval $[0,1]$ retracts to X_c. Thus, the natural inclusion $X_1\subset X_{[0,1]}$ induces a homomorphism $\iota:\tilde{\mathrm{H}}_n(X_1,\mathbb{Z})\to\tilde{\mathrm{H}}_n(X_c,\mathbb{Z})$ whose kernel $\ker(\iota)$ is rank 1 module $\mathbb{Z}$ (since the Hessian of $f_{\underline{t}}$ at the critical point is non-degenerate).

Definition Let the setting be as above. A base e (up to sign) of the kernel $\ker(\iota)$ in $\tilde{\mathrm{H}}_n(X_1,\mathbb{Z})$ is called a *vanishing cycle* along the path g. We denote by R_f the set of all vanishing cycles running all possible paths g and the critical values c.

Let γ be a path in D_ε which starts at the base point 1 and move along g close to the critical value c and then turns once around c counter-clockwisely, and then return to 1 along g. This path induces the monodromy $\rho(\gamma)\in\mathrm{Aut}(\tilde{\mathrm{H}}_n(X_1,\mathbb{Z}))$, whose action on $u\in\tilde{\mathrm{H}}_n(X_1,\mathbb{Z})$ is described by the following Picard-Lefschetz formula:

$$\rho(\gamma)(u) \;=\; u-(-1)^{\frac{n(n-1)}{2}}(u,e)e$$

where $(\cdot,\cdot):\tilde{\mathrm{H}}_n(X_1,\mathbb{Z})\times\tilde{\mathrm{H}}_n(X_1,\mathbb{Z})\to\mathbb{Z}$ is the intersection form on the middle homology group (see Footnote 35). If n is even, it is symmetric and $(e,e)=(-1)^{n/2}2$ so that $\rho(\gamma)$ is a *reflection action* with respect to the vector e, denoted by w_e.

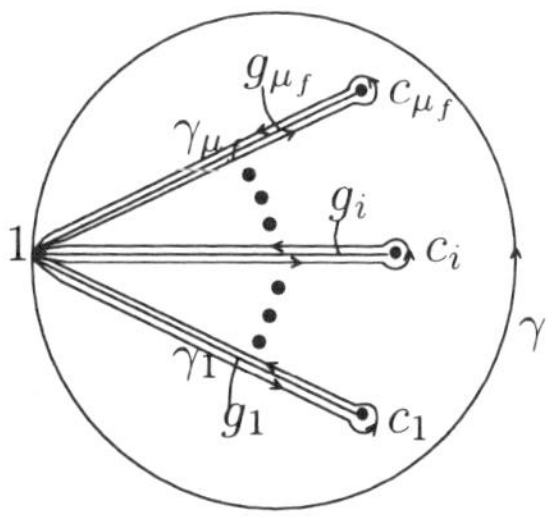

Table 4.

Now, we describe the *distinguished basis* of the middle homology group $\tilde{\mathrm{H}}_n(X_1,\mathbb{Z})$, depending on two choices: i) to give a numbering of the critical values, say $c_1,\cdots,c_{\mu_f}$, of $f_{\underline{t}}$, ii) to choose μ_f paths $g_1,\cdots,g_\mu$ in D_ε such that a) each g_i is a path connecting 1 with c_i as above, which is not self-intersecting, b) distinct paths g_i and g_j are intersecting only at 1, and c) the passes $g_1, \cdots, g_\mu$ are starting at the point 1 in the linear order $1, \ldots, \mu_f$ counter-clock wisely (see Table 4).

Fact-Definition. Under the above the setting, the set $e_1, \cdots, e_{\mu_f}$ of vanishing cycles (up to choices of sign) associated to the paths $g_1, \cdots, g_{\mu_f}$ form an ordered basis of $\tilde{\mathrm{H}}_n(X_1, \mathbb{Z})$, called a *distinguished basis* (see [Br3], [Le1], [Gab1], [Eb1])

Monodromy. Let γ be the path starting at 1 turning once around the boundary of D_ε counter-clock wisely and comes back to 1. The monodromy of this path $\mathbf{c} := \rho(\gamma) \in \mathrm{Auto}(\tilde{\mathrm{H}}_n(X_1, \mathbb{Z}))$ is called the *Milnor monodromy.* Since γ is homotopic to the product $\gamma_1 \cdots \gamma_{\mu_f}$ of paths γ_i (see Table 4), we express the monodromy $\mathbf{c}$:

$$\mathbf{c} \;=\; w_{e_1} \cdots w_{e_{\mu_f}}$$

as a product of reflections associated to a distinguished basis $e_1, \cdots, e_{\mu_f}$.

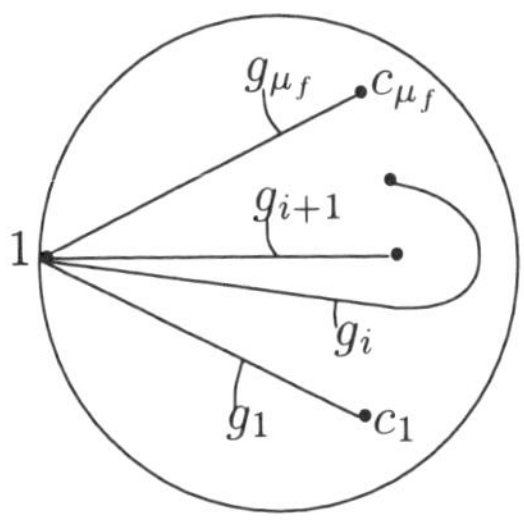

Table 5.

Braid group B_{μ_f} action on distinguished basis: First, we remark that the homotopy classes of the paths $\gamma_1, \cdots, \gamma_{\mu_f}$ give a free generator system of the group $\pi_1(D_\varepsilon \setminus \{c_1, \cdots, c_{\mu_f}\}, 1)$. Thus the choice of the paths $g_1, \cdots, g_{\mu_f}$, up to homotopy, corresponds to a choice of a free generator system of the free group. On the other hand, the braid group B_{μ_f} acts on the set of free generator systems, as usual as follows: for $1 \le i < \mu_f$, define an action σ_i : $\gamma_1, \cdots, \gamma_{\mu_f} \mapsto \gamma_1, \cdots, \gamma_{i-1}, \gamma_i\gamma_{i+1}\gamma_i^{-1}, \gamma_i, \gamma_{i+2}, \cdots, \gamma_{\mu_f}$. This causes an action of σ_i on paths $g_1, \cdots, g_{\mu_f}$ to those given in Table 5. and on the distinguished basis $e_1, \cdots, e_{\mu_f}$ to the distinguished basis $e_1, \cdots, e_{i-1}, w_{\gamma_i}(e_{i+1}), e_i, e_{i+2}, \cdots, e_{\mu_f}$. One can immediately verify that σ_i $(1 \le i < \mu_f - 1)$ satisfy Artin braid relations (see [Ar]) so that we obtain *a braid group action on the set of distinguished basis.*

Remark 2. Even if we start with a globally defined weighted homogeneous polynomial f of positive weights, in order to construct the fibration $f_{\underline{t}} : X \to D_\varepsilon$ above, we need to shrink the domain of $f_{\underline{t}}$ suitably by a use of δ and ε as above. In fact, if one of the coordinate t_i has negative weight (c.f. §11,b),4)), the embedding of a Milnor fiber $X_{\underline{t}}$ into the global affine surface $\hat{X}_{\underline{t}} := \{\underline{x} \in \mathbb{C}^{n+1} \mid F(\underline{x}, \underline{t}) = 0\}$ induces a non-trivial extension $\tilde{\mathrm{H}}_n(X_{\underline{t}}, \mathbb{Z}) \subset \tilde{\mathrm{H}}_n(\hat{X}_{\underline{t}}, \mathbb{Z})$. The extension is achieved by adding the lattice of the vanishing cycles "coming from ∞" and is expected to play key role in analytic theory of primitive forms (see [Sa19]§6 Conjecture and Problem I').

Remark 3. In mathematical physics, hypersurface singularity is studied under the name of Landau-Ginzburg model.

§6. Simply elliptic singularities

We return to the main stream of our considerations in the present paper: to seek for a connection of primitive forms with Lie theory.

In the year 1974, the author [Sa2] came up with a new class of normal surface singularities, which are "located on the boundary" of the deformation space of simple singularities. They are called the *simply elliptic singularities*, which include the following three types of hypersurfaces:

Type	equation f_W	$E \cdot E$	(μ_+, μ_0, μ_-)
$\tilde{E}_6$ or $E_6^{(1,1)}$	$x^3+y^3+z^3+\lambda xyz$	-3	0, 2, 6
$\tilde{E}_7$ or $E_7^{(1,1)}$	$x^4+y^4+z^2+\lambda xyz$	-2	0, 2, 7
$\tilde{E}_8$ or $E_8^{(1,1)}$	$x^6+y^3+z^2+\lambda xyz$	-1	0, 2, 8

Table 6.

The simple elliptic singularities X_0 are characterized from two different view points: a) by the resolution of the singularity X_0: *a normal singular point 0 of a surface X_0 is simply elliptic if and only if*, by definition, *the exceptional set $E=\pi^{-1}(0)$ of the minimal resolution $\pi:\tilde{X}_0 \to X_0$ of the singularity contains only a single elliptic curve*, and b) by deformation of the singularity: *a singular point 0 of a hypersurface surface X_0 is either simple or simply elliptic if and only if any singularity in a local deformation of X_0 admits a weighted homogeneous structure.*[12]

[12]Let us explain what do we mean by 1. "singularity in a local deformation of X_0", and 2. "weighted homogeneous structure" on a singularity X_0.

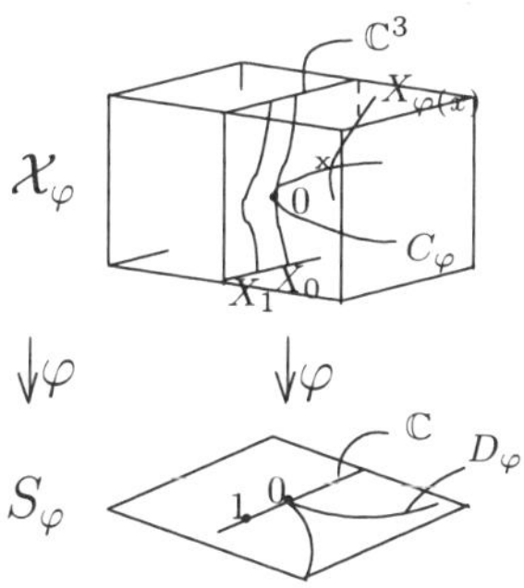

Local deformation of X_0

1. Recall §5 the universal unfolding $F(\underline{x},\underline{t})$ defined in a neighborhood $\tilde{\mathcal{U}}$ of the origin of $\mathbb{C}^{n+1} \times \mathbb{C}^{\mu_f}$. Then, it defines a local analytic flat family of analytic varieties $\varphi : \mathcal{X}_\varphi \to S_\varphi$ where $\mathcal{X}_\varphi := \{(\underline{x},\underline{t}) \in \tilde{\mathcal{U}} \mid F(\underline{x},\underline{t}) = 0\}$, S_φ is a neighborhood of the origin of $\mathbb{C}^{\mu_f}$, and φ is the projection to the second factor. The fiber $\varphi^{-1}(0)$ over 0 is nothing but the original singular surface X_0 so that the family $\{X_{\underline{t}} := \varphi^{-1}(\underline{t})\}_{\underline{t}\in S_\varphi}$ is called the semi-universal deformation of the singularity X_0 ([K-S], [Sch]). One can show that the critical set C_φ of the map φ is (locally near at the origin 0) a smooth subvariety of dimension $\mu_f - 1$, which is finite over S_φ so that the image $D_\varphi := \varphi(C_\varphi)$ is (locally near at 0) is a hypersurface in S_φ, called the

Here in the case of simply elliptic singularity, a) the resolution diagram in the sense of Du Val consists only of a single elliptic curve E and Lie theoretic data are hardly seen, in contrast with the case of the simple singularity. However, b) they show a new relation (in a symbolical level) with Lie theory through deformation theory as follows: *in the local deformation (see* 1. *of* Footnote 12*) of an elliptic singularity of type* $\tilde{\Gamma} \in \{\tilde{E}_6, \tilde{E}_7, \tilde{E}_8\}$[13]*, only an elliptic singularity of the same type* $\tilde{\Gamma}$ *or a simple singularity can appear. The simple singularity of type* Γ *can appear if and only if* Γ *is a subdiagram of* $\tilde{\Gamma}$. This fact was explained soon after its finding by use of the lattice $(\mathrm{H}_2(X_1, \mathbb{Z}), I)$ (here, $I = -(\cdot,\cdot)$, see Footnote 35).[14] Thus, for a simply elliptic singularity X_0, a relationship with Lie theory begun to appear from the lattice of the smoothing X_1, instead of the resolution $\tilde{X}_0$. Do we need to change our view point? [15] We shall come back to this question of "change of view-points" later when we discuss $*$-duality in §14 and 15.

---

discriminant of φ. Then, for any point $x \in C_\varphi$, the variety $X_{\varphi(x)} = \varphi^{-1}(\varphi(x))$ is singular at the point x. This is a singularity in a local deformation of X_0. As we saw already, for a generic point $x \in C_\varphi$, $(X_{\varphi(x)}, x)$ is an ordinary double point (i.e. Morse singularity).

2. Let X_0 be a hypersurface in a neighborhood of the origin 0 of $\mathbb{C}^{n+1}$ defined by an analytic equation $f(\underline{x}) = 0$ with an isolated singular point at 0. We say that X_0 admits a weighted homogeneous structure at 0 if there is a local analytic coordinate change at 0 such that the defining equation $f(\underline{x})$ is transformed to a weighted homogeneous polynomial $P(\underline{x})$ (i.e. $P(\underline{x}) = \sum_{a_0 i_0 + \cdots a_n i_n = h} c_{i_0 \cdots i_n} x_0^{i_0} \cdots x_n^{i_n}$ for some positive integers $a_0, \cdots, a_n$ and h). Then, the following i), ii) and iii) are equivalent [Sa1]: i) X_0 admits an weighted homogeneous structure, ii) The sequence: $0 \to \mathbb{C} \to \mathcal{O}_{X_0,0} \xrightarrow{d} \Omega^1_{X_0,0} \to \cdots \xrightarrow{d} \Omega^{n+1}_{X_0,0} \to 0$ is exact, where $(\Omega^{\cdot}_{X_0,0}, d)$ is the Poincaré complex over X_0 at 0, and iii) f belongs to the ideal $\left(\frac{\partial f(\underline{x})}{\partial x_0}, \cdots, \frac{\partial f(\underline{x})}{\partial x_n}\right)$ in the local ring $\mathbb{C}\{\underline{x}\}$.

[13]The names $\tilde{E}_i$ are taken from that of the affine Coxeter diagrams (Table 3) for the reason explained in this section. They are nowadays called also $E_i^{(1,1)}$ for the reason explained in the next §7.

[14]This is shown by using the fact that the lattice $(\mathrm{H}_2(X_1, \mathbb{Z}), I)$ is isomorphic to $Q_{\tilde{\Gamma}} \oplus \mathbb{Z}$ (see [Ga], [Eb1,2]) where $Q_{\tilde{\Gamma}}$ is the affine root lattice of a type $\tilde{E}_6, \tilde{E}_6$ and $\tilde{E}_8$. See next §7.

[15]This question is supported by the fact that the period domain for the period map $\int \zeta$ of the primitive form is determined from the lattice $\mathrm{H}_2(X_1, \mathbb{Z})$ [Sa7], [Sa14]II.

§7. Vanishing cycles for simple and simply elliptic singularities

In order to sharpen the new view point, i.e. to study the lattice $(\mathrm{H}_2(X_1,\mathbb{Z}),I)$ of the middle homology group of the smoothing X_1 of singular surface X_0, we consider a particular subset $R\subset \mathrm{H}_2(X_1,\mathbb{Z})$, the set of *vanishing cycles* introduced in §5 (c.f. [Sa15](5.2),(5.3)). From this view point, let us state some consequences of Brieskorn's description [Br4] on simple singularities:

1) *The minimal resolution $\tilde{X}_0$ and the smoothing X_1 of a simple singularity X_0 of type Γ are homeomorphic. Hence one obtains an isomorphism of lattices:*

$$*)\qquad \mathrm{H}_2(X_1,\mathbb{Z}) \quad\simeq\quad \mathrm{H}_2(\tilde{X}_0,\mathbb{Z})\ .$$

Here, the homotopy type of the homeomoprhims, and hence the isomorphism of lattices $*$) depend on the Weyl group of type Γ. In fact, the ambiguity of the isomorphism can be resolved (up to an outer automorphism of the Weyl group) by choosing the base point 1 in the totally real region of the deformation parameter space S_φ (see Footnote 16).

2) *The set of vanishing cycles R in $\mathrm{H}_2(X_1,\mathbb{Z})$ (see §5) forms a finite root system of type Γ, and $\mathrm{H}_2(X_1,\mathbb{Z})$ is identified with the root lattice Q_Γ of the root system.*

3) *The homology classes $[E_i]\in \mathrm{H}_2(\tilde{X}_0,\mathbb{Z})$ $(i=1,\cdots,l)$ of the exceptional curves E_i in the resolution $\tilde{X}_0$ are mapped by the homomorphism $*$) to a simple root basis Γ of the root system R, which are also distinguished basis in the sense in §5.*[16]

If X_0 is a simply elliptic singularity, none of 1), 2) or 3) holds. However, 2) suggests *to regard the set of vanishing cycles in $\mathrm{H}_2(X_1,\mathbb{Z})$ for a Milnor fiber X_1 of an elliptic singularity as a generalization of root systems.* In fact, we can generalize the root systems[17] by removing

---

[16]The paths $g_1,\cdots,g_{\mu_f}$ in S_φ (Footnote 12), with whom associated distinguished basis $e_1,\cdots,e_{\mu_f}$ is the simple root basis, is given in [Sa20] §4.3 Figure 6. and Theorems 4.1 and 4.2, using semi-algebraic geometry of the real discriminant $D_{\varphi,\mathbb{R}}$ of the universal deformation of the simple singularity. Furthermore, the associated paths γ_i $i=1,\cdots,\mu$ (Table 4) generate the fundamental group $\pi_1(S_\varphi\backslash D_\varphi,1)$ and satisfy Artin braid relations of type Γ so that the fundamental group becomes an Artin group ([Br5] [B-S]). Then, *the intersection matrix $(I(e_i,e_j))_{ij=1,\cdots,\mu}$ is shown to become the Cartan matrix of type Γ by solving the braid relations where $\gamma_1,\cdots,\gamma_\mu$ are substituted by Picard-Lefschetz formula for $\rho(\gamma_1),\cdots,\rho(\gamma_\mu)$ in §5.*

[17]A subset R of a real vector space equipped with a symmetric form I is called a (generalized) root system if $\mathbb{Z}R$ is a full lattice, $2I(\alpha,\beta)/I(\beta,\beta)\in\mathbb{Z}$ and $\alpha-2I(\alpha,\beta)/I(\beta,\beta)\beta\in R$ for $\forall\alpha,\beta\in R$, and irreducible in a suitable sense

the finiteness axiom from the classical one for a finite root system [Bou] Chap. VI §1 so that the set of vanishing cycles for any even dimensional hypersurface isolated singularity becomes a generalized root system. In particular, the set of vanishing cycles for a simply elliptic singularity is characterized as an *elliptic root system*, that is, a root system belonging in a semipositive lattice with radical of rank 2 (see [Sa14] I).

However, by the lack of 1) and 3) for the case of simple singularity, we cannot find a generalization of "the simple root basis" of the elliptic root system naively from the resolution of X_0. Also, no geometric method to choose one particular distinguished basis (see §5) is knowm.[18] However, we choose some root basis arithmetically[19] such that the elliptic Coxeter-Killing transformation defined as a product of reflections associated with the basis is of finite order. As in the case of classical finite root systems, we associate a diagram, called an elliptic diagram, to the basis (see Footnote 41). Some of the simply-laced elliptict diagrams are given in following Table 7.

([Sa14]I). A root system is finite or affine if I is positive definite or semidefinite and rank(radical)=1, respectively. A root system is called *elliptic* if I is positive semidefinite and rank(radical)=2. The set of vanishing cycles for a simply elliptic singularity of type $\tilde{E}_6, \tilde{E}_7$ or $\tilde{E}_8$ is the elliptic root system of type $E_6^{(1,1)}, E_7^{(1,1)}$ or $E_8^{(1,1)}$.

[18]Gabrielov [Gab2] (Fig. 10 and 11.) obtained the diagrams in Table 7. for certain distinguished basis as one of the possible choices after the braid group action under the guiding principle to find the diagrams containing small number of triangles. On the other hand, in the simple singularity case, the semialgebraic geometry of the discriminant ([Sa20]) can yield the distinguished basis which corresponds to the simple root basis of the finite root system (see also A'Campo's [AC]). There seems a gap between topology and semi-algebraic geometry.

[19]There does not exist elliptic Weyl chambers and, hence, there seemed no a priori definition of a simple basis for an elliptic root system (see [Klu]). However, the elliptic diagram in Table 7. is defined by duplicating the vertex of the affine diagram at the largest exponent (see [Sa14]I(8.6)). We define the elliptic Coxeter-Killing transformation $\mathbf{c}_e$ as the product of reflections (acting on $\mathrm{H}_2(X_1,\mathbb{Z})$) attached to the vertices of the elliptic diagram (in a suitable order). Then one has: i) $\mathbf{c}_e$ is of finite order h, and the eigenvalues of $\mathbf{c}_e$ determine the exponents of the elliptic root system (see §8 and Table 9), ii) the eigenvector of $\mathbf{c}_e$ belonging to the eigenvalue 1 is regular in the elliptic Cartan algebra $\mathfrak{h}_e$ with respect to the elliptic Weyl group W_e and iii) the universal central extension $\tilde{W}_e$ of W_e is generated by a lift $\tilde{\mathbf{c}}_e^h$. Using i), ii) and iii), a flat structure on the quotient space $\tilde{\mathfrak{h}}_e//\tilde{W}_e$ is constructed ([Sa15]II, [Sat,1,2]).

Table 7. Simply laced Elliptic diagrams of Codim=1 ([Sa14] I, Table 1).

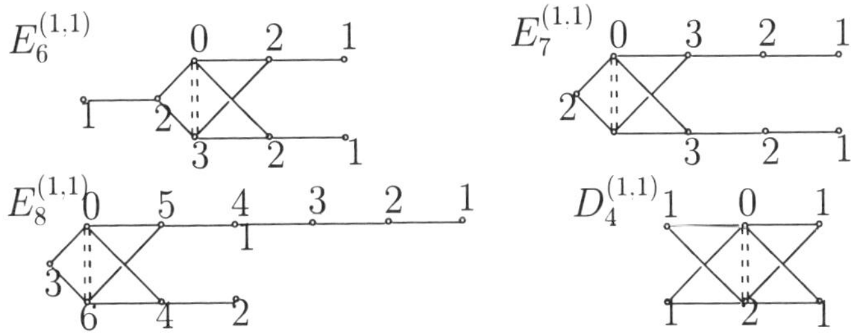

The numbers attached at vertices are the exponents of the root system (see §7).

The diagrams plays basic role, as in the finite root system case, in describing the elliptic root systems [Sa14]I, elliptic Weyl groups [ibid]III, elliptic Lie algebras [S-Y]. The construction of the primitive forms from the elliptic Lie algebras is a work in progress.[20]

§8. Exponents and weight systems

In this section, we first introduce the *exponents* for a finite or elliptic root system, which play important role in the classical and elliptic Lie theory[21]. Then, we try to extend the definition of exponents for a generalized root system, and meet with a problem of "choice of the phases"

[20]In [S-Y] the following three algebras are shown to be isomorphic: a) an algebra generated by vertex operators [Bo1] for all elliptic real roots, b) an algebra generated by the Chevalley triplets attached to the elliptic diagram (Table 7) satisfying certain generalized Serre relations, and c) an amalgamation of an affine algebra and a Heisenberg algebra. An algebra isomorphic to any one of them is called an elliptic algebra. It is also a universal central extension of a 2-toroidal algebra. We remark that the elliptic root systems and the Lie algebras are found also from the representation theory of tubular algebras (see Y. Lin and L. Peng [L-P,1&2]). Works on highest weight representations and Chevalley type invariant theory for an elliptic algebra and group are in progress (see Footnote 2). Due to the existence of the regular element (see Footnote 19), several properties similar to classical algebraic groups and its invariant theory hold for the elliptic Lie algebras and its adjoint groups. These facts supports the program that the elliptic primitive forms are constructed on the elliptic Lie algebras (see references in Footnote 2).

[21]The exponents are equal to the degrees of basic $\mathfrak{g}$- or W-invariants and play basic roles in Lie theory (see [Ko],[Sp],[St1]), and also in the study of the flat structures ([Sa16],[Sa14]II,[Sa7]).

of the exponents. In order to solve the problem, we are lead to introduce a new concept: the *regular system of weights*.

First, we recall a definition of exponents for a finite or elliptic root system. In both cases, we define a Coxeter-Killing transformation as a product **c**, in a suitable order, of reflection actions on the lattice $H_2(X_1, \mathbb{Z})$ attached to a simple root basis (recall §5). The **c** is of finite order h (called the *Coxeter number*, see §19 Remark)[22]. Then the exponents $m_1, \cdots, m_\mu$ are integers such that $\exp(2\pi\sqrt{-1}\frac{m_i}{h})$ $(i=1,\cdots,\mu)$ are the eigenvalues of **c** (see [Bou]Ch.v,n°6.2 and [Sa14] I (9.7) Lemma A.iii)). However, this determines the exponents only up to modulo h. In case of finite root systems and elliptic root systems, one poses further the constraint on the range $0 \le m_i \le h$ and on the symmetricity $m_i + m_{\mu-i+1} = h$ for $i = 1, \cdots, \mu$. Under these constraints, we determine uniquely the exponents as in the next tables.

Type	$(a,b,c;h)$	exponents
A_l $(l \ge 1)$	$(1,b,c;l+1)$	$1,2,\ldots,l$ $(b+c=l+1)$
D_l $(l \ge 3)$	$(2,l-2,1-1;2(l-1))$	$1,3,5,\ldots,2l-3,l-1$
E_6	$(3,4,6;12)$	$1,4,5,7,8,11$
E_7	$(4,6,9;18)$	$1,5,7,9,11,13,17$
E_8	$(6,10,15;30)$	$1,7,11,13,17,19,23,29$

Table 8.

Type	$(a,b,c:h)$	exponents
$E_6^{(1,1)}$	$(1,1,1:3)$	$0,1,1,1,2,2,2,3$
$E_7^{(1,1)}$	$(1,1,2:4)$	$0,1,1,2,2,2,3,3,4$
$E_8^{(1,1)}$	$(1,2,3:6)$	$0,1,2,2,3,3,4,4,5,6$

Table 9.

We try further to introduce the exponents through Coxeter-Killing transf. (Milnor Monodromy) for root systems of singularities (since they are necessary data for primitive forms; see discussions below). In fact,

---

[22]The Coxeter-Killing transformation has distinguished properties: i) **c** is of finite order h, ii) the primitive hth roots of unity (or, 1 for the case of an elliptic root system) are eigenvalues of **c**, and iii) the eigenvectors of **c** belonging to them are *regular* (i.e. they are not fixed by the Weyl group and the adjoint group of the Lie algebra, [Col], [Bou] chap.V§6 n°2, [Sa14]II §10 Lemma B). This existence of regular eigenvectors is basic for the construction of the adjoint quotient morphism $\mathfrak{g} \to \mathfrak{g}/\!\!/Ad(\mathfrak{g}) \simeq \mathfrak{h}/\!\!/W$ ([Ko],[Sp],[St1]) and of the flat structure on $\mathfrak{h}/\!\!/W$ ([Sa16], [Sa14]II).

we shall obtain in §18 quite interesting class of generalized root systems of Witt index 2 together with some distinguished root basis. However, we meet here at present a subtle problem: *phase of exponents*, which lead the author to introduce the concept of the weight system below. To explain the problem concretely, we cite some results from later sections as follows.

1. Consider a polynomial in LHS of Table 10 in §13. The zero loci of the polynomial in $\mathbb{C}^3$ defines a hypersurface X_0 with an isolated singular point at the origin.

2. The generalized root system (= the set of vanishing cycles) in $\mathrm{H}_2(X_1, \mathbb{Z})$ in the middle homology group of a Milnor fiber X_1 of X_0 has a root basis whose associated diagram is given in Table 12 (where p, q, r, called the Gabrielov#, are given in Table 13).

3. Define the Coxeter-Killing transformation $\mathbf{c}$ as the product of reflection actions on $\mathrm{H}_2(X_1, \mathbb{Z})$ associated with the vertices of the diagram in a suitable order. Then, $\mathbf{c}$ is of finite order h and the characteristic polynomial of $\mathbf{c}$ is given in the form (15) for a suitable choice of a system of integers m_i called exponents given in Table 10.

4. Observes that m_i's in Table 10 is exceeding the interval $[0, h]$. Thus, *the Coxeter-Killing transformation is unable to determine their phases* $(:= [m_i/h])$ for these new class of root systems. On the other hand, these m_i/h without the ambiguity "modulo 1" are well defined directly from a choice of a primitive form. [23]

Concern: The root system with basis may not have sufficient data to determine the phases of exponents and to construct the primitive forms.

We shall discuss again on this issue (see §14 Remark 7). This fact, due to the important role of exponents [Sa7][Sai], leads the author to handle them directly (but not through eigenvalues of Coxeter-Killing transformation) as follows.

Consider the generating function (called a characteristic function) [24] for each type of exponents in Tables 8 and 9.

$$\chi(T) := T^{m_1} + T^{m_2} + \cdots + T^{m_\mu}. \tag{1}$$

[23]The proportions m_i/h are eigenvalues of an operator N in the flat structure associated to a primitive form, and are called exponents of the flat structure ([Sa4] and [Sa7] (3.3) Definition). Therefore, we should have stated more exactly that, conjecturally, there exist a primitive form (constructed from the Lie algebra which we shall study) such that the associated flat structure determines the set of exponents m_i/h.

[24]It is introduced as the Fourier transform of the distribution of the exponents (see [Sa4] (3.1.1) and [Sa7] (3.3.14)) in order to study the zero-loci of χ.

Then, these generating functions for the finite and elliptic root systems of types A_l, D_l, E_6, E_7, E_8 and $E_6^{(1,1)}, E_7^{(1,1)}, E_8^{(1,1)}$ have a decomposition[25] of the form:

$$(2) \qquad \chi(T) \ = \ T^{-h}\frac{(T^h - T^a)(T^h - T^b)(T^h - T^c)}{(T^a - 1)(T^b - 1)(T^c - 1)}$$

where a, b, c, called weights, are integers and h, called the Coxeter number, is the order of the Coxeter element $\mathbf{c}$ such that

$$(3) \qquad 0 < a, b, c < h \quad \text{and} \quad \gcd(a, b, c) = 1.$$

Note that the set of weights a, b, c are uniquely determined from the characteristic function $\chi(T)$, except for the type A_{h-1}.[26] See Tables 8 and 9 for explicit lists of $(a, b, c; h)$. The generating function (1) of exponents for a finite or an elliptic root system are characterized by the factorization (2) without a pole as follows. Consider abstractly a system:

$$(4) \qquad W \ := \ (a, b, c; h)$$

of 4 integers satisfying (3) (and additionally, $a=1$ if $b+c=h$ called type A_{h-1}), and call it a *weight system*, where a, b, c are called the *weights* and h is called the *Coxeter number.*

Fact 1. ([Sa11]Theorem 2) *If the function* χ_W (2) *for* W *has no poles, then it is equal to a generating function* (1) *of exponents either for a finite root system of type* A_l, D_l, E_6, E_7, E_8 *or for an elliptic root system of type* $E_6^{(1,1)}, E_7^{(1,1)}, E_8^{(1,1)}$.

Let us call the rational function $\chi_W := \chi$ in (2) the *characteristic function* associated to W, and call a weight system W is *simple* (resp. *elliptic*) if its characteristic function χ_W is equal to a generating function (1) for a finite root system (resp. elliptic root system) (explicitly, see Table 8 and 9). [27]

---

[25]In the present paper, we are interested in only the cases when all roots of $\chi(T) = 0$ are on the unit circle. But, this is not the case in general for a general primitive form (see [Sa4]).

[26]The characteristic function for the type A_{h-1} is expressed as $\chi_{A_{h-1}}(T) = T + \cdots + T^{h-1} = \frac{T^h - T}{T-1} = T^{-h}\frac{(T^h-T)(T^h-T^b)(T^h-T^c)}{(T-1)(T^b-1)(T^c-1)}$ for $a = 1$ and for any integer b, c with $b + c = h$.

[27]To be exact, one should add the diagram for $D_4^{(1,1)}$ (recall Table 7.) in the list. A diagram is called simply-laced if it does not contain a multiple edges. Any other diagrams for simple (or, elliptic) root system is obtained by the foldings of these simply-laced diagrams.

Before analyzing the characteristic function χ_W further, we state another fact, which gives a geometric meaning to the weights a, b, c and to the Coxeter number h in case of a simple and elliptic weight system (see Table 2 and 4 for a proof):

Fact 2. *A simple polynomial $f_{\tilde{G}}(x, y, z)$ in* Table 2 *(resp. an equation for an elliptic singularity in* Table 6*) is a weighted homogeneous polynomial of degree h with the weights a, b, c on the variables x, y, z for a simple (resp. elliptic) weight system $(a, b, c; h)$. The simple weight system determines the simple polynomial, up to a homogeneous coordinate change, uniquely. The elliptic weight system determines the equation up to one parameter (=the modulus parameter of elliptic curves).*

§9. Triangle Δ of weight system, geometry and algebra

Summarizing the results of previous sections, we obtain the following triangle among three mathematical objects: weight system, geometry and algebra:

$$\begin{array}{ccc} & \{\text{ Simple weight systems }\} & \\ (5) & \Downarrow\!\!\!\!\swarrow \qquad\qquad \nwarrow\!\!\!\!\Uparrow & \\ \{\text{ Kleinean groups }\} \Rightarrow & \left\{\begin{array}{c}\text{Simple Lie algebras}\\ \text{with simple root basis}\end{array}\right\}. & \end{array}$$

Here, the three arrows are constructed as follows.
1) The correspondence $\swarrow\!\!\!\!\swarrow$ (denoted by $\Phi_{\swarrow\!\!\!\swarrow}$) is given by the pair of the fundamental group $\pi_1(X_0 \setminus \{0\})$ for the hypersurface X_0 defined by the polynomial in Table 2 and its action to the covering space $\tilde{X}_0$ (use §1 Theorem, Fact 2 and a theorem due to Mumford [Mu1], see Appendix).

2) The correspondence $\Rightarrow$ (denoted by $\Phi_{\Rightarrow}$) is given in three different ways (depending on the view points), all of which give the same result:

a) Use the Du Val diagram for the simple singularity (§1 and 2) and obtain the diagram of the simple root basis of the simple Lie algebra,

b) Use the set of vanishing cycles for the singularity (§5) and obtain the set of real roots of the simple Lie algebra,

c) Use the McKay correspondence ([Mc], see Appendix) and obtain the Dynkin graph for the simple Lie algebra.

Here, the first two approaches a) and b) are equivalent due to Brieskorn's theorem (recall §7 1),2) and 3)). The third approach c) gives the dual basis of the basis given by a) with respect to the Killing form (see Appendix), but is more direct algebraic construction.

3) The correspondence $\Nwarrow$ (denoted by $\Phi_{\Nwarrow}$) is given by the decomposition (2) of the generating function (1) of the exponents (Table 7) of the root system of the simple Lie algebra.

By a direct inspection of the cases, we see that *a composition of the three arrows* $\Phi_{\Swarrow}$, $\Phi_{\Rightarrow}$ *and* $\Phi_{\Nwarrow}$ *starting at any corner of the triangle* (5) *is an identity.*[28] Here we stress that the key step among the three arrows is the horizontal correspondence $\Phi_{\Rightarrow}$. The others are rather straight forward. As a consequence of this observation, we conclude that

The datum of the set of exponents for a finite root system, which, a priori, is a very small part of the information of the root system, is sufficient to recover the whole root system and the simple Lie algebra. In the same way, the datum of a system W of weights (4) *is sufficient to reconstruct the simple Lie algebra.*

A similar triangle as (5) holds for the triple of elliptic weight systems, Heisenberg groups of rank 2 ([Sa14] II, Appendix) and elliptic Lie algebras ([Sa14] IV). This supports the construction of the elliptic primitive forms and the flat structures from the elliptic Lie algebras. This motivates the author to generalize the triangle by starting with a wider class of weight systems and search for corresponding Lie algebras.

We propose to use the top corner of the triangle (5) as the key to uncover a new class of objects: consider any system W (4) of 4 integers, *relaxing the condition on $\chi_W(T)$* (2) *to be a polynomial to to be a Laurent polynomial. Then, associated to the new weight system, we look for new geometric objects in the left corner and new algebras in the right corner, respectively.* That is: we try to recover the triangle:

$$(6)\qquad \begin{array}{ccc} & \{\text{ Weight system } W \text{ }\} & \\ \Swarrow & & \Nwarrow \\ \{\text{Geometry of } X_W\} & \Longrightarrow & \{\text{Algebra } \mathfrak{g}_W\} \end{array}$$

with *the goal to construct primitive forms and their associated period mappings and automorphic forms* (see [Sa19] for the details on the goal). Actually, without this setting of the goal, the objects and the correspondences in the triangle (6) are ambiguous (see §12). Note that each corner of the triangle is not a category and the correspondences $\Rightarrow$, $\Swarrow$ and $\Nwarrow$

[28] A similar triangle is obtained by replacing the three corners by {elliptic weight systems}, {Heisenberg groups of rank 2 with the extension classes -3,-2,-1} and {Elliptic Lie algebras of type $E_6^{(1,1)}$,$E_7^{(1,1)}$, $E_8^{(1,1)}$ with their simple basis}, where we choose the correspondence b) as for the arrow $\Rightarrow$.

are not functors. However, *we expect a sort of "functoriarity" (yet to be defined) due to the deformation relations among X_W's.*

§10. Top corner of the triangle: regular systems of weights

We start anew by introducing the concept of a regular system of weights.[29]

Definition. A weight system $W = (a,b,c;h)$ (4) satisfying (3) is called *regular* if the function $\chi_W(T) := T^{-h}\frac{(T^h-T^a)(T^h-T^b)(T^h-T^c)}{(T^a-1)(T^b-1)(T^c-1)}$ is a Laurent polynomial in $\mathbb{Z}[T,T^{-1}]$.

We give two basic properties of a regular system of weights in the following Fact 3. and in Fact 4. in the next section. The two properties are equivalent to the definition of the regular systems of weights, and they already attribute to the properties in the right and left corners of the triangle (6), respectively.

We first discuss about the new definition of exponents.

Fact 3.([Sa11]Theorem 1) *A weight system W (4) is regular, if and only if there exist integers $m_1, \cdots, m_\mu$ with $\mu = \mu_W = \frac{(h-a)(h-b)(h-c)}{abc}$ called the rank of W, such that χ_W (2) is developed into the sum of monomials of the form* (1).

We call $m_1, \cdots, m_\mu$ the *exponents* of W[30], which we order: $m_1 \leq \cdots \leq m_\mu$ linearly. By use of the functional equality $T^h\chi_W(T^{-1}) = \chi_W(T)$, one has the duality of exponents:

$$m_i + m_{\mu-i+1} \;=\; h \qquad (i = 1, \cdots, \mu). \tag{7}$$

A fact which is not used in the present paper but shall be of basic importance (see Footnote 22, ii)), is that there exists always an exponent prime to h [Sa,13,18].

The advantage to start from a weight system is that the exponents are a priori defined without an ambiguity of their phases (i.e. $[m_i/h] \in$

[29]This is slightly modified ([Yas]) from the original definition [Sa11]: $\chi_W(T)$ has a pole at most only at $T=0$. Using a relation: $T^h\chi_W(T^{-1}) = \chi_W(T)$, the two definitions are equivalent.

[30]In order to agree with the classical convention in Lie theory (e.g. [Bou]), we have called the integers m_i exponents. However, from a view point of the flat structure on S_φ (recall Footnote 23), one should better call the rational numbers m_i/h exponents. This view point becomes important again, when we consider the category of graded matrix factorizations §16.

$\mathbb{Z}$). The smallest exponent $\min\{m_1, \cdots, m_\mu\}$ is given and denoted by

$$\varepsilon_W \ := \ a+b+c-h. \tag{8}$$

Actually, §8 Fact 1. implies that if $\varepsilon_W > 0$ (resp. $=0$), then automatically one has $\varepsilon_W = 1$ (resp. $=0$) and W is a simple (resp. elliptic) weight system, whose exponents coincide with the exponents of the corresponding finite or elliptic root system.

For each negative integer $\varepsilon < 0$, there always exist a finite number of regular systems of weights having ε as the smallest exponent (see [Sa12, Sa17] Appendix 1,2. for many interesting examples of W with $\varepsilon_W < 0$). In particular, there exist 14+8 regular systems of weights for the case $\varepsilon_W = -1$ having no 0 exponents (see Table 10), on which we shall discuss more in details in the present paper.

We are now to analyze the other corners of the triangle (6). Recall that the finite or elliptic root system cannot be directly constructed from the weight system, but we needed to turn the triangle (5) counter-clockwisely. Similarly, we start with analyzing the left corner of (6) in the next section.

§11. Left corner of the triangle: a geometry of X_W

Finding the objects in the left corner of the triangle (6) and $\Phi_{\swarrow}$ follows from the following characterization Fact 4. of the regularity of a weight system W.

For any given weight system $W = (a, b, c; h)$, consider a weighted homogeneous polynomial

$$f_W(x,y,z) \ := \ \textstyle\sum_{ai+bj+ck=h} c_{ijk} x^i y^j z^k. \tag{9}$$

Fact 4. ([Sa11]Theorem 3) *The weight system W (4) is regular, if and only if there exists a polynomial f_W of the form (9) such that the quotient ring:*

$$J_W \ := \ \mathbb{C}[x,y,z]/\left(\frac{\partial f_W}{\partial x}, \frac{\partial f_W}{\partial y}, \frac{\partial f_W}{\partial z}\right), \tag{10}$$

called the Jacobi ring of f_W, is of finite rank μ_W over $\mathbb{C}$.

"If" part of the statement is trivial. Actually, any polynomial (9) with generic coefficients carries this property.

In fact, Fact 4. is trivially equivalent to that the hypersurface

$$X_{W,0} \ := \ \{(x,y,z) \in \mathbb{C}^3 \mid f_W(x,y,z) = 0\} \tag{11}$$

has an isolated singular point at the origin, i.e. $X_{W,0}$ is smooth except at the origin $0 \in X_{W,0}$, due to the Nullstellensatz of Hilbert.

Let us call f_W in Fact 4. a *polynomial of type W*. We employ the hypersurface $X_{W,0}$ (11) with an isolated singular point at 0 and admitting a $\mathbb{C}^\times$-action[31]:

$$\lambda \in \mathbb{C}^\times \ : \ (x,y,z) \ \mapsto \ (\lambda^a x, \lambda^b y, \lambda^c z)$$

as for the object in the left corner of the triangle (6). Following the history in §2-7, we analyze $X_{W,0}$ from two a) algebraic and b) topological view points.

a) **Orbi-bundle $K_{C_W}^{\frac{1}{\varepsilon_W}}$ over the curve C_W.**

There are many studies on surface singularities with a good $\mathbb{C}^\times$-action (e.g. [Dol1,2,3,4], [Pin4,5], [Sa11,12,16], [Wa,1,2]). We recall a few results of them, which are necessary in our purpose. First, we remark that the smoothness of $X_0\setminus\{0\}$ implies that the quotient variety

$$(12) \qquad C_W \ := \ (X_{W,0} \setminus \{0\})/\mathbb{C}^\times \ = \ \mathrm{Proj}(\mathbb{C}[x,y,z]/(f_W(x,y,z))$$

is a smooth curve. However, the $\mathbb{C}^\times$-bundle $X_{W,0}\setminus\{0\} \overset{\mathbb{C}^\times}{\to} C_W$ has some finite number of singular fibers (i.e. fixed by some non-trivial finite subgroups, called isotropy groups, of $\mathbb{C}^\times$). In this sense, C_W carries also a structure of an *orbifold curve* (to be precise, an *algebraic stack*). The pair $(g : p_1, \cdots, p_r)$ of the genus g of the curve C_W and the *set*, called the signature set, of the orders of the isotropy groups:

$$(13) \qquad A(W) \ = \ \{p_1, \cdots, p_r\}$$

is called the *signature* of the orbifold ([F-K]pp.182-190). In fact, we have

Fact 5. ([Sa11]Theo.6) *The genus g of the curve C_W is equal to the multiplicity $a_0 := \#\{1 \le i \le \mu \mid m_i = 0\}$ of exponents equal to 0.*

The signature set $A(W)$, up to some $p_i = 1$, is explicitly determined from the weights W arithmetically.[32]

The orbifold Euler number: $2 - 2g + \sum(1/p_i - 1)$ is positive, 0 or negative according to whether ε_W is positive, 0 or negative. Accordingly, the orbifold universal covering of C_W is either $\mathbb{P}_1$, the complex plane $\mathbb{C}$

---

[31]The action is said good since the the exponents of the action a, b, c are positive (or, equivalently, the coordinate ring $R_W := \mathbb{C}[x,y,z]/(f_W)$ is non-negatively graded.

[32]The genus and the signature set of the orbi-curve C_W is explicitly give as follows.

$$a_0 := \#\{(i,j,k) \in \mathbb{Z}^3_{\ge 0} \mid ai + bj + ck = h\} = \#\{1 \le i \le \mu \mid m_i = 0\},$$
$$A(W) := \{a_i \mid \ a_i \not| h, 1 \le i \le 3\} \amalg \{\ \gcd(a_i, a_j) * (m(a_i, a_j; h) - 1), \ 1 \le i < j \le 3\}$$

where $\{a_1, a_2, a_3\} = \{a, b, c\}$ and $m(a,b;h) = \#\{(u,v) \in \mathbb{Z}^2_{\ge 0} \mid au + bv = h\}$.

or the complex upper half plane $\mathfrak{H} := \{z \in \mathbb{C} \mid \mathrm{Im}(z) > 0\}$. Then, for a weight system W with $\varepsilon_W \neq 0$, we have the description of the $\mathbb{C}^\times$-bundle: $X_{W,0} \setminus \{0\} \to C_W$ ([Dol3]Prop.1, [Sa11](5.5)Lemma)

Fact 6. *Let W be a regular system of weights. According as $\varepsilon_W > 0$ or < 0, one has the following natural commutative diagrams, respectively.*

$$(14) \qquad \begin{array}{ccc} K_{\mathbb{P}^1}^{\frac{1}{\varepsilon_W}} \setminus \text{0-section} & \xrightarrow[/\tilde{\Gamma}_W]{} & X_{W,0} \setminus \{0\} \\ \downarrow /\mathbb{C}^\times & & \downarrow /\mathbb{C}^\times \\ \mathbb{P}^1 & \xrightarrow[/\tilde{\Gamma}_W]{} & C_W \end{array} \quad \text{and} \quad \begin{array}{ccc} K_{\mathfrak{H}}^{\frac{1}{\varepsilon_W}} \setminus \text{0-section} & \xrightarrow[/\tilde{\Gamma}_W]{} & X_{W,0} \setminus \{0\} \\ \downarrow /\mathbb{C}^\times & & \downarrow /\mathbb{C}^\times \\ \mathfrak{H} & \xrightarrow[/\tilde{\Gamma}_W]{} & C_W. \end{array}$$

Here, 1) $K_{\mathbb{P}^1}^{\frac{1}{\varepsilon_W}}$ and $K_{\mathfrak{H}}^{\frac{1}{\varepsilon_W}}$ is a ε_Wth root of the canonical bundle of $\mathbb{P}_1$ or $\mathfrak{H}$, respectively, and 2) $\tilde{\Gamma}_W$ is a co-compact discrete subgroup of $SU(2)$ or $PSL(2,\mathbb{R})$, whose actions on $\mathbb{P}^1$ or $\mathfrak{H}$ are liftable to the bundles (Footnote 33), respectively.

The action of $\tilde{\Gamma}_W$ on $\mathbb{P}_1$ or $\mathfrak{H}$ may have fixed points such that the quotient map $/\tilde{\Gamma}_W$ gives the orbifold universal covering of C_W. That is: the signature of the group $\tilde{\Gamma}_W$ ([Ma]) coincides with that $(a_0 : A(W))$ of the orbifold curve C_W.[33]

These imply that C_W in a Deligne-Mumford stack. They give the "algebraic data" of the geometry of $X_{W,0}$ for $\varepsilon_W \neq 0$.[34]

Example. Case $\varepsilon_W > 0$ (i.e. W is a simple weight system in Table 8). Then, we naturally have $K_{\mathbb{P}^1}^{\frac{1}{\varepsilon_W}} \setminus \{0\} \simeq \mathbb{C}^2 \setminus \{0\}$, and the $\tilde{\Gamma}_W$ action in LHS is identified with the Kleinean group $\tilde{G}$-action in RHS (recall §1). I.e. the liftablity condition in Fact 6. is automatically satisfied). The induced action of $\tilde{\Gamma}_W$ on $\mathbb{P}_1 = (\mathbb{C}^2 \setminus \{0\})/\mathbb{C}^\times$ is identified with the

---

[33] We have a similar geometry for $\varepsilon_W = 0$. Namely, the three simply elliptic singularities of types $\tilde{E}_6, \tilde{E}_7$ and $\tilde{E}_8$ are quotients of the trivial $\mathbb{C}^\times$ bundle over $\mathbb{C}$ by an action of a Heisenberg group of rank 2 of characteristic class -3,-2 and -1, respectively ([Sa2],[Sa14]II Appendix).

[34] To be exact, there remains still the problem to characterize (or, to list up) the pair $(\tilde{\Gamma}_W, \varepsilon)$ of a number $\varepsilon \in \mathbb{Z}_{<0}$ and a co-compact Fuchsian group $\tilde{\Gamma}_W \subset PSL(2,\mathbb{R})$ such that the action of $\tilde{\Gamma}_W$ on $\mathfrak{H}$ is liftable to that on $K_{\mathfrak{H}}^{1/\varepsilon}$. This condition on $(\tilde{\Gamma}_W, \varepsilon)$ (in order to obtain a Gorenstein normal surface singularity $K_{\mathfrak{H}}^{1/\varepsilon} // \tilde{\Gamma}_W^*$) is equivalent to finding a splitting factor $\tilde{\Gamma}_W^*$ in $\tilde{\Gamma}_d$ of the central extension $1 \to \mathbb{Z}/\varepsilon\mathbb{Z} \to \tilde{\Gamma}_d \to \tilde{\Gamma}_W \to 1$ (see [Sa12] (5.2)(5.3) and (5.4)). To list the cases when $K_{\mathfrak{H}}^{1/\varepsilon} // \tilde{\Gamma}_W^*$ is a hypersurface requires further works (e.g. [Dol2] [Wa1]) which is generally unsolved yet. To remain in the category of Gorenstein singularities seems theoretically easier and natural.

regular polyhedral group G-action on S^2. So, there are three singular orbits {centers of faces of the polyhedron}, {centers of edges of the polyhedron} and {vertices of the polyhedron} of fixed points on $\mathbb{P}_1$. Therefore, the signature set $A(W)$ (13) consists of the three numbers p, q, r in Table 1.

Similarly, in case $\varepsilon_W < 0$, the Fuchsian group $\tilde{\Gamma}_W$ has elliptic fixed points in $\mathfrak{H}$, whose orbits correspond in 1:1 to the elements of $A(W)$.

b) **Generalized root system and Coxeter-Killing transformation.**

We discuss about some of the topological data obtained from the semi-universal deformation (also, called universal unfolding) of $X_{W,0}$.

1) Generalized root system

Let us denote by Q_W the lattice $(\mathrm{H}_2(X_{W,1}, \mathbb{Z}), I = -(\cdot,\cdot))$ of vanishing cycles[35] and by R_W the set of vanishing cycles for f_W (which depend only on W but not on a choice of f_W). As is explained already, it is easy to see that R_W *satisfies the axiom of generalized root system having* Q_W *as its root lattice in the sense* [Sa14] I. The following 2) and 3) describe some strong properties carried by R_W. However, we do not know a characterization of a root system which arises as the set of vanishing cycles associated to a singular point.

2) Coxeter-Killing transformation

The Milnor monodromy induces an automorphism $\mathbf{c}$ of finite order h of the lattice Q_W, which we shall call also the Coxeter-Killing transformation of the root system R_W. Using the weighted homogeneity of the defining equation f_W, it is easy to see that the characteristic polynomial $\det(\lambda \cdot \mathrm{id}_Q - \mathbf{c})$ is given by

$$(15) \qquad \varphi_W(\lambda) \;=\; \prod_{i=1}^{\mu}(\lambda - \exp(2\pi\sqrt{-1}\tfrac{m_i}{h})) \;\in\; \mathbb{Z}[\lambda].$$

[35]The middle homology group $\mathrm{H}_2(X_{W,1}, \mathbb{Z})$ admits the symmetric bilinear form, called the intersection form, $(u, v) := \langle u, P(v)\rangle$ obtained from the Poincare duality $P : \mathrm{H}_2(X_{W,1}, \mathbb{Z}) \to \mathrm{H}^2(X_{W,1}, \mathbb{Z})$. In the above definiton of the lattice Q_W, we put the minus sign factor in order to adjust with the classical convention in the Killing form that $I(e, e) = 2$ for any vanishing cycle e. The signature (μ_-, μ_0, μ_+) of I is given by $\mu_- = \#\{1 \le i \le \mu \mid m_i < 0 \text{ or } h < m_i\}$, $\mu_0 = \#\{1 \le i \le \mu \mid m_i = 0 \text{ or } h\} = 2a_0$, $\mu_+ = \#\{1 \le i \le \mu \mid 0 < m_i < h\}$, ([Sai]). Then the *Witt index* (=the maximal rank of totally isotropic subspace) of $\mathrm{H}_2(X_{W,1}, \mathbb{Z})$= $\mu_0 + \mu_- = \#\{$exponents exceeding the interval $(0, h)\}$ is always *even*. This fact supports the existence of the Coxeter-Killing transformation of finite order and to ask for Chevalley type invariant theory to the algebra $\mathfrak{g}_W$ in §12 iv).

The set $\{\exp(2\pi\sqrt{-1}m_i/h) \mid i=1,\cdots,\mu\}$ is closed under the action of the Galois group over $\mathbb{Q}$.[36] Recall that:

Fact 7. ([Sa, 13(Theorem 1), 18(Theorem 5.1)]) *Let us denote by $e_W(h)$ the multiplicity of the hth primitive roots of unity in the roots of the equation $\varphi_W(\lambda) = 0$. Then, for any regular system of weights W, one has $e_W(h) > 0$.*

Remark 4. In the classical simple Lie algebra case, the eigenvector of the Coxeter-Killing transformation belonging in to the hth primitive root of unity (in the Cartan subalgebra of $\mathfrak{g}_W$) is regular with respect to the adjoint action of the simple Lie group and that of the Weyl group. This gives a key role to the vector in the invariant theory by Kostant [Ko], Springer [Sp], Steinberg [St1] as well as in the construction of the primitive form and the flat structure [Sa18].

3) Root basis

Any distinguished basis $(e_1,\cdots,e_{\mu_W})$ (recall §5) gives a root basis of the root system R_W in the sense: i) $R_W = \cup_{i=1}^{\mu_W}\langle w_{e_1},\cdots,w_{e_{\mu_W}}\rangle \cdot e_i$, and ii) the Coxeter-Killing transformation is given by the product $w_{e_1}\cdots,w_{e_{\mu_W}}$. This implies: iii) $Q_W=\oplus_{i=1}^{\mu_W}\mathbb{Z}e_i$ and iv) $\langle w_{e_1},\cdots,w_{e_{\mu_W}}\rangle$ coincides with the group generated by reflections for all $e\in R_W$ (=the Weyl group of the root system R_W).

As we saw already in §5, the braid group of μ_W-strings acts on the set of distinguished basis. It is desirable to find some "simple" basis for the root system R_W by the use of the action. There are several works in the direction by Gabrielov [Gab 1,2], Ebeling [Eb 1,2], Kluitman [Klu] and others. However, purely topological data of the braid group action alone seems insufficient to choose some distinguished ones. On the other hand, one may still have a hope to choose some particular basis, either by a use of semi-algebraic geometry of the discriminant of the family $\mathcal{X}_\varphi \to S_\varphi$ (see Footnote 12 and [Sa20]), or by the algebraic approach a) by a use of the orbifold structure on C_W given in the first half of the present §. The study of this subject belongs still to a future work.

4) Cycles from ∞.

We already discussed about the cycles from infinity in §5 Remark2. Under the setting of a regular system of weights W, let us discuss again about it.

Let us define explicitely a universal unfolding of f_W by

$$F(\underline{x},\underline{t}) \ := \ f_W(\underline{x}) + \varphi_1(\underline{x})\cdot t_1 + \varphi_2(\underline{x})\cdot t_2 + \cdots + 1\cdot t_{\mu_W}$$

[36] This is shown as follows. Substitute any power of $\exp(2\pi\sqrt{-1}/h)$ in (1). (2) implies that it is a rational number.

where $\varphi_1, \varphi_2, \cdots, \varphi_\mu \equiv 1$ are weighted homogeneous polynomials in $\mathbb{C}[x, y, z]$ (with respect to the weights (a, b, c)) such that their images in the Jacobi ring J_W (10) gives its $\mathbb{C}$-basis. Clearly, the function in a neighborhood of origin gives the universal unfolding in the sense explained in §5. However, we remark that $F(\underline{x}, \underline{t})$ is affine globally defined, where, by putting $\deg(t_i) := h - \deg(\varphi_i) = m_i + \varepsilon_W$ $(1 \le i \le \mu_W)$, it is a weighted homogeneous polynomial. The lowest degree coordinate is t_1 and its degree is equal to $2\varepsilon_W$. That is, the unfolding parameter $\underline{t}$ gets negative weights if (and only if) $\varepsilon_W < 0$. Consider the affine global family of affine surfaces: $\hat{\varphi}_W : \hat{\mathcal{X}}_W \to S_W$, where $\hat{\mathcal{X}}_W := \{(\underline{x}, \underline{t}) \in \mathbb{C}^3 \times \mathbb{C}^{\mu_W} \mid F(\underline{x}, \underline{t}) = 0\}$, $S_W := \mathbb{C}^{\mu_W}$ and $\hat{\varphi}$ is the projection to the second factor. The discriminants of $\hat{\varphi}_W$ is a divisor of S_W and decomposes into a union of $D_{W,+}$, $D_{W,0}$ and $D_{W,-}$ according as the behavior of the vanishing cycles vanishing at the components (see [Sa19]II §6). Then, as was shown in §5, the lattice Q_W of middle homology group of the Milnor fiber is generated by the vanishing cycles which are degenerating to the discriminant $D_{W,+}$. Then, the extension $\hat{Q}_W := (\mathrm{H}_2(\hat{X}_{W,\underline{t}}, \mathbb{Z}), -I)$ for a generic parameter value $\underline{t}$ such that the coordinate component $t_1 \neq 0$ is a orthogonal direct sum of the lattice Q_W with the lattice Q_W^∞ generated by the vanishing cycles which are degenerating to the discriminant $D_{W,-}$. It is expected that the periods of the cycles in Q_W^∞ give the denominators for primitive forms ([Sa19]II §6 Conjecture and Problem).

Remark 5. The concept of the generalized root system of vanishing cycles and the braid group action on its basis may better be lifted to a categorical level due to the recent developments of the study of Floer homology groups of Lagrangean subvarieties in symplectic varieties [Sei].[37]

Remark 6. As we shall see in §16, for weight systems W having its $*$-dual, the lattices Q_W and Q_W^∞ are expected to have a categorical construction as the K-groups of the category of the graded and un-graded matrix factorizations, respectively, where the Coxeter-Killing transf. is defined as the A-R translation.

§12. Right corner of the triangle: an algebra $\mathfrak{g}_W$

We now come to the main question of the present paper:

[37]A comprehensive treatment of this subject shall appear in: K. Fukaya, Y.-G. Oh, H. Ohta, K. Ono: Lagrangian intersection Floer theory - anomaly and obstruction, in preparation.

Question. *For any regular system of weights W, define the correspondences $\Phi_{\Rightarrow}$ and $\Phi_{\Nwarrow}$ which make the triangle* (6) *commutative. Precisely, construct the algebra $\mathfrak{g}_W$ from the data of the geometry of X_W, satisfying the following conditions* i) - vi).

Then, we automatically have $\Phi_{\Nwarrow}$ and commutativity of the triangle.

We impose some working hypothetical conditions i)-vi) on the algebra $\mathfrak{g}_W$; otherwise the question is ambiguous. Under these constraints, we expect a sort of functoriality and uniqueness for the correspondence $\Phi_{\Rightarrow}$ (recall §9).[38]

i) The algebra $\mathfrak{g}_W$ should be a simple Lie algebra for the case $\varepsilon_W > 0$ and a elliptic Lie algebra for the case $\varepsilon_W = 0$.

ii) The algebra $\mathfrak{g}_W$ should carry an integrability structure, generalizing the Jacobi identity for Lie algebras (i.e. $\mathfrak{g}_W$ should be the prolongation of X_W with respect to the equations for a primitive form; see the last paragraph in §4).

iii) $\mathfrak{g}_W$ should contain an abelian subalgebra $\mathfrak{h}_W$ isomorphic to $\mathrm{Hom}(Q_W, \mathbb{C})$ (which we may call the Cartan-Killing subalgebra of $\mathfrak{g}_W$). The adjoint action of $\mathfrak{h}_W$ on $\mathfrak{g}_W$ induces the root space decomposition of $\mathfrak{g}_W$ so that R_W should be the set of real roots (i.e. a root $\alpha \in Q_W$ such that $I(\alpha,\alpha) > 0$), whose multiplicities are equal to 1. The real root spaces $\mathfrak{g}_{W,\alpha}$ for $\alpha \in R_W$ generate the algebra $\mathfrak{g}_W$.

iv) Depending on a choice (*Note* 3. below), one should have a family of Chevalley type invariant theories for the adjoint group G_W action on $\mathfrak{g}_W$ and the adjoint quotient morphism with the identification of the quotient varieties $\tilde{\mathfrak{g}}_W//\mathrm{Ad}(\tilde{\mathfrak{g}}_W) \simeq \tilde{\mathfrak{h}}_W//\tilde{W}_W$. Here, $\tilde{\mathfrak{g}}_W$, $\tilde{\mathfrak{h}}_W$ and $\tilde{W}_W$ are suitable *hyperbolic extensions* of $\mathfrak{g}_W$, $\mathfrak{h}_W$ and W_W, if the Killing form I has a degeneration .[39]

---

[38]Beside the classical construction of semi-simple Lie algebras, there are several new approaches, e.g. using vertex operators [Bo1], or using Ringel-Hall algebras [P-X], as was discussed in Preface. However, in connection with our final goal (the construction of primitive forms), we would like to be cautious in choosing the type of construction.

[39]One supporting reason for this condition is the following fact ([Sa13](2.2) Theorem1, [Sa17] Theorem5.1 and (5.6)): *for any regular system of weights, there always exists an exponent which is prime to the Coxeter number h.* This generalizes the existence of an eigenvalue of a primitive hth root of unity of the Coxeter-Killing transformation $\mathbf{c}$ in the classical case [Col] [Bou]Ch.v§6 Theorem 1. This is a key fact for the construction of the adjoint quotient morphism and for the global construction of the flat structure (see Footnotes 18,19).

v) The universal unfolding $\mathcal{X}_W \to S_W$ of the singularity $X_{W,0}$ should be embedded into the adjoint quotient map $\mathfrak{g}_W \to \tilde{\mathfrak{h}}_W//\tilde{W}_W$ (c.f. §4 when $\varepsilon_W > 0$). The relative (with respect to the adjoint quotient map) symplectic form on $\mathfrak{g}_W$ (Kostant-Kirirov form when $\varepsilon_W > 0$) induced from the involutive structure given in ii) should (up to a unit factor) induce a primitive form on the family $\mathcal{X}_W \to S_W$, whose exponents (recall Footnote 23) coincide with the exponents of the weight system W (up to the factor h).

vi) The flat structure on the quotient variety $\tilde{\mathfrak{h}}_W//\tilde{W}_W$ (c.f. [Sa16], [Sa14] II) and the flat structure on S_W defined from the theory of primitive forms [Sa7] should be identified by the isomorphism in iv). This, in particular, requires that the set of exponents for the primitive form on $X_{W,0}$ should coincides with the set of exponents associated to the flat structure of the algebra $\mathfrak{g}_W$.

The last condition vi) implies that the generating function (1) ([Sa7] (3.3.14)) of the exponents for the flat structures of the algebra $\mathfrak{g}_W$ decomposes as in (2), and defines the weight system $W = (a, b, c; h)$, which we had at the beginning. That is, *the correspondence* $\Phi_{\Nwarrow}$ *of the triangle* (6) *is defined by the use of the decomposition of the generating function* (1) *of the exponents of the algebra. Then, the composition* $\Phi_{\Nwarrow} \circ \Phi_{\Rightarrow} \circ \Phi_{\Swarrow}$ *is the identity on the top of the triangle* (6). Thus, we shall obtain a family of primitive forms having the exponents given at the beginning by a regular system of weights, when the problem is solved.

Obviously, the simple Lie algebra $\mathfrak{g}_W$ of type W for a simple weight system W satisfies all conditions i)-vi). The elliptic algebra $\mathfrak{g}_W$ for an elliptic weight system W satisfies i), ii) and iii), and the flat structure on $\tilde{\mathfrak{h}}_W//\tilde{W}_W$ has been constructed. However, the construction of the adjoint quotient space $\tilde{\mathfrak{g}}_W//\mathrm{Ad}(\tilde{\mathfrak{g}}_W)$ is still a work in progress (see Footnote 20).

For general weight system W, we introduce in §16 a category $\mathrm{HMF}^{gr}_{A_W}(f_W)$, which is expected to give three constructions of Lie algebras. We ask to clarify the relationship among the constructions, and whether they satisfy i)-vi) (up to the $*$-duality which we shall introduce in §14) (see Problem at the end of §18).

On the other hand, elliptic root systems have a radical of rank 2. Then, depending on the choice of its rank 1 subspace, called a marking, one defines the extensions $\tilde{\mathfrak{g}}_W$, $\tilde{\mathfrak{h}}_W$ and $\tilde{W}_W$ (see [S-Y],[Sa14]I,II,[S-T]). These extensions, called hyperbolic, are necessary for the construction of the flat structure [Sa14]II as well as in the representation theory and invariant theory of the elliptic algebra.

Note. 1. If the Killing form I on the root lattice $Q_W = \mathrm{H}_2(X_{W,1}, \mathbb{Z})$ degenerates ($\Leftrightarrow$ the genus a_0 of the curve C_W is positive, see Fact 5, Footnotes 32 and 35), then the algebra $\mathfrak{g}_W$ may have a "radical" (corresponding to the moduli parameter of the curve C_W). In that case, as for $\mathfrak{g}_W$, we assign the universal algebra (i.e. the one having the largest radicals) for the unicity of the notation $\mathfrak{g}_W$.

2. The other problem in answering **Question** is, which view-point a) or b) in §9 do we generalize? It seems likely that, in the above iii), the two view points a) and b) give two different root systems and two different algebras. Let us tentatively denote by $\Phi\overset{a)}{\Rightarrow}$ the correspondence using the algebraic geometric data of the singularity $X_{W,0}$ and by $\Phi\overset{b)}{\Rightarrow}$ the correspondence using the topological data of the deformations of $X_{W,0}$. In fact, these two different view points are, nowadays, called mirror symmetric to each other (see [Kon1], [Yau] for mirror symmetry in general). There is a duality operation on the set of weight systems, called the $*$-duality, which conjecturally exchanges the two approaches (see §14 **Addition to Question**). Then, the conditions iv) and v) on the period map seem to choose $\Phi\overset{b)}{\Rightarrow}$ for the correspondence $X_W \Rightarrow \mathfrak{g}_W$.

3. The denominator of an elliptic primitive form depends on a choice of a primitive element in the radical of the root lattice ([Sa7] (3.1) Example), which determines the polarization (marking [Sa14] I) of the elliptic root system. Similarly, the primitive form for the 14 exceptional unimodular singularities is conjectured to be a proportion of a form with its integral over the cycle coming from infinity (see [Sa19]II 6. Conjecture, §11b) 4) and Footnote 49).

§13. Strange duality of Arnold

In order to get an insight to the Question in §12 and also to sharpen it by Addition to Question in the next §14, we look closely at the case $\varepsilon_W = -1$ in this section where the singularities are called exceptional unimodular singularties. We recall the *strange duality* among the 14 cases due to Arnold [Ar4].

There are 14+8+9 regular systems of weights of $\varepsilon_W = -1$, where the first 14+8 cases are genus $a_0 = 0$ and the remaining 9 cases are positive genus $a_0 > 0$. The multiplicity $e_W(h)$ of the first 14 weight systems is equal to 1 and that of the next 8 weight systems are either equal to 2 or 3. Accordingly, the signature set $A(W)$ (Footnote 32) consists of 3 elements for the first 14 cases, and of 4 or 5 elements for the 8 cases (where f_W depends on parameter(s)). (see [Sa11, Tables 3,4 and 5] for details on the geometry of them in the sense of §11). In the present paper, we study only the 14+8 cases where genus a_0 is zero.

Table 10. 14+8 regular systems of weights of genus $a_0 = 0$ and $\varepsilon_W = -1$

Polynomial f_W	$(a, b, c; h)$	exponents
$x^7 + y^3 + z^2$:	$(6, 14, 21; 42)$	$-1, 5, 11, 13, 17, 19, 23, 25, 29, 31, 37, 43$
$yx^5 + y^3 + z^2$:	$(4, 10, 15; 30)$	$-1, 3, 7, 9, 11, 13, 15, 17, 19, 21, 23, 27, 31$
$x^4 z + y^3 + z^2$:	$(3, 8, 12; 24)$	$-1, 2, 5, 7, 8, 10, 11, 13, 14, 16, 17, 19, 22, 25$
$x^5 + xy^3 + z^2$:	$(6, 8, 15; 30)$	$-1, 5, 7, 11, 13, 15, 17, 19, 23, 25, 31$
$yx^4 + xy^3 + z^2$:	$(4, 6, 11; 22)$	$-1, 3, 5, 7, 9, 11, 11, 13, 15, 17, 19, 23$
$x^3 z + xy^3 + z^2$:	$(3, 5, 9; 18)$	$-1, 2, 4, 5, 7, 8, 9, 10, 11, 13, 14, 16, 19$
$x^5 + y^2 z + z^2$:	$(4, 5, 10; 20)$	$-1, 3, 4, 7, 8, 9, 11, 12, 13, 16, 17, 21$
$yx^4 + y^2 z + z^2$:	$(3, 4, 8; 16)$	$-1, 2, 3, 5, 6, 7, 8, 9, 10, 11, 13, 14, 17$
$x^4 + y^3 + xz^2$:	$(6, 8, 9; 24)$	$-1, 5, 7, 8, 11, 13, 16, 17, 19, 25$
$x^3 y + y^3 + xz^2$:	$(4, 6, 7; 18)$	$-1, 3, 5, 6, 7, 9, 11, 12, 13, 15, 19$
$x^3 z + y^3 + xz^2$:	$(3, 5, 6; 15)$	$-1, 2, 4, 5, 5, 7, 8, 10, 10, 11, 13, 16$
$x^4 + y^2 z + z^2 x$:	$(4, 5, 6; 16)$	$-1, 3, 4, 5, 7, 8, 9, 11, 12, 13, 17$
$x^3 y + y^2 z + z^2 x$:	$(3, 4, 5; 13)$	$-1, 2.3.4.5.6.7.8.9.10.11.14$
$x^4 + yz(y - z)$:	$(3, 4, 4; 12)$	$-1, 2, 3, 3, 5, 6, 6, 7, 9, 9, 10, 13$
$y(y - x^3)(y - \lambda x^3)$:	$(2, 6, 9; 18)$	$-1, 1, 3, 5, 5, 7, 7, 9, 9, 11, 11, 13, 15, 17, 19$
$xy(y - x^2)(y - \lambda x^2) + z^2$:	$(2, 4, 7; 14)$	$-1, 1, 3, 3, 5, 5, 7, 7, 7, 9, 9, 11, 11, 13, 15$
$y(y - x^2)(y - \lambda x^2) + xz^2$:	$(2, 4, 5; 12)$	$-1, 1, 3, 3, 4, 5, 5, 7, 7, 8, 9, 9, 11, 13$
$(y^2 - x^3)(y^2 - \lambda x^3) + z^2$:	$(2, 3, 6; 12)$	$-1, 1, 2, 3, 4, 5, 5, 6, 7, 7, 8, 9, 10, 11, 13$
$x(z - x^2)(z - \lambda x^2) + y^2 z$:	$(2, 3, 4; 10)$	$-1, 1, 2, 3, 3, 4, 55, 6, 7, 7, 8, 9, 11$
$x^3 y + z(z - y)(z - \lambda y)$:	$(2, 3, 3; 9)$	$-1, 1, 2, 2, 3, 4, 4, 5, 5, 6, 7, 7, 8, 10$
$xy(x - y)(y - \lambda_1 x)(y - \lambda_2 x) + z^2$:	$(2, 2, 5; 10)$	$-1, 1, 1, 3, 3, 3, 5, 5, 5, 5, 7, 7, 7, 9, 9, 11$
$y(y - x)(y - \lambda_1 x)(y - \lambda_2 x) + xz^2$:	$(2, 2, 3; 8)$	$-1, 1, 1, 2, 3, 3, 3, 4, 5, 5, 5, 6, 7, 7, 9$

Here λ, λ_1 and λ_2 are parameters $\neq 0, 1$ and $\lambda_1 \neq \lambda_2$.

In order to construct $\Phi_{\Rightarrow}$, remember a) and b) of §9, 1) for the case $\varepsilon_W = 1$. An immediate analogy of $\Phi^{a)}_{\Rightarrow}$ is to study the resolution of the singularity $X_{W,0}$. The minimal resolution $\pi : \tilde{X}_{W,0} \to X_{W,0}$ of the singularity is determined by [Dol1] as follows: the exceptional set $\pi^{-1}(0) \subset \widetilde{X}_0$ of the minimal resolution is a union of 4-rational curves E_0, E_1, E_2 and E_3, which intersect transversely as illustrated in Table 11 and are self-intersecting as

$$-1 = E_0^2, \quad -p = E_1^2, \quad -q = E_2^2, \quad -r = E_3^2$$

where p, q, r are positive integers such that $(0 : p, q, r)$ is the signature of the orbifold curve C_W (§11 Fact 5., [Dol 1,2,3,4], [Pin4,5],[Sa 11,12]). The signature set $A(W) = \{p, q, r\}$ (13), in this particular case, was called the *Dolgachev numbers* [Ar3] [Dol1], which are listed in Table 13.

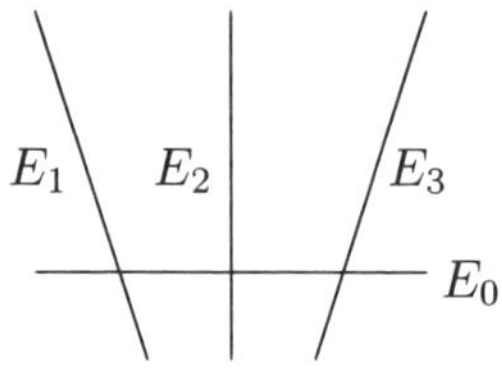

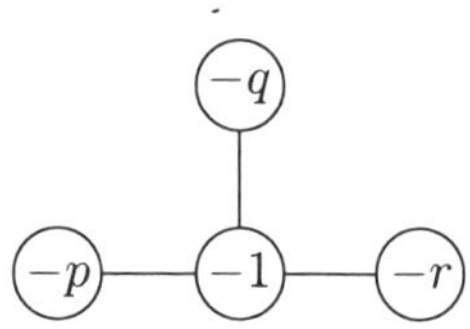

Table 11. Exceptional set Resolution diagram

On the other hand, as for the correspondence $\Phi^{b)}_{\Rightarrow}$, one should know the set of vanishing cycles in the lattice $\mathrm{H}_2(X_{W,1},\mathbb{Z})$, whose signature is $(2,0,*)$ (Footnote 35.). The *distinguished basis* of the lattices were studied by the authors A.M. Gabrielov [Gab1,2] and W. Ebeling [Eb1,2]. In particular, they found certain "simple"[40] distinguished basis for each of the 14 exceptional singularities, which is expressed by the following diagram[41]:

Table 12. Distinguished basis for the exceptional unimodular singularities.

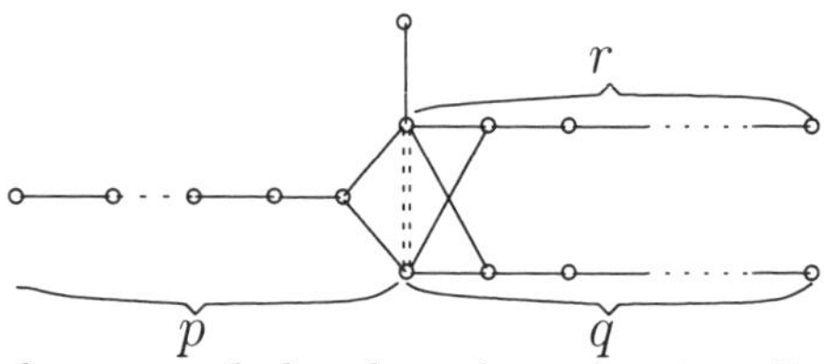

where the length p,q,r of the three branches is called the *Gabrielov numbers* [Ar3].[42]

[40]Here, we mean by "simple" the following: 1) the vertices of the diagram is a $\mathbb{Z}$-basis of the lattice $H_2(X,\mathbb{Z})$, 2) a product in suitable order of the reflections on the lattice attached to the vertices of the diagram (i.e. a Coxeter-Killing transformation) is of finite order h and its eigenvalues are $\exp(2\pi\sqrt{-1}m_i/h)$ for the exponents m_i, 3) consider the group W acting on the lattice generated by the reflections attached to the vertices of the diagram. Then the set of the vanishing cycles is equal to the union of the W-orbits of the vertices of the diagram.

[41]Let $e_1,\cdots,e_\mu$ be a basis of a lattice Q_W such that $I(e_i,e_i)=2$ for $i=1,\cdots,\mu$. Then, we associate a *diagram* to the basis as follows: to each basis element e_i for $i=1,\cdots,\mu$, we associate ith vertex of the diagram. Between ith and jth vertices of the diagram, we draw $-I(e_i,e_j)$ edges if $I(e_i,e_j)<0$, $I(e_i,e_j)$ dotted edges if $I(e_i,e_j)>0$ and no edges if $I(e_i,e_j)=0$.

[42]There is a strong reason to suspect that the diagram should be (a part of) the correspondence $\Phi_{\Rightarrow}$ for the 14 weight systems, since it partially answers to the questions iv) and v) in §12 as follows. Let $\xi,\bar{\xi}$ be the eigenvectors of the Coxeter-Killing transformation belonging to the eigenvalues $\exp(\pm 2\pi\sqrt{-1}/h)$. Then each belongs to the two connected component of the cone $\{x\in Q_W\otimes_{\mathbb{Z}}\mathbb{C}\mid I(x,\bar{x})<0, I(x,x)=0\}$ over a symmetric domain of type IV and is regular

Then, Arnold [Ar3] observed the following duality and called it the **strange duality**: *there exists an involutive bijection σ on the set of 14 exceptional singularities, by which Dolgachev numbers and Gabrielov numbers interchange.*

In the next table, we indicate the involution σ by the two-sided arrows $\leftrightarrow$ between the weight systems corresponding to the singularities.

	Weights	$A(W)$=Dolgachev#	Gabrielov#	$\varphi_W(\lambda)$
↺	(3, 4, 4; 12)	4, 4, 4	4, 4, 4	$\frac{(\lambda^{12}-1)(\lambda^4-1)}{(\lambda^3-1)(\lambda-1)}$
↺	(3, 4, 5; 13)	3, 4, 5	3, 4, 5	$\frac{(\lambda^{13}-1)}{(\lambda-1)}$
	(4, 5, 6; 16)	2, 5, 6	3, 4, 4	$\frac{(\lambda^{16}-1)(\lambda^4-1)}{(\lambda^8-1)(\lambda-1)}$
↺	(3, 5, 6; 15)	3, 3, 6	3, 3, 6	$\frac{(\lambda^{15}-1)(\lambda^3-1)}{(\lambda^5-1)(\lambda-1)}$
	(4, 6, 7; 18)	2, 4, 7	3, 3, 5	$\frac{(\lambda^{18}-1)(\lambda^3-1)}{(\lambda^9-1)(\lambda-1)}$
	(6, 8, 9; 24)	2, 3, 9	3, 3, 4	$\frac{(\lambda^{24}-1)(\lambda^4-1)(\lambda^3-1)}{(\lambda^{12}-1)(\lambda^8-1)(\lambda-1)}$
	(3, 4, 8; 16)	3, 4, 4	2, 5, 6	$\frac{(\lambda^{16}-1)(\lambda^2-1)}{(\lambda^4-1)(\lambda-1)}$
↺	(4, 5, 10; 20)	2, 5, 5	2, 5, 5	$\frac{(\lambda^{20}-1)(\lambda^5-1)(\lambda^2-1)}{(\lambda^{10}-1)(\lambda^4-1)(\lambda-1)}$
	(3, 5, 9; 18)	3, 3, 5	2, 4, 7	$\frac{(\lambda^{18}-1)(\lambda^2-1)}{(\lambda^6-1)(\lambda-1)}$
↺	(4, 6, 11; 22)	2, 4, 6	2, 4, 6	$\frac{(\lambda^{22}-1)(\lambda^2-1)}{(\lambda^{11}-1)(\lambda-1)}$
	(6, 8, 15; 30)	2, 3, 8	2, 4, 5	$\frac{(\lambda^{30}-1)(\lambda^5-1)(\lambda^2-1)}{(\lambda^{15}-1)(\lambda^{10}-1)(\lambda-1)}$
	(3, 8, 12; 24)	3, 3, 4	2, 3, 9	$\frac{(\lambda^{24}-1)(\lambda^3-1)(\lambda^2-1)}{(\lambda^8-1)(\lambda^6-1)(\lambda-1)}$
	(4, 10, 15; 30)	2, 4, 5	2, 3, 8	$\frac{(\lambda^{30}-1)(\lambda^3-1)(\lambda^2-1)}{(\lambda^{15}-1)(\lambda^6-1)(\lambda-1)}$
↺	(6, 14, 21; 42)	2, 3, 7	2, 3, 7	$\frac{(\lambda^{42}-1)(\lambda^7-1)(\lambda^3-1)(\lambda^2-1)}{(\lambda^{21}-1)(\lambda^{14}-1)(\lambda^6-1)(\lambda-1)}$

Table 13. The strange duality and the $*$-duality.

The strange duality captured the attention of many authors and was interpreted by Dolgachev, Nikulin and Pikham in terms of duality between algebraic cycles and transcendental cycles on certain K3 surfaces [Pin1,2]. Further generalizations of the duality were studied by several authors [N-G][Pin4,5][Lo4][E-W].

In §14, we induce the strange duality from the $*$-duality of weight systems [Sa17], which is interpreted as a mirror symmetry [Ta1].

§14. $*$-duality of regular systems of weights

We introduce one key operation $*$ of the present paper: the $*$-duality on regular systems of weights [Sa17]. It induces the strange duality in the arithmetical level, and induces, a much wider class of dualities among weight systems beyond the strange duality.

---

w.r.t. the Weyl group (i.e. does not belong to any reflection hyperplane ([Sa15] (5.6) Lemma 2). We remark also that the diagram defines a splitting hyperbolic plane of the lattice Q_W.

Recall the *characteristic polynomial* $\varphi_W(\lambda)$ (15) of a regular system of weights W. Since $\varphi_W \in \mathbb{Z}[\lambda]$ (recall §11 b) and Footnote 36) and is a cyclotomic polynomial, there exists a unique expression:

$$(16) \qquad \varphi_W(\lambda) \;=\; \prod_{i|h} (\lambda^i - 1)^{e_W(i)}$$

for some integer $e_W(i) \in \mathbb{Z}$ for all $i \in \mathbb{Z}_{>0}$ with $i|h$, where h is the Coxeter number of W.

Definition. A regular system of weights W' is called $*$-*dual* to W if 1) its Coxeter number h' coincides with that h of W, and 2) one has the duality relation:

$$(17) \qquad e_W(i) + e_{W'}(h/i) \;=\; 0 \qquad \text{for } {}^\forall i \in \mathbb{Z}_{>0}.$$

Here, we put $e_W(i) = e_{W'}(i) = 0$ for an integer i with $i \not| h$.

Example. 1. The characteristic polynomial for the type E_8 decomposes as

$$\varphi_{E_8}(\lambda) \;=\; \frac{(\lambda^{30}-1)(\lambda^5-1)(\lambda^3-1)(\lambda^2-1)}{(\lambda-1)(\lambda^6-1)(\lambda^{10}-1)(\lambda^{15}-1)}.$$

Then $e_{E_8}(30) + e_{E_8}(1) = e_{E_8}(5) + e_{E_8}(6) = e_{E_8}(3) + e_{E_8}(10) = e_{E_8}(2) + e_{E_8}(15) = 0$. This implies E_8 is selfdual. This is a special case of the next 2.

2. All regular weight systems W with $\varepsilon_W > 1$ (i.e. simple weight systems) are selfdual ([Sa17] Theo.7.10.1). This fact resemble the result of Brieskorn in §7, 1). However, the $*$-duality, in general, implies neither of the the homeomorphisms $\tilde{X}_{W,0} \simeq X_{W^*,1}$ nor $\tilde{X}_{W^*,0} \simeq X_{W,1}$ (see the examples below). Therefore, it seems interesting to ask what the natural generalization of [O-O] is for the $*$-dual pair?

3. Any regular system of weights W of rank μ_W equal to 24 is selfdual. It is curious to observe that there are 11 such weight systems with $\varepsilon_W < 0$, and the set of their characteristic polynomials is exactly equal to the set of all selfdual characteristic polynomials of the conjugacy classes of the Conway group $\cdot 0$ ([Sa17] Appendix 1) except for the four 6A, 10A, 15D and 18A.

We have the following uniqueness of the $*$-dual weight system.

Theorem. ([Sa17] The.7.8) 1. *For a regular system of weights W if there exists a $*$-dual weight system, then it is uniquely determined from W, which we denote by W^*. By definition, we have $(W^*)^* = W$.*

2. *The smallest exponent of W^* is equal to that of W, $\varepsilon_W = \varepsilon_{W^*}$.*

3. *The multiplicities $e_W(h)$ and $e_{W^*}(h)$ are equal to 1.*

In general, there does not exist a dual weight system W^* for a given regular system of weights W (eg. if the multiplicity $e_W(h)$ is larger or equal than 2, then W cannot have the dual weight system), but if the $*$-dual for W exists, then it is purely arithmetically determined from W. In fact, we have a complete list of dual pair of regular systems of weights ([Sa17] The.7.9). As a consequence, we can prove the following.

Fact 7. ([Sa17] Theo.7.10.2 & §12) *Any of the 14 regular systems of weights W in* Table 10 *(i.e. $\varepsilon_W = -1$, $a_0 = 0$ and $e_W(h) = 1$) is $*$-dual to a weight system in the same* Table. *Further, if W and W^* are dual, then $\mu_W + \mu_{W^*} = 24$*[43] *and*

$$\text{Dolgachev } \# \text{ of } W = \text{Gabrielov } \# \text{ of } W^*$$
$$\text{Dolgachev } \# \text{ of } W^* = \text{Gabrielov } \# \text{ of } W$$

That is: the strange duality of Arnold is induced from the $*$-duality.

Remark 7. Whether W^* is dual to W or not is determined only by the characteristic polynomials φ_{W^*} and φ_W, and hence, in view of (15), it is determined only by the exponential: $\exp(2\pi\sqrt{-1}m_i/h)$ $(i=1,\cdots,\mu)$ and $\exp(2\pi\sqrt{-1}m_i^*/h)$ $(i = 1,\cdots,\mu^*)$. That is, *the information of the phases $[m_i/h]$, $[m_i^*/h]$ of the exponents are unnecessary to determine the $*$-duality.*

This brings us to a puzzle: we had mentioned (§8 Concern) that the eigenvalues $\exp(2\pi\sqrt{-1}m_i/h)$ of a Coxeter-Killing transformation may not be sufficient to recover the phases of the exponents. This was the main reason why we introduced the concept of regular systems of weights in §10 (but not a root system with a simple basis) as our starting point, since a regular system of weights carries the full data of the set of exponents. From this starting point, we arrive at a result that the phase is unnecessary for the definition of duality among them.

The author does not have a good answer to this puzzle. The only fact, we can mention here is that *a regular system of weights W, which admits the dual W^*, has a peculiarity such that the datum of the set of exponentials $\{\exp(2\pi\sqrt{-1}m_i/h) \mid i=1,\cdots,\mu\}$ is sufficient to recover $\{m_i \mid i=1,\cdots,\mu\}$* (see the proof of [Sa17] The.7.9).

Namely, the uniqueness of the dual weight system can be shown briefly as follows: if a weight system W admits a dual weight system, then the characteristic polynomial $\varphi_W(\lambda)$ is reduced (i.e. $e_W(i) \in \{0, \pm 1\}$ for $i \in \mathbb{Z}_{>0}$). This is equivalent to $e_W(h) = 1$ and we call such W *simple.* On the other hand, a simple weight system W is arithmetically

[43] In the original proof [Sa17] Theorem 7.10, 2., the rank relation was not stated explicitly.

reconstructed from its characteristic polynomial φ_W ([ibid] The.6.3). This proves the uniqueness of W (and W^*).

Perhaps, the above puzzle is closely related to another puzzle: the $*$-duality is described purely in terms of arithmetic whereas the strange duality in general exchanges the algebraic and the transcendental structures in geometry.

Remark 8. It seems an interesting and important problem to find a reasonable extension of regular systems of weights, which is closed under the $*$-duality. For instance, the end remark of Footnote 34 suggests that Gorenstein surface singularities with good $G \times \mathbf{A}^{\times}$-action should be included (see [Ta1]). However, in the present paper, we do not go into any details of the question. Instead, we proceed here as if we were already in the extended category which is closed under the $*$-duality, and ask the following follow-up to the question in §12.

For two decades, inspired from mathematical physics, one observes "symmetry relations" called *mirror symmetry* between some symplectic topological varieties, called the A-model side and some algebraic (or complex analytic) varieties, called the B-model side. Mirror symmetry is formulated at different levels: from identities of numerical invariants of the varieties to the equivalence of categories associated to the varieties. In the present paper, we do not go into any details of the subject but just refer the reader to some of the literature (see for instance [Kon1],[Yau],[H-V]). It is expected that the models on both sides finally should induce an isomorphic flat structures (recall the condition vi) in §12). Mirror symmetry on topological Landau-Ginzburg orbifold model (which corresponds to the singularity theory in mathematics) is described by Kawai-Yang [K-Y] in terms of the duality of orbifoldized Poincaré polynomials. Therefore, it is natural to ask whether (and this was actually proven by A. Takahashi [Ta1]) *the $*$-duality of weight systems is equivalent to mirror symmetry in the Landau-Ginzburg orbifold model in mathematical physics.*[44]

Having these background, we ask the following mathematical question.

Addition to §12 Question. Does there exist an involutive correspondence $\mathfrak{g} \mapsto \mathfrak{g}^*$ on the set of algebras in the right corner of the triangle (6) so that one has the isomorphism: $\mathfrak{g}_{W^*} \simeq (\mathfrak{g}_W)^*$? That is: does

[44]Accordingly, the definition of the $*$-duality for the weight systems of type V in the classification of [Sa17] §5 is modified.

there exist a $*$-duality in the right corner of the triangle (6) making the following diagram commutative?

(18)

$$\begin{array}{ccc} \{\text{Regular systems of weights } W\} & \Leftrightarrow & \{\text{Algebras } \mathfrak{g}_W\} \\ \Updownarrow * & & \Updownarrow * \\ \{\text{Regular systems of weights } W^*\} & \Leftrightarrow & \{\text{Algebras } \mathfrak{g}_{W^*} \simeq (\mathfrak{g}_W)^*\} . \end{array}$$

This means that we seek duality operations in each corner of the triangle (6) (i.e. the $*$-duality on the top, the mirror symmetry in the left and the new conjectural $*$-duality in the right) so that the arrows are compatible with them. Likewise, the domain of the definition of the $*$-duality in the top is restricted, so a similar constraint on the domain of the definition in the RHS might exist. Note also that the $*$-duality in the RHS does not keep the rank μ_W (of the Killing-subalgebra $\mathfrak{h}_W$, recall §12 iii)), but is rather complementary in the sense that $\mu_W + \mu_{W^*} = 24$ in case of $\varepsilon_W = -1$.

What seems remarkable here is the fact that the $*$-duality in the RHS exchanges the algebras which are constructed from algebraic data with that from transcendental data, whereas the $*$-duality in the LHS is purely arithmetically defined. In §15, we shall discuss the duality at a categorical level.

η-product. In the rest of this section, we give a digression on the reformulation of the $*$-duality in terms of eta products ([Sa17] §13).

1. Let $\eta(\tau) := q^{1/24}\prod_{n=1}^{\infty}(1-q^n)$ (where $q^a := \exp(2\pi\sqrt{-1}a\tau)$ be Dedekind eta function. To the product (16), we associate a product

$$\eta_W(\tau) \ := \ \prod_{i|h} \eta(i\tau)^{e_W(i)}. \tag{19}$$

Assertion ([Sa17](13.3)) *Two weight systems W and W^* are dual to each other if and only if one of the following (equivalent) relations holds:*

$$\begin{aligned} \eta_W(-1/h\tau)\cdot\eta_{W^*}(\tau) &= (\tau/\sqrt{-1})^{a_0}/\sqrt{d_{W^*}}, \\ \eta_{W^*}(-1/h\tau)\cdot\eta_W(\tau) &= (\sqrt{-1}/\tau)^{a_0}/\sqrt{d_W}, \end{aligned}$$

where d_W is the discriminant defined by $\prod_{i|h} i^{e_W(i)}$.

2. We observe the following behavior of the coefficients of the expansion of $\eta_W(\tau)$ in the powers of q (called the Fourier coefficients of $\eta_W(\tau)$ at ∞).

i) Fourier coefficients of the eta-products of type A_l ($l \geq 1$), D_l ($l \geq 4$) and E_l ($l = 6, 7, 8$) are positive and are exponentially growing.

ii) Fourier coefficients of the eta-products of type $D_4^{(1,1)}, E_6^{(1,1)}, E_7^{(1,1)}$ or $E_8^{(1,1)}$ are non-negative and are polynomially growing. For example:

$$\begin{aligned}\eta_{E_8^{(1,1)}}(12\tau) &:= \frac{\eta(30\cdot 12\tau)\eta(5\cdot 12\tau)\eta(3\cdot 12\tau)\eta(2\cdot 12\tau)}{\eta(12\tau)\eta(6\cdot 12\tau)\eta(10\cdot 12\tau)\eta(15\cdot 12\tau)}\\ &= q^5+q^{17}+q^{29}+q^{41}+q^{53}+2q^{65}+q^{89}+q^{101}+q^{113}+2q^{125}+q^{137}+q^{149}+\cdots.\end{aligned}$$

Here, i) is trivially checked but ii) requires some work [Sa14]V, where the Melin transform $L_W(s)$ of η_W is expressed by the L-functions of the cyclotomic field $\mathbb{Q}(\sqrt{-1}), \mathbb{Q}(\sqrt{-3}), \mathbb{Q}(\sqrt{-1},\sqrt{-2})$ or $\mathbb{Q}(\sqrt{-1},\sqrt{-3})$ with abelian Galois group according to W is of type $D_4^{(1,1)}, E_6^{(1,1)}, E_7^{(1,1)}$ or $E_8^{(1,1)}$. Then, their L-functions have quadratic expressions by Dirichlet L-functions. For example: using some Dirichlet characters ε, χ on $\mathbb{Z}/12\mathbb{Z}$, the L-function of type $E_8^{(1,1)}$ is expressed as

$$L_{E_8^{(1,1)}}(s) = \frac{1}{4}\prod_{p\in P_1}\frac{1}{(1-p^{-s})^2}\prod_{p\in P_\varepsilon\cup P_{\varepsilon\chi}}\frac{1}{1-p^{-2s}}\left\{\prod_{p\in P_\chi}\frac{1}{(1-p^{-1})^2}-\prod_{p\in P_\chi}\frac{1}{(1+p^{-1})^2}\right\}.$$

A direct inspection on this Euler product shows the non-negativity of all Dirichlet coefficients of them.

For each elliptic root system, we associate the eta-product (19) using the decomposition (16) of the characteristic polynomial of its Coxeter-Killing transformation. Then,

Fact 8. *The Fourier coefficients are non-negative if and only if the root system is of type $D_4^{(1,1)}, E_6^{(1,1)}, E_7^{(1,1)}$ or $E_8^{(1,1)}$, which are exactly the types of simply-laced elliptic root systems admitting the flat structure* (compare [Sa14]V, Theorem with [Sa14]II, §11 Theorem.)

Finally, we remark that a stronger form[45] of the following statement was conjectured in [Sa17](Conjecture 13.5) and is proven by S. Yasuda [Yas].

Fact 9. *Let us define the dual rank ν_W of W by $\nu_W := -\sum_{i|h} e_W(h/i)i$ ($\nu_W = \mu_{W^*}$ if the $*$-dual of W exists). Then, all Fourier coefficients of $\eta_W(\tau)$ at ∞ are non-negative integers if and only if $\nu_W \geq 0$.*

In particular, if a weight system W admits the $*$-dual, then all Fourier coefficients of η_W are non-negative.

Question: Interpret the Fourier series of η_W as the generating function of counting of some objects either from the geometry of X_W or from the algebra $\mathfrak{g}_W$.

[45] An eta product $\eta(h\tau)^\nu\eta_W(\tau)$ has non-negative Fourier coefficients if and only if $\nu_W \geq \nu$.

§15. Towards algebraic construction of the correspondence $\Phi_{\Rightarrow}$

We return to the question of constructing the correspondence $\Phi_{\Rightarrow}$ posed in §12. According the program §12 iii), the algebra $\mathfrak{g}_W$ should have the root space decomposition with respect to the adjoint action of the Cartan-Killing subalgebra $\mathfrak{h} \simeq \mathrm{H}^2(X_1, \mathbb{C})$ with the generalized root system R_W of the vanishing cycles in $\mathrm{H}_2(X_1, \mathbb{Z})$ (§5,7). So, our first task should be to give a good description of the set of vanishing cycles R_W and to find a good basis for it.

For the 14 cases of $\varepsilon_W = -1$, we explained already that the diagram in Table 12. due to Gabrielov is a good candidate for the simple root basis for the generalized root system (recall Footnotes 40 and 42). However, the diagrams were obtained after several braid actions on the basis of the lattice of vanishing cycles ([Gab2], see also [Eb1]). It seems as if the diagrams are obtainded ad hoc, and hard to find an explanation on their naturarity and necessity from purely topological machinery alone.

On the other hand, once we introduce the use of $*$-duality in §14, the situation changes drastically. Namely, owing to §14 Fact 7., the Gabrielov numbers of the diagram of the 14 weight systems W with $\varepsilon_W = -1$ are given by the signature set $A(W^*)$ ((13) and Footnote 32) of the $*$-dual weight system W^*. That is, the Gabrielov number for W is determined "intrinsically" by two arithmetic steps: step 1. calculate the $*$-dual weight system W^* from W ([Sa17] The.7.9) and step 2. calculate the signature set $A(W^*)$ of W^* ([Sa11] The.6), which can be done without any ambiguity. That is: the diagram in Table 12. for W is, at least in its numerical level, determined from the algebraic approach through the $*$-dual W^*. Actually, the same phenomenon occurred already for the simple singularities, where $\varepsilon_W > 0$ and the weight system W is self-dual (§14 Example 2) and then the signature set $A(W^*)$ ($= A(W)$) gives the branch lengths of the diagram of the simple basis of the finite root system R_W (recall §7, 3), §11 a) Example and Table 3).

These facts led the author to ask the following question: [46]

Problem: Construct the root system R_W and its basis through the algebraic approach $\Phi^{a)}_{\Rightarrow}(X_{W^*})$ instead of the topological approach $\Phi^{b)}_{\Rightarrow}(X_W)$.

---

[46] Problem ([Sa15], in English translation p.124). Construct directly from the system of weights $(a, b, c; h)$, without pathing through the homology group of the Milnor fiber, arithmetically or combinatorially, the generalized root system (Q, I, R, c) given above. That is to say, give a basis $\alpha_1, \cdots, \alpha_\mu$ and their inner products $I(\alpha_i, \alpha_j)$ $(1 \le i \le j \le \mu)$ directly from the data $(a, b, c; h)$.

The numerical data $A(W^*)$ alone are not sufficient, and we need to find a structural approach to reconstruct the root system R_W and its basis. Based on recent developments in mathematical physics[47], A. Takahashi [Ta2] gave an answer to the first part of the problem (i.e. the construction of the root system R_W). He conjectured that the K-group of the category of graded matrix factorizations for f_{W^*} should be isomorphic to the lattice of vanishing cycles $(\mathrm{H}_2(X_1,\mathbb{Z}), I)$. He has shown that the category for the case of the polynomial $f_{A_l}=x^{l+1}$ of type A_l is derived equivalent to the category of modules over the path algebra for the Dynkin quiver of type A_l, so that the set of indecomposable objects in the K-group gives the set of roots of type A_l, and he further conjectured that this should hold for all the other simple polynomials.

In the following three sections 16, 17 and 18, we report the results of some joint works of H. Kajiura, A. Takahashi and the author along these line and on its further development. We introduce in §16 the homotopy category $\mathrm{HMF}^{gr}_{A_W}(f_W)$ of graded matrix factorizations for f_W, in three different formulations.

In §17, we study the category for a simple weight system W for $\varepsilon_W=1$, and show that it is generated by a strongly exceptional collection (see §16, 4. for a definition) whose associated quiver is a classical Dynkin quiver of the type $W=W^*$ [K-S-T 1]. Then, due to a classical result by Gabriel [Ga], the set of indecomposable objects in the category form the classical finite root system in the associated Grothendieck group (= K-group), as was expected.

In §18, we study [K-S-T 2] the category for a weight system W of $14+8$ weight systems of $\varepsilon_W=-1$ with $a_0=0$. We show that it is generated by a strongly exceptional collection whose associated quiver is of the form Table 14, where the set of lengths of branches of the quiver is given by the signature set $A(W)$ (13) of the weight system W. We show further that *the path algebra for the quiver with relations is isomorphic to the finite dimensional algebra consisting of morphisms among the objects of the exceptional collection.* Then, owing to a result of Bondal-Kapranov [B-K], the category is equivalent to the bounded

---

[47] A hint was given by the Gonzalez-Verdier interpretation of c) ([G-V], see Appendix), where the dual basis of the simple root basis was constructed by certain vector bundles on $\tilde{X}_{W,0}$. Then, the derived category of the abelian category of coherent sheaves was acknowledged in the recent development in mirror symmetry of D-branes due to Kapustin-Li [K-L 1,2], Hori-Walcher [H-W] and Walcher [Wal].

The category of graded D-branes of type B in Landau-Ginzburg models was formulated by D. Orlov [Orl2] as the triangulated category of the singularity X.

derived category of that of modules over the path algebra of wild type with two relations. In particular, in the 14 exceptional modular cases, in view of the $*$-duality (§14, Fact 7.) and by the comparison of Table 12 with 14, *the Grothendieck group* $\mathrm{K}_0(\mathrm{HMF}^{gr}_{A_W}(f_W))$ *is isomorphic to the lattice of vanishing cycles* $\mathrm{H}_2(X_{W^*,1},\mathbb{Z})$ *for the $*$-dual weight system* W^* *of* W. "*Whether the generalized root system* R_{W^*} *in* $\mathrm{H}_2(X_{W^*,1},\mathbb{Z})$ *(i.e. the set of vanishing cycles, see* §11,b), 1) *) is exactly the image of the set of exceptional indecomposable objects in* $\mathrm{HMF}^{gr}_{A_W}(f_W)$ *or not*" is an open and interesting question.[48]

The above results on the category of graded matrix factorization for $\varepsilon_W=\pm 1$ seem to suggest that the category $\mathrm{HMF}^{gr}_{A_W}(f_W)$ for W with $a_0=0$ may possibly have certain canonical strongly exceptional collections, which are liftings, at the categorical level, of an answer to the latter half of Problem in Footnote 46.

§16. The category of graded matrix factorizations

In this section, we introduce the triangulated category $\mathcal{T}_W$ associated with a regular system of weights W in three equivalent forms: by the homotopy category of the graded matrix factorizations,[49] by the stable category of maximal Cohen-Macaulay modules, and by the category of singularities ([Bu],[Orl1],[Ta2]). We discuss some basic properties of the category such as Serre duality, the generation of the category,

---

[48]One should lift the question into the categorical level as follows: since R_{W^*} is a union of the Weyl group orbits of a distinguished basis due to the irreducibility of the discriminant D_φ (Footnote 12), and a distinguished basis is the image of the objects of an exceptional collection, we ask "whether any exceptional indecompsable object in $\mathrm{HMF}^{gr}_{A_W}(f_W)$ is obtained by a successive application of mutations on the objects of the exceptional collection or not?".

[49]The concept of a matrix factorization is introduced by D. Eisenbud [Ei] in order to describe the *two periodic resolutions of maximal Cohen-Macaulay modules*. It was applied in the study of hypersurface isolated singularities ([Kn1,2],[Gr],[Sch]). It obtained a new impetus through mathematical physics ([K-L], [H-W]) and found new application to the categorification of link invariants ([K-R]). From a graded matrix factorization, forgetting about its grading one obtains a ungraded Marx factorization. This induces a comparison of the categories of graded and ungraded Matrix factorizations. This forgetful functor induces the embedding of the corresponding K-groups, which should conjecturally mirror dual to the embedding of the lattice of vanishing cycles to that of *cycles coming from infinity*. However, in the present paper, we shall not discuss this subject further (see §11 b) 4) and §12 Note 3.).

exceptional collections and Auslander-Reiten translation. For basic terminology and concepts in category theory, one is referred to [Ke].

Let $W=(a,b,c;h)$ be a regular system of weights. We regard the polynomial ring

$$A_W \ := \ \mathbb{C}[x,y,z]$$

to be graded by the weight $\deg(x)=2a/h$, $\deg(y)=2b/h, \deg(z)=2c/h$.[50] Fix a polynomial $f_W \in A_W$ of type W (9), which is of degree 2. Put [51]

$$R_W \ := \ A_W/(f_W) \ = \ \mathbb{C}[x,y,z]/(f_W).$$

Obvious remarks are that A_W is a regular ring and R_W is a Gorenstein ring. By definition, both A_W and R_W are graded rings graded by $\frac{2}{h}\mathbb{Z}_{\geq 0}$. In the present paper, by a graded module M over A_W or R_W, we always mean a module which is graded by $\frac{2}{h}\mathbb{Z}$, i.e. $M=\oplus_{d\in\frac{2}{h}\mathbb{Z}}M_d$. A graded homomorphism $f: M \to N$ of degree a between graded modules is defined as usual a homomorphism with $f(M_d)\subset N_{d+a}$ for any $d \in \frac{2}{h}\mathbb{Z}$. We denote by gr-$A_W$ or gr-R_W the category of *finitely generated* graded A_W or R_W-modules, respectively, whose morphisms are homogeneous of degree 0. We denote by τ the degree shift operator on the set of graded modules to itself defined by $(\tau M)_d = M_{d+\frac{2}{h}}$. For a morphism f, we associate the same morphism $\tau(f):\tau M\to\tau N$.

For $M, N \in$ gr-A_W, the module $\mathrm{Hom}_{A_W}(M,N)$ of all A_W-homomorphisms naturally belongs to gr-A_W by letting $\mathrm{Hom}_{A_W}(M,N)_{\frac{2n}{h}} := \mathrm{Hom}_{\text{gr-}A_W}(M,\tau^n N)$. The same statement replacing A_W by R_W holds also.

1. The homotopy category of graded matrix factorizations for f_W.

Definition. i) A graded matrix factorization for f_W is a system

$$M \ := \ (P_0 \underset{p_1}{\overset{p_0}{\rightleftarrows}} P_1)$$

where P_1, P_2 are graded free A_W-modules of finite rank and p_0, p_1 are graded A_W-homomorphisms such that $p_0p_1 = f_W\cdot id_{P_0}$, $p_1p_0=f_W\cdot id_{P_1}$ and $\deg(p_0)=0$, $\deg(p_1)=2$. The set of all graded matrix factorizations for f_W is denoted by

$$\mathrm{MF}^{gr}_{A_W}(f_W) \ := \ \{\text{graded matrix factorizations for } f_W\}.$$

---

[50]In order to compare with the conventions of matrix factorizations, we have to duplicate the grading compared with that for the flat structure. Hence, one should note that $\deg(f_W)=2$.

[51]The reader is notified with the fact that there is an unfortunate coincidence of this notation with that for the set of vanishing cycles in §11 b) 1).

ii) A graded homomorphism from $M=(P_0 \underset{p_1}{\overset{p_0}{\rightleftarrows}} P_1)$ to $M'=(P'_0 \underset{p'_1}{\overset{p'_0}{\rightleftarrows}} P'_1)$ is a pair $\Phi=(\phi_0,\phi_1):(P_0,P_1)\to(P'_0,P'_1)$ of graded A_W-homomorphisms homogeneous of degree 0 making the following diagram commutative.

$$
\begin{array}{ccccccccc}
 & \xrightarrow{p_1} & P_0 & \xrightarrow{p_0} & P_1 & \xrightarrow{p_1} & P_0 & \xrightarrow{p_0} & \\
\cdots & & \downarrow{\scriptstyle \phi_0} & & \downarrow{\scriptstyle \phi_1} & & \downarrow{\scriptstyle \phi_0} & & \cdots \\
 & \xrightarrow{p'_1} & P'_0 & \xrightarrow{p'_0} & P'_1 & \xrightarrow{p'_1} & P'_0 & \xrightarrow{p'_0} &
\end{array}
$$

The set of all graded homomorphisms is denoted by $\mathrm{Hom}_{\mathrm{MF}^{gr}_{A_W}(f_W)}(M, M')$.

iii) We denote also by $\mathrm{MF}^{gr}_{A_W}(f_W)$ the additive category of all matrix factorizations with respect to above defined homomorphisms.

Definition. We denote by $\mathrm{HMF}^{gr}_{A_W}(f_W)$ the homotopy category of $\mathrm{MF}^{gr}_{A_W}(f_W)$. That is, the objects of $\mathrm{HMF}^{gr}_{A_W}(f_W)$ are the same as $\mathrm{MF}^{gr}_{A_W}(f_W)$. The module of homomorphisms is defined as the quotient space by the homotopy equivalence

$$\mathrm{Hom}_{\mathrm{HMF}^{gr}_{A_W}(f_W)}(M,M') := \mathrm{Hom}_{\mathrm{MF}^{gr}_{A_W}(f_W)}(M,M')/\sim$$

where a morphism $\Phi=(\phi_0,\phi_1)$ is homotopic to zero, denoted by $\Phi\sim 0$, if there exists A_W-homomorphisms $h_0 : P_0 \to P'_1$ and $h_1 : P_1 \to P'_0$ with $\deg(h_0) = -2$ and $\deg(h_1) = 0$ such that $\phi_0 = p'_1 h_0 + h_1 p_0$ and $\phi_1 = p'_0 h_1 + h_0 p_1$.

$$
\begin{array}{ccccccccc}
 & \xrightarrow{p_1} & P_0 & \xrightarrow{p_0} & P_1 & \xrightarrow{p_1} & P_0 & \xrightarrow{p_0} & \\
\cdots & \swarrow & \phi_0\downarrow & \swarrow{\scriptstyle h_1} & \phi_1\downarrow & \swarrow{\scriptstyle h_0} & \phi_0\downarrow & \swarrow & \cdots \\
 & \xrightarrow{p'_1} & P'_0 & \xrightarrow{p'_0} & P'_1 & \xrightarrow{p'_1} & P'_0 & \xrightarrow{p'_0} &
\end{array}
$$

Example. The $A_W \underset{f}{\overset{1}{\rightleftarrows}} A_W$ and $A_W \underset{1}{\overset{f}{\rightleftarrows}} \tau^h A_W$ are matrix factorizations which are homotopic to 0, since we have the following commutative diagram:

$$
\begin{array}{ccccccccc}
\to & A_W & \xrightarrow{f} & A_W & \xrightarrow{1} & A_W & \xrightarrow{f} \\
\swarrow{\scriptstyle 1} & \downarrow{\scriptstyle 1} & \swarrow{\scriptstyle 0} & \downarrow{\scriptstyle 1} & \swarrow{\scriptstyle 1} & \downarrow{\scriptstyle 1} & \swarrow{\scriptstyle 0} \\
\to & A_W & \xrightarrow{f} & A_W & \xrightarrow{1} & A_W & \to
\end{array}
$$

Any 0-object M (i.e. $1_M \sim 0$) in the category $\mathrm{HMF}^{gr}_{A_W}(f_W)$ is a direct sum of copies of some τ-powers shifts of $(A_W \underset{f}{\overset{1}{\rightleftarrows}} A_W)$ and $(A_W \underset{1}{\overset{f}{\rightleftarrows}} A_W)$.

Definition. (Shift functors) We introduce two auto-equivalence functors:

$$\tau(P_0 \underset{p_1}{\overset{p_0}{\rightleftarrows}} P_1) := (\tau P_0 \underset{\tau p_1}{\overset{\tau p_0}{\rightleftarrows}} \tau P_1), \qquad \tau(\phi_0, \phi_1) = (\tau\phi_0, \tau\phi_1),$$

$$T(P_0 \underset{p_1}{\overset{p_0}{\rightleftarrows}} P_1) := (P_1 \underset{-\tau^h p_0}{\overset{-p_1}{\rightleftarrows}} \tau^h P_0), \quad T(\phi_0, \phi_1) = (-\phi_1, -\tau^h \phi_0).$$

By definition, they satisfy an obvious but basic relation:

$$T^2 = \tau^h \tag{20}$$

Here are some elementary properties of the category $\mathrm{HMF}_A^{gr}(f_W)$.

i) $\mathrm{HMF}_A^{gr}(f_W)$ *is a Krull-Schmidt category: i.e. if* $e \in End_{\mathrm{HMF}_R^{gr}(f)}(M)$ *for an object* M *is idempotent* $e^2 = e$, *then there exist an object* M' *and morphisms* $\Phi' : M' \to M$ *and* $\Phi : M \to M'$ *such that* $\Phi' \circ \Phi = e$ *and* $\Phi \circ \Phi' = \mathrm{id}_{M'}$.

ii) $\mathrm{HMF}_A^{gr}(f_W)$ *is of Ext-finite type: i.e.* $\oplus_{n\in\mathbb{Z}}\mathrm{Hom}(M, T^n N)$ *is finite dimensional vector space for all objects* M *and* N *of the category.*

Sketch of proof. The direct sum $\oplus_{n\in\mathbb{Z}}\mathrm{Hom}_{\mathrm{HMF}_A^{gr}(f_W)}(M, \tau^n M')$ is a finitely generated A_W-module. Since the sum is annihilated by multiplications by $\partial_x f_W, \partial_y f_W, \partial_z f_W$, it is a finite module over $J_W := A_W/(\partial_x f_W, \partial_y f_W, \partial_z f_W)$. Since f_W is of type W and in view of §11 Fact 4., it is of finite rank over $\mathbb{C}$. □

Definition. (Mapping cone) For any morphism $\Phi = (\phi_0, \phi_1) \in \mathrm{Hom}_{\mathrm{MF}_R^{gr}(f)}(M, M')$, we introduce the *mapping cone* $C(\Phi) \in \mathrm{MF}_R^{gr}(f)$ as follows.

$$\left(\begin{matrix} P_1 & \begin{pmatrix} -p_1 & 0 \\ \phi_1 & p'_0 \end{pmatrix} & \tau^h P_0 \\ \oplus & \rightleftarrows & \oplus \\ P'_0 & \begin{pmatrix} -\tau^h p_0 & 0 \\ \tau^h \phi_0 & p'_1 \end{pmatrix} & P'_1 \end{matrix} \right)$$

and obtains a sequence: $*)\quad M \xrightarrow{\Phi} M' \xrightarrow{inclusion} C(\Phi) \xrightarrow{-projection} TM$. Then, we have the following general fact (c.f. [G-M], [K-S], [B-K2], [Ta2]).

Theorem. *The additive category* $\mathrm{HMF}_{A_W}^{gr}(f_W)$ *endowed with the shift function* T *and distinguished triangles isomorphic to* $*)$ *for all morphisms* Φ *forms an enhanced triangulated category of Ext-finite type.*

See [B-K2] for a definition of the *enhanced triangulated category.*

2. The stable category of maximal Cohen-Macaulay modules over R_W.

Definition. A graded module $M \in$ gr-R_W is a maximal Cohen-Macaulay module over R_W if $\operatorname{depth}(M) = \dim(R_W)$ $(=: d = 2)$ $(\Leftrightarrow \operatorname{Ext}^i_R(R_W/(x,y,z), M) = 0$ for $i < d = 2)$. The full subcategory of gr-R_W consisting of all graded maximal Cohen-Macaulay modules over R_W is denoted by $\mathrm{CM}^{\mathrm{gr}}(R_W)$.

For an element M of gr-R_W and $n \geq d$, the n-th syzygy $syz^n(M)$ (:= nth kernel of a graded free resolution of M) up to a free module factor becomes a maximal Cohen-Macaulay module and doubly periodic in n. Hence, one sees $\mathrm{CM}^{\mathrm{gr}}(R_W)$ is a Frobenius category (i.e. it has enough injective and projective objects which coincide to each other). Then, *the stable category* $\underline{\mathrm{CM}}^{\mathrm{gr}}(R_W)$, *defined below, becomes a triangulated category* (c.f. [Ke]): the objects of the stable category $\underline{\mathrm{CM}}^{\mathrm{gr}}(R_W)$ is the same as $\mathrm{CM}^{\mathrm{gr}}(R_W)$ and the space of morphisms is given by

$$\underline{\mathrm{Hom}}_{\text{gr-}R_W}(M,N) := \mathrm{Hom}_{\text{gr-}R_W}(M,N)/\mathcal{P}(M,N),$$

where $\mathcal{P}(M,N)$ is the subspace of $\mathrm{Hom}_{\text{gr-}R_W}(M,N)$ consisting of morphisms which factor through projective modules.

Fact 10. *For a graded matrix factorization $M \in \mathrm{MF}^{\mathrm{gr}}_{A_W}(f_W)$, we associate a maximal Cohen-Macaulay module* $\operatorname{coker}(P_1 \xrightarrow{p_1} P_0) \in \mathrm{CM}^{\mathrm{gr}}(R_W)$ *over R_W. This correspondence induces an equivalence of the triangulated categories:*

$$\mathrm{HMF}^{\mathrm{gr}}_{A_W}(f_W) \quad \simeq \quad \underline{\mathrm{CM}}^{\mathrm{gr}}(R_W). \tag{21}$$

The advantage of the category $\underline{\mathrm{CM}}^{\mathrm{gr}}(R_W)$ is that it easily admits the concepts: *Auslander-Reiten triangles* and *Serre duality*, which we explain below. For details on the subject, the reader is referred to textbooks, e.g. [Hap], [Yos].

We first define the Auslander transpose $\mathrm{tr}(M)$ (up to free module factor) of $M \in$gr-R_W by putting $\mathrm{tr}(M) := \mathrm{Coker}({}^t f)$ where $F_1 \xrightarrow{f} F_0 \to M \to 0$ is a finite presentation of M and ${}^t f$ is the contragradient homomorphism of f. Let us denote by $\mathrm{syz}^d(\mathrm{tr}(M))$ the reduced dth syzygy of $\mathrm{tr}(M)$ obtained by avoiding all graded free summands from a dth syzygy of $\mathrm{tr}(M)$. Then, the Auslander-Reiten translation, or A-R translation, $\tau_{AR}(M) \in$ gr-R_W is defined by

$$\textbf{A-R translation}: \qquad \tau_{AR}(M) \ := \ \mathrm{Hom}_{R_W}(\mathrm{syz}^d(\mathrm{tr}(M)), \mathrm{K}_{R_W})$$

where $\mathrm{K}_{R_W} = \mathrm{Res}_{X_{W,0}} \begin{bmatrix} A_W\, dxdydz \\ f_W \end{bmatrix} = \tau^{-\varepsilon_W} R_W$ is the canonical module of $R_W = A_W/(f_W)$. If M is a maximal Cohen-Macaulay module without a free direct summand, then we easily see that $\mathrm{syz}^2(\mathrm{tr}(M)) \simeq \mathrm{Hom}_{R_W}(M, R_W)$, and, hence,

$$\tau_{AR}(M) \ \simeq \ \tau^{-\varepsilon_W} M. \tag{22}$$

The auto-equivalence of the category $\underline{\mathrm{CM}}^{\mathrm{gr}}(R_W)$ induced by τ_{AR} is denoted again by τ_{AR}. In view of the relation (20), we have the following relation:

$$\tau_{AR}^h \ = \ (T^2)^{-\varepsilon_W}.$$

The following duality was shown by Auslander and Reiten [A-R3]:

$$\mathrm{Ext}^d_{\mathrm{gr}\text{-}R_W}(\underline{\mathrm{Hom}}_{R_W}(M,N), \mathrm{K}_{R_W}) \ \simeq \ \mathrm{Ext}^1_{\mathrm{gr}\text{-}R_W}(N, \tau_{AR}(M))$$

for $M, N \in \underline{\mathrm{CM}}^{\mathrm{gr}}(R_W)$. This, in particular, implies the following

Serre duality : $\mathrm{Hom}_{\mathbb{C}}(\underline{\mathrm{Hom}}_{\mathrm{gr}\text{-}R_W}(M,N), \mathbb{C}) \ \simeq \ \underline{\mathrm{Hom}}_{\mathrm{gr}\text{-}R_W}(N, SM)$

as a bi-functorial isomorphism of vector spaces for $M, N \in \underline{\mathrm{CM}}^{\mathrm{gr}}(R_W)$, where S is an auto-equivalence of the category $\underline{\mathrm{CM}}^{\mathrm{gr}}(R_W)$, called Serre functor [B-K1], defined by

$$S := T\tau_{AR}. \tag{23}$$

As a consequence of Serre duality, one can show that, for any indecomposable object Z of $\underline{\mathrm{CM}}^{\mathrm{gr}}(R_W)$, there exists the *AR-triangle of Z* in the following sense: let $Z \stackrel{w}{\to} T\tau_{AR}(Z)$ be the morphism, which, by Serre duality, corresponds to the dual of the identity element in $\mathrm{Hom}_{\mathbb{C}}(\underline{\mathrm{Hom}}_{\mathrm{gr}\text{-}R_W}(Z,Z), \mathbb{C})$. Then, *there exists an object $AR(Z)$ and the triangle, called A-R triangle, in $\underline{\mathrm{CM}}^{\mathrm{gr}}(R_W)$:*

A-R triangle : $\tau_{AR}(Z) \ \stackrel{u}{\to} \ AR(Z) \ \stackrel{v}{\to} \ Z \ \stackrel{w}{\to} \ T\tau_{AR}(Z)$

such that, for any morphism $g : W \to Z$ in $\underline{\mathrm{CM}}^{\mathrm{gr}}(R_W)$ which is not a split epimorphism, there exists $h : W \to AR(Z)$ with $vh = g$.

3. The category of the singularity $X_{W,0} := \mathrm{Spec}(R_W)$.

Definition. ([Orl1]) The triangulated category of the singularity $X_{W,0}$ is

$$D^{\mathrm{gr}}_{Sg}(R_W) := D^b(\mathrm{gr}\text{-}R_W)/D^b(\mathrm{gr\ proj}\text{-}R_W)$$

where $D^b(\mathrm{gr}\text{-}R_W)$ is the bounded derived category of the abelian category gr-R_W with the natural triangulated structure and $D^b(\mathrm{gr}\ \mathit{proj}\text{-}R_W)$ is its full triangulated subcategory consisting of objects which are isomorphic to the bounded complexes of projectives. Actually, the subcategory is the derived category of the exact category of graded projective modules [Ke], and is called the subcategory of perfect complexes.

Since R_W defines the hypersurface $X_{W,0}$ and is Gorenstein, we have

Theorem. (Buchweitz [Bu], Orlov [Orl2]§1.3) *The natural inclusion map* $\mathrm{CM}^{\mathrm{gr}}(R_W) \to \mathrm{gr}\text{-}R_W$ *induces the equivalence of triangulated categories:*

$$\underline{\mathrm{CM}}^{\mathrm{gr}}(R_W) \quad \simeq \quad D^{\mathrm{gr}}_{Sg}(R_W). \tag{24}$$

Orlov [Orl2] gave further a comparison Theorem of $D^{\mathrm{gr}}_{Sg}(R_W)$ with the quotient abelian category $\mathrm{qgr}\text{-}R_W := \mathrm{gr}\text{-}R_W/\mathrm{tors}\text{-}R_W$ where $\mathrm{tors}\text{-}R_W$ is the full subcategory of $\mathrm{gr}\text{-}R_W$ consisting of all finite dimensional R_W-modules. Actually, in case when C_W is a rational curve, we may regard it as a weighted projective line in the sense of Geigle and Lenzing [G-L 1]. Then $\mathrm{qgr}\text{-}R_W$ is derived equivalent to the category of coherent sheaves on the weighted projective line [G-L 2].[52]

4. The triangulated category $\mathcal{T}_W$ associated with a regular system of weights W.

Owing to (21) and (24), we introduce an enhanced triangulated category

$$\mathcal{T}_W \ := \ \mathrm{HMF}^{\mathrm{gr}}_{A_W}(f_W) \ \simeq \ \underline{\mathrm{CM}}^{\mathrm{gr}}(R_W) \ \simeq \ D^{\mathrm{gr}}_{Sg}(R_W) \tag{25}$$

associated to a regular system of weights W up to equivalences. The advantage of the third expression is that we have the following generation theorem ([K-S-T 2]), which we shall use in the proof of our main theorem in §18.

Theorem. *Let $\mathcal{T}$ be a right-admissible full triangulated subcategory of $\mathcal{T}_W$ satisfying:*

i) The shift functor τ induces an auto-equivalence of $\mathcal{T}$.

ii) There is an object of $\mathcal{T}$ which is isomorphic to the pure complex of the torsion (sky-scraper) module $R_W/(x,y,z)$ in $\mathcal{T}_W$.

Then the natural inclusion $\mathcal{T} \subset \mathcal{T}_W$ induces the triangulated equivalence.

Here, a subcategory $\mathcal{T}'$ of a triangulated category $\mathcal{T}$ is called *right-admissible* if, for any object X of $\mathcal{T}$, there exist $N \in \mathcal{T}'$, $M \in \mathcal{T}'^{\perp} := \{M \in \mathcal{T} \mid \mathrm{Hom}_{\mathcal{T}}(N,M) = 0 \ \forall N \in \mathcal{T}'\}$ and a triangle: $N \to X \to M \to TN$ in $\mathcal{T}$.

For a later use, we recall some terminologies and results from [Bon].

[52] [G-L 2] treats the case corresponding to the regular weight systems with $\varepsilon_W = 1$. A general proof which covers the case for any regular weight system of genus 0 shall appear in: H. Kajiura, K. Saito and A. Takahashi: Weighted projective lines associated to regular systems of weights, in preparation.

Definition. i) An object E in a triangulated category $\mathcal{T}$ over $\mathbb{C}$ is called *exceptional* if $\mathrm{Hom}_{\mathcal{T}}(E, T^pE) \simeq \mathbb{C}$ if $p = 0$ and 0 if $p \neq 0$.

ii) An *exceptional collection* is a sequence $(E_1, \cdots, E_\mu)$ of exceptional objects satisfying $\mathrm{Hom}_{\mathcal{T}}(E_i, T^pE_j) = 0$ for $\forall p \in \mathbb{Z}$ and $0 \leq j < i \leq \mu$.

iii) An exceptional collection $(E_1, \cdots, E_\mu)$ is called *strongly exceptional* if $\mathrm{Hom}_{\mathcal{T}}(E_i, T^pE_j) = 0$ for all $1 \leq i, j \leq \mu$ and $p \neq 0$.

iv) For an exceptional collection $\mathcal{E} := (E_1, \cdots, E_\mu)$, we denote by $\langle\mathcal{E}\rangle := \langle E_1, \cdots, E_\mu\rangle$ the smallest triangulated full subcategory containing $E_1, \cdots, E_\mu$. We say that $\mathcal{E}$ generates $\mathcal{T}$ if $\langle\mathcal{E}\rangle$ is equivalent to $\mathcal{T}$.

v) For a strongly exceptional collection $\mathcal{E} := (E_1, \cdots, E_\mu)$, let us introduce a finite dimensional agebra

$$\mathrm{Hom}(\mathcal{E}, \mathcal{E}) \ := \ \oplus_{0 \leq i,j \leq \mu} \mathrm{Hom}_{\mathcal{T}}(E_i, E_j)$$

and call it the *hom-algebra* of the collection $\mathcal{E}$.

Theorem. ([Bon],[B-K2]) *Let $\mathcal{T}$ be an enhanced triangulated category of Ext-finite type, and let $\mathcal{E}$ be a strongly exceptional collection. Then, $\langle\mathcal{E}\rangle$ is right admissible and is, as an enhanced triangulated category, equivalent to the bounded derived category*

$$D^b(\text{mod-Hom}(\mathcal{E}, \mathcal{E})). \tag{26}$$

5. K-group and Auslander-Reiten translation of $\mathcal{T}_W$.

In this paragraph, we show that the Auslander-Reiten translation induces an automorphism of the K-group of the category, which is expressed as the product of reflections. This expression is presumably the mirror dual of the expresion given in §5 of the Milnor monodromy $\mathbf{c}$ by the product of reflections.

For a triangulated category $\mathcal{T}$, let $\mathrm{K}_0(\mathcal{T})$ be its Grothendieck group (or K-group), i.e. the quotient group of the free abelian group generated by the equivalence classes $[X]$ of objects X of $\mathcal{T}$ divided by the submodule generated by $[X]+[Z]-[Y]$ for all triangles $X \to Y \to Z \to TX$. We denote by $[X]$ the image element in K_0 of X. If a set $E_1, \cdots, E_\mu$ of objects generates the triangulated category, then their images $[E_1], \cdots, [E_\mu]$ generates the K-group over $\mathbb{Z}$.

The shift functor T on $\mathcal{T}$ induces an action $[T] = -\mathrm{id}_{\mathrm{K}_0(\mathcal{T})}$ on $\mathrm{K}_0(\mathcal{T})$, since $[X]+[TX]=0$ for any object X because of the triangle $X \xrightarrow{1} X \to 0 \to TX$. In particular T^2 induces identity on the K-group

The Auslander-Reiten translation τ_{AR} is an auto-equivalence of the triangulated category, so that it induces an automorphism of the group $\mathrm{K}_0(\mathcal{T})$, denoted by $[\tau_{AR}]$. For the category $\mathcal{T}_W$ associated to a regular weight system W, it is of finite order h, since, using the expression (22) and the fact (20), we calculate as

$$[\tau_{AR}]^h = [T^{-2\varepsilon_W}] = (-id_{\mathrm{K}_0(\mathcal{T}_W)})^{-2\varepsilon_W} = id_{\mathrm{K}_0(\mathcal{T}_W)}.$$

If $\mathcal{T}$ is of Ext-finite type over $\mathbb{C}$, the Euler pairing is defined by

$$\chi(X,Y) \ := \ \textstyle\sum_{n\in\mathbb{Z}}(-1)^n \hom_{\mathcal{T}}(X,T^nY)$$

for any two objects X and Y of $\mathcal{T}$. Because of the (co-)homological property of $\hom_{\mathcal{T}}$, it induces a bilinear form on $\mathrm{K}_0(\mathcal{T})$, which we denote again by χ. We equip the K-group with the symmetric bilinear form (e.g. see [Ri1] 2.4) [53]:

$$I(e,f) \ := \ \chi(e,f) \ + \ \chi(f,e). \tag{27}$$

for $e, f \in \mathrm{K}_0(\mathcal{T})$. We remark that if $e = [E]$ where E is an exceptional object of $\mathcal{T}$, then $\chi(e,e) = 1$ and, hence, $I(e,e) = 2$. Then, similarly to Picard-Lefschetz formula in §5, we can define the reflection $w_e \in O(\mathrm{K}_0(\mathcal{T}), I)$ by letting

$$w_e(u) \ := \ u - I(u,e)e \qquad \text{for } u \in \mathrm{K}_0(\mathcal{T}).$$

The $[\tau_{AR}]$ preserves the bilinear form χ, i.e. $[\tau_{AR}] \in O(\mathrm{K}_0(\mathcal{T}), I)$. Let us express now $[\tau_{AR}]$ as a product of reflections on $\mathrm{K}_0(\mathcal{T})$.

Let $\mathcal{E} := (E_1, \cdots, E_\mu)$ be a strongly exceptional collection of $\mathcal{T}$. Assume that $E_1, \cdots, E_\mu$ generate $\mathcal{T}$ and, hence $[E_1], \cdots, [E_\mu]$ is a basis of $\mathrm{K}_0(\mathcal{T})$. Associated to $[E_1], \cdots, [E_\mu]$, we consider two basis: $f_1, \cdots, f_\mu$ and $g_1, \cdots, g_\mu$ of $\mathrm{K}_0(\mathcal{T})$ defined by the following relations:

$$[E_i] \ = \ \textstyle\sum_{j=1}^{\mu} \chi(E_i,E_j)f_j \ = \ \sum_{j=1}^{\mu} g_j\chi(E_j,E_i) \tag{28}$$

Here, we remark that the matrix $\chi_{\mathcal{E}} := (\chi(E_i,E_j))_{ij=1,\cdots,\mu}$ is an upper triangular matrix with 1 at each diagonal entry so that χ is invertible. Let us denote by $C_{\mathcal{E}} = (C_{\mathcal{E},ij})_{ij=1,\cdots,\mu}$ the inverse matrix $\chi_{\mathcal{E}}^{-1}$ (which is also an upper triangular integral matrix). In fact, using the mapping cone constructions, one can find objects F_i and G_i in $\mathcal{T}$ such that $f_i = [F_i]$ and $g_i = [G_i]$ for $i = 1, \cdots, \mu$.[54] The intersection matrix of them are given

[53]For the purpose of the period map for odd dimensional Milnor fiber, we need to study the skew symmetric bilnear form: $I_{odd}(e,f) := \chi(e,f) - \chi(f,e)$ (see [Sa18]§6 (6.2.2)), [Sa19] latter half of §4). However, we shall not treat them in the present paper.

[54]Actually, these objects F_i and G_i are constructed by use of mutations and are shown to be exceptional objects ([Bon]).

by $C_{\mathcal{E}}$ as follows:
(29)

$$\chi([F_i],[F_j]) \;=\; \chi([G_i],[G_j]) \;=\; C_{\mathcal{E},ji} \qquad \text{for } i,j=1,\cdots,\mu.$$

Since $I([F_i],[F_i])=I([G_i],[G_i])=2$ for $1\le i\le\mu$, we define reflections $w_{[F_1]},\cdots,w_{[F_\mu]}$ and $w_{[G_1]},\cdots,w_{[G_\mu]}$. Then, one can easily verify the formula:

Fact 11. *Let $(E_1,\cdots,E_\mu)$ be a strongly exceptional collection, then the transformation $[\tau_{AR}]$ is expressed as the product of reflections associated to the basis:*

$$(30) \qquad [\tau_{AR}] \;=\; w_{[F_1]}\cdots w_{[F_\mu]} \;=\; w_{[G_1]}\cdots w_{[G_\mu]}\;.$$

6. Quiver and path algebra associated with $\mathcal{E}$.

In this paragraph, associated with a strongly exceptional collection, we consider a slight generalization of a quiver, and then, associated to the (generalized) quiver, we introduce a path algebra with relations, which we shall use in §17 and 18 (see [Ri1] for quivers and path algebras).

Let $\mathcal{E}=(E_1,\cdots,E_\mu)$ be a strongly exceptional collection of a triangulated category $\mathcal{T}$. Then, we associated a quiver $\Delta_{\mathcal{E}}$ given by a pair

$$(31) \qquad \Delta_{\mathcal{E}} \;=\; (\Delta_0,\Delta_1),$$

where $\Delta_0=\{v_1,\cdots,v_\mu\}$ is a set of μ elements, called the *vertices*, and Δ_1, called the set of *allows*, is a multi-set of triplet $(v_i,v_j,\epsilon)\in\Delta_0\times\Delta_0\times\{\pm\}$ where $(v_i,v_j,+)$ appears in Δ_1 only when $i\neq j$, $C_{\mathcal{E},ij}<0$ and $-C_{\mathcal{E},ij}$-times, and $(v_i,v_j,-)$ appears in Δ_1 only when $i\neq j$, $C_{\mathcal{E},ij}>0$ and $C_{\mathcal{E},ij}$-times. We regard $(v_i,v_j,+)\in\Delta_1$ as an arrow (with positive sign) from the vertex v_i to the vertex v_j, and similarly $(v_i,v_j,-)\in\Delta_1$ as a dotted arrow from v_i to v_j.

Remark 16.1. If one forgets the directions of the arrows from the quiver $\Delta_{\mathcal{E}}$ and leaves only lines or dotted lines together with the vertices, then one obtains automatically the intersection diagram Γ of the symmetric bilinear form I with respect to the basis $[F_1],\cdots,[F_\mu]$ or $[G_1],\cdots,[G_\mu]$ of $\mathrm{K}_0(\mathcal{T})$, i.e. Γ is the intersection diagram for the symmetrization of the matrix (29)) (for instance, [Sa14]I (8.2)).

Associated with the above given quiver $\Delta_{\mathcal{E}}$ (31), the path algebra

$$(32) \qquad \mathbb{C}(\Delta_{\mathcal{E}},R)$$

with relations R is defined as follows. Let $\Delta_1=\Delta_1^+\amalg\Delta_1^-$ be the decomposition of the set of arrows into those of positive and negative signs.

We regard $\Delta_{\mathcal{E}}^{+} := (\Delta_0, \Delta_1^{+})$ as the quiver in the classical sense (e.g. [Ri1] 2.1), then by concatenating arrows, one defines paths and the path algebra $\mathbb{C}\Delta^{+}$ as usual ([ibid]). Let

$$R : \Delta_1^{-} \longrightarrow \mathbb{C}\Delta_{\mathcal{E}}^{+}$$

be a map such that the image of an arrow $(v_i, v_j, -)$ belongs in the subspace $(v_j \mid v_j) \cdot \mathbb{C}\Delta_{\mathcal{E}}^{+} \cdot (v_i \mid v_i)$ spanned by all paths from v_i to v_j (here, we denote by $(v \mid v)$ the path of length 0 at a vertex v). Then, we put

$$\mathbb{C}(\Delta_{\mathcal{E}}, R) := \mathbb{C}\Delta_{\mathcal{E}}^{+} \,/\, (\mathbb{C}\Delta_{\mathcal{E}}^{+} \cdot R(\Delta_1^{-}) \cdot \mathbb{C}\Delta_{\mathcal{E}}^{+}), \tag{32}$$

where $\mathbb{C}\Delta_{\mathcal{E}}^{+} \cdot R(\Delta_1^{-}) \cdot \mathbb{C}\Delta_{\mathcal{E}}^{+}$ is the both-sided ideal of the path algebra $\mathbb{C}\Delta_{\mathcal{E}}^{+}$ generated by the image set $R(\Delta_1^{-})$. We call $\mathbb{C}(\Delta_{\mathcal{E}}, R)$ the *path-algebra with relations* R.

Remark 16.2. Assigning to each arrow $(v_i, v_j; +) \in \Delta_1^{+}$ a morphism $f_{ij} \in \mathrm{Hom}_{\mathcal{T}}(E_i, E_j)$, we can define a ring homomorphism:

$$\mathbb{C}(\Delta_{\mathcal{E}}, R) \longrightarrow \mathrm{Hom}(\mathcal{E}, \mathcal{E}) := \oplus_{0 \le i,j \le \mu} \mathrm{Hom}_{\mathcal{T}}(E_i, E_j)$$

for a suitable choice of relations R. In general, the homomorphism can neither be isomorphic nor induce derived equivalence for any choices of f_{ij} and R.

Example. Let us consider a strongly exceptional collection $\mathcal{E} = (E_1, E_2, E_3)$ such that $\chi_{\mathcal{E}} = \begin{bmatrix} 1 & 1 & 1 \\ 0 & 1 & 1 \\ 0 & 0 & 1 \end{bmatrix}$ and $C_{\mathcal{E}} = \begin{bmatrix} 1 & -1 & 0 \\ 0 & 1 & -1 \\ 0 & 0 & 1 \end{bmatrix}$. Then the associated quiver is a Dynkin quiver $\Delta_{\mathcal{E}} = \circ \to \circ \to \circ$ of type A_2 and $\mathbb{C}\Delta_{\mathcal{E}}$ is a path algebra of type A_2. On the other hand, there are two cases of the structure of the hom-algebra $\mathrm{Hom}(\mathcal{E}, \mathcal{E}) := \oplus_{1 \le i \le j \le 3} \mathrm{Hom}(E_i, E_j)$ depending on whether the product $\mathrm{Hom}(E_1, E_2) \times \mathrm{Hom}(E_2, E_3) \to \mathrm{Hom}(E_1, E_3)$ is a) non-zero or b) zero. Then the homomorphism $\mathbb{C}\Delta_{\mathcal{E}} \to \mathrm{Hom}(\mathcal{E}, \mathcal{E})$ assigning the two arrows in $\Delta_{\mathcal{E}}$ to the base of $\mathrm{Hom}(E_1, E_2)$ and $\mathrm{Hom}(E_2, E_3)$, respectively, is isomorphic in the case a), but neither isomorphic nor derived equivalent in the case b).

§17. The category of matrix factorizations: the case $\varepsilon_W = 1$.

In this section, we study the category $\mathcal{T}_W = \mathrm{HMF}^{\mathrm{gr}}_{A_W}(f_W)$ associated with a weight system W with $\varepsilon_W = 1$. Recall that, in this case, the weight systems are classified into types A_l $(l \ge 1)$, D_l $(l \ge 4)$, E_6, E_7 and E_8 (see Table 8), and that the associated polynomials f_W of type W are called the simple polynomials (see Table 2).

Then the following theorem is proven in [K-S-T 1] (c.f. also [Ue]).

Theorem. *Let W be a regular system of weights of type ADE. For any Dynkin quiver $\vec{\Delta}$ of type W (see Note below), there exists a unique strongly exceptional collection $\mathcal{E}_{\vec{\Delta}}$ of the category $\mathcal{T}_W$* (25) *such that*

i) *the $\mathcal{E}_{\vec{\Delta}}$ generate the triangulated category $\mathcal{T}_W$,*

ii) *the quiver associated with the collection $\mathcal{E}_{\vec{\Delta}}$ is isomorphic to $\vec{\Delta}$,*

iii) *the path algebra $\mathbb{C}\vec{\Delta}$ is isomorphic to the hom-algebra $\mathrm{Hom}_{\mathcal{T}_W}(\mathcal{E}_{\vec{\Delta}}, \mathcal{E}_{\vec{\Delta}})$.*

Note. By a Dynkin quiver, we mean an oriented Dynkin diagram of type ADE.

Sketch of proof. According to the works [Ei], [A-R1] and [Au], the Auslander-Reiten quiver for the triangulated category $\mathrm{HMF}_{\hat{\mathcal{O}}}(f_W)$ of ungraded matrix factorizations over the local rings $\mathcal{O}$ and $\hat{\mathcal{O}}$ are well-known to be isomorphic to the both side oriented Dynkin quiver $\overleftrightarrow{\Delta}$ of type W. We consider the natural forgetful functor: $\mathrm{HMF}^{\mathrm{gr}}_{A_W}(f_W) \to \mathrm{HMF}_{\hat{\mathcal{O}}}(f_W)$ forgetting the gradings on matrix factorizations. Then, by "lifting" the results on $\mathrm{HMF}_{\hat{\mathcal{O}}}(f_W)$ back to the graded category together with the knowledge of the Serre duality, in [K-S-T 1], we determine the list of all indecomposable objects and all irreducible morphisms in $\mathrm{HMF}^{\mathrm{gr}}_{A_W}(f_W)$. Using these data, we can verify directly the existence (up to τ-shift) of a strongly exceptional collection $\mathcal{E}_{\vec{\Delta}}$ of $\mathrm{HMF}^{\mathrm{gr}}_{A_W}(f_W)$, and of the natural isomorphism: $\mathbb{C}\vec{\Delta} \simeq \mathrm{Hom}(\mathcal{E}_{\vec{\Delta}}, \mathcal{E}_{\vec{\Delta}})$ (i.e. the non-vanishing of compositions of morphisms corresponding to concatenations of arrows in $\vec{\Delta}$).□

Applying a theorem of Bondal-Kaplanov to the enhanced category $\mathrm{HMF}^{\mathrm{gr}}_{A_W}(f_W)$, we see the equivalence among the triangulated categories:

$$D^b(\text{mod-}\mathbb{C}\vec{\Delta}) \simeq D^b(\text{mod-}\mathrm{Hom}(\mathcal{E}_{\vec{\Delta}}, \mathcal{E}_{\vec{\Delta}})) \simeq \mathrm{HMF}^{\mathrm{gr}}_{A_W}(f_W). \tag{33}$$

Combining with the well known results on the representations of the hereditary algebra $\mathbb{C}\vec{\Delta}$ (c.f. [Ga], [Ri1], [Hap]), we obtain the following expected results.

Corollary. *Let the setting be as in Theorem. Then,* i) *the K-group $\mathrm{K}_0(\mathcal{T}_W)$ of $\mathcal{T}_W$ is isomorphic to the root lattice of type $W = W^*$,* ii) *the image set in $\mathrm{K}_0(\mathcal{T}_W)$ of indecomposable objects form the root system R_{W^*} of type W^*, and iii) the image in $K_0(\mathcal{T}_W)$ of a strongly exceptional collection $\mathcal{E}_{\vec{\Delta}}$ forms a simple root basis of the root system R_{W^*}.*

Remark. As in the A_l case [Ta2], a stability condition (Bridgeland [Bri 1]) can be naturally given by the grading of matrix factorizations.

The abelian category associated to the stability condition (as a full subcategory of $\mathrm{HMF}^{\mathrm{gr}}_{A_W}(f_W)$) is equivalent to the category mod-$\mathbb{C}\vec{\Delta}$ of finite modules over the path algebra of a Dynkin quiver $\vec{\Delta}$ of the principal orientation introduced in [Sa21].

§18. The category of matrix factorizations: the case $\varepsilon_W = -1$

In this section, we describe the category $\mathcal{T}_W = \mathrm{HMF}^{\mathrm{gr}}_{A_W}(f_W)$ associated with a regular system of weights W with $\varepsilon_W{=}-1$ and $a_0{=}0$. Recall that the orbifold curve C_W (12) is of genus a_0 so that we are considering the case of rational orbifold curves. There are 14+8 such weight systems, which are listed in Table 10. The associated polynomials f_W of type W are also listed in Table 10, where we remark that, in the first 14 weight systems, there are 3 orbifold points on the curve C_W so that the polynomial f_W contains no moduli parameter, whereas, in the latter 8 weight systems case, there are either 4 or 5 orbifold points on the curve C_W so that the polynomial f_W contains either one or two moduli parameters λ or λ_1, λ_2, respectively.

In order to recall Theorem in [K-S-T 2] 5.4, we introduce some particular quiver $\Delta_{A(W)}$ depending only on the signature set $A(W)$ (13) (see Footnote 32) for the orbifold structure on C_W. Slightly more generally, let us define

Definition. Let $$A = \{\alpha_1, \cdots, \alpha_r\}$$
be a multi-set of r positive integers for some $r \in \mathbb{Z}_{\geq 0}$. Then the quiver Δ_A of type A is defined by the following figure and data.

Table 14. $\Delta_A = (\Pi_A, E_A)$

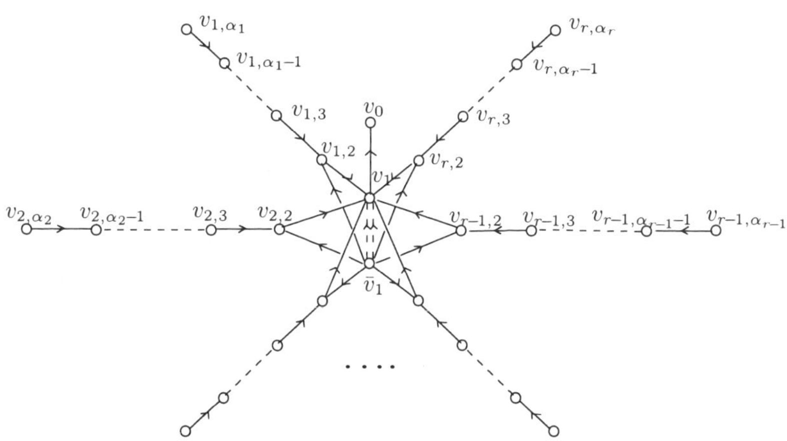

where the set of vertices and the set of arrows are given as follows:

$$\begin{array}{rcl} \Pi_A & := & \{v_0,\ v_1,\ \overline{v}_1\} \amalg \amalg_{i=1}^r\{v_{i,2},\ \cdots,v_{i,\alpha_i}\}, \\ E_A & := & \{(v_1,v_0;+),\ (\overline{v}_1,v_1;-)_1,\ (\overline{v}_1,v_1;-)_2\} \\ & & \amalg\ \amalg_{i=1}^r \{(v_{i,2},v_1;+),\ \cdots,(v_{i,\alpha_i},v_{i,\alpha_i-1};+),\ (\overline{v}_1,v_{i,2};+)\}. \end{array}$$

Remark 9. We have only two negatively signed arrows between the vertices $\bar{v}_1$ and v_1. They are distinguished by the subscripts 1 and 2 as $(\overline{v}_1,v_1;-)_1$ and $(\overline{v}_1,v_1;-)_2$. They shall later turn to relations in the path algebra.

Before we state the main theorem, we introduce one more numerical invariant: the *dual rank* ν_W for any regular weight system W. It is defined by using exponents $e_W(i)$ defined at Preface of the paper, as

$$\nu_W := -\textstyle\sum_{j|h} j \cdot e_W(h/j). \tag{34}$$

It is introduced [Sa17] (7.2) as the rank of W^* (if it exists). Actually, we prove the formula (whose proof will appear elsewhere):

$$\nu_W = \textstyle\sum_{i=1}^r(\alpha_i-1)+2(1-a_0)-\varepsilon_W$$

where $A(W)=\{\alpha_1,\cdots,\alpha_r\}$ is the signature set of W (see Footnote 31).

Remark 10. In this section, we have $\varepsilon_W=-1$ and $a_0=0$. So the formula reduces to

$$\nu_W = \textstyle\sum_{i=1}^r(\alpha_i-1)+3.$$

Then, one observes that the first term of this formula coincides with the number of vertices on the r branches of the diagram Δ_A and the last term 3 coincides with the number of vertices on the central axis of the diagram Δ_A.

The same interpretation is possible for the case of the previous section §17, where one has $\varepsilon_W=1$ and $a_0=0$ so that one has $\nu_W = \sum_{i=1}^r(\alpha_i-1)+1$. Then this formula again describes the number of vertices in a Dynkin diagram. However, in the case when the weight systems are selfdual, rank and dual rank coincide with each other, and it is unnecessary to introduce such dual rank.

Theorem. *Let W be a regular system of weights with $\varepsilon_W=-1$ and $a_0=0$. We fix a polynomial f_W of type W. Let $\mathcal{T}_W$ (25) be triangulated category associated to f_W. Then, there exists a strongly exceptional collection $\mathcal{E}_{\Delta_A}=(E_1,\cdots,E_{\nu_W})$ of the category $\mathcal{T}_W$ satisfying the following properties.*

i) *The* $\mathcal{E}_{\Delta_A}$ *generate the triangulated category* $\mathcal{T}_W$.

ii)) *The quiver associated with the collection* $\mathcal{E}_{\Delta_A}$ *is equal to* Δ_A (Table 14), *where* A *is equal to the signature set* $A(W)$ *of* W.

iii) *If the* $(v_i, v_j; +)$ *is a real arrow of* Δ_A, *then* $\mathrm{Hom}_{\mathcal{T}_W}(E_i, E_j)$ *is a vector space of rank* $-C_{\mathcal{E},ij}(=1)$. *If, further, the arrow* $(v_i, v_j; +)$ *is on the branches of* Δ_A, *then* $\mathrm{Hom}_{\mathcal{T}_W}(E_i, E_j)$ *is spanned by an irreducible homomorphism.*

iv) *The assignments*

$$(v_i, v_j, +) \longmapsto f_{ij}$$

of a base f_{ij} *of* $\mathrm{Hom}_{\mathcal{T}_W}(E_i, E_j)$ *to each arrow* $(v_i, v_j; +)$ *of* $\Delta_{\mathcal{E}}^{+}$ *together with suitable choices, depending on* f_W *and* f_{ij}, *of the relations*

$$(35) \qquad \begin{array}{rcl} R((\overline{v}_1, v_1; -)_1) & = & \sum_{i=1}^{r} \lambda_{1,i}\ (\overline{v}_1, v_{i,2}; +) \circ (v_{i,2}, v_1; +), \\ R((\overline{v}_1, v_1; -)_2) & = & \sum_{i=1}^{r} \lambda_{2,i}\ (\overline{v}_1, v_{i,2}; +) \circ (v_{i,2}, v_1; +), \end{array}$$

induce an isomorphism:

$$(36) \qquad \mathbb{C}(\Delta_{A(W)}, R) \ \simeq \ \mathrm{Hom}_{\mathcal{T}_W}(\mathcal{E}, \mathcal{E})$$

between the path algebra (32) *and the hom-algebra (recall* §16 **4.** Theorem*).*

Combining the isomorphism (36) with the theorem of Bondal-Kaplanov (see §16 **4.**) on the enhanced category $\mathrm{HMF}^{\mathrm{gr}}_{A_W}(f_W)$, we obtain:

Corollary. *We have the equivalence between the triangulated categories:*

$$D^b(\mathrm{mod}\text{-}\mathbb{C}(\Delta_{A(W)}, R)) \ \simeq \ \mathrm{HMF}^{\mathrm{gr}}_{A_W}(f_W).$$

Recall that the signature set $A(W)$ for the 14 weight systems coincides with the set of Dolgachev numbers (§13), and that it is equal to the set of Gabrielov numbers (recall Table 12, 13) for the $*$-dual weight system W^* (§14, Fact 7.).

Recall the basis f_i (or g_i) defined by the formula (28) of the K-group of the category $D^b(\mathrm{mod}\text{-}\mathbb{C}(\Delta_{A(W)}, R))$. In view of the definition (27) of the bilinear form on the K-group and the intesection number (29) among the basis elements, we see that the K-group, as a lattice, coincides with the lattice associated with the Gabrielov diagram (Table 12) for the dual weight system W^*. That is, we have the isomorphism of lattices equipped with symmetric bilinear forms:

$$(37) \qquad \mathrm{K}_0(\mathcal{T}_W) \ \simeq \ \mathrm{H}_2(X_{W^*,1}, \mathbb{Z})$$

of the K-group of the category for the weight system W and the middle homology group of the Milnor fiber (see Footnote 35) of the dual weight system W^*. In this sense, mirror symmetry at the homology group level is confirmed. However, the characterization of the subset in the LHS corresponding to the set of vanishing cycles R_{W^*} in the RHS is unknown. We ask whether it is the set of images of indecomposable exceptional objects in $\mathcal{T}_W$ or not (see Footnote 48).

As was discussed in the Preface, there are three Lie algebras associated to the 14 regular weight systems W (which admit a $*$-dual weight system W^*):

i) the algebra $\mathfrak{g}_{W^*}$ defined by the Chevalley generators and generalized Serre relations [S-Y] (4.1.1) for the Cartan matrix associated to the diagram $\Delta_{A(W)}$,

ii) the algebra $\mathfrak{g}'_{W*}$ generated by the vertex operators e^α for roots $\alpha \in R_{W^*}$ in the Lie algebra $V_{K_0(\mathcal{T}_W)}/DV_{K_0(\mathcal{T}_W)}$ ([Bo1], [S-Y](3.2.1)) for the lattice $K_0(\mathcal{T}_W)$,

iii) the algebra $\mathfrak{g}''_{W^*}$ constructed by Ringel-Hall construction ([To], [P-X], [X-X-Z]) for the derived category $D^b(\text{mod-}\mathbb{C}(\Delta_{A(W)}, R))$ of the path algebra $\mathbb{C}(\Delta_{A(W)}, R)$.

The following question is the last question of the present paper.

Problem. Clarify the relationship among these three Lie algebras. Are they isomorphic to each other? Do any of (or the covering of) these algebras satisfies the requirements posed by Question in §12 and by Addition to Question in §14?

Remark. For the 14 exceptional weight systems W, the (conjectural) period domain for the period map for the primitive form of type W is introduced [Lo6], [Sa22] (c.f. [Ao]) as

$$\mathcal{B}_V \ := \ \{ \ \varphi \in \mathrm{Hom}_{\mathbb{R}}(V, \mathbb{C}) \mid \ker(\varphi) < 0\}$$

where $V := (Q_W \otimes \mathbb{R}, I)$ is the real vector space equipped with a quadratic form I of the signature $(l+2, 0, 2)$, and $\ker(\varphi) < 0$ means that the restriction of I to the subspace $\ker(\varphi)$ is negative definite. It is interesting to clarify the relationship of the period domain $\mathcal{B}_V$ for W with the space of stability conditions (Bridgeland [Bri 1,2,3],[H-M-S]) for the category $\mathcal{T}_{W^*}$ through the identification $\mathrm{K}_0(\mathcal{T}_{W^*}) \simeq (Q_W, -I)$ due to the above Theorem. The ring of "automorphic forms" (in suitable sense, c.f. [Ao]) on $\mathcal{B}_V$ with respect to the group W_W is expected to carry the flat structure (c.f. §12 Question vi)). For some recent developments on the geometry of the modular varieties for the orthogonal groups $O(2,n)$, we refer to [Bo1, Bo2, Bo3], [G-H-S 1, G-H-S 2, G-H-S 3] and [Gr].

§19. Appendix. McKay correspondence and its Inverse.

1. McKay correspondence (1979) [Mc].

We recall McKay correspondence in its original form [Mc]. For its further understanding from a categorical view point, see [B-K-R].

Let ρ be the faithful representation of the Kleinean group $\tilde{G}$ into $SU(2)$. Let $\{\rho_0 = 1, \rho_2, \ldots, \rho_n\}$ be the set of isomorphism classes of all irreducible representations of $\tilde{G}$. Consider the decomposition

$$\rho \otimes \rho_j \quad = \quad \sum_{i=\phi}^{n} n_{ij}\rho_i \qquad (j = 0, \ldots, n)$$

for some nonnegative integers $n_{ij} \in \mathbb{Z}_{\geq 0}$. Then, one has:

0) $n_{ij} \in \{0, 1\}$
i) $n_{ii} = 0 \quad (i = 0, \ldots, n)$
ii) $n_{ij} = n_{ji}$
iii) $\tilde{C} := 2I_{n+1} - (n_{ij})_{i,j=0}^{n}$ is negative semi-definite with 1-dimensional kernel.

Actually, from these properties, it is not hard see that $\tilde{C}$ is an affine Cartan matrix of one of types $\tilde{A}_l$, $\tilde{D}_l$ or $\tilde{E}_6, \tilde{E}_7$ or $\tilde{E}_8$, and that the matrix C obtained by deleting column and low for the trivial representation is a Cartan matrix of one of types A_l, D_l, E_6, E_7 or E_8 (see Table 3.). The correspondence:

(MC): $\qquad \tilde{G} \longmapsto \Gamma :=$ the graph associated to C

induces the bijection, called the McKay correspondence:

$$\{\text{Kleinean groups}\} \xrightarrow{\sim} \{\text{Simply laced Coxeter-Dynkin graphs of finite type}\}$$

McKay wrote [Mc] "Would not the Greeks appreciate the result that the simple Lie algebras may be derived from Platonic solids?".

2. Gonzalez-Verdier interpretation of McKay correspondence

The work by Gonzalez-Verdier [G-V] says that the representations ρ_i are interpreted as vector bundles $\tilde{V}_i$ on the resolution $\tilde{X}_0$ of the singularity X_0. Then, the 1-st Chern classes $c_1(\tilde{V}_i)$ of the vector bundles form the dual basis to the homology classes of the exceptional curves $[E_i]$ in $\tilde{X}_0$. That is: $c_1(\tilde{V}_i)$ form the fundamental weight for the simple root system.

Let $\rho_i : G \to GL(V_i)$ be an irred. repr. of G. So G acts on $\mathbb{C}^2 \times V_i$ diagonally. Then the diagram (not precise)

$$\begin{array}{ccc} (\mathbb{C}^2 \times V_i)/G & \leftarrow & \widetilde{V}_i \\ \downarrow & & \downarrow \\ \mathbb{C}^2/G \simeq X_0 & \leftarrow & \tilde{X}_0 \end{array}$$

defines an irreducible vector bundle $\widetilde{V}_i$ on $\widetilde{X}_0$

Theorem (Gonzalez-Verdier 1984). *The first Chern class $c_1(\widetilde{V}_i)$ of $\widetilde{V}_i$ defines a divisor (a smooth curve) in $\widetilde{X_0}$, which is transversal to exactly one irreducible component, say E_i, of $E = \pi^{-1}(0)$. That is: $c_1(\widetilde{V}_1), \cdots, c_1(\widetilde{V}_l) \in \mathrm{H}^2(\widetilde{X}_0, \mathbb{C})$ forms the dual basis of $[E_1], \ldots, [E_l] \in \mathrm{H}_2(\widetilde{X}_0, \mathbb{Z}) \simeq Q := \bigoplus_{i=1}^{l} \mathbb{Z}\rho_i^v$.*

Table 15. The first Chern classes of irreducible vector bundles over $\widetilde{X}_{A_4,0}$.

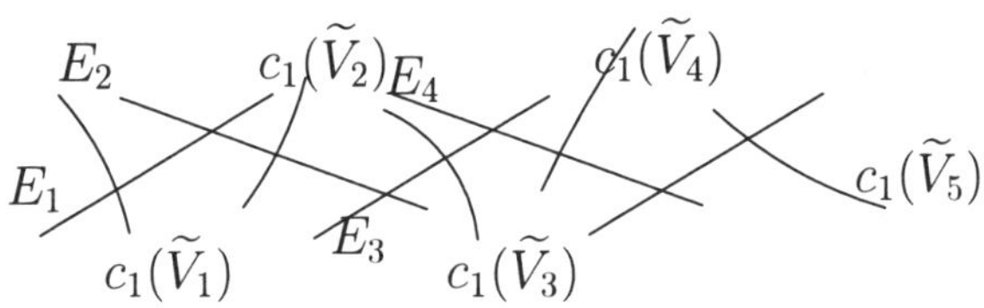

3. The inverse of McKay correspondence: $\Gamma \mapsto W_\Gamma \mapsto \langle A(W_\Gamma)\rangle$.

Let us construct conceptually the inverse of the McKay correspondence (MC) (through regular systems of weights) without using the classification.

Let a simply-laced Dynkin diagram Γ (or, equivalently a Cartan matrix C of finite type) be given. The data determine the Coxeter-Killing transformation $\mathbf{c}$ and using its eigenvalues, as we did in §8, we obtain the system of exponents $m_1, \cdots, m_\mu$. Then, as was discussed in §8, the generating function (1) of the exponents decomposes as (2) so that we obtain a simple weight system $W = W_\Gamma$.

How to recover the Kleinean group $\tilde{G}$ from a simple weight system W?

Let W be a simple weight system (i.e. $\varepsilon_W > 0$, see §8 Fact 1). Let f_W be the simple polynomial of the type W (9), and let us consider the associated hypersurface $X_{W,0}$ (11) (the simple singularity). Due to Fact 2 in §8 and Theorem in §1, the fundamental group of $X_{W,0} \setminus \{0\}$ is nothing but the isomorphic to the Kleinean group to define the simple singularity. on the other hand, we can determine the fundamental group purely arithmetically as follows.

Fact 12. *Let W be either a simple weight system or one of the 14 non-degenerate weight systems with $\varepsilon_W = -1$ and $a_0 = 0$. Then, we have the following isomorphism:*

$$\pi_1(X_{W,0} \setminus \{0\}, *) \simeq \langle A(W)\rangle,$$

where we recall that $A(W)$ is the signature set of W (see §11 a) Fact 5.*) and that* $\langle\{p,q,r\}\rangle := \langle p,q,r\rangle$ *denotes the group defined in* §1.

Proof. Combining a result of Mumford, which we quote below, and the description of the singularity $X_{W,0}$ in §11 a) Fact 6. (see also its following Example), we get the result. □

Theorem (Mumford 1961). *Let X_0 be a two dimensional normal singularity, and let $\tilde{X}_0 \to X_0$ be a resolution of the singularity such that the exceptional set $E := \pi^{-1}(0)$ is a union of $\mathbb{P}^1$ such that the intersection diagram is a tree. Then, by the use of the data of the tree (details are omitted), one can write down $\pi_1(X_0 \backslash \{0\}, *)$ by suitable generators and relations, explicitly.*

References

[Ac] N. A'Campo, Le nombre de Lefschetz d'une monodromie, Indag. Math., **35** (1973), 113–118.

[Ao] H. Aoki, Automorphic forms on the expanded symmetric domain of type IV, Publ. Res. Inst. Math. Sci., **35** (1999), 263–283.

[Ar1] V. I. Arnol'd, Normal forms of functions near degenerate critical points, the Weyl groups A_k, D_k, and E_k, and Lagrangian singularities, Funct. Anal. Appl., **6** (1972), 254–272.

[Ar2] ———, Remarks on the stationary phase method and Coxeter numbers, Russian Math. Surveys, **28** (1973), no. 5, 19–48.

[Ar3] ———, Critical points of smooth functions, Proc. Internat. Congr. Math., Vancouver, 1974, **1**, pp. 19–39.

[Ar4] ———, Critical points of smooth functions and their normal forms, Russian Math. Surveys, **30** (1975), no. 5, 1–75.

[Art] M. Artin, On isolated rational singularities of surfaces, Amer. J. Math., **88** (1966), 129–136.

[A-W] M. Artin and J.-L. Verdier, Reflective modules over rational double points, Math. Ann., **270** (1985), 79–82.

[Au] M. Auslander, Rational singularities and almost split sequences, Trans. Amer. Math. Soc., **293** (1986), 511–531.

[A-R1] M. Auslander and I. Reiten, Almost split sequences for rational double points, Trans. Amer. Math. Soc., **302** (1987), 87–99.

[A-R2] M. Auslander and I. Reiten, Cohen-Macaulay modules for graded Cohen-Macaulay rings and their completions, Commutative algebra, Berkeley, CA, 1987, Math. Sci. Res. Inst. Publ., **15**, Springer-Verlag, 1989, pp. 21–31.

[A-R3] M. Auslander and I. Reiten, Almost split sequences for $\mathbb{Z}$-graded rings, Singularities representation of algebras, and vector bundles, Lambrecht, 1985, Lecture Notes in Math., **1273**, Springer-Verlag, 1987, pp. 232-243.

[Bon] A. Bondal, Representations of associative algebras and coherent sheaves, Izv. Akad. Nauk SSSR Ser. Mat., **53** (1989), 25–44; translation in Math. USSR-Izv., **34** (1990), 23–42; Helices, representations of quivers and Koszul algebras, Helices and vector bundles, 75–95, London Math. Soc. Lecture Note Ser., **148**, Cambridge Univ. Press, Cambridge, 1990.

[B-K1] A. Bondal and M. Kapranov, Representable functors, Serre functors and Mutations, IZv. Akad. Nauk SSSR Ser. Mat., **53** (1989), 1183–1205; English transl. in Math. USSR Izv., **35** (1990), 519–541.

[B-K2] A. Bondal and M. Kapranov, Enhanced triangulated categories, Math. USSR Sbornik, **70** (1991), 93–107.

[B-O] A. Bondal and D. Orlov, Semiorthogonal decomposition for algebraic varieties, arXiv:math.AG/9506012.

[Bo1] R. Borcherds, Vertex algebras, Kac-Moody algebras and the Monster, Proc. Nat. Acad. Sci. USA, **83** (1986), 3068–3071.

[Bo2] R. Borcherds, Generalized Kac-Moody algebras, J. Algebra, **115** (1988), 501–512.

[Bo3] R. Borcherds, Automorphic forms on $O_{s+2,2}(\mathbb{R})$ and infinite products, Invent. Math., **120** (1995), 161–213.

[Bou] N. Bourbaki, Éléments de mathématique, Fasc. XXXIV, Groupes et algèbres de Lie, Chs. 4–6. Hermann, Paris 1968.

[Br1] E. Brieskorn, Rationale Singularitäten komplexer Flächen, Invent. Math., **4** (1968), 336–358.

[Br2] ———, Die Auflösung der rationalen Singularitäten holomorpher Abbildungen, Math. Ann., **178** (1968), 255–270.

[Br3] ———, Die Monodromie der isolierten Singularitäten von Hyperflächen, Manuscripta Math., **2** (1970), 103–161.

[Br4] ———, Singular elements of semi-simple algebraic groups, Actes du Congrès International des Mathématiciens, Nice, 1970, **2**, Gauthier-Villars, Paris, 1971, pp. 279–284.

[Br5] ———, Die Fundamentalgruppe des Raumes der regulären Orbits einer endlichen komplexen Spiegelungsgruppe, Invent. Math., **12** (1971), 57–61.

[Br6] ———, Sur les groupes de tresses (d'après V. I. Arnol'd), Séminaire Bourbaki 1971/72, Exposé 401, Lecture Notes in Math., **317**, Springer-Verlag, 1973, pp. 21–44.

[Br7] ———, Milnor lattices and Dynkin diagrams, in [Or], part 1, pp. 153–165.

[Br8] ———, Die Milnorgitter der exzeptionellen unimodularen Singularitäten, Bonner Math. Schriften, **150** (1983).

[B-S] E. Brieskorn and K. Saito, Artin-Gruppen und Coxeter-Gruppen, Invent. Math., **17** (1972), 245–271.

[Bri 1] T. Bridgeland, Stability conditions on triangulated category, math.AG/0212237.

[Bri 2] T. Bridgeland, Stability conditions and Kleinian singularities, math.AG/0508257.

[Bri 3] T. Bridgeland, Stability conditions on K3 surfaces, math.AG/0307164.

[B-K-R] T. Bridgeland, A. King and M. Reid, The McKay correspondence as an equivalence of derived categories, J. Amer. Math. Soc., **14** (2001), 535–554 (electronic).

[Bu] R.-O. Buchweitz, Maximal Cohen-Macaulay modules and Tate-Cohomology over Gorenstein rings, preprint, 1986.

[B-E-H] R.-O. Buchweitz, D. Eisenbud and J. Herzog, Cohen Macaulay Modules on Quadrics, with an appendix by R.-O. Buchweitz, Lect. Notes in Math., **1273**, Springer-Verlag, 1987, pp. 58–116.

[Ca] E. Cartan, a) Sur la structure des groupes de transformations finis et continus (thèse), Paris (Nony), 1894 (=Œuvres complètes, Paris, (Gauthier-Villars), 1952, t. I_1, 137–287); b) Sur la réduction à sa forme canonique de la structure d'un groupe de transformations fini et continu, Amer. J. Math., t. XVIII (1896), 1–46 (=Œuvres complètes, t. I_1, 293–353); c) Le principe de dualité et la théorie des groupes simples et semi-simples, Bull. Sci. Math., t. XLIX (1925), 361–374 (=Œuvres complètes, t. I_1, 555–568); d) La géométrie des groupes simples, Ann. di Mat. (4), t. IV (1927), 209–256 (=Œuvres complètes, t. I_2, 793–840); e) Sur certaines formes riemanniennes remarquables des géométries à groupe fondamental simple, Ann. Ec. Norm. Sup. (3), t. XLIV (1927), 345–467 (=Œuvres complètes, t. I_2, 867–989); f) Complément au mémoire ⟨Sur la géométrie des groupes simples⟩, Ann. di Mat. (4), t. V (1928), 253–260 (=Œuvres complètes, t. I_2, 1003–1010).

[Ch] C. Chevalley, Invariants of finite groups generated by reflexions, Amer. J. Math., **77** (1955), 778–782.

[Col] A. J. Coleman, The Betti numbers of the simple Lie groups, Canad. J. Math., **10** (1958), 349–356.

[Co1] H. S. M. Coxeter, a) Groups whose fundamental regions are simplexes, J. Lond. Math. Soc., t. VI (1931), 132–136; b) The polytopes with regular prismatic figures, Proc. Lond. Math. Soc. (2), t. XXXIV (1932), 126–189; c) Discrete groups generated by reflections, Ann. of Math. (2), t. XXXV (1934), 588–621; d) The complete enumeration of finite groups of the form $R_i^2 = (R_i \cdot R_j)^{k_{ij}} = 1$, J. Lond. Math. Soc., t. X (1935), 21–25.

[Co2] H. S. M. Coxeter, The product of generators of a finite group generated by reflections, Duke Math. J., **18** (1951), 765–782.

[C-M] H. S. M. Coxeter and W. Moser, Generators and relations for discrete groups, 2nd ed., Springer-Verlag, 1965.

[De1] P. Deligne, Les immeubles des groupes de tresses généraliesés, Invent. Math., **17** (1972), 273–302.

[De2] ———, Théorie de Hodge. I, II, III, Proc. Internat. Congr. Math., Nice, 1970, **1**, pp. 425–430; Inst. Hautes Études Sci. Publ. Math., **40** (1971), 5–57; ibid., **44** (1974), 5–77.

[De3] ———, A letter to E. Looijenga, dated 9. 3. 74, published in the Anhang of [Klu].

[Dem] M. Demazure, Surfaces de Del Pezzo. I–V, Séminaire sur les Singularités des Surfaces 1976/77, Lecture Notes in Math., **777**, Springer-Verlag, 1980, pp. 21–69.

[Do] B. Dobrovin, Geometry of 2D-Topological Field Theory, Lecture Notes in Math., **1620** (1996), pp. 120–348.

[Dol1] I. Dolgachev, Quotient-conical singularities on complex surfaces, Funkcional Anal. i Priložen., Translated in Funct. Anal. Appl., **8** (1974), 160–161.

[Dol2] ———, Automorphic forms and quasihomogeneous singularities, Funkcional Anal. i Priložen., **9** (1975), 67–68; translated in Funct. Anal. Appl., **9** (1975), 149–151.

[Dol3] ———, On the link space of a Gorenstein quasihomogeneous surface singularity, Math. Ann., **265** (1983), 529–540.

[Dol4] ———, Weighted projective varieties, In: Group actions and Vector Fields, Proceedings, Vancouver 1981 (ed. J. Carrell), Lecture Note in Math., **956**, Springer-Verlag, 1982, pp. 34–71.

[Dol5] ———, Mirror symmetry for lattice polarized K3 surfaces, J. Math. Sci., **81** (1986), 2599–2630.

[Du] P. D. Val, On isolated singularities which do not affect the conditions of adjunction. I, II, III, Proc. Cambridge Philos. Soc., **30** (1934), 453–459, 460–465, 483–491.

[Dun] V. Nguyen-Dung, The fundamental groups of the spaces of regular orbits of the affine Weyl groups, Topology, **22** (1983), 425–435.

[Dur] A. H. Durfee, Fifteen characterizations of rational double points and singular critical points, Enseignement Math. (2), **25** (1979), 131–163.

[D-X] B. Deng and J. Xiao, On Ringel-Hall algebras, Representations of finite dimensional algebras and related topics in Lie theory and geometry, Fields Inst. Commun., **40**, Amer. Math. Soc., Providence, RI, 2004, pp. 319–348.

[Eb1] W. Ebeling, Quadratische Formen und Monodromiegruppen von Singularitäten, Math. Ann., **255** (1981), 463–498.

[Eb2] ———, An arithmetic characterisation of the symmetric monodromy groups of singularities, Invent. Math., **77** (1984), 85–99.

[Eb3] ———, The Poincaré series of a quasihomogeneous surface singularity, preprint, 2000.

[E-W] W. Ebeling and T. Wall, Kodaira Singularities and an Extension of Arnold'd Strange Duality, Compositio Math., **56** (1985), pp. 3–77.

[E-G] W. Ebeling and S. M. Gusein-Zades, Monodromies and Poincaré series of quasihomogeneous complete intersections, Abh. Math. Sem. Univ. Hamburg, **74** (2004), 175–179.

[Ei] D. Eisenbud, Homological algebra on a complete intersection, with an application to group representations, Trans. Amer. Math. Soc., **260** (1980), 35–64.

[Fr] I. B. Frenkel, Representations of Kac-Moody Algebras and Dual Resonance Models, Lect. Appl. Math., **21** (1985), 325–353.

[F-K] R. Fricke and F. Klein, Vorlesungen über die Theorie der automorphen Funktionen, **I**, Teubner, Leipzig, 1897.

[Ga] P. Gabriel, Unzerlegbare Darstellungen I, Manuscripta Math., **6** (1972), 71–163.

[Gab1] A. M. Gabrièlov, Intersection matrices for certain singularities, Funct. Anal. Appl., **7** (1973), 182–183.

[Gab2] ———, Dynkin diagrams for unimodal singularities, Funct. Anal. Appl., **8** (1974), 192–196.

[G-V] G. Gonzalez-Sprinberg and J.-L. Verdier, Construction géométrique de la correspondence de McKay, Ann. Sci. École Norm. Sup. (4), **16** (1983), 409–449.

[G-L 1] W. Geigle and H. Lenzing, A class of weighted projective curves arising in representation theory of finite dimensional algebras, In: Singularities, Representations of Algebras and Vector Bundles, Lecture Notes in Math., **1273**, Springer-Verlag, 1987, pp. 265–297.

[G-L 2] W. Geigle and H. Lenzing, Perpendicular categories with applications to representations and sheaves, J. Algebra, **144** (1991), 273–234.

[G-M] S. I. Gelfand and Y. I. Manin, Methods of Homological algebra, Vol. 1: Introduction to Cohomology Theory and Derived Categories, Nauka, Moscow, 1988, (in Russian); Homological Algebra in Algebra V, Springer-Verlag, 1994, encycl. Math. Sci., **38**, pp. **1**–222.

[G-M] G.-M. Greuel and H. Knörrer, Einfache Kurven Singularitäten und torsionfreie Moduln, Math. Ann., **270** (1985), 417–425.

[G-H-S 1] V. A. Gritsenko, K. Hulek and G. K. Sankaran, The Kodaira dimension of the moduli of K3 surfaces Invent. Math., **169** (2007), 519–567.

[G-H-S 2] V. A. Gritsenko, K. Hulek and G. K. Sankaran, The Hirzebruch-Mumford Volume for the Orthogonal Group and Applications Documenta Math., **12** (2007), 215–241.

[G-H-S 3] V. A. Gritsenko, K. Hulek and G. K. Sankaran, Hirzebruch-Mumford proportionality and locally symmetric varieties of orthogonal type, preprint, math.AG/0609774.

[Gr] V. Gritsenko, A solution of one automorphic problem of K. Saito, to appear in Koukyu-roku, RIMS, 2008.

[H-L] H. Hamm and L. D. Tráng, Un théorème de Zariski du type de Lefschetz, Ann. Sci. École Norm. Sup. (4), **6** (1973), 317–366.

[H-R] D. Happel and C. Ringel, The derived category of a tubular algebra, Lecture Notes in Math., **1273**, Springer-Verlag, 1986, pp. 156–180.

[Hap] D. Happel, Triangulated categories in the representation theory of finite-dimensional algebras, London Math. Soc. Lecture Note Ser., **119**, Cambridge Univ. Press, Cambridge, 1988, x+208p.

[Ha] T. Hawkins, Wilhelm Killing and the Structure of Lie Algebras, Arch. Hist. Exact Sci., **26** (1982), 127–192.

[H-S 1] S. Helmke and P. Slodowy, On Unstable Principal Bundles over Elliptic Curves, preprint, April 2000.

[H-S 2] S. Helmke and P. Slodowy, Loop groups, elliptic singularities and principal bundles over elliptic curves, MathSci. MR2055872.

[He] C. Hertling, Frobenius manifolds and moduli spaces for singularities, Cambridge Univ. Press, 2002.

[H-W] K. Hori and J. Walcher, F-term equations near Gepner points, J. High Energy Phys., (1):008, 23 pp. (electronic), 2005.

[H-I-V] K. Hori, A. Iqbal and C. Vafa, D-branes and mirror symmetry, hep-th/0005247, 2000.

[H-V] K. Hori and C. Vafa, Mirror symmetry, hep-th/0002222, 2000.

[H-M-S] D. Huybrechts, E. Macri and P. Stellari, Stability conditions for generic K3 categories, math.AG/0608430.

[I-M] N. Iwahori and H. Matsumoto, On some Bruhat decomposition and the structure of the Hecke rings of p-adic Chevalley groups, Inst. Hautes Études Sci. Publ. Math., **25** (1965), 5–48.

[I-U-U] A. Ishii, K. Ueda and H. Uehara, Stability conditions on A_n-singularities, math.AG/0609551.

[I-U] A. Ishii and H. Uehara, Autoequivalences of derived categories on the minimal resolutions of A_n-singularities on surfaces, J. Differential Geom., **71** (2005), 385–435.

[K-P] V. Kac and D. H. Peterson, Infinite-dimensional Lie algebras, theta functions and modular forms, Adv. in Math., **50** (1984), 125–264.

[K-L] A. Kapustin and Y. Li, D-branes in Landau-Ginzburg models and algebraic geometry, J. High Energy Phys., (12):005, 44p. (electronic), 2003.

[K-R] M. Khovanov and L. Rozansky, Matrix factorizations and link homology, arXiv:math. QA/0401268 v2.

[K-S] M. Kashiwara and P. Schapira, Sheaves on manifolds, Grundlehren, **292**, Springer-Verlag, 1990.

[K-S-T 1] H. Kajiura, K. Saito and A. Takahashi, Matrix Factorizations and Representations of Quivers II: type ADE case, math.ag/0511155.

[K-S-T 2] H. Kajiura, K. Saito and A. Takahashi, Category of Matrix Factorizations for Exceptional Singularities, preprint, RIMS-1600.

[K-W] M. Kato and S. Watanabe, The flat coordinate system of the rational double point of E_8 type, Bull. Coll. Sci. Univ. Ryukyus, **32** (1981), 1–3.

[K-Y] T. Kawai and S. K. Yang, Duality ofbifoldized elliptic genera, Prog. Theo. Phys. Supple., **118** (1995), 277 p.

[Ke] B. Keller, Derived categories and their uses, Chapter of the Handbook of algebra, **1**, (ed. M. Hazewinkel), Elsevier, 1996.

[Ki] W. Killing, Die Zusammensetzung der stetigen endlichen Transformationsgruppen: I) Math. Ann., t. XXXI (1888), 252–290; II) ibid., t. XXXIII (1889), 1–48; III) ibid., t. XXXIV (1889), 57–122; IV) ibid., t. XXXVI (1890), 161–189.

[Kl1] F. Klein, Vorlesungen über das Ikosaeder und die Auflösung der gleichung vom fünften gerade, B. G. Teubner, Stuttgart und Leipzig, 1993.

[Kl2] F. Klein, Vorlesungen über die Entwicklung der Mathematik im 19. Jahrhundert, I, II, Springer-Verlag, 1929.

[Klu] P. Kluitman, Geometrische Basen des Milnorgitters einer einfach elliptische Singularität, Diplomarbeit, Bonn, 1983.

[Kn1] F. Knörrer, Group representations and resolution of rational double points, Finite Groups—Coming of Age, Montréal, 1982, Contemp. Math., **45**, Amer. Math. Soc., 1985, pp. 175–222.

[Kn2] H. Knörrer, Cohen-Macauley modules on hypersurface singularities I, Invent. Math., **88** (1987), 153–164.

[Ko] B. Kostant, The principal three-dimensional subgroup and the Betti numbers of a complex simple group, Amer. J. Math., **81** (1959), 973–1032.

[Kob] M. Kobayashi, Duality of weights, mirror symmetry and Arnold's strange duality, algeom/9502004 (1995).

[Kon1] M. Kontsevich, Homological algebra of mirror symmetry, Proceedings of ICM, Zurich, 1994, Basel: Birkhäuser, 1995, pp. 120–139.

[Kr] P. Kronheimer, The construction of ALE spaces as hyper-Kähler quotients, J. Differential Geom., **29** (1989), 665–683.

[La] K. Lamotke, Die Homologie isolierte Singularitäten, Math. Z., **143** (1975), 27–44.

[Lau] H. Laufer, On minimally elliptic singularities, Amer. J. Math., **99** (1977), 1257–1295.

[Le1] D. T. Le, Singularités isolées des hypersurfaces complexes, Ph. D. thesis, 1969, Acta Sc. Vietnamicarum, Hanoi, **7** (1972), 24–33.

[L-P1] Y. Lin and L. Peng, Elliptic Lie algebras and tubular algebras, Advances in Math., **196** (2005), 487–530.

[L-P2] Y. Lin and L. Peng, Representation Theory of Finite Dimensional Algebras and Relationship with Lie Theory, Lecture Note at RIMS, 2005, to appear in Kokyuu-roku (2007).

[Lo1] E. Looijenga, On the semi-universal deformation of a simple elliptic singularity. II, Topology, **17** (1978), 23–40.

[Lo2] ———, Root systems and elliptic curves, Invent. Math., **38** (1976), 17–32.

[Lo3] ———, Invariant theory for generalized root systems, Invent. Math., **61** (1980), 1–32.

[Lo4] ———, Rational surfaces with an anti-canonical cycle, Ann. of Math., **114** (1981), 267–322.

[Lo5] ———, The smoothing components of a triangle singularity. I, in [Or], part 2, 173–184.

[Lo6] ———, The smoothing components of a triangle singularity. II, Math. Ann., **269** (1984), 357–387.

[Lo7] ———, A new compactification theory, preprint, 1985.

[Mac] E. Macri, Some examples of moduli spaces of stability conditions on derived categories, arXiv:math.AG/0411613v2.

[Ma] W. Magnus, NonEuclidean tesselations and their groups, Academic Press, 1974.

[Man] Y. Manin, Three constructions of Frobenius manifolds: a comparative study, Asian J. Math.

[Mat] A. Matsuo, Summary of the Theory of Primitive Forms, In: Topological Field Theory, Primitive Forms and Related Topics, Progr. Math., **160**, Birkhäuser, 1998, pp. 337–363.

[Mc] J. McKay, Graphs, Singularities and Finite Groups, Proc. Sympos. Pure Math., **37**, 1980.

[Mi] J. Milnor, Singular points of complex hypersurfaces, Ann. of Math. Stud., **61**, Princeton Univ. Press, 1968.

[M-O] J. Milnor and P. Orlik, Isolated singularities defined by weighted homogeneous polynomials, Topology, **9** (1970), 385–393.

[Mu1] D. Mumford, The topology of normal singularities of an algebraic surface and a criterion for simplicity, Inst. Hautes Études Sci. Publ. Math., **9** (1961), 5–22.

[Mu2] ———, Abelian varieties, Oxford Univ. Press, 1970.

[Mu3] ———, Tata Lectures on Theta II, Jacobian theta functions and Differential equations, Birkhäuser, 1984.

[Na] H. Nakajima, Quiver varieties and Kac-Moody algebras, Duke Math. J., **91** (1998), 515–560.

[N] I. Naruki, Cross ratio variety as a moduli space of cubic surface, Proc. London Math. Soc. (3), **45** (1980), 1–30.

[No] M. Noumi, Expansion of the solution of a Gauss-Manin system at a point of infinity, Tokyo J. Math., **7** (1984), 1–60.

[Od1] T. Oda, K. Saito's period map for holomorphic functions with isolated critical points, In: Algebraic Geometry, Sendai, 1985, Adv. Stud. Pure Math., **10**, North-Holland, 1987, pp. 592–648.

[Od2] T. Oda, Introduction to Algebraic Singulariteis (with Appendix by Taichi Ambai), preprint, Tohoku Univ., 198?.

[Od:3] T. Oda, Convex bodies and algebraic geometry, Springer-Verlag, 1988.

[O-K] T. Oda and N. Katz, On the differentiation of De Rham cohomology classes with respect to parameters, J. Math. Kyoto Univ., **8** (1968), 199–213.

[O-O] H. Ohta and K. Ono, Simple singularities and topology of symplectic filling 4-manifold, Comment. Math. Helv., **74** (1999), 575–590.

[Or] P. Orlik (editor), Singularities, Proc. Sympos. Pure Math., **40**, parts 1, 2, Amer. Math. Soc., 1983.

[O-S] P. Orlik and L. Solomon, Combinatorics and topology of complements of hyperplanes, Invent. Math., **56** (1980), 167–189.

[O-W] P. Orlik and P. Wagreich, Isolated singularities with C^*-actions, Ann. of Math. (2), **93** (1971), 205–228.

[Orl1] D. Orlov, Triangulated Categories of Singularities and D-branes in Landau-Ginzburg Models, Tr. Mat. Inst. Steklova, **246** (2004); Algebr. Geom. Metody, Svyazi i Prilozh., 240–262; translation in Proc. Steklov Inst. Math., **2004**, no. 3 (246), 227–248, math.AG/0302304.

[Orl2] D. Orlov, Derived categories of coherent sheaves and triangulated categories of singularities, math.AG/0503632.

[P-X] L. Peng and J. Xiao, Triangulated categories and Kac-Moody algebras, Invent. Math., **140** (2000), 563–603.

[Pi] E. Picard, Sur les fonctions de deux variables indépendantes analogues aux fonctionons modulaires, Acta Math., **2** (1883), 114–135.

[Pin1] H. Pinkham, Singularités exceptionelles, la dualité étrange d'Arnol'd et les surfaces K-3, C. R. Acad. Sci. Paris Sér. A, **284** (1977), 615–618.

[Pin2] ———, Groupe de monodromie des singularités unimodulaires exceptionelles, C. R. Acad. Sci. Paris Sér. A, **284** (1977), 1515–1518.

[Pin3] ———, Simple elliptic singularities, Del Pezzo surfaces and Cremona transformations, Several Complex Variables (ed. R. O. Wells, Jr.), Proc. Sympos. Pure Math., **30**, part 1, Amer. Math. Soc., 1977, pp. 69–71.

[Pin4] ———, Normal surfaces singularities with C^*-action, Math. Ann., **227** (1977), 183–193.

[Pin5] ———, Deformation of normal surface singularities with C^*-action, Math. Ann., **232** (1978), 65–84.

[Ri1] C. M. Ringel, Tame algebras and integral quadratic forms, Lecture Notes in Math., **1099**. Springer-Verlag, 1984, xiii+376 p.

[Ri2] C. M. Ringel, Hall algebras, In: Topics in Algebra, Banach Center Publ., **26** (1990), 433–447.

[Ri3] C. M. Ringel, Hall algebras and quantum groups, Invent. Math., **101** (1990), 583–592.

[Ri4] C. M. Ringel, Hall algebras revisited, Israel Math. Conference Proc., **7** (1993), 171–176.

[Sab] C. Sabbah, Déformations isomonodromiques et variétés de Frobenius, 2002, EDP Sciences.

[Sa1] K. Saito, Quasihomogene isolierte Singularitäten von Hyperflächen, Invent. Math., **14** (1971), 123–142.

[Sa2] ———, Einfach elliptische Singularitäten, Invent. Math., **23** (1974), 289–325.

[Sa3] ———, Theory of logarithmic differential forms and logarithmic vector fields, J. Fac. Sci. Univ. Tokyo Sect. I A Math., **27** (1980), 265–291.

[Sa4] ———, The zeroes of characteristic function χ_f for the exponents of hypersurface isolated singular point, In: Proceedings of a Symposium Algebraic Varieties and Analytic Varieties, Tokyo, 1981, Adv. Stud. Pure Math., **1**, North-Holland, 1983, pp. 195–217.

[Sa5] ———, On the identification of intersection forms on the middle homology with the flat function via period mapping, Proc. Japan Acad. Ser. A, **58** (1982), 196–199.

[Sa6] ———, On the periods of primitive integrals. I, preprint, **412**, Res. Inst. Math. Sci., Kyoto Univ., 1982.

[Sa7] ———, Period mapping associated to a primitive form, Publ. Res. Inst. Math. Sci., **19** (1983), 1231–1264.

[Sa8] ———, The higher residue pairings $K_F^{(k)}$ for a family of hypersurface singular points, in [Or], part 2, pp. 441–463.

[Sa9] ———, A new relation among Cartan matrix and Coxeter matrix, J. Algebra, **105** (1987), 149–158.

[Sa10] ———, On theta invariants for extended affine root systems and the moduli spaces for simple elliptic singularities, RIMS Kokyuroku, **535** (1984), 1–23. (Japanese)

[Sa11] ———, Regular systems of weights and associated singularities, In: Complex Analytic Singularities, Tsukuba, 1984, Adv. Stud. Pure Math., **8**, North-Holland, 1986, pp. 479–526.

[Sa12] ———, Algebraic surfaces for regular systems of weights, Algebraic Geometry and Commutative Algebra (in Honor of Masayoshi Nagata), Kinokuniya, 1987, pp. 517–614.

[Sa13] ———, On the existence of exponents prime to the Coxeter number, J. Algegra, **114** (1988), 333–356.

[Sa14] ———, Extended affine root systems. I. (Coxeter transformations) Publ. Res. Inst. Math. Sci., **21** (1985), 75–179; II. (Flat invariants) Publ. Res. Inst. Math. Sci., **26** (1990), 15–78; V (Elliptic Eta-Products and Their Dirichlet Seires) Centre de recherches Mathématiques, CRM Proceedings and Necture Notes, **30** (2001), 185–222.

[Sa15] ———, Around the Theory of the Generalized Weight System: Relations with Singularity Theory, the Generalized Weyl Group and Its Invariant Theory, Etc., (Japanese) Sugaku, **38** (1986), 97–115, 202–217; translation in English, Amer. Math. Soc. Transl. (2), **183** (1998), 101–143.

[Sa16] ———, On a linear Structure of the Quotient Variety by a finite Reflection Group, Publ. Res. Inst. Math. Sci., **29** (1993), 535–579.

[Sa17] ———, Duality for Regular Systems of Weights, Asian. J. Math., **2** (1998), 983–1048.

[Sa18] ———, Uniformization of the orbifold of a finite reflection group, In: Frobenius Manifolds, (eds. C. Hertling and M. Marcolli), Aspect Math., **E36** (2004), 265–320.

[Sa19] ———, Primitive Automorphic Forms, Mathematics Unlimited - 2001 and Beyond, edited by Engquist & Schmidt, Springer-Verlag, 2001, 1003–1018.

[Sa20] ———, Polyhedra Dual to the Weyl Chamber Decomposition: A Précis, Publ. RIMS, Kyoto Univ., **40** (2004), 1337–1384.

[Sa21] ———, Principal Γ-cone for a tree Γ, Adv. Math., **212** (2007), 645–668.

[Sa22] ———, Period domain $\mathcal{B}_V$ associated with indefinite quadratic form I, Seminar note, 5 August 1991, in Japanese, English translation by H. Aoki (June 2001), to appear.

[S-T] K. Saito and T. Takebayashi, Extended affine root systems III (Elliptic Weyl Groups), Publ. Res. Inst. Math. Sci., **33** (1997), 301–329.

[S-Y] K. Saito and D. Yoshii, Extended affine root systems IV (Elliptic Lie Algebras), Publ. Res. Inst. Math. Sci., **36** (2000), 385–421.

[S-Y-S] K. Saito, T. Yano and J. Sekiguchi, On a certain generator system of the ring of invariants of a finite reflection group, Comm. Algebra, **8** (1980), 373–408.

[Sai] M. Saito, On the structure of Brieskorn lattices, Ann. Inst. Fourier (Grenoble), **39** (1989), 27–72.

[Sat1] I. Satake, Automorphism of the Extended Affine Root System and Modular Property for the Flat Theta Invariants, Publ. Res. Inst. Math. Sci., **31** (1995), 1–32.

[Sat2] ———, Flat Structure for the simple Elliptic Singularities of type $\widetilde{E}_6$, Proceedings of 37th Taniguchi Symposium, Birkhäuser, 1998.

[Sc] H. A. Schwarz, Über diejenigen Falle, in welche die Gaussische hypergeometrische Reihe eine algebraische Funktion ihres vierten Elementes darstellt, J. Reine Angew. Math., **75** (1872), 292–335.

[Sch] F. O. Schryer, Finite and countable CM-representation type, Singularities, representation of algebras, and vector bundles, Lambrecht, 1985, Lecture Notes in Math., **1273**, Springer-Verlag, 1987, pp. 9–34.

[Se1] J.-P. Serre, Lie algebras and Lie groups, lectures given at Harvard University, 1964, Benjamin, Inc. New York Amsterdam, 1965.

[Se2] ———, A course in arithmetic, Springer-Verlag, 1973.

[Sei] P. Seidel, Vanishing cycles and mutation, European Congress of Mathematics, Vol. II, Barcelona, Progr. Math., **202** (2001), 65–85.

[Sh] I. G. Shcherbak, Algebras of automorphic forms with three generators, Funct. Anal. Appl., **12** (1978), 156–158.

[Shi] H. Shiga, One attempt to the K3 modular function. I, II, Ann. Scuola Norm. Sup. Pisa Cl. Sci. (4), **6** (1979), 609–635; (4) **8** (1981), 157–182.

[Si] C. L. Siegel, Topics in complex function theory, Vols. I, II, III, Wiley-Interscience, 1969.

[Sl1] P. Slodowy, Simple singularities and simple algebraic groups, Lecture Notes in Math., **815**, Springer-Verlag, 1980.

[Sl2] ———, A character approach to Looijenga's invariant theory for generalized root systems, Compositio Math., **55** (1985), 3–32.

[Sl3] ———, Singularitäten, Kac-Moody-Lie algebren, assoziierte Gruppen und Verallgemeinerungen, Habilitationsschrift, Bonn, 1984.

[Sl4] ———, Beyond Kac-Moody algebras and inside, Can. Math. Soc. Proc., **5** (1986), 361–371.

[Sp] A. T. Springer, Regular elements of finite reflection groups, Invent. Math., **25** (1974), 159–198.

[St1] R. Steinberg, Regular elements of semisimple algebraic groups, Inst. Hautes Études Sci. Publ. Math., **25**, 1965.

[St2] R. Steinberg, Classes of elements of semisimple algebraic groups, Proc. of the Int. of Math., Moscow, 1966, 277–289.

[St3] R. Steinberg, Subgroups of SU(2), preprint, UCLA, 1981.

[Ta1] A. Takahashi, K. Saito's Duality for Regular Weight Systems and Duality for Orbifoldized Poincaré Polynomials, Commun. Math. Phys., **205** (1999), 571–586.

[Ta2] ———, Matrix Factorizations and Representations of Quivers I, math.AG/0506347.

[Ta3] ———, Seiberg-Witten Differential as a Primitive Form, Progress of Theoretical Physics Supplement, **135** (1999), 109–117.

[Te] H. Terao, Generalized exponents of a free arrangement of hyperplanes and Shepherd-Todd-Brieskorn formula, Invent. Math., **63** (1981), 159–179.

[Th] R. Thom, Stabilité structurelle et morphogénèse, Bejamin, Reading, Mass.

[To] A. N. Todorov, Applications of the Kähler-Einstein-Calabi-Yau metric to moduli of K3 surfaces, Invent. Math., **61** (1980), 251–266.

[Toe] B. Toën, Derived Hall algebras, Duke Math. J., **135** (2006), 587–615.

[Tu] G. N. Turina, Flat locally semi-universal deformations of isolated singulariteis of complex spaces, Izv. Akad. Nauk S. S. S. R. Ser. Mat., **33** (1969), 1026–1058.

[Ue] K. Ueda, A remark on a Theorem of Kajiura, Saito and Takahashi, Appendix to [K-S-T 1].

[Va] V. S. Varadarajan, On the ring of invariant polynomials on a semisimple Lie algebras, Amer. J. Math., **90** (1968), 308–317.

[Wae] B. L. van der Waerden, Die Klassification der einfachen Lieschen Gruppen, Math. Zeitschr., t. XXXVII (1933), 446–462.

[Wal] J. Walcher, Stability of Landau-Ginzburg branes, hep-th/0412274.

[Wa1] P. Wagreich, Algebra of automorphic forms with few generators, Trans. Amer. Math. Soc., **262** (1980), 367–389.

[Wa2] ———, The structure of quasihomogeneous singularities, in [Or], part 2, pp. 593–624.

[We1] A. Weil, Variétés kähleriennes, Hermann, 1958.

[We2] ———, Elliptic functions according to Eisenstein and Kronecker, Springer-Verlag, 1976.

[We] H. Weyl, Theorie der Darstellung kontinuierlicher halb-einfacher Gruppen durch lineare Transformationen, Math. Zeitschr., t. XXIII (1925), 271–309, t. XXIV (1926), 328–395 et 789–791 (= Selecta, Basel-Stuttgart (Birkhäuser), 1956, 262–366).

[Wi] E. Witt, Spiegelungsgruppen und Aufzählung halbeinfacher Liescher Ringe, Abh. Math. Sem. Hamb. Univ., t. XIV (1941), 289–322.

[X-X] J. Xiao and F. Xu, Hall algebras associated triangulated categories, math.QA/0608144.

[X-X-Z] J. Xiao, F. Xu and G. Zhang, Derived categories and Lie algebras, math.QA/0604564.

[Yah] T. Yahiro, Hyper-Kaehler metric, Master thesis, Feburally 1991, RIMS.

[Ya1] H. Yamada, Lie group theoretic construction of period mappings, Math. Z., **220** (1995), 231–255.

[Ya2] ———, Elliptic Root Systems and Elliptic Artin Group, Publ. Res. Inst. Math. Sci., **36** (2000), 111–138.

[Ya3] ———, On a characterization of an unstale principal G-bundle on an elliptic curve (in Japanese), Koukyuroku, **1501** (Recent topics in Real and Complex Singularities) (2006), 79–95.

[Yan] T. Yano, Flat coordinate system for the deformation of type E_6, Proc. Japn Acad. Ser. A Math. Sci., **57** (1982), 412–414.

[Yas] S. Yasuda, Non-negativity of Fourier coefficients of eta-product associated to regular systems of weights, submitted.

[Yau] S.-T. Yau, Essays on Mirror Manifolds, International Press Co., 1992.

[Yo] M. Yoshinaga, On the relative de Rham cohomology of the adjoint quotient map (in Japanese), Master thesis, February 2001, RIMS.

[Yos] Y. Yoshino, Cohen-Macaulay modules over Cohen-Macaulay rings, London Math. Soc. Lecture Note Ser., **146**, Cambridge Univ. Press, Cambridge, 1990. viii+177 pp.

Research Institute for Mathematical Sciences
Kyoto University
Kyoto 606-8502 JAPAN
E-mail address: saito@kurims.kyoto-u.ac.jp

Advanced Studies in Pure Mathematics 50, 2008
Algebraic Geometry in East Asia — Hanoi 2005
pp. 177–215

Elliptic parameters and defining equations for elliptic fibrations on a Kummer surface

Masato Kuwata and Tetsuji Shioda

Abstract.

We pose the problem to determine explicit defining equations of various elliptic fibrations on a given $K3$ surface, and study the case of the Kummer surfaces of the product of two elliptic curves.

§1. Introduction

1.1. Problem setting

Let X be a $K3$ surface defined over a base field k, and let $k(X)$ denote its function field. Suppose $f : X \to \mathbf{P}^1$ is an elliptic fibration on X with a section O. Then it defines a non-constant function $u = f(x)$ $(x \in X)$, and hence an element $u \in k(X)$. We call u the *elliptic parameter* for the elliptic fibration f. (Actually u is unique only up to the linear fractional transformations, but to fix the idea, we always choose one u. Note that the subfield $k(u)$ of $k(X)$ is uniquely defined by f).

Now let E denote the generic fiber of f. Then E is an elliptic curve defined over $k(u)$ such that the function field $k(u)(E)$ is isomorphic to $k(X)$ as the extensions of k.

Problem 1. *Given a $K3$ surface X/k and an elliptic fibration f, determine (i) the elliptic parameter u for f, (ii) the defining equation of the elliptic curve $E/k(u)$, and (iii) the Mordell-Weil lattice (MWL) $E(k(u))$.*

Problem 2. *Given a $K3$ surface X/k, determine all the (essentially distinct) elliptic parameters.*

Received June 21, 2006.
Revised February 2, 2007.
2000 *Mathematics Subject Classification.* 14J27, 14J28.

Problem 2 is a combination of Problem 1 and the following standard problem:

Problem 3. *Given a $K3$ surface X/k, classify the elliptic fibrations $f : X \to \mathbf{P}^1$ up to isomorphisms.*

1.2. Main results

In this paper, we focus on the case of Kummer surfaces $X = \mathrm{Km}(A)$, where $A = C_1 \times C_2$ is a product of two elliptic curves, and assume k is an algebraically closed field of characteristic different from 2.

In this case, Problem 3 has been solved by Oguiso [8] under the assumption

(#) C_1, C_2 are not isogenous to each other and $k = \mathbf{C}$ (the field of complex numbers).

Namely he classifies the configuration of singular fibers on such a Kummer surface X into eleven types $\mathcal{J}_1, \dots, \mathcal{J}_{11}$, and determines the number of the isomorphism classes for each type.

Our main results can be stated as follows: we solve Problem 1 for each type of Oguiso's list (without assuming (#)), and thus solve Problem 2 under the assumption (#). More details will be given in §1.5 and 1.7 after we fix the notation and review some known cases.

1.3. Notation

By a (-2)-curve we mean a smooth rational curve on X whose self-intersection number is -2. (It is called a "nodal curve" in Oguiso [8].) It is known (cf. [4]) that all irreducible components of a reducible fiber in an elliptic fibration are (-2)-curves.

We have a configuration of twenty-four (-2)-curves on X, called the *double Kummer pencil* (see Fig. 1, cf. [10]). It consists of the 16 exceptional curves A_{ij} arising from the minimal resolution $X \to A/\iota_A$, plus the 8 curves F_i, G_j obtained as the image of $v_i \times C_2$ or $C_1 \times v'_j$ under the rational map $A \to S$. Here $\{v_i\}$ (or $\{v'_i\}$) denote the 2-torsion points of C_1 (resp. C_2) ($i, j \in I = \{0, 1, 2, 3\}$), and ι_A denotes the inversion automorphism of A. These curves will be referred to as the *basic curves* below.

Suppose that the elliptic curve C_i is defined by the Legendre form

$$C_i : y_i^2 = x_i(x_i - 1)(x_i - \lambda_i) \qquad \lambda_i \neq 0, 1.$$

We order the 2-torsion points by $v_1 = (0,0), v_2 = (1,0), v_3 = (\lambda_1, 0)$, with v_0 denoting the origin of C_1; similarly for v'_j and C_2.

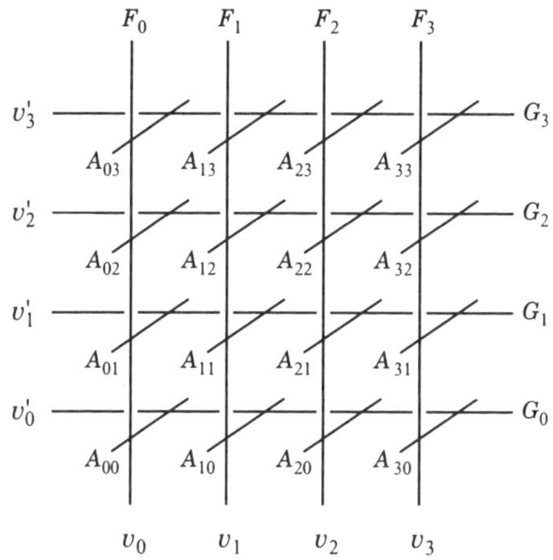

Fig. 1. double Kummer pencil

The function field $k(X)$ is equal to the subfield of the function field $k(A) = k(x_1, y_1, x_2, y_2)$ consisting of the elements invariant under the inversion $(x_1, y_1, x_2, y_2) \mapsto (x_1, -y_1, x_2, -y_2)$, namely we have

$$k(X) = k(x_1, x_2, t), \quad t = \frac{y_2}{y_1},$$

where x_1, x_2 and t are naturally regarded as functions on X, satisfying the relation

$$x_1(x_1 - 1)(x_1 - \lambda_1)t^2 = x_2(x_2 - 1)(x_2 - \lambda_2). \tag{1.1}$$

1.4. Examples

We start from the most classical and elementary example:

Example 1.1 (Kummer pencils). The projection of A to the first factor induces an elliptic fibration $\pi_1 : X \to \mathbf{P}^1$ with four singular fibers of type I_0^*:

$$\Phi_i = 2F_i + \sum_j A_{ij}$$

(see Fig. 2). This π_1 and the similar π_2 (obtained from the second projection) are respectively called the first or second Kummer pencil on X. The elliptic parameter for π_1 (or π_2) is obviously given by the

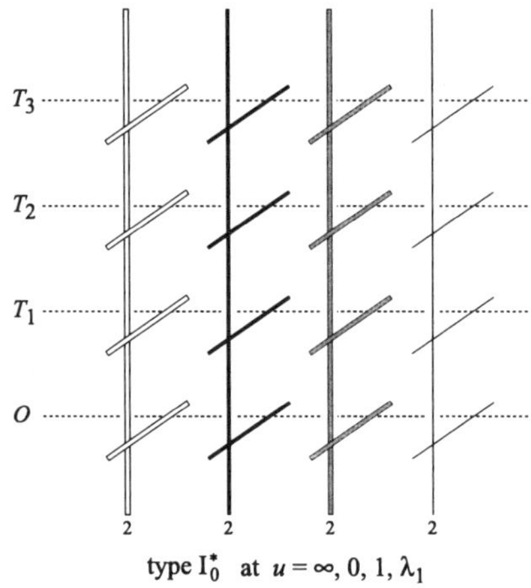

Fig. 2. Kummer pencil (type $\mathscr{J}_4$)

function x_1 (resp. x_2) in $k(X)$. (This belongs to type $\mathscr{J}_4$ in [8], and π_1 and π_2 are the two representatives of isomorphism classes, if C_1, C_2 are not isogenous.)

The defining equation of the generic fiber over $k(x_1)$ is easily obtained (see §2.3), which is isomorphic to the constant curve C_2 over the quadratic extension $k(x_1, y_1) = k(C_1)$ of $k(x_1)$. The Mordell-Weil lattice is isomorphic to the *lattice* $\mathrm{Hom}(C_1, C_2)$ with norm $\varphi \mapsto \deg(\varphi)$ up to torsion (see [14, Prop.3.1]).

The next is the motivating example for studying the elliptic parameters and the problems posed in §1.1 in general.

Example 1.2 (Inose's pencils). Using the twenty-four basic curves, we can find two disjoint divisors of Kodaira type IV*. Namely, take the following divisors shown in Fig. 3:

$$\begin{cases} \Psi_1 = G_1 + G_2 + G_3 + 2(A_{01} + A_{02} + A_{03}) + 3F_0, \\ \Psi_2 = F_1 + F_2 + F_3 + 2(A_{10} + A_{20} + A_{30}) + 3G_0, \end{cases}$$

There is an elliptic fibration, called *Inose's pencil*, having these divisors as the singular fibers over $u = 0$ and $u = \infty$, as first shown by Inose [3]. The elliptic parameter for this is given by the function $u = t(= y_2/y_1) \in k(X)$, and the generic fiber $E/k(t)$ is isomorphic to the cubic curve

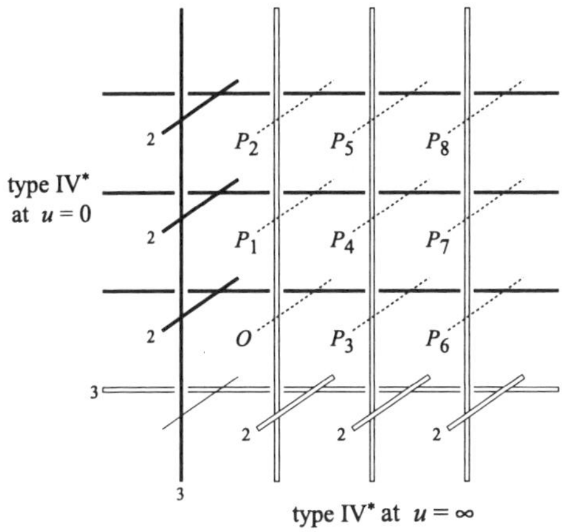

Fig. 3. Inose's pencil (type $\mathcal{J}_3$)

defined by the equation (1.1) in the projective plane with inhomogeneous coordinates x_1, x_2. (This belongs to type $\mathcal{J}_3$ in [8].)

It should be remarked that Kuwata [6] has succeeded in constructing, by the use of Inose's pencil, some elliptic $K3$ surfaces with high Mordell-Weil rank which have an explicit defining equation. For example, the base change $t = s^3$ gives rise to the elliptic curve $E/k(s)$ which has the highest possible rank $r = 18$ (for $k = \mathbf{C}$) provided that C_1 and C_2 are mutually isogenous but non-isomorphic elliptic curves with complex multiplications. We refer to Kuwata [6] and Shioda [12], [14] for more details including the defining equation of E in the Weierstrass form as well as the structure of MWL; see also §2.2.

Example 1.3. Besides the Kummer pencils (Example 1.1), the elliptic pencil on the Kummer surface $X = \mathrm{Km}(C_1 \times C_2)$ which has been studied first is perhaps the one introduced in Shioda-Inose [10]. It has II^*, I_0^*, I_0^* as reducible singular fibers (for general values of λ_1 and λ_2). This has type $\mathcal{J}_9$ in [8] (see Fig. 16). Via the base change of degree 2, it gives rise to an elliptic $K3$ surface with two II^* fibers, which plays an important role in the theory of singular $K3$ surfaces [10] and which has been reconsidered by Morrison [7] in a more general situation. It turns out that the elliptic parameter and the defining equation for this type $\mathcal{J}_9$ is the hardest case treated in this paper (see §5.3).

1.5. Results

In the following Table 1, we give a summary of the elliptic parameters and the structure of the MWL for each type $\mathcal{J}_n$, to be constructed in the subsequent sections.

The first column shows the type $\mathcal{J}_n$ of elliptic fibration following Oguiso's notation (cf. [8]). The second column shows the configuration of singular fibers in the *generic case*, which means that λ_1 and λ_2 are algebraically independent elements of k over $\mathbf{Q}_0$, where $\mathbf{Q}_0$ is the prime field in k. The third column shows the structure of MWL of the generic fiber E over $k(u)$, again in the generic case. The last column gives the elliptic parameter which can be used for any $\lambda_1, \lambda_2 (\neq 0, 1)$.

The explicit form of defining equations should be found in the text, since it is not suitable to tabulate here. We note that each of these defining equations has coefficients in $\mathbf{Q}_0(\lambda_1, \lambda_2)(u)$, where u is the elliptic parameter.

We see from the table that the elliptic parameters for $\mathcal{J}_n$ for $n = 1, 2, 3$ are of the form $u = t\varphi(x_1, x_2)$ with $\varphi(x_1, x_2) \in k(x_1, x_2)$, while those for $\mathcal{J}_n$ for $n > 3$ are contained in $k(x_1, x_2)$.

1.6. Basic strategy of construction

Theoretically, constructing an elliptic fibration on a $K3$ surface is to find a divisor that has the same type as a singular fiber in the Kodaira's list (cf. [4] [9]). In practice, however, we need to find two divisors, one for the fiber at $u = 0$, and the other for the fiber at $u = \infty$, to write down an actual elliptic parameter. This is where the difficulty is.

Once an elliptic parameter is found, we want to find a change of variables that converts the defining equation to a Weierstrass form. In most cases we encounter an equation of the form $y^2 =$ (quartic polynomial). We then use a standard algorithm to transform it to a Weierstrass form (see for example Cassels [1], or Connell [2]).

Some elliptic fibrations have nontrivial Mordell-Weil group. To determine the structure of Mordell-Weil lattice, we can use the method in [11] since we understand very well the intersection between the section and the components of singular fibers. Alternatively, we can compute the height pairing using the algorithm in [5] once we establish the conversion to the Weierstrass form. Note that [11] and [5] use different normalization of the height pairing, and they differ by a multiple of 2. In this article we adopt the normalization used in [11].

Type	Singular fibers	MWL	Elliptic parameter u
$\mathscr{J}_1$	$2\mathrm{I}_8 + 8\mathrm{I}_1$	$\mathbf{Z}^2 \oplus \mathbf{Z}/2\mathbf{Z}$	$\dfrac{tx_1}{x_2}$
$\mathscr{J}_2$	$\mathrm{I}_4 + \mathrm{I}_{12} + 8\mathrm{I}_1$	$A_2^*[2] \oplus \mathbf{Z}/2\mathbf{Z}$	$\dfrac{t(x_1-\lambda_1)(x_1-x_2)}{x_2(x_2-1)}$
$\mathscr{J}_3$	$2\mathrm{IV}^* + 8\mathrm{I}_1$	$(A_2^*[2])^2$	t
$\mathscr{J}_4$	$4\mathrm{I}_0^*$	$(\mathbf{Z}/2\mathbf{Z})^2$	x_i
$\mathscr{J}_5$	$\mathrm{I}_6^* + 6\mathrm{I}_2$	$(\mathbf{Z}/2\mathbf{Z})^2$	$\dfrac{(x_1-x_2)\big(\lambda_2(x_1-\lambda_1)+(\lambda_1-1)x_2\big)}{(\lambda_2 x_1-x_2)\big(x_1-\lambda_1+(\lambda_1-1)x_2\big)}$
$\mathscr{J}_6$	$2\ \mathrm{I}_2^* + 4\mathrm{I}_2$	$(\mathbf{Z}/2\mathbf{Z})^2$	$\dfrac{x_1}{x_2}$
$\mathscr{J}_7$	$\mathrm{I}_4^* + 2\mathrm{I}_0^* + 2\mathrm{I}_1$	$\mathbf{Z}/2\mathbf{Z}$	$\dfrac{(x_2-\lambda_2)(x_1-x_2)}{(x_2-1)(\lambda_2 x_1-x_2)}$
$\mathscr{J}_8$	$\mathrm{III}^* + \mathrm{I}_2^* + 3\mathrm{I}_2 + \mathrm{I}_1$	$\mathbf{Z}/2\mathbf{Z}$	$-\dfrac{(x_2-\lambda_2)(x_1-x_2)}{\lambda_2(\lambda_2-1)x_1(x_1-1)}$
$\mathscr{J}_9$	$\mathrm{II}^* + 2\mathrm{I}_0^* + 2\mathrm{I}_1$	$\{0\}$	$\dfrac{(x_2-\lambda_2)(x_1-x_2)\big(\lambda_2 x_1(x_1-1)+(\lambda_1-1)(x_2-1)(\lambda_2 x_1-x_2)\big)}{(x_2-1)(\lambda_2 x_1-x_2)\big(\lambda_2 x_1(x_1-1)+(\lambda_1-1)(x_2-\lambda_2)(x_1-x_2)\big)}$
$\mathscr{J}_{10}$	$\mathrm{I}_8^* + \mathrm{I}_0^* + 4\mathrm{I}_1$	$\{0\}$	$\dfrac{(x_2-\lambda_2)(x_1-x_2)\big((\lambda_1-1)(x_2-1)(\lambda_2 x_1-x_2)+\lambda_2 x_1(x_1-1)\big)}{x_2(x_2-1)(x_1-1)(\lambda_2 x_1-x_2)}$
$\mathscr{J}_{11}$	$2\mathrm{I}_4^* + 4\mathrm{I}_1$	$\{0\}$	$\dfrac{x_2(x_2-\lambda_2)(x_1-x_2)}{x_1(x_2-1)(\lambda_2 x_1-x_2)}$

Table 1. Results

1.7. Remark

Fix a type $\mathscr{J}_n$ $(n = 1, \ldots, 11)$. As noted in §1.5, each of the defining equation of $E/k(u)$ constructed in this paper has the coefficients in $\mathbf{Q}_0(\lambda_1, \lambda_2)(u)$, where u is the elliptic parameter, and λ_1 (resp. λ_2) is the Legendre parameter for C_1 (resp. C_2). Given C_i, there are in general six choices of λ_i (i.e., six different level 2-structures on C_i). We have verified that, by different choices of λ_1 or λ_2, we obtain as many nonisomorphic E's belonging to the same type $\mathscr{J}_n$, as predicted by Oguiso's result ([8, Table B, p. 652]), and thus solved Problem 2 when C_1 and C_2 are not isogenous. The proof for this will be omitted in this paper, but we write down the results in a single special case where we take $C_1 : y_1^2 = x_1^3 - 1$ and $C_2 : y_2^2 = x_2^3 - x_2$ (see §6).

This paper is organized as follows.

CONTENTS

§2. Elliptic parameters for $\mathscr{J}_1$, $\mathscr{J}_3$, $\mathscr{J}_4$, and $\mathscr{J}_6$

In this section we construct elliptic fibrations that have two singular fibers consisting only of the twenty-four basic curves. We use combinations of the following divisors of typical functions (cf. Examples in §1.4):

$$(x_1) = 2F_1 + A_{10} + A_{11} + A_{12} + A_{13} - (2F_0 + A_{00} + A_{01} + A_{02} + A_{03}),$$

$$(x_2) = 2G_1 + A_{01} + A_{11} + A_{21} + A_{31} - (2G_0 + A_{00} + A_{10} + A_{20} + A_{30}),$$

$$(t) = G_1 + G_2 + G_3 + 2(A_{01} + A_{02} + A_{04}) + 3F_0 \\ - (F_1 + F_2 + F_3 + 2(A_{10} + A_{20} + A_{30}) + 3G_0).$$

2.1. $\mathscr{J}_1$

An elliptic parameter for the type $\mathscr{J}_1$ fibration is given by

$$u = \frac{tx_1}{x_2}.$$

It is easy to verify that the divisor of u is given by

$$(u) = F_0 + F_1 + G_2 + G_3 + A_{02} + A_{03} + A_{12} + A_{13} \\ - (G_0 + G_1 + F_2 + F_3 + A_{20} + A_{21} + A_{30} + A_{31}),$$

which is indicated in Fig. 4. Choosing A_{00} as the 0-section of the group

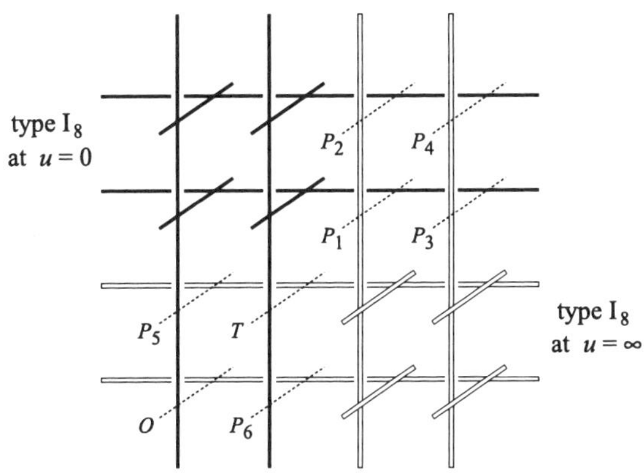

Fig. 4. $\mathscr{J}_1$

structure, we obtain the Weierstrass equation of the elliptic fibration

$$Y^2 = X^3 + \big((\lambda_1 - 1)^2 u^4 - 2(\lambda_1 + 1)(\lambda_2 + 1)u^2 + (\lambda_2 - 1)^2\big)X^2 + 16\lambda_1\lambda_2 u^4 X,$$

where the change of variables is given by

$$X = \frac{4t^2x_1^3}{x_2}, \quad Y = \frac{4t^2x_1^3\big(t^2x_1(x_1^2 - \lambda_1) + x_2(x_2^2 - \lambda_2)\big)}{x_2^3}.$$

Its discriminant is given by

$$\Delta(u) = 2^{12}\lambda_1^2\lambda_2^2\, u^8\, d(u)d(-u),$$

where $d(u)$ is a polynomial of degree 4 in u:

$$d(u) = (\lambda_1 - 1)^2 u^4 + 4(\lambda_1 - 1) u^3 - 2(\lambda_1\lambda_2 + \lambda_1 + \lambda_2 - 3) u^2 + 4(\lambda_2 - 1) u + (\lambda_2 - 1)^2.$$

[The discriminant of $d(u)$ vanishes if and only if $\lambda_1 = \lambda_2$. If $\lambda_1 = \lambda_2$, the elliptic fibration has two I_2 fibers for general λ_1.]

The curve A_{11} corresponds to the 2-torsion section $T = (0,0)$. The correspondence between the curves and the sections are as follows:

$$\begin{array}{lll}
A_{22} & \leftrightarrow & P_1 = \big(4u^2, -4u^2((\lambda_1 - 1)u^2 + \lambda_2 - 1)\big) \\
A_{23} & \leftrightarrow & P_2 = \big(4\lambda_2 u^2, -4\lambda_2 u^2((\lambda_1 - 1)u^2 - \lambda_2 + 1)\big) \\
A_{32} & \leftrightarrow & P_3 = \big(4\lambda_1 u^2, 4\lambda_1 u^2((\lambda_1 - 1)u^2 - \lambda_2 + 1)\big) \\
A_{33} & \leftrightarrow & P_4 = \big(4\lambda_1\lambda_2 u^2, 4\lambda_1\lambda_2 u^2((\lambda_1 - 1)u^2 + \lambda_2 - 1)\big) \\
A_{01} & \leftrightarrow & P_5 = \big(4\lambda_2, 4\lambda_2((\lambda_1 + 1)u^2 - \lambda_2 - 1)\big) \\
A_{10} & \leftrightarrow & P_6 = \big(4\lambda_1 u^4, -4\lambda_1 u^4((\lambda_1 + 1)u^2 - \lambda_2 - 1)\big)
\end{array}$$

These sections satisfy the following relations.

$$P_3 = P_2 + T, \qquad P_4 = P_1 + T,$$
$$P_5 = P_1 + P_2, \qquad P_6 = P_5 + T.$$

The Mordell-Weil group is generated by T, P_1 and P_2 in the general case where C_1 and C_2 are not isogenous. The height matrix with respect to $\{P_1, P_2\}$ is shown to be

$$\begin{pmatrix} 1 & 0 \\ 0 & 1 \end{pmatrix}.$$

2.2. $\mathscr{J}_3$

As we have seen in Example 1.2 (§1.4),

$$u = t$$

gives an elliptic parameter of type $\mathscr{J}_3$. We regard (1.1) as a cubic curve in x_1 and x_2 with coefficients in $k(u) = k(t)$. We choose $(x_1, x_2) = (0,0)$ as the origin of the group structure. The Weierstrass form is given by

$$Y^2 = X^3 + 4(\lambda_1 + 1)(\lambda_2 + 1)u^2 X^2 + 16u^4\big((\lambda_1\lambda_2 + 1)(\lambda_1 + \lambda_2 + 1) - 1\big)X + 16u^4\big((\lambda_1(\lambda_1 - 1)u^2 + \lambda_2(\lambda_2 - 1))^2 + 4\lambda_1\lambda_2(\lambda_1 + \lambda_2)u^2\big).$$

(This is relatively simple, but the intermediate calculations are rather complicated.) The change of variables between two forms of equations is given by

$$
\begin{aligned}
X &= \frac{4\big(\lambda_2(x_1-1)(x_1-\lambda_1)+\lambda_1(x_2-1)(x_2-\lambda_2)-\lambda_1\lambda_2\big)\,t^2}{x_1x_2},\\
Y &= \frac{8(x_2-1)(x_2-\lambda_2)\big(\lambda_2(\lambda_1+1)x_1+\lambda_1(\lambda_2+1)x_2-\lambda_1\lambda_2\big)\,t^2}{x_1^2x_2}\\
&\quad+\frac{4\lambda_1\big((\lambda_1+1)x_1-2\lambda_1\big)\,t^4}{x_1}+\frac{4\lambda_2\big((\lambda_2+1)x_2-2\lambda_2\big)\,t^2}{x_2}.
\end{aligned}
$$

The discriminant is of the form $u^8d(u)$, where $d(u)$ is an irreducible polynomial of degree 8. Besides two IV* fibers, the elliptic fibration has eight I_1 fibers in the generic case. These eight I_1 fibers can degenerate in four different ways; $2\,\mathrm{I}_2+4\,\mathrm{I}_1$, $4\,\mathrm{I}_2$, 4 II or 2 IV. For more detail, see Prop. 5.1 in [14].

There are eight other A_{ij}'s which define sections; the correspondence between these curves and the sections is as follows:

$$
\begin{aligned}
A_{12} &\leftrightarrow P_1=\big(4u^2(\lambda_1^2u^2-\lambda_2(\lambda_1+1)),\\
&\qquad\qquad -4u^2(2\lambda_1^3u^4-\lambda_1(\lambda_1+1)(2\lambda_2-1)u^2+\lambda_2(\lambda_2-1))\big)\\
A_{13} &\leftrightarrow P_2=\left(\frac{4u^2(\lambda_1^2u^2-\lambda_2^2(\lambda_1+1))}{\lambda_2^2},\right.\\
&\qquad\qquad \left.-\frac{4u^2(2\lambda_1^3u^4+\lambda_1\lambda_2^2(\lambda_1+1)(\lambda_2-2)u^2-\lambda_2^4(\lambda_2-1))}{\lambda_2^3}\right)\\
A_{21} &\leftrightarrow P_3=\big(-4(\lambda_1(\lambda_2+1)u^2-\lambda_2^2),\\
&\qquad\qquad -4(\lambda_1(\lambda_1-1)u^4-\lambda_2(\lambda_2+1)(2\lambda_1-1)u^2+2\lambda_2^3)\big)\\
A_{22} &\leftrightarrow P_4=\big(-4\lambda_1\lambda_2u^2,-4u^2(\lambda_1(\lambda_1-1)u^2+\lambda_2(\lambda_2-1))\big)\\
A_{23} &\leftrightarrow P_5=\big(-4\lambda_1u^2,-4u^2(\lambda_1(\lambda_1-1)u^2-\lambda_2(\lambda_2-1))\big)\\
A_{31} &\leftrightarrow P_6=\left(-\frac{4(\lambda_1^2(\lambda_2+1)u^2-\lambda_2^2)}{\lambda_1^2},\right.\\
&\qquad\qquad \left.\frac{4(\lambda_1^4(\lambda_1-1)u^4-\lambda_1^2\lambda_2(\lambda_1-2)(\lambda_2+1)u^2-2\lambda_2^3)}{\lambda_1^3}\right)\\
A_{32} &\leftrightarrow P_7=\big(-4\lambda_2u^2,4u^2(\lambda_1(\lambda_1-1)u^2-\lambda_2(\lambda_2-1))\big)\\
A_{33} &\leftrightarrow P_8=\big(-4u^2,4u^2(\lambda_1(\lambda_1-1)u^2+\lambda_2(\lambda_2-1))\big)
\end{aligned}
$$

These sections satisfy the following relations:

$$\begin{aligned} P_1 &= P_5 + P_8, & P_2 &= P_4 + P_7, \\ P_3 &= P_7 + P_8, & P_6 &= P_4 + P_5. \end{aligned}$$

We can show that P_4, P_8, P_5, and P_7 generate the Mordell-Weil group in the generic case. The height matrix with respect to the basis $\{P_4, P_8, P_5, P_7\}$ is

$$\begin{pmatrix} \frac{4}{3} & \frac{2}{3} & 0 & 0 \\ \frac{2}{3} & \frac{4}{3} & 0 & 0 \\ 0 & 0 & \frac{4}{3} & \frac{2}{3} \\ 0 & 0 & \frac{2}{3} & \frac{4}{3} \end{pmatrix}.$$

This is the direct sum of two copies of $A_2^*[2]$, the dual lattice of A_2 scaled by 2.

2.3. $\mathscr{J}_4$

The elliptic parameter for the fibration π_1 in Example 1.1 is given by

$$u = x_1,$$

while the elliptic parameter for π_2 is given by $u = x_2$. For π_1, the change of variables

$$\begin{aligned} X &= u(u-1)(u-\lambda_1)x_2, \\ Y &= u^2(u-1)^2(u-\lambda_1)^2\, t, \end{aligned}$$

converts the equation (1.1) to

$$Y^2 = X\bigl(X - u(u-1)(u-\lambda_1)\bigr)\bigl(X - \lambda_2\, u(u-1)(u-\lambda_1)\bigr).$$

The curve G_0 is the 0-section. Other sections are:

$$\begin{aligned} G_1 &\quad\leftrightarrow\quad T_1 = (0,0), \\ G_2 &\quad\leftrightarrow\quad T_2 = (u(u-1)(u-\lambda_1), 0), \\ G_3 &\quad\leftrightarrow\quad T_3 = (\lambda_2\, u(u-1)(u-\lambda_1), 0). \end{aligned}$$

Similar results hold for π_2.

2.4. $\mathscr{J}_6$

The divisor of the function x_1/x_2 is given by

$$\left(\frac{x_1}{x_2}\right) = 2(F_1 + A_{10} + G_0) + A_{12} + A_{13} + A_{20} + A_{30} - \big(2(F_0 + A_{01} + G_1) + A_{02} + A_{03} + A_{21} + A_{31}\big).$$

This is the difference of two disjoint divisors of type I_2^*, and thus

$$u = \frac{x_1}{x_2}$$

is an elliptic parameter of type $\mathscr{J}_6$.

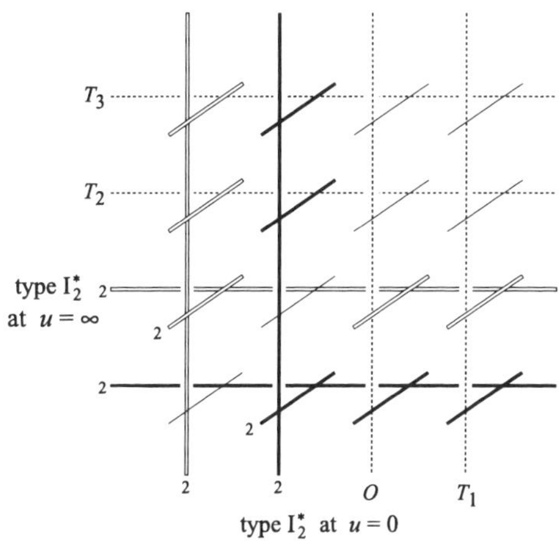

Fig. 5. $\mathscr{J}_6$

In order to write down a Weierstrass equation using the curve F_2 as the 0-section, we put

$$X = \frac{x_1(x_1 - \lambda_1)(x_1 - x_2)(\lambda_2 x_1 - x_2)}{(x_1 - 1)x_2^3},$$

$$Y = \frac{(\lambda_1 - 1)\, t\, x_1^3(x_1 - \lambda_1)(x_1 - x_2)(\lambda_2 x_1 - x_2)}{(x_1 - 1)x_2^5}.$$

Then we obtain the Weierstrass equation

$$Y^2 = X\big(X - u(u-1)(\lambda_2 u - \lambda_1)\big)\big(X - u(u - \lambda_1)(\lambda_2 u - 1)\big).$$

Its discriminant is given by

$$\Delta(u) = 16u^8(\lambda_1 - 1)^2(\lambda_2 - 1)^2(u-1)^2(u-\lambda_1)^2(\lambda_2 u - 1)^2(\lambda_2 u - \lambda_1)^2.$$

Besides two I_2^* fibers at $u = 0$ and ∞, there are four I_2 fibers at $u = 1, \lambda_1, 1/\lambda_2$ and λ_1/λ_2. This elliptic surface has the following three 2-torsion sections:

$$\begin{aligned} F_3 &\leftrightarrow T_1 = (0,0), \\ G_2 &\leftrightarrow T_2 = \bigl(u(u-\lambda_1)(\lambda_2 u - 1), 0\bigr), \\ G_3 &\leftrightarrow T_3 = \bigl(u(u-1)(\lambda_2 u - \lambda_1), 0\bigr), \end{aligned}$$

Note that A_{22}, A_{23}, A_{32}, and A_{33} are components of four I_2 fibers. The other components of these four I_2 fibers are new (-2)-curves not among the basic curves, which will be clarified in §3.2.

§3. More (-2)-curves

In order to describe elliptic parameters for other types, we need more (-2)-curves than the basic curves. When we constructed elliptic parameters of type $\mathcal{J}_6$ just above, we obtained some new (-2)-curves as components of I_2 fibers. In this section we give a systematic way to obtain such (-2)-curves.

For our purpose, it is convenient to regard $X = \mathrm{Km}(C_1 \times C_2)$ as a double cover of the product of projective lines: $\mathbf{P}^1 \times \mathbf{P}^1 = \{(x_1 : z_1), (x_2 : z_2)\}$. Let $p_i : C_i \to \mathbf{P}^1$ $(i = 1, 2)$ be the projection given by

$$\begin{array}{cccc} p_i : & C_i & \longrightarrow & \mathbf{P}^1 \\ & (x_i : y_i : z_i) & \longmapsto & \begin{cases} (x_i : z_i) & \text{if } z_i \neq 0 \\ (1 : 0) & \text{if } z_i = 0 \end{cases} \end{array}$$

Then the map $p_1 \times p_2 : A = C_1 \times C_2 \to \mathbf{P}^1 \times \mathbf{P}^1$ factors through $\bar{\pi} : A/\iota_A \to \mathbf{P}^1 \times \mathbf{P}^1$. Let π be the morphism of degree two from X to $\mathbf{P}^1 \times \mathbf{P}^1$ that makes the following diagram commutative:

$$\begin{array}{ccccc} & & X & & \\ & & \downarrow & \overset{\pi}{\searrow} & \\ A & \longrightarrow & A/\iota_A & \underset{\bar{\pi}}{\longrightarrow} & \mathbf{P}^1 \times \mathbf{P}^1 \end{array}$$

We denote by R_{ij} the point in $\mathbf{P}^1 \times \mathbf{P}^1$ that is the image of the exceptional curve A_{ij} by π. To obtain more (-2)-curves, we look for curves in $\mathbf{P}^1 \times \mathbf{P}^1$ which lift to a (-2)-curve via the map π.

3.1. (1, 1)-curves

Let L be a curve in $\mathbf{P}^1 \times \mathbf{P}^1$ defined by a bihomogeneous equation of bidegree $(1,1)$:

$$ax_1x_2 + bx_1z_2 + cz_1x_2 + dz_1z_2 = 0.$$

We call such a curve $(1,1)$-curve for short. By an abuse of notation, we denote the image of F_i and G_i under $\pi : S \to \mathbf{P}^1 \times \mathbf{P}^1$ by the same letters F_i and G_i, respectively. For example, F_1 is the curve with the equation $x_1 = 0$, and G_3 with $x_2 - \lambda_2 z_2 = 0$, etc.

Let L be a $(1,1)$-curve in $\mathbf{P}^1 \times \mathbf{P}^1$. Its pullback $\pi^{-1}(L)$ ramifies at the intersection of L and F_i or G_j, except when the intersection point falls on $R_{ij} = F_i \cap G_j$.

Lemma 4. *Let L be a $(1,1)$-curve. Then,*

(1) *If L passes three of sixteen R_{ij}'s, then $\pi^{-1}(L)$ is a curve of genus* 0.

(2) *If L passes two out of sixteen R_{ij}'s, then $\pi^{-1}(L)$ is a curve of genus* 1.

Proof. In general a $(1,1)$-curve L intersects with $\sum F_i$ (resp. $\sum G_j$) at four points. If L passes three of sixteen R_{ij}'s, then it intersects with F_i one more time and G_j one more time outside R_{ij}. This implies that $\pi^{-1}(L)$ ramifies at two points. By Hurwitz's theorem $\pi^{-1}(L)$ is a curve of genus 0. Similarly, if L passes two out of sixteen R_{ij}'s, $\pi^{-1}(L)$ ramifies at four points, and it is a curve of genus 1. Q.E.D.

A $(1,1)$-curve is uniquely determined by a set of three points in a general position. If we choose $R_{i_0j_0}$, $R_{i_1j_1}$, $R_{i_2j_2}$ so that no two of them are on the same F_i or G_j, then they are in general position. Let i_3 and j_3 be the missing indices. In other words, we choose i_3 and j_3 such that $\{i_0, i_1, i_2, i_3\} = \{j_0, j_1, j_2, j_3\} = \{0,1,2,3\}$. Under the condition that the two elliptic curves C_1 and C_2 are not isomorphic, the $(1,1)$-curve passing through $R_{i_0j_0}$, $R_{i_1j_1}$, and $R_{i_2j_2}$ does not pass $R_{i_3j_3}$. Thus, choosing $R_{i_0j_0}$, $R_{i_1j_1}$, $R_{i_2j_2}$ we obtain a $(1,1)$-curve whose pullback by π is an irreducible (-2)-curve in X. We denote such a $(1,1)$-curve by $L_{i_0j_0,i_1j_1,i_2j_2}$, and its pullback by $\tilde{L}_{i_0j_0,i_1j_1,i_2j_2}$. There are ninety-six such (-2)-curves. Also note that $\tilde{L}_{i_0j_0,i_1j_1,i_2j_2}$ intersects twice with each of $A_{i_0j_0}$, $A_{i_1j_1}$, and $A_{i_2j_2}$.

The $(1,1)$-curve $L_{00,11,22}$ passes through R_{00}, R_{11}, R_{22}. It is given by the bihomogeneous equation $x_2z_1 - x_1z_2 = 0$. In the sequel we write it in the affine form $x_2 - x_1 = 0$ for simplicity. $\tilde{L}_{00,11,22}$ is denoted by A^{44} in Oguiso [8], which appears in the $\mathcal{J}_2$ fibration. We denote it by B_{33} to make it consistent with our notation, indicating that it intersects with F_3 and G_3 outside A_{33}. Note, however, that there are six (-2)-curves of the form $\tilde{L}_{i_0j_0,i_1j_1,i_2j_2}$ that intersect with F_3 and G_3.

Fig. 6 shows the curve B_{33} in the affine space $\mathbf{A}_{x_1} \times \mathbf{A}_{x_2} \times \mathbf{A}_t$. As a matter of fact, if we substitute x_2 by x_1 in (1.1), the equation factorizes into

$$x_1(x_1 - 1)(x_1t^2 - x_1 - t^2\lambda_1 + \lambda_2) = 0,$$

which implies that the intersection between $x_2 - x_1 = 0$ and the affine Kummer surface (1.1) has three irreducible components, namely A_{11}, A_{22}, and B_{33}. We also see that a parametrization of B_{33} is given by

$$(x_1, x_2, t) = \left(\frac{\lambda_1 s^2 - 1}{s^2 - 1}, \frac{\lambda_1 s^2 - 1}{s^2 - 1}, s \right).$$

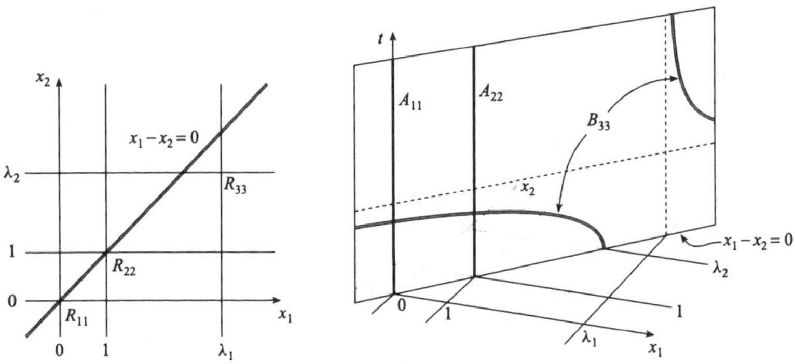

Fig. 6. (-2)-curve B_{33}

The zero divisor of the function $x_2 - x_1 \in k(x_1, x_2, t)$ is $A_{11} + A_{22} + B_{33}$, while the polar divisor is of the form $D_1 + D_2 + rA_{00}$, where

$$D_1 = 2F_0 + A_{00} + A_{01} + A_{02} + A_{03},$$
$$D_2 = 2G_0 + A_{00} + A_{10} + A_{20} + A_{30}.$$

Since A_{00} intersects twice with the divisor $A_{11} + A_{22} + B_{33}$, the intersection number $A_{00} \cdot (D_1 + D_2 + rA_{00})$ must be 2, which implies $r = -1$.

This shows

$$(x_2 - x_1) = A_{11} + A_{22} + B_{33} \\ - (2F_0 + 2G_0 + A_{00} + A_{01} + A_{02} + A_{03} + A_{10} + A_{20} + A_{30}).$$

This and similar calculations of divisors are used to find the elliptic parameter with a prescribed divisor in §4 and §5.

3.2. I_2 fibers of type $\mathscr{J}_6$ fibration

The elliptic parameter $u = x_1/x_2$, which is of type $\mathscr{J}_6$, defines a pencil of $(1,1)$-curves $x_1 - ux_2 = 0$. The general fiber of this elliptic fibration is the pullback of a $(1,1)$-curve passing through R_{00} and R_{11}. If $x_1 - ux_2 = 0$ passes through a third R_{ij}, then its pullback is a singular fiber (see Fig. 7). Four fibers of type I_2, which are mentioned in §2.4

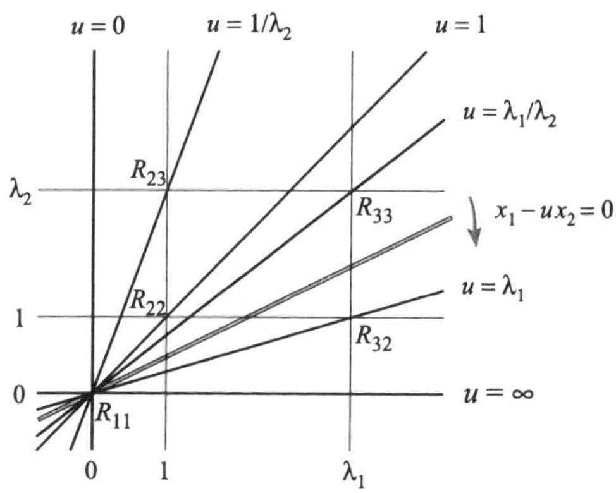

Fig. 7. pencil of $(1,1)$-curves

arise as follows:

$$\begin{array}{llll} \tilde{L}_{00,11,23} + A_{23} & \text{at } u = 1/\lambda_2, & B_{33} + A_{22} & \text{at } u = 1, \\ \tilde{L}_{00,11,33} + A_{33} & \text{at } u = \lambda_1/\lambda_2, & \tilde{L}_{00,11,32} + A_{32} & \text{at } u = \lambda_1. \end{array}$$

3.3. Notation

Even though the notation "B_{33}" is ambiguous as we mentioned earlier, it is quite convenient. We thus use the following notation in the sequel:

(3.1)

$$B_{32} = \tilde{L}_{00,11,23} : \lambda_2 x_1 - x_2 = 0, \qquad B_{33} = \tilde{L}_{00,11,22} : x_1 - x_2 = 0,$$
$$B_{22} = \tilde{L}_{00,11,33} : \lambda_2 x_1 - \lambda_1 x_2 = 0, \quad B_{23} = \tilde{L}_{00,11,32} : x_1 - \lambda_1 x_2 = 0, .$$

Later in §4.3 and §5.2, we introduce more (-2)-curves of this type, B_{31}, B_{12} and B_{13}.

§4. Elliptic parameters for $\mathscr{J}_2$, $\mathscr{J}_7$, $\mathscr{J}_8$ and $\mathscr{J}_{11}$

4.1. $\mathscr{J}_2$

Using B_{33}, we can construct an elliptic parameter of type $\mathscr{J}_2$. In fact, the divisor

$$\Psi_{2,0} = F_3 + A_{33} + G_3 + B_{33}$$

is a divisor of type I_4, and it does not intersect with the divisor of type I_{12} given by

$$\Psi_{2,\infty} = F_0 + A_{02} + G_2 + A_{12} + F_1 + A_{10} + G_0 + A_{20} + F_2 + A_{21} + G_1 + A_{01}$$

(see Fig. 8 below). It turns out that the divisor of the function

$$u = \frac{t(x_1 - \lambda_1)(x_1 - x_2)}{x_2(x_2 - 1)}$$

is $\Psi_{2,0} - \Psi_{2,\infty}$, and it is an elliptic parameter of type $\mathscr{J}_2$. Choosing A_{30} as the 0-section, we obtain the Weierstrass equation

$$Y^2 = X^3 + \big(u^4 + 2\,(2\lambda_1\lambda_2 - \lambda_1 - \lambda_2 + 2)\,u^2 + (\lambda_2 - \lambda_1)^2\big)X^2 - 16\lambda_1\lambda_2(\lambda_1 - 1)(\lambda_2 - 1)\,u^2 X,$$

where the change of variables is given by

$$X = -\frac{4\lambda_1(\lambda_1 - 1)(x_1 - x_2)(x_2 - \lambda_2)}{x_1(x_1 - 1)},$$
$$Y = -\frac{4\lambda_1(\lambda_1 - 1)(x_1 - x_2)(x_2 - \lambda_2)(2x_1 - 2x_2 - \lambda_1 + \lambda_2)}{x_1(x_1 - 1)} + \frac{4\lambda_1(\lambda_1 - 1)(x_1 - x_2)^2(x_2 - \lambda_2)^3(2x_1x_2 - x_1 - x_2)}{t^2 x_1^3 (x_1 - 1)^3}.$$

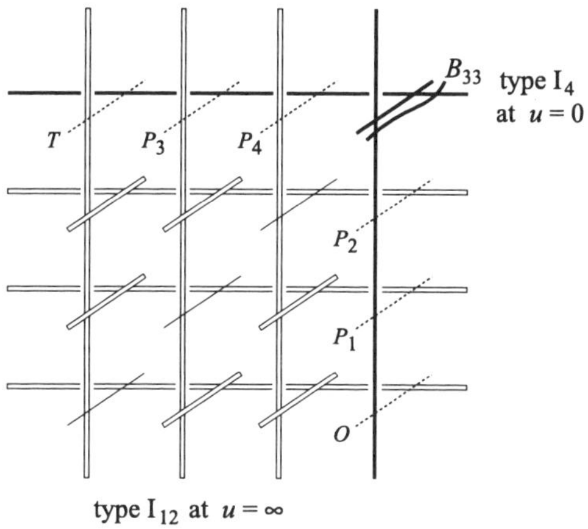

Fig. 8. $\mathscr{J}_2$

The discriminant of the fibration is of the form $u^4 d(u)$, where $d(u)$ is a polynomial of degree 8. The discriminant of $d(u)$ vanishes if and only if

$$\lambda_2 = \lambda_1, 1-\lambda_1, \frac{1}{\lambda_1}, \text{or } \frac{\lambda_1}{\lambda_1 - 1}.$$

The curve A_{03} corresponds to the 2-torsion section $T=(0,0)$. The correspondence between the curves and the sections is as follows:

$$\begin{aligned}
A_{31} &\leftrightarrow P_1 = \big(4\lambda_1\lambda_2, 4\lambda_1\lambda_2(u^2+\lambda_1+\lambda_2)\big)\big) \\
A_{32} &\leftrightarrow P_2 = \big(4(\lambda_1-1)(\lambda_2-1), \\
&\qquad -4(\lambda_1-1)(\lambda_2-1)(u^2-\lambda_1-\lambda_2+2)\big)\big) \\
A_{13} &\leftrightarrow P_3 = \big(-4u^2(\lambda_1-1)(\lambda_2-1), \\
&\qquad 4(\lambda_1-1)(\lambda_2-1)u^2(u^2+\lambda_1+\lambda_2)\big)\big) \\
A_{23} &\leftrightarrow P_4 = \big(-4\lambda_1\lambda_2u^2, -4\lambda_1\lambda_2u^2(u^2-\lambda_1-\lambda_2+2)\big)\big)
\end{aligned}$$

These sections satisfy the following relations.

$$P_3 = P_1 + T, \qquad P_4 = P_2 + T.$$

The Mordell-Weil group is generated by T, P_1 and P_2 in the general case where C_1 and C_2 are not isogenous. The height matrix with respect to

$\{P_1, P_2\}$ is shown to be

$$\begin{pmatrix} \frac{4}{3} & \frac{2}{3} \\ \frac{2}{3} & \frac{4}{3} \end{pmatrix}.$$

Thus the Mordell-Weil lattice is isomorphic to $A_2^*[2]$.

4.2. $\mathscr{J}_7$

Using the curves B_{33} and B_{32} introduced in §3.3, we can form two disjoint divisors of type I_0^*:

$$\begin{aligned} \Psi_{7,0} &= 2G_3 + A_{03} + A_{13} + A_{33} + B_{33}, \\ \Psi_{7,\infty} &= 2G_2 + A_{02} + A_{12} + A_{32} + B_{32}. \end{aligned}$$

Looking for the function whose divisor is $\Psi_{7,0} - \Psi_{7,\infty}$, we obtain the elliptic parameter

$$u = \frac{(x_2 - \lambda_2)(x_1 - x_2)}{(x_2 - 1)(\lambda_2 x_1 - x_2)}. \tag{4.1}$$

The divisor of the function

$$u - 1 = -\frac{(\lambda_2 - 1)\, x_2(x_1 - 1)}{(x_2 - 1)(\lambda_2 x_1 - x_2)}$$

is given by the following divisor consisting only of the basic curves:

$$A_{01} + A_{31} + 2G_1 + 2A_{21} + 2F_2 + 2A_{20} + 2G_0 + A_{10} + A_{30}.$$

This is a singular fiber of type I_4^* (see Fig. 9). Thus, the elliptic parameter given by (4.1) is of type $\mathscr{J}_7$.

The change of variables

$$\begin{aligned} X &= \frac{\lambda_2 u(u-1)^2 x_1}{x_2} \\ &= \frac{\lambda_2(\lambda_2 - 1)^2 x_1(x_1 - 1)^2 x_2(x_2 - \lambda_2)(x_1 - x_2)}{(x_2 - 1)^3(\lambda_2 x_1 - x_2)^3}, \\ Y &= \frac{\lambda_2(\lambda_2 - 1)u^2(u-1)^2}{t} \\ &= \frac{\lambda_2(\lambda_2 - 1)^3(x_1 - 1)^2 x_2^2(x_2 - \lambda_2)^2(x_1 - x_2)^2}{t(x_2 - 1)^4(\lambda_2 x_1 - x_2)^4}, \end{aligned}$$

converts (1.1) to the Weierstrass equation

$$Y^2 = X^3 - u(u-1)\big((\lambda_1\lambda_2 + 1)u - \lambda_1 - \lambda_2\big)X^2 + \lambda_1\lambda_2 u^2(u-1)^4 X.$$

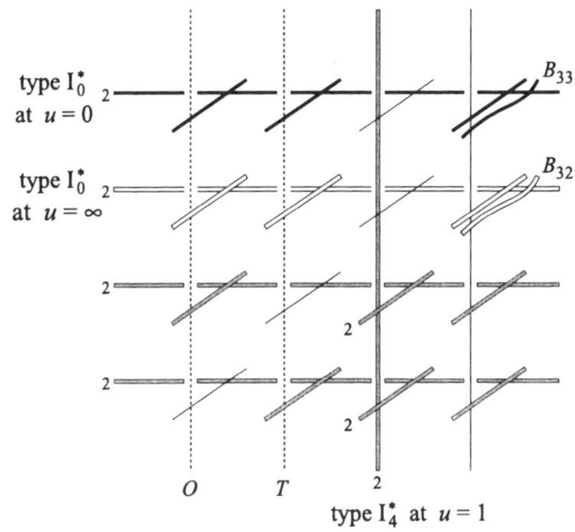

Fig. 9. $\mathscr{J}_7$

Its discriminant is of the form $u^6(u-1)^{10}d(u)$, where $d(u)$ is a polynomial of degree 2.

Generically, it has only one section other than 0-section:

$$F_2 \quad \leftrightarrow \quad T = (0,0).$$

4.3. $\mathscr{J}_8$

To find an elliptic parameter of type $\mathscr{J}_8$, we need to construct a I_2^* fiber. For this, we can make use of B_{33} once again. The divisor

$$\Psi_{8,0} = A_{01} + A_{02} + 2F_0 + 2A_{03} + 2G_3 + A_{33} + B_{33}$$

is of type I_2^* and it does not intersect with the divisor

$$\Psi_{8,\infty} = A_{12} + 2F_1 + 3A_{10} + 4E_0 + 3A_{20} + 2F_2 + A_{21} + 2A_{30},$$

which is of type III^*. We look for a function whose divisor is $\Psi_{8,0} - \Psi_{8,\infty}$, and we obtain the elliptic parameter of type $\mathscr{J}_8$

$$u = -\frac{(x_2 - \lambda_2)(x_1 - x_2)}{\lambda_2(\lambda_2 - 1)x_1(x_1 - 1)}.$$

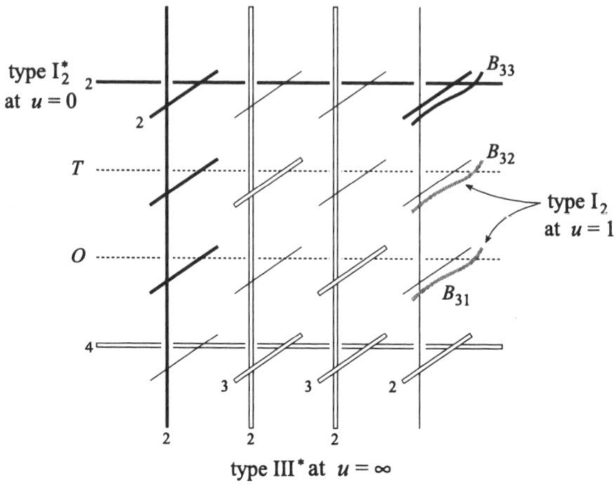

Fig. 10. $\mathscr{J}_8$

Let B_{31} be the (-2)-curve $\tilde{L}_{00,13,22} : (\lambda_2 - 1)x_1 + x_2 - \lambda_2 = 0$. Then B_{32} and B_{31} form a fiber of type I_2 at $u = 1$. Also A_{32} and the pullback of a certain $(2, 2)$-curve form another fiber of type I_2 at $u = 1/(\lambda_1\lambda_2)$, while A_{31} together with the pullback of a certain $(2, 2)$-curve form the third fiber of type I_2 at $u = (\lambda_1 - 1)^{-1}(\lambda_2 - 1)^{-1}$. The change of variables

$$X = u\big((\lambda_1 - 1)(\lambda_2 - 1)u - 1\big)\frac{(x_2 - 1)(\lambda_2 x_1 - x_2)}{(\lambda_2 - 1)x_2(x_1 - 1)},$$
$$Y = -u^3\big((\lambda_1 - 1)(\lambda_2 - 1)u - 1\big)\frac{\lambda_2(x_2 - 1)(\lambda_2 x_1 - x_2)}{t\, x_2(x_1 - 1)},$$

converts (1.1) to the Weierstrass equation

$$Y^2 = X^3 - u\big((2\lambda_1\lambda_2 - \lambda_1 - \lambda_2 + 2)u - 2\big)X^2 - u^2(u-1)(\lambda_1\lambda_2 u - 1)\big((\lambda_1 - 1)(\lambda_2 - 1)u - 1\big)X.$$

Its discriminant is

$$\Delta(u) = 16u^8(u-1)^2(\lambda_1\lambda_2 u - 1)^2\big((\lambda_1 - 1)(\lambda_2 - 1)u - 1\big)^2 \times \big(4\lambda_1\lambda_2(\lambda_1 - 1)(\lambda_2 - 1)u + (\lambda_1 - \lambda_2)^2\big).$$

[If $\lambda_2 = -\lambda_1, 2 - \lambda_2$, or $\lambda_1/(2\lambda_1 - 1)$, this elliptic fibration has fiber of type III for general λ_1.]

Generically, it has only one section other than 0-section:

$$G_2 \quad \leftrightarrow \quad T = (0, 0).$$

4.4. $\mathscr{J}_{11}$

Modifying the divisors appearing in the type $\mathscr{J}_7$ fibration we constructed in §4.2, we form two divisors

$$\Psi_{11,0} = A_{31} + A_{21} + 2G_1 + 2A_{01} + 2F_0 + 2A_{03} + 2G_3 + A_{33} + B_{33},$$
$$\Psi_{11,\infty} = A_{30} + A_{20} + 2G_0 + 2A_{10} + 2F_1 + 2A_{12} + 2G_2 + A_{32} + B_{32}.$$

They are of type I_4^* and they do not intersect with each other.

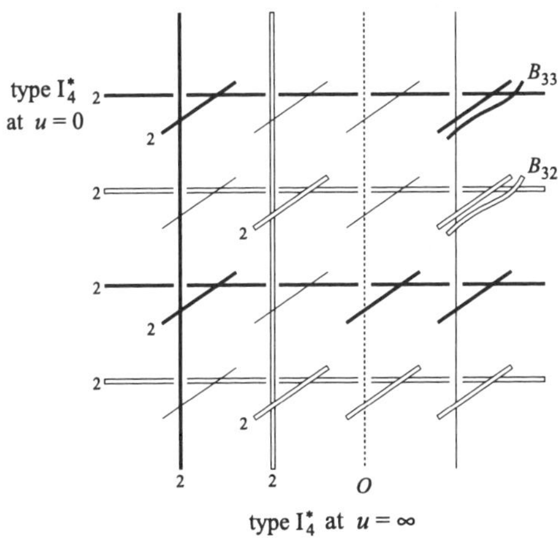

Fig. 11. $\mathscr{J}_{11}$

We look for a function whose divisor is $\Psi_{11,0} - \Psi_{11,\infty}$, and we obtain the elliptic parameter of type $\mathscr{J}_{11}$:

$$u = \frac{x_2(x_2 - \lambda_2)(x_1 - x_2)}{x_1(x_2 - 1)(\lambda_2 x_1 - x_2)}.$$

The change of variables

$$
\begin{aligned}
X &= u\frac{(\lambda_1-1)(x_2-\lambda_2)(x_1-x_2)}{x_1(x_1-1)} \\
&= \frac{(\lambda_1-1)\,x_2(x_2-\lambda_2)^2(x_1-x_2)^2}{x_1^2(x_1-1)(x_2-1)(\lambda_2 x_1-x_2)}, \\
Y &= u^2\frac{(\lambda_1-1)(x_2-\lambda_2)^2(x_1-x_2)^2}{t\,x_1^2(x_1-1)^2} \\
&= \frac{(\lambda_1-1)\,x_2^2(x_2-\lambda_2)^4(x_1-x_2)^4}{t\,x_1^4(x_1-1)^2(x_2-1)^2(\lambda_2 x_1-x_2)^2}
\end{aligned}
$$

converts (1.1) to the Weierstrass equation

$$
\begin{aligned}
&Y^2 = X^3 + \bigl(\lambda_1 u^2 - (2\lambda_1\lambda_2 - \lambda_1 - \lambda_2 + 2)u + \lambda_2\bigr)\,u\,X^2 \\
&+ (\lambda_1-1)(\lambda_2-1)\bigl((\lambda_1\lambda_2+1)u - 2\lambda_2\bigr)\,u^3\,X + \lambda_2(\lambda_1-1)^2(\lambda_2-1)^2\,u^5.
\end{aligned}
$$

Its discriminant is of the form $u^{10}d(u)$, where $d(u)$ is a polynomial of degree 4. The discriminant of $d(u)$ is too complicated to write down here. However, a simple search reveals that there are cases where four I_1 fibers degenerate even when C_1 and C_2 are not isogenous.

Remark. Suppose that the characteristic of the base field is 0.

(1) If $\lambda_1 = -1$ and $\lambda_2 = 9 \pm 4\sqrt{5}$, then the fibration has one I_2 fibers and one type II fiber. In this case j-invariant of C_1 is 1728 and that of C_2 is $78608 = 2^4 17^3$. They are not isogenous, and they can be defined over $\mathbf{Q}$.

(2) If $\lambda_1 = -1$ and $\lambda_2 = \pm\sqrt{-1}$, then the fibration has two type II fibers. In this case j-invariant of C_1 is 1728 and that of C_2 is 128. They are not isogenous, and they can be defined over $\mathbf{Q}$.

§5. (2,2)-curves and $\mathcal{J}_5$, $\mathcal{J}_9$ and $\mathcal{J}_{10}$

5.1. (2,2)-curves

Now the pullbacks of (1,1)-curves are not enough to construct all the elliptic fibrations in Oguiso's list. A pullback of a (2,2)-curve is a candidate for missing (-2)-curves. A nonsingular (2,2)-curve in $\mathbf{P}^1 \times \mathbf{P}^1$ is a curve of genus 1, and thus, we first look for (2,2)-curves with a node. Then we try to impose conditions such that their pullbacks are (-2)-curves. Here, we do not try to make a systematic search as before.

Actually, we can construct an elliptic fibration of type $\mathcal{J}_5$ using only pullbacks of (1,1)-curves and the basic curves. As a by-product, however, we obtain some new (-2)-curves which are pullbacks of $(2,2)$-curves. Such curves have a node at R_{11}. They are given by an equation of the form

$$a\,x_1^2 z_2^2 + b\,x_2 x_1 z_2 z_1 + c\,x_2^2 z_1^2 + d\,x_1 z_2^2 z_1 + e\,x_2 z_2 z_1^2 + f\,z_2^2 z_1^2 = 0.$$

The fact that it has a node at R_{11} corresponds to the fact that the equation does not have the terms $x_1^2 x_2^2$, $x_1^2 x_2 z_2$, and $x_1 z_1 x_2^2$. In order to obtain such a $(2,2)$-curve, we need to specify six points among R_{ij} $(1 \le i, j \le 4)$ such that no three among them are on the same F_i or G_j. We use such curves to construct an elliptic fibration of type $\mathscr{J}_9$ and $\mathscr{J}_{10}$.

5.2. $\mathscr{J}_5$

An elliptic fibration of type $\mathscr{J}_5$ has six I_2 fibers together with one I_6^* fiber. In order to write down an elliptic parameter for $\mathscr{J}_5$, we need to identify these six I_2 fibers.

Let B_{33} and B_{32} be the (-2)-curves introduced in §3.3. Consider two more (-2)-curves of this type:

$$\begin{aligned} B_{12} &:= \tilde{L}_{00,23,31} : \lambda_2(x_1 - \lambda_1) + (\lambda_1 - 1)x_2 = 0, \\ B_{13} &:= \tilde{L}_{00,22,31} : x_1 - \lambda_1 + (\lambda_1 - 1)x_2 = 0. \end{aligned}$$

Looking at Fig. 12, we see that B_{33} and B_{12} intersect each other only at two points above the intersection of lines $x_1 - x_2 = 0$ and $\lambda_2(x_1 - \lambda_1) + (\lambda_1 - 1)x_2 = 0$. Thus, the divisor $B_{33} + B_{12}$ is a singular fiber of type I_2. Similarly, $B_{32} + B_{13}$ is another singular fiber of type I_2. Furthermore, $B_{33} + B_{12}$ and $B_{32} + B_{13}$ do not intersect each other since the image of these curves in $\mathbf{A}_{x_1} \times \mathbf{A}_{x_2}$ intersect only at R_{ij} (see Fig. 12 below).

Computing the divisors $(x_1 - x_2)$, $\big(\lambda_2(x_1 - \lambda_1) + (\lambda_1 - 1)x_2\big)$, $(\lambda_2 x_1 - x_2)$ and $\big(x_1 - \lambda_1 + (\lambda_1 - 1)x_2\big)$, we see that

$$u = \frac{(x_1 - x_2)\big(\lambda_2(x_1 - \lambda_1) + (\lambda_1 - 1)x_2\big)}{(\lambda_2 x_1 - x_2)\big(x_1 - \lambda_1 + (\lambda_1 - 1)x_2\big)}$$

is an elliptic parameter of type $\mathscr{J}_5$. We have

$$u - 1 = \frac{-\lambda_1(\lambda_2 - 1)\,x_2(x_1 - 1)}{(\lambda_2 x_1 - x_2)\big(x_1 - \lambda_1 + (\lambda_1 - 1)x_2\big)},$$

which shows that the fiber at $u = 1$ is a singular fiber of type I_6^*. Each of the divisors A_{12}, A_{13}, A_{32} and A_{33} is a component of a singular fiber

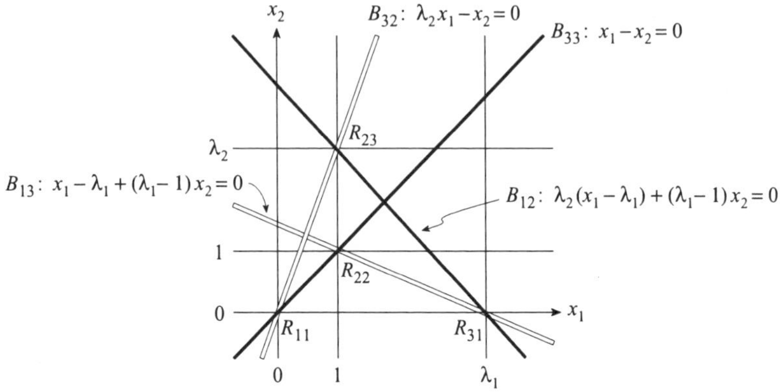

Fig. 12. $(1,1)$-curves

of type I_2. The other (-2)-curves are pullbacks of $(2,2)$-curves. For example, the singular fiber at $u = \lambda_1\lambda_2 - \lambda_1 + 1$ consists of A_{12} and the pullback of the $(2,2)$-curve given by

$$\lambda_2 x_1(x_1 - 1) + (\lambda_1 - 1)(x_2 - 1)(\lambda_2 x_1 - x_2) = 0.$$

In order to obtain a Weierstrass equation using the curve G_0 as the 0-section, we first put

$$X_0 = \frac{(x_1 - x_2)(x_1 - \lambda_1)}{x_1\big((x_1 - \lambda_1) + (\lambda_1 - 1)x_2\big)},$$
$$Y_0 = \frac{\lambda_1(\lambda_1 - 1)\, t\, x_2(x_1 - 1)(x_1 - \lambda_1)(x_2 - x_1)}{x_1(\lambda_2 x_1 - x_2)\big((x_1 - \lambda_1) + (\lambda_1 - 1)x_2\big)^2},$$

and then put

$$X = \lambda_1(\lambda_2 - 1)(u - 1)(u - \lambda_1\lambda_2 + \lambda_1 - 1)\big((\lambda_1\lambda_2 - \lambda_1 - \lambda_2)u + \lambda_2\big)X_0,$$
$$Y = \lambda_1^2(\lambda_2 - 1)^2(u - 1)^2(u - \lambda_1\lambda_2 + \lambda_1 - 1)\big((\lambda_1\lambda_2 - \lambda_1 - \lambda_2)u + \lambda_2\big)Y_0.$$

Then (X, Y) satisfy the Weierstrass equation

$$Y^2 = X(X - \alpha)(X - \beta),$$

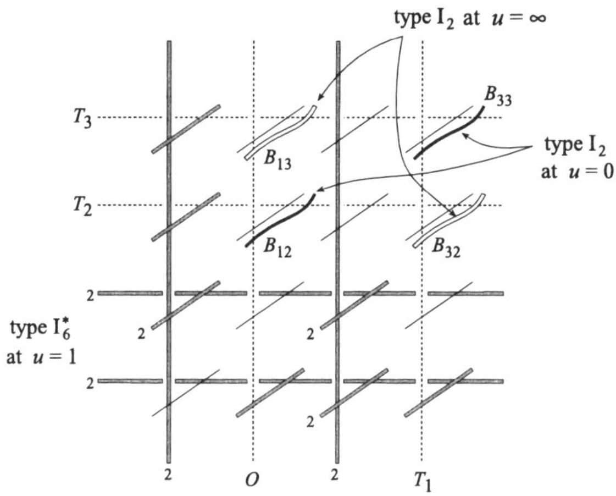

Fig. 13. $\mathscr{J}_5$

where

$$\alpha = -\lambda_1(\lambda_2-1)(u-1)\big((\lambda_1\lambda_2-1)u-\lambda_1+1\big)\big((\lambda_1\lambda_2-\lambda_1-\lambda_2)u+\lambda_2\big),$$
$$\beta = \lambda_1(\lambda_2-1)u(u-1)(u-\lambda_1\lambda_2+\lambda_1-1)\big((\lambda_1\lambda_2-\lambda_2)u-\lambda_1+\lambda_2\big).$$

The discriminant of this fibration is given by

$$\begin{aligned}\Delta(u) = 16\lambda_1^6\lambda_2^2(\lambda_1-1)^2u^2(u-1)^{12}\\ \times (u-\lambda_1\lambda_2+\lambda_1-1)^2\big((\lambda_1\lambda_2-1)u-\lambda_1+1\big)^2\\ \times \big((\lambda_1\lambda_2-\lambda_2)u-\lambda_1+\lambda_2\big)^2\big((\lambda_1\lambda_2-\lambda_1-\lambda_2)u+\lambda_2\big)^2.\end{aligned}$$

The Mordell-Weil group of this elliptic surface has the following three sections:

$$\begin{aligned} F_3 &\leftrightarrow T_1=(0,0),\\ G_2 &\leftrightarrow T_2=(\alpha,0),\\ G_3 &\leftrightarrow T_3=(\beta,0).\end{aligned}$$

5.3. $\mathscr{J}_9$

In order to construct an elliptic fibration of type $\mathscr{J}_9$, we need to find a divisor of type I_0^* different from the ones appearing in $\mathscr{J}_4$ or $\mathscr{J}_7$. To do so we look for a (-2)-curve P_{33} such that $2G_3 + A_{03} + A_{33} + B_{33} + P_{33}$ is of type I_0^*. We can show that P_{33} cannot be a pullback of a $(1,1)$-curve; if that were the case, B_{33} and P_{33} would have to intersect each other. Thus, we look for a $(2,2)$-curve whose double cover serves as P_{33}.

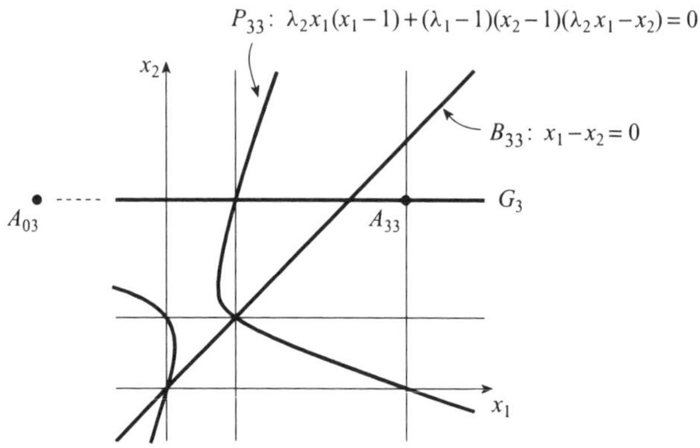

Fig. 14. fiber at $u = 0$

It turns out that the pullback of the $(2,2)$-curve

$$\lambda_2 x_1(x_1 - 1) + (\lambda_1 - 1)(x_2 - 1)(\lambda_2 x_1 - x_2) = 0 \tag{5.1}$$

can be used as P_{33}. This curve is a component of a I_2 fiber of the elliptic fibration of type $\mathscr{J}_5$ which we constructed in the previous subsection. The $(2,2)$-curve (5.1) has a node at R_{00}, and passes through $R_{11}, R_{12}, R_{22}, R_{23}$, and R_{31}. Fig. 14 shows the projection of the (-2)-curves contained in the divisor $\Psi_{9,0} = 2G_3 + A_{03} + A_{33} + B_{33} + P_{33}$. (The projection of A_{03} is R_{03}, which is a point at infinity.)

Similarly, let P_{32} be the pullback of the $(2,2)$-curve

$$\lambda_2 x_1(x_1 - 1) + (\lambda_1 - 1)(x_2 - \lambda_2)(x_1 - x_2) = 0.$$

Then, the divisor $\Psi_{9,\infty} = 2G_2 + A_{02} + A_{32} + B_{32} + P_{32}$ is again of type I_0^*, which does not intersect with $\Psi_{9,0}$. Fig. 15 shows the curves contained in the divisor $\Psi_{9,\infty}$. Looking for the function having the

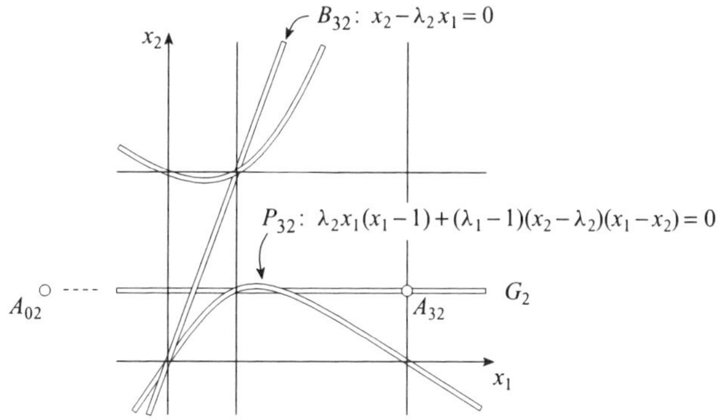

Fig. 15. fiber at $u = \infty$

divisor $\Psi_{9,0} - \Psi_{9,\infty}$, we find the elliptic parameter u given by

(5.2)
$$u = \frac{(x_2 - \lambda_2)(x_1 - x_2)\big(\lambda_2 x_1(x_1 - 1) + (\lambda_1 - 1)(x_2 - 1)(\lambda_2 x_1 - x_2)\big)}{(x_2 - 1)(\lambda_2 x_1 - x_2)\big(\lambda_2 x_1(x_1 - 1) + (\lambda_1 - 1)(x_2 - \lambda_2)(x_1 - x_2)\big)}.$$

We have

$$u - 1 = \frac{-\lambda_2(\lambda_2 - 1)x_1 x_2 (x_1 - 1)^2}{(x_2 - 1)(\lambda_2 x_1 - x_2)\big(\lambda_2 x_1(x_1 - 1) + (\lambda_1 - 1)(x_2 - \lambda_2)(x_1 - x_2)\big)}.$$

The zero divisor of this function $u - 1$ is given by the following divisor consisting only of the basic curves:

$$A_{01} + 2G_1 + 3A_{21} + 4F_2 + 5A_{20} + 6G_0 + 3A_{30} + 4A_{10} + 2F_1.$$

This is a singular fiber of type II* (see Fig. 16). Thus, the elliptic parameter given by (5.2) is of type $\mathscr{J}_9$.

Our next task is to write down a Weierstrass equation. If we regard (5.2) as the defining equation of a curve in $\mathbf{P}^1 \times \mathbf{P}^1$ defined over $k(u)$, then we can show that this curve is a curve of genus 0, and thus it can be parametrized. In fact, we can parametrize x_1 and x_2 satisfying (5.2) using the parameter

$$\xi = -\frac{(x_2 - 1)(\lambda_2 x_1 - x_2)}{\lambda_2 x_1 x_2}.$$

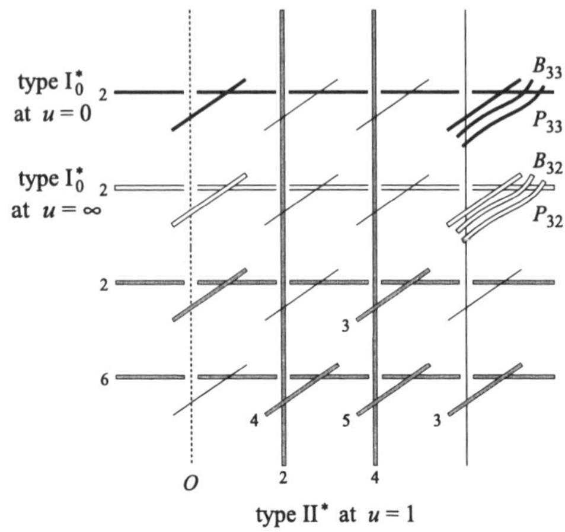

Fig. 16. $\mathscr{J}_9$

Actual parametrizations of x_1 and x_2 are complicated and we omit here. Substituting x_1 and x_2 in the equation (1.1) by these parametrizations, we obtain an equation of a curve of genus 1 with variables in (ξ, t) defined over $k(u)$. This equation turns out to be a quadratic equation in t, and it is easily converted to a Weierstrass equation. Combining all these, we obtain the change of variables

$$X_0 = -\frac{\lambda_2(\lambda_2 - 1)x_1(x_1 - 1)}{(x_2 - 1)(\lambda_2 x_1 - x_2)}, \quad Y_0 = \frac{\lambda_2(\lambda_2 - 1)}{t}$$

that converts (1.1) to a twisted form of Weierstrass equation

$$\begin{aligned} u(u-1)Y_0^2 = X_0^3 &+ (\lambda_1\lambda_2 - 2\lambda_1 - 2\lambda_2 + 1)(u-1)X_0^2 \\ &- (\lambda_1 + \lambda_2 - 1)(\lambda_1\lambda_2 - \lambda_1 - \lambda_2)(u-1)^2 X_0 \\ &\qquad - \lambda_1\lambda_2(\lambda_1 - 1)(\lambda_2 - 1)(u-1)^2. \end{aligned}$$

By letting

$$X = u(u-1)X_0, \quad Y = u^2(u-1)^2 Y_0,$$

we obtain the following Weierstrass equation:

$$\begin{aligned}Y^2 = X^3 &+ (\lambda_1\lambda_2 - 2\lambda_1 - 2\lambda_2 + 1)\, u\, (u-1)^2 X^2 \\ &- (\lambda_1 + \lambda_2 - 1)(\lambda_1\lambda_2 - \lambda_1 - \lambda_2)\, u^2 (u-1)^4 X \\ &- \lambda_1\lambda_2(\lambda_1 - 1)(\lambda_2 - 1)\, u^3 (u-1)^5.\end{aligned}$$

Its discriminant is of the form $u^6(u-1)^{10}d(u)$, where $d(u)$ is a polynomial of degree 2. The discriminant of $d(u)$ is given by

$$16\lambda_1^2\lambda_2^2(\lambda_1-1)^2(\lambda_2-1)^2(\lambda_1^2-\lambda_1+1)^3(\lambda_2^2-\lambda_2+1)^3.$$

If either λ_1 or λ_2 is a sixth root of unity, then two I_1 fibers of the fibration degenerate to form a type II fiber.

5.4. $\mathscr{J}_{10}$

In order to construct an elliptic fibration of type $\mathscr{J}_{10}$, we must find yet another divisor of type I_0^*. The divisor $\Psi_{9,0} = 2G_3 + A_{03} + A_{33} + B_{33} + P_{33}$ is a divisor of type I_0^* appearing in the elliptic fibration constructed in the previous subsection. Since neither B_{33} nor P_{33} intersects with A_{13}, we see that

$$\Psi_{10,0} = 2G_3 + A_{13} + A_{33} + B_{33} + P_{33}$$

is also a divisor of type I_0^*. We then find a divisor of type I_6^* that does not intersect with $\Psi_{10,0}$:

$$\begin{aligned}\Psi_{10,\infty} = B_{32} &+ A_{32} + 2G_2 + 2A_{02} + 2F_0 + 2A_{01} \\ &+ 2G_1 + 2A_{21} + 2F_2 + 2A_{20} + 2G_0 + A_{10} + A_{30}\end{aligned}$$

(see Fig.17). Looking for the function having the divisor $\Psi_{10,0} - \Psi_{10,\infty}$, we find the elliptic parameter of type $\mathscr{J}_{10}$ given by

(5.3)
$$u = \frac{(x_2-\lambda_2)(x_1-x_2)\big((\lambda_1-1)(x_2-1)(\lambda_2x_1-x_2)+\lambda_2x_1(x_1-1)\big)}{x_2(x_2-1)(x_1-1)(\lambda_2x_1-x_2)}.$$

The curve in $\mathbf{P}^1 \times \mathbf{P}^1$ over $k(u)$ defined by (5.3) is a curve of genus 0. As in the case of $\mathscr{J}_9$, the parameter

$$\xi = \frac{(x_1-x_2)(x_2-\lambda_2)}{(x_1-1)x_2},$$

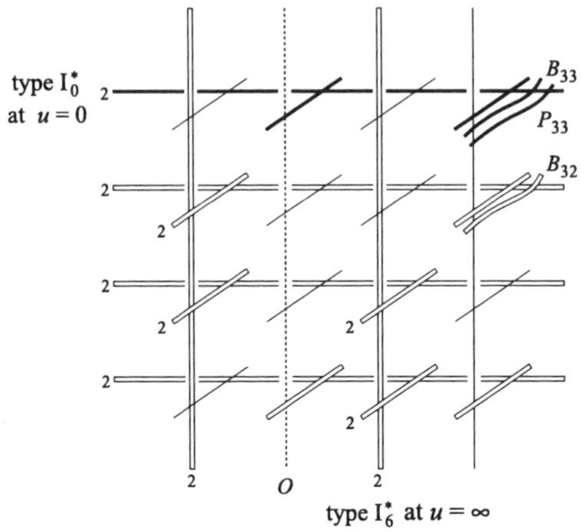

Fig. 17. $\mathscr{I}_{10}$

can be used to parametrize this curve. We can proceed in a similar manner to the case of $\mathscr{I}_9$ and we obtain the change of variables

$$X_0 = \frac{\lambda_1(\lambda_1-1)(x_2-1)(\lambda_2 x_1 - x_2)}{x_1(x_1-1)},$$
$$Y_0 = \frac{\lambda_1(\lambda_1-1)(x_2-1)^2(\lambda_2 x_1 - x_2)^2}{t x_1^2(x_1-1)^2},$$

which converts (1.1) to

$$\begin{aligned} uY_0^2 = X_0^3 &- (u+\lambda_1+\lambda_2-1)(u-\lambda_1\lambda_2+\lambda_1+\lambda_2)X_0^2 \\ &+ \lambda_1\lambda_2(\lambda_1-1)(\lambda_2-1)(2u-\lambda_1\lambda_2+2\lambda_1+2\lambda_2-1)X_0 \\ &+ \lambda_1^2\lambda_2^2(\lambda_1-1)^2(\lambda_1-1)^2. \end{aligned}$$

Putting

$$X = uX_0, \quad Y = u^2 Y_0,$$

we obtain the Weierstrass equation

$$\begin{aligned} Y^2 = X^3 &+ u(u-\lambda_1-\lambda_2+1)(u+\lambda_1\lambda_2-\lambda_1-\lambda_2)X^2 \\ &+ \lambda_1\lambda_2(\lambda_1-1)(\lambda_2-1)u^2(2u+\lambda_1\lambda_2-2\lambda_1-2\lambda_2+1)X \\ &+ \lambda_1^2\lambda_2^2(\lambda_1-1)^2(\lambda_1-1)^2u^3. \end{aligned}$$

Its discriminant is of the form $u^6 d(u)$, where $d(u)$ is a polynomial of degree 2. We can show that $d(u)$ can have a multiple root without C_1 and C_2 being isogenous.

§6. Full list of the defining equations in a special case

In this section, we take as C_1 and C_2 the most familiar elliptic curves

$$C_1 : y_1^2 = x_1^3 - x_1, \quad C_2 : y_2^2 = x_2^3 - 1, \tag{6.1}$$

and write down the full list of the defining equations of mutually non-isomorphic elliptic fibrations on the Kummer surface $S = \mathrm{Km}(C_1 \times C_2)$ in characteristic 0. Although they are very special among elliptic curves (e.g. automorphisms or complex multiplications), corrseponding Kummer surface S serves as a more or less "typical" case, since C_1, C_2 are not isogenous to each other.

In this case, the number $N(n)$ of nonisomorphic elliptic fibrations on S of type $\mathcal{J}_n$ has been determined by Oguiso as follows:

$$N(n) = 1 \quad \text{for } n = 2, 3, 5, 8, 9, 10,$$

and

$$N(n) = 2 \quad \text{for } n = 1, 4, 6, 7, 11.$$

(See Oguiso [8, p. 652]. We note that this number N is *not* typical among all non-isogenous curves, as shown there.)

Now observe that the values of Legendre parameter λ_i for the present C_i are as follows:

$$\lambda_1 = -1,\ 2 \text{ or } 1/2, \quad \lambda_2 = -\omega \text{ or } -\omega^2,$$

where ω is a cubic root of unity. In the following, we write down the $N = N(n)$ defining equations for each type $\mathcal{J}_n$. When $N = 1$, we give essentially the same equation as the one constructed in the previous sections, except that we make some coordinate change when it makes the equation look simpler. When $N > 1$, we make the same construction as before using a suitable equivalent value of λ_i. We briefly indicate how to verify that the resulting defining equations are not isomorphic to each other.

6.1. $\mathcal{J}_1$

$$y^2 = x\left(x^2 + (u^4 + 1)x + 4u^4\right), \tag{6.2}$$

$$J = \frac{1}{108}\frac{(u^8 - 10u^4 + 1)^3}{u^8(u^8 - 14u^4 + 1)}.$$

$$(6.3) \qquad y^2 = x\big(x^2 + (u^4 + 6(2\omega + 1)u^2 + 1)x - 32u^4\big),$$

$$J = \frac{1}{6912}\frac{\big(u^8 + 12(2\omega + 1)u^6 - 10u^4 + 12(2\omega + 1)u^2 + 1\big)^3}{u^8\big(u^8 + 12(2\omega + 1)u^6 + 22u^4 + 12(2\omega + 1)u^2 + 1\big)}.$$

Both the equations (6.2) and (6.3) have two I_8 fibers at $u = 0$ and ∞ and eight I_1 fibers. Suppose they define isomorphic elliptic curves over $k(u)$. Then there must be a linear transformation of u fixing 0 and ∞ which sends one J into the other, J denoting the classical absolute invariant of the generic fibre (normalized so that $J = 1$ for $y^2 = x^3 - x$). But this is impossible, as the positions of the eight I_1 fibers are determined by the simple poles of J and they cannot be transformed by such a linear transformation. This proves that the two elliptic fibrations are not isomorphic to each other.

6.2. $\mathscr{J}_2$

$$(6.4) \qquad y^2 = x\big(x^2 - (3u^4 + 6u^2 - 1)x + 32u^6\big),$$

$$J = \frac{1}{6912}\frac{(9u^8 - 60u^6 + 30u^4 - 12u^2 + 1)^3}{u^{12}(u^4 - 10u^2 + 1)(9u^4 - 2u^2 + 1)}.$$

6.3. $\mathscr{J}_3$

$$(6.5) \qquad y^2 = x^3 + u^4(u^4 + 1), \quad J = 0.$$

6.4. $\mathscr{J}_4$

$$(6.6) \qquad y^2 = x^3 - (u^3 - 1)^2 x, \ (u = x_1), \quad J = 1.$$

$$(6.7) \qquad y^2 = x^3 - (v^3 - v)^3, \ (v = x_2), \quad J = 0.$$

These are the two obvious elliptic fibrations on S induced by the projections $C_1 \times C_2 \to C_1$ or C_2.

6.5. $\mathscr{J}_5$

$$y^2 = x(x-4)\bigl(x+2u(u^2+3u+3)\bigr). \tag{6.8}$$

$$J = \frac{1}{27}\frac{(u^6+6u^5+15u^4+20u^3+15u^2+6u+4)^3}{u^2(u+2)^2(u^2+u+1)^2(u^2+3u+3)^2}.$$

6.6. $\mathscr{J}_6$

$$y^2 = x(x+2u^2)\bigl(x-u(u^2-u+1)\bigr), \tag{6.9}$$

$$J = \frac{1}{27}\frac{(u^4+5u^2+1)^3}{u^2(u^4+u^2+1)^2}.$$

$$y^2 = x(x-\omega u^2)\bigl(x+u(2u-1)(u+\omega^2)\bigr), \tag{6.10}$$

$$J = \frac{4}{27}\frac{\omega\bigl(2u^2-(\omega+2)u-\omega^2\bigr)^3\bigl(2u^2-2(\omega+2)u-\omega^2\bigr)^3}{u^2(u-1)^2(2u-1)^2(u+\omega^2)^2(2u+\omega^2)^2}.$$

The Legendre parameters we employed for the first equation (6.9) are $\lambda_1 = -1$, $\lambda_2 = -\omega$, while those for the second one (6.10) are $\lambda_1 = 2$, $\lambda_2 = -\omega$. That the two equations define nonisomorphic elliptic fibrations can be checked in the same way as the case for $\mathscr{J}_1$ above.

6.7. $\mathscr{J}_7$

$$y^2 = x\bigl(x^2-u(u+1)(u+2)x+u^2(u+2)^2\bigr), \tag{6.11}$$

$$J = \frac{4}{27}\frac{(u^2+2u-2)^3}{(u-1)(u+3)}.$$

$$y^2 = x\bigl(x^2+\omega u(u-1)(u-3\omega-2)x+2\omega^2u^2(u-1)^2\bigr). \tag{6.12}$$

$$J = \frac{1}{27}\frac{(u^2-2(3\omega+2)u+3\omega-11)^3}{(u^2-2(3\omega+2)u+3\omega-13)}.$$

In this case, we can check that there is a linear transformation of u sending the first J into the second one. However it does not preserve the position of singular fibres which can be seen from the discriminants (but not from the absolute invariants). Hence (6.11) and (6.12) are not isomorphic.

6.8. $\mathscr{J}_8$

$$y^2 = x\bigl(x^2 + u(3u+2)x + 1 + 3u + 3u^2 + 2u^3\bigr), \tag{6.13}$$

$$J = \frac{4}{27}\frac{(6u^3 - 3u - 1)^3}{u^2(2u+1)^2(u^2+u+1)^2(8u+3)}.$$

6.9. $\mathscr{J}_9$

$$y^2 = x^3 + u(u^2-4)^3, \quad J = 0. \tag{6.14}$$

6.10. $\mathscr{J}_{10}$

$$y^2 = x^3 + u^2(u+3)x^2 + u^2(-2u^2 - 2u + 3)x + u^4(u-1). \tag{6.15}$$

We omit J.

6.11. $\mathscr{J}_{11}$

$$\begin{aligned} y^2 = x^3 &- 27u^2(u^4 + 6u^3 + 5u^2 - 6u + 1)x \\ &- 54u^3(u^2+1)(u^4 + 9u^3 + 20u^2 - 9u + 1). \end{aligned} \tag{6.16}$$

$$\begin{aligned} y^2 = x^3 &- 27u^2(4u^4 - 12u^3 + 10(\omega+1)u^2 - 6\omega u + \omega)x \\ &+ 27u^3\bigl(16u^6 - 72u^5 + 6(19+10\omega)u^4 \\ &\quad - 63(1+2\omega)u^3 + 3(-9+10\omega)u^2 + 18u - 2\bigr). \end{aligned} \tag{6.17}$$

We omit J, but it can be checked that the two elliptic fibrations are not isomorphic to each other by a similar argument as before.

Thus we have listed the defining equations of elliptic fibrations (with a section) on the Kummer surface $S = \mathrm{Km}(C_1 \times C_2)$ with C_i given by (6.1) over an algebraically closed field k of characteristic 0. Needless to say that the function field $k(x, y, u)$ defined by each of the equations (6.2) through (6.17) is isomorphic to one and the same function field $k(S)$, which is the extension $k(x_1, x_2, t)$ with $t = y_1/y_2$ determined by (6.1).

§7. Closing remark

In closing this paper, it should be remarked that the problems posed in the Introduction (§1.1) should be interesting and worth considering for more general $K3$ surfaces.

Even in the case of Kummer surfaces, we could ask such questions as follows:

Problem 5. *Study Problems 1 and 2 for the Kummer surface* $X = \mathrm{Km}(A)$, *when A is the Jacobian variety of a genus two curve.*

For this, the so-called 16_6-configuration of thirty-two (-2)-curves on X should play an important role in place of the twenty-four basic curves used in this paper. A special case has been treated in Shioda [13].

According to Weil [15], a principally polarized abelian surface is either the Jacobian variety of a genus two curve or a product of two elliptic curves. Beyond the case of principally polarized abelian surfaces, we ask:

Problem 6. *Find at least one elliptic parameter for the Kummer surface* $X = \mathrm{Km}(A)$ *when A is a generic member in a family of polarized abelian surfaces.*

The coefficients in the defining equation (especially the discriminant) for such should be related to some modular forms or theta-functions of interest.

Acknowledgements We would like to thank Professor Fumio Sakai of Saitama University who arranged a Seminar by the second author about "Elliptic parameters" in September 2005, which resulted in this fruitful joint work with the first author who attended the Seminar. We thank Matthias Schütt for careful reading of the manuscript. We are most grateful to the referee for making very sharp and constructive comments on the manuscript, who has requested the three things: (i) to make the paper free from mistakes (since many standard calculations are nicely omitted), (ii) to add a few words about the case of quasi-elliptic fibrations, and (iii) to include "a full account in one non-trivial case" after saying that Remark 1.7 is very impressive.

In answering these requests, (i) we have tried our best to minimize the possible errors (but still some errors might have crept in during correction). (ii) In $\mathrm{char}\, k = 3$, it is possibe that some of the defining equations becomes a quasi-elliptic fibration. We should leave this question to the interested reader, but let us say this: if for example we let $\lambda_1 = \lambda_2 = -1$ in $\mathrm{char}\, k = 3$, then $\mathcal{J}_3$-fibration gives the same equation as the $\mathscr{J}_3$-fibration in §6 (for char $k \neq 3$), which is indeed quasi-elliptic in char $k = 3$. (iii) We have added a new section §6 in the revised version to respond to this request.

The second-named author ackowledges the support from Grant-in-Aid for Scientific Research No.17540044.

References

[1] J. W. S. Cassels, Lectures on elliptic curves, London Math. Soc. Stud. Texts, **24**, Cambridge Univ. Press, Cambridge, 1991.

[2] I. Connell, Addendum to a paper of K. Harada and M.-L. Lang: Some elliptic curves arising from the Leech lattice, J. Algebra, **125** (1989), 298–310, J. Algebra, **145** (1992), 463–467.

[3] H. Inose, Defining equations of singular $K3$ surfaces and a notion of isogeny, In: Proceedings of the International Symposium on Algebraic Geometry, Kyoto Univ., Kyoto, 1977, Kinokuniya, Tokyo, 1978, pp. 495–502.

[4] K. Kodaira, On compact analytic surfaces. II, III, Ann. of Math. (2), **77** (1963), 563–626; ibid., **78** (1963), 1–40.

[5] M. Kuwata, The canonical height and elliptic surfaces, J. Number Theory, **36** (1990), 201–211.

[6] ———, Elliptic $K3$ surfaces with given Mordell-Weil rank, Comment. Math. Univ. St. Paul., **49** (2000), 91–100.

[7] D. R. Morrison, On $K3$ surfaces with large Picard number, Invent. Math., **75** (1984), 105–121.

[8] K. Oguiso, On Jacobian fibrations on the Kummer surfaces of the product of nonisogenous elliptic curves, J. Math. Soc. Japan, **41** (1989), 651–680.

[9] I. I. Pjateckiĭ -Šapiro and I. R. Šafarevič, Torelli's theorem for algebraic surfaces of type K3, Izv. Akad. Nauk SSSR Ser. Mat., **35** (1971), 530–572; English tansl., Math. USSR, Izv., **5** (1971), 547–588.

[10] T. Shioda and H. Inose, On singular $K3$ surfaces, Complex Analysis and Algebraic Geometry, Iwanami Shoten, Tokyo, 1977, pp. 119–136.

[11] T. Shioda, On the Mordell-Weil lattices, Comment. Math. Univ. St. Paul., **39** (1990), 211–240.

[12] ———, A note on $K3$ surfaces and sphere packings, Proc. Japan Acad. Ser. A Math. Sci., **76** (2000), 68–72.

[13] ———, Classical Kummer surfaces and Mordell-Weil lattices, Algebraic geometry, Contemp. Math., **422**, Amer. Math. Soc., Providence, RI, 2007, pp. 213–221.

[14] ———, Correspondence of elliptic curves and Mordell-Weil lattices of certain elliptic $K3$'s, Algebraic Cycles and Motives, vol. 2, London Math. Soc. Lecture Note Ser., **344**, Cambridge Univ. Press, Cambridge, 2007, pp. 319–339.

[15] A. Weil, Zum Beweis des Torellischen Satzes, Nachr. Akad. Wiss. Göttingen. Math.-Phys. Kl. IIa., **1957** (1957), 33–53.

Masato Kuwata
Faculty of Economics
Chuo University
742-1 Higashi-Nakano, Hachioji-shi
Tokyo 192-0393 JAPAN
E-mail address: kuwata@tamacc.chuo-u.ac.jp

Tetsuji Shioda
Department of Mathematics
Rikkyo University
3-34-1 Nishi-Ikebukuro, Toshima-ku
Tokyo 171-8501 JAPAN
E-mail address: shioda@rkmath.rikkyo.ac.jp

Advanced Studies in Pure Mathematics 50, 2008
Algebraic Geometry in East Asia — Hanoi 2005
pp. 217–236

Characterization of the rational homogeneous space associated to a long simple root by its variety of minimal rational tangents

Jaehyun Hong and Jun-Muk Hwang

Abstract.

Let $S = G/G'$ be a rational homogeneous space defined by a complex simple Lie group G and a maximal parabolic subgroup G'. For a base point $s \in S$, let $\mathcal{C}_s \subset \mathbb{P}T_s(S)$ be the variety of minimal rational tangents at s. In the study of rigidity of rational homogeneous spaces, the following question naturally arises. Let X be a Fano manifold of Picard number 1 such that the variety of minimal rational tangents at a general point $x \in X$, $\mathcal{C}_x \subset \mathbb{P}T_x(X)$, is isomorphic to $\mathcal{C}_s \subset \mathbb{P}T_s(S)$. Is X biholomorphic to S? An affirmative answer has been given by Mok when S is a Hermitian symmetric space or a homogeneous contact manifold. Extending Mok's method further and combining it with the theory of differential systems on S, we will give an affirmative answer when G' is associated to a long simple root.

§1. Introduction

Let X be a Fano manifold of Picard number 1. An irreducible component $\mathcal{K}$ of the space of rational curves on X is called a minimal dominating rational component, if for a general point $x \in X$, the subvariety $\mathcal{K}_x$ consisting of members passing through x is non-empty and projective. The tangent directions at x of members of $\mathcal{K}_x$ form a subvariety $\mathcal{C}_x$ of $\mathbb{P}T_x(X)$, called the variety of minimal rational tangents at x, and the closure of the union of $\mathcal{C}_x$ as x varies over general points of X gives the subvariety $\mathcal{C} \subset \mathbb{P}T(X)$, called the variety of minimal rational tangents (associated with $\mathcal{K}$). It is generally believed that the projective

---

Received February 3, 2006.

2000 *Mathematics Subject Classification.* 53C15, 32M10, 14J45.

Key words and phrases. rational homogeneous space, Cartan connection, minimal rational curve.

Jun-Muk Hwang was supported by the Korea Research Foundation Grant funded by the Korean Government(MOEHRD)(KRF-2006-341-C00004).

geometry of $\mathcal{C}_x \subset \mathbb{P}T_x(X)$ controls the geometry of the Fano manifold X. There are many results manifesting this philosophy which were surveyed in [Hw01]. In this paper, we will give another example of this philosophy in the study of rational homogeneous spaces, namely, the projective varieties homogeneous under the action of a complex semi-simple Lie group.

Recall that when S is a rational homogeneous space of Picard number 1, there is a unique choice of a minimal dominating rational component and the varieties of minimal rational tangents at two different points are isomorphic. In the study of rigidity of rational homogeneous spaces, the following conjecture naturally arises. This is Conjecture 2.2 in [Hw06].

Conjecture *Let S be a rational homogeneous space of Picard number 1 and $\mathcal{C}_s \subset \mathbb{P}T_s(S)$ be the variety of minimal rational tangents at a base point $s \in S$. Let X be a Fano manifold of Picard number 1 and $\mathcal{C}_x \subset \mathbb{P}T_x(X)$ be the variety of minimal rational tangents at a general point $x \in X$ associated to a minimal dominating rational component $\mathcal{K}$. Suppose that $\mathcal{C}_s \subset \mathbb{P}T_s(S)$ and $\mathcal{C}_x \subset \mathbb{P}T_x(X)$ are isomorphic as projective subvarieties. Then X is biholomorphic to S.*

We can write $S = G/G'$ for a complex simple Lie group G and a maximal parabolic subgroup G'. As explained in Section 2, the subgroup G' is determined by a choice of a simple root α of G. The main goal of this paper is to prove Conjecture for $S = G/G'$ when G' is associated to a long simple root.

Main Theorem *Let $S = G/G'$ where G' is a maximal parabolic subgroup associated to a long root and let $\mathcal{C}_s \subset \mathbb{P}T_s(S)$ be the variety of minimal rational tangents at a base point $s \in S$. Let X be a Fano manifold of Picard number 1 and $\mathcal{C}_x \subset \mathbb{P}T_x(X)$ be the variety of minimal rational tangents at a general point $x \in X$ associated to a minimal dominating rational component $\mathcal{K}$. Suppose that $\mathcal{C}_s \subset \mathbb{P}T_s(S)$ and $\mathcal{C}_x \subset \mathbb{P}T_x(X)$ are isomorphic as projective subvarieties. Then X is biholomorphic to S.*

There are three classes of rational homogeneous spaces associated to long simple roots. When the isotropy representation of G' on $T_s(S)$ is irreducible, S is a Hermitian symmetric space. When the isotropy representation on $T_s(S)$ has an irreducible subspace of codimension 1, S is a homogeneous contact manifold. The third class consisting of the remaining homogeneous spaces is distinguished from the first two classes from the fact that the automorphism group of the homogeneous space in this class is determined by the basic linear differential system on them (cf. Section 2).

For Hermitian symmetric spaces and homogeneous contact manifolds, Main Theorem was proved by N. Mok in [Mo]. The essential part of Mok's argument is to show that if $\mathcal{C}_x \subset \mathbb{P}T_x(X)$ is isomorphic to $\mathcal{C}_s \subset \mathbb{P}T_s(S)$ at a general point $x \in X$, then it is true for every point x on a general member of $\mathcal{K}$. This is equivalent to extending the corresponding geometric structure up to codimension 1. After this is done, he used the result of [HM97] for Hermitian symmetric spaces and the result of [Ho] for homogeneous contact manifolds to conclude that the geometric structure defined by $\mathcal{C}$ is isomorphic to that of S.

To prove Main Theorem for the third class of homogeneous spaces associated to long simple roots, we have to do two things. The first is to extend Mok's argument to the homogeneous spaces in question. The second is to establish an analog of [HM97] and [Ho] for the homogeneous spaces in the third class. In both parts, a crucial role is played by the theory of linear differential systems on homogeneous spaces. This theory plays a minor role in [Ho], [HM97] and [Mo], because for Hermitian symmetric spaces or homogeneous contact manifolds, the basic linear differential system is either trivial or very simple. But it is an essential component in the current paper, as was the case for other works on the third class of homogeneous spaces, as explained in the introduction of [HM02].

Main Theorem verifies Conjecture except for symplectic Grassmannians and two F_4-homogeneous spaces. For these remaining cases, our method cannot be applied at all. These cases remain a challenge for future research.

We will start with reviewing the basics of the theory of linear differential systems in Section 2. It turns out that this theory plays such a decisive role in the problem that Main Theorem can be proved easily except for quadric Grassmannians of isotropic 3-spaces. We will collect some basic properties of these homogeneous spaces in Section 4, after reviewing some known results about the variety of minimal rational tangents in Section 3. The analog of Mok's argument, extending the geometric structure up to codimension 1, is proved in Section 5 and the analog of [HM97] and [Ho] is proved in Section 6.

We would like to thank Keizo Yamaguchi for beneficial discussions.

§2. Review of Tanaka's theory of linear differential systems

We recall the theory of linear differential systems on rational homogeneous spaces, due to Tanaka [Ta].

First, let us recall a few basic definitions. A linear differential system on a complex manifold is just a subbundle of the tangent bundle. Given a subbundle D of the tangent bundle $T(M)$ of a complex manifold M, the k-th *weak derived system* $\mathcal{D}^k$ of D is a subsheaf of $T(M)$ defined inductively by $\mathcal{D}^{-1} = \mathcal{D}$ and $\mathcal{D}^k = \mathcal{D}^{k+1} + [\mathcal{D}, \mathcal{D}^{k+1}]$ for $k < -1$, where $\mathcal{D}$ is the sheaf of sections of D. We say that D is *regular at a point* $x \in M$, or equivalently, $x \in M$ is a *regular point* of D, if the k-th derived system $\mathcal{D}^k$ of D is a subbundle D^k of $T(M)$ in a neighborhood of x for every $k \leq -1$. The *symbol algebra* of D at a regular point $x \in M$ is defined as the graded nilpotent Lie algebra

$$\mathbf{sym}_x(D) := D_x^{-1} + D_x^{-2}/D_x^{-1} + \cdots + D_x^{-\mu}/D_x^{-\mu+1}$$

where μ is the largest integer satisfying $D^{-\mu} \neq D^{-\mu+1}$ and the Lie bracket is induced by the bracket of local vector fields in a neighborhood of x.

Now fix a complex simple Lie algebra $\mathfrak{g}$. Choose a Cartan subalgebra $\mathfrak{h}$ and the root system $\Phi \subset \mathfrak{h}^*$ of $\mathfrak{g}$ with respect to $\mathfrak{h}$. Fix a system of simple roots $\{\alpha_1, \dots, \alpha_l\}$ and a distinguished choice of a simple root α, say, $\alpha = \alpha_i$. Given an integer $k, -\mu \leq k \leq \mu$, we define Φ_k as the set of all roots $\sum_{j=1}^{l} m_j \alpha_j$ with $m_i = k$. Here μ is the largest integer such that $\Phi_\mu \neq 0$. For $\beta \in \Phi$, let $\mathfrak{g}_\beta$ be the corresponding root space. Define

$$\begin{aligned} \mathfrak{g}_0 &= \mathfrak{h} \oplus \bigoplus_{\beta \in \Phi_0} \mathfrak{g}_\beta \\ \mathfrak{g}_k &= \bigoplus_{\beta \in \Phi_k} \mathfrak{g}_\beta, \quad k \neq 0. \end{aligned}$$

The decomposition $\mathfrak{g} = \bigoplus_{k=-\mu}^{\mu} \mathfrak{g}_k$ gives a graded Lie algebra structure on $\mathfrak{g}$. Define

$$\begin{aligned} \mathfrak{g}' &= \mathfrak{g}_0 \oplus \mathfrak{g}_1 \oplus \cdots \oplus \mathfrak{g}_\mu \\ \mathfrak{m} &= \mathfrak{g}_{-1} \oplus \cdots \oplus \mathfrak{g}_{-\mu}. \end{aligned}$$

We say that $\mathfrak{g}'$ is the maximal parabolic subalgebra associated to the simple root α and $\mathfrak{m}$ is the nilpotent graded Lie algebra of type $(\mathfrak{g}, \alpha)$. Here we follow the convention of [Ya] for the sign of roots in the definition of the parabolic subalgebra. This is opposite to the choice in [HM02].

Now let G be a complex Lie group with Lie algebra $\mathfrak{g}$ and G' be the complex Lie subgroup with Lie algebra $\mathfrak{g}'$. The quotient space G/G' is called the rational homogeneous space of type $(\mathfrak{g}, \alpha)$. The quotient map $G \to G/G'$ defines a G'-principal bundle on G/G'. The vector bundle associated to this principal bundle by the natural G'-representation on $\mathfrak{m}$

$$\rho : G' \longrightarrow \mathrm{GL}(\mathfrak{m}), \quad \rho(a)v := \mathrm{Ad}(a)v \mod \mathfrak{g}' \text{ for } v \in \mathfrak{m}$$

is the tangent bundle $T(G/G')$. Let $E \subset T(G/G')$ be the subbundle corresponding to the G'-invariant subspace $\mathfrak{g}_{-1}$ of $\mathfrak{m}$. This E will be called the *standard differential system* on G/G'. The symbol algebra $\mathbf{sym}_s(E)$ of E at each point $s \in G/G'$ is isomorphic to the nilpotent graded Lie algebra $\mathfrak{m}$.

When $\mu = 1$, namely, $E = T(G/G')$, the homogeneous space G/G' is a Hermitian symmetric space. When $\mu = 2$ and $\dim \mathfrak{g}_{-2} = 1$, namely, when $E \subset T(G/G')$ is of corank 1, G/G' is a homogeneous contact manifold. We will exclude these two cases. We will say that $(\mathfrak{g}, \alpha)$ is neither of symmetric type nor of contact type.

The fundamental question regarding the basic differential system E is the following equivalence problem. Let D be a linear differential system on a complex manifold M and $x \in M$ be a regular point of D. We say that D is a *regular differential system of type* $\mathfrak{m}$ at x if $\mathbf{sym}_y(D)$ is isomorphic to $\mathfrak{m}$ for each point y in a neighborhood of x. The equivalence problem asks for a regular differential system D of type $\mathfrak{m}$ on a manifold M, when we can find a biholomorphic map $\varphi : U \to U'$ between a neighborhood U of $x \in M$ and a neighborhood U' of $\varphi(x)$ in G/G' such that its differential $d\varphi : T(U) \to T(U')$ sends D onto E. An answer to this question was given in [Ta]. We will review this result.

Before going into details, let us remark that the assumption that $(\mathfrak{g}, \alpha)$ is neither of symmetric type nor of contact type guarantees that $\mathfrak{g}$ is the prolongation of $\mathfrak{m}$ by Theorem 5.2 of [Ya] and thus a $G_0^\sharp$- structure of type $\mathfrak{m}$ that Tanaka considered in [Ta] is nothing but a regular differential system of type $\mathfrak{m}$.

A *Cartan connection* of type G/G' on a manifold M of dimension $= \dim G/G'$ is a principal G'-bundle P on M with a $\mathfrak{g}$-valued 1-form ω on P satisfying the following conditions:

(1) For a tangent vector v of P, $\omega(v) = 0$ implies $v = 0$.

(2) $\omega(A^*) = A$ for all $A \in \mathfrak{g}'$ where A^* denotes the natural vector field on P induced by A via the G'-action on P.

(3) $R_a^*\omega = \mathrm{Ad}(a^{-1})\omega$ for $a \in G'$ where R_a denotes the right G'-action on P.

For example, the principal G'-bundle given by the canonical projection $G \to G/G'$ with the Maurer-Cartan form ω_G of the Lie group G is a Cartan connection on the homogeneous space G/G'.

The $\mathfrak{g}$-valued 2-form $\Omega = d\omega + \frac{1}{2}[\omega, \omega]$ is called the *curvature form* of (P, ω). If the curvature form Ω vanishes then (P, ω) is locally isomorphic to (G, ω_G)(Theorem 5.1 of [Sh]). The curvature form Ω can be considered as a function K on P, to be called the *curvature function*, in the following way. This proposition is just Lemma 2.2 and Lemma 2.3 in [Ta].

Proposition 2.1. *Let (P, ω) be a Cartan connection of type G/G'.*
(1) There is a unique function $K : P \to \mathfrak{g} \otimes \wedge^2 \mathfrak{m}^*$ such that

$$\Omega = \frac{1}{2} K(\omega_- \wedge \omega_-)$$

where ω_- is the $\mathfrak{m}$-component of ω with respect to the decomposition $\mathfrak{g} = \mathfrak{g}' \oplus \mathfrak{m}$.
(2) For $z \in P$, $a \in G'$ and $v_1, v_2 \in \mathfrak{m}$,

$$K(za)(v_1, v_2) = Ad(a^{-1})K(z)(\rho(a)v_1, \rho(a)v_2),$$

where $\rho : G' \to GL(\mathfrak{m})$ is the adjoint representation modulo $\mathfrak{g}'$.

To get a more refined condition for the vanishing of K, we define the harmonic space $H(\mathfrak{m}, \mathfrak{g})$. The coboundary operator ∂ of the complex

$$\to \mathfrak{g} \otimes \wedge^q \mathfrak{m}^* \to \mathfrak{g} \otimes \wedge^{q+1} \mathfrak{m}^* \to$$

is defined by

$$\begin{aligned} \partial c(v_1 \wedge \cdots \wedge v_{q+1}) = & \sum_{1 \le i \le q+1} (-1)^{i+1} [v_i, c(v_1 \wedge \cdots \wedge \hat{v}_i \wedge \cdots \wedge v_{q+1})] \\ + & \sum_{1 \le i < j \le q+1} (-1)^{i+j} c([v_i, v_j] \wedge v_1 \wedge \cdots \wedge \hat{v}_i \wedge \cdots \wedge \hat{v}_j \wedge \cdots \wedge v_{q+1}) \end{aligned}$$

for $c \in \mathfrak{g} \otimes \wedge^q \mathfrak{m}^*$ and $v_i \in \mathfrak{m}$. Let ∂^* be the adjoint operator with respect to the inner product induced by the Killing form. Then we have the orthogonal decomposition

$$\begin{aligned} \mathfrak{g} \otimes \wedge^q \mathfrak{m}^* &= \mathrm{Ker}(\partial\partial^* + \partial^*\partial) \oplus \mathrm{Im}(\partial\partial^* + \partial^*\partial) \\ &= (\mathrm{Ker}\partial \cap \mathrm{Ker}\partial^*) \oplus \mathrm{Im}\partial \oplus \mathrm{Im}\partial^* \end{aligned}$$

so that the cohomology space $H^q(\mathfrak{m}, \mathfrak{g}) = \mathrm{Ker}\partial / \mathrm{Im}\partial$ is isomorphic to the harmonic space $\mathrm{Ker}(\partial\partial^* + \partial^*\partial) = \mathrm{Ker}\partial \cap \mathrm{Ker}\partial^*$. From now on we will identify them and consider the cohomology space $H^q(\mathfrak{m}, \mathfrak{g})$ as a subspace of $\mathfrak{g} \otimes \wedge^q \mathfrak{m}^*$. Under this identification, we will denote by $H(K)$, the $H^2(\mathfrak{m}, \mathfrak{g})$-component of the curvature function K.

Decompose the complex $\mathfrak{g} \otimes \mathfrak{m}^* \to \mathfrak{g} \otimes \wedge^2 \mathfrak{m}^* \to \mathfrak{g} \otimes \wedge^3 \mathfrak{m}^*$ as the direct sum of the subcomplexes

$$\begin{aligned} C^{r+1,1} &= \bigoplus_{i<0} \mathfrak{g}_{r+i+1} \otimes \mathfrak{g}_i^* \to C^{r,2} = \bigoplus_{i,j<0} \mathfrak{g}_{r+i+j+1} \otimes (\mathfrak{g}_i \wedge \mathfrak{g}_j)^* \\ &\to C^{r-1,3} = \bigoplus_{i,j,k<0} \mathfrak{g}_{r+i+j+k+1} \otimes (\mathfrak{g}_i \wedge \mathfrak{g}_j \wedge \mathfrak{g}_k)^*. \end{aligned}$$

We denote by $H^{r,2}(\mathfrak{m}, \mathfrak{g})$ the cohomology of this subcomplex. Then

$$H^2(\mathfrak{m}, \mathfrak{g}) = \bigoplus_r H^{r,2}(\mathfrak{m}, \mathfrak{g}).$$

Let K^r be the $C^{r,2}$-component of K. A *normal* Cartan connection of type G/G' is a Cartan connection (P, ω) of type G/G' on M such that the curvature function K satisfies

(1) $K^r = 0$ for $r < 0$ (admissibility)

(2) $\partial^* K^r = 0$ for $r \geq 0$ (normality).

The existence of a normal Cartan connection which governs the geometric structure associated with a regular differential system of type $\mathfrak{m}$ is guaranteed by the following, which is in Theorem 2.7 of [Ta].

Proposition 2.2. *Let (M, D) be a regular differential system of type $\mathfrak{m}$. Then there is a normal Cartan connection (P, ω) of type G/G' such that*

(1) $D^{-k}/D^{-(k-1)}$ is the vector bundle associated to P with the fiber $\mathfrak{g}_{-k}$ where the unipotent part of G' acts on $\mathfrak{g}_{-k}$ trivially and

(2) if the curvature function K vanishes then (M, D) is locally isomorphic to the standard differential system $(G/G', E)$.

The existence of a normal Cartan connection simplifies the curvature computation greatly as follows, which is just Corollary of Theorem 2.9 in [Ta]

Proposition 2.3. *For a normal Cartan connection (P, ω), the curvature function K vanishes if and only if its harmonic part $H(K)$, which is a $\oplus_{r \geq 0} H^{r,2}(\mathfrak{m}, \mathfrak{g})$-valued function on P, vanishes.*

Regarding the cohomology group $H^{r,2}(\mathfrak{m}, \mathfrak{g})$, we have the following result.

Proposition 2.4. *Let $(\mathfrak{g}, \alpha)$ be a complex simple Lie algebra with a choice of a long simple root, which is neither of symmetric type nor of contact type. Then*

(1) $H^{r,2}(\mathfrak{m}, \mathfrak{g}) = 0$ for all $r \geq 0$ except when $(\mathfrak{g}, \alpha)$ is either (B_ℓ, α_3) with $\ell \geq 4$ or (D_ℓ, α_3) with $\ell \geq 5$ and

(2) when $(\mathfrak{g}, \alpha)$ is either (B_ℓ, α_3) with $\ell \geq 4$ or (D_ℓ, α_3) with $\ell \geq 5$, $H^{r,2}(\mathfrak{m}, \mathfrak{g}) = 0$ for all $r > 0$ and $H^{0,2}(\mathfrak{m}, \mathfrak{g})$ is contained in $\mathfrak{g}_{-1} \otimes \wedge^2 \mathfrak{g}^*_{-1} \subset C^{0,2}$.

Proof. (1) follows from the vanishing results in Proposition 5.5 of [Ya]. In fact, Yamaguchi listed all non-vanishing $H^{r,2}(\mathfrak{m}, \mathfrak{g})$ with $r \geq 0$ for any parabolic subalgebra of a simple Lie algebra. In his list, when $\mathfrak{g}$ is an exceptional Lie algebra, the cases with a single root α are either of symmetric type or of contact type. When $\mathfrak{g}$ is of type A or C, and α is a long root, $(\mathfrak{g}, \alpha)$ is of symmetric type. Thus we are left with $\mathfrak{g}$ of B or D type only. From Yamaguchi's list, we can easily check that α must be α_3 to have non-zero $H^{r,2}(\mathfrak{m}, \mathfrak{g})$ for some $r \geq 0$.

For (2), we need to recall some terms. Let θ be the maximal root of $\mathfrak{g}$. Let x_β denote a root vector of a root β. Given a collection $\Psi = \{\beta_1, \ldots, \beta_q\}$ of positive roots, define

$$x_\Psi := x_{\beta_1} \wedge \cdots \wedge x_{\beta_q}.$$

Let σ_j be the element in the Weyl group associated to the simple root α_j. By Kostant's Theorem on the Lie algebra cohomology (Theorem 5.14 of [Ko] or p. 471 of [Ya]), there is a collection, say Σ, of elements of the Weyl group and a collection Φ_σ of positive roots such that $H^q(\mathfrak{m}, \mathfrak{g})$ is a direct sum of irreducible $\mathfrak{g}_0$-modules with the highest weight vector $x_{-\sigma(\theta)} \otimes x_{\Phi_\sigma}$ for $\sigma \in \Sigma$. Proposition 5.5 of [Ya] says that if

$$x_{-\sigma(\theta)} \otimes x_{\Phi_\sigma} \in H^{r,2}(\mathfrak{m}, \mathfrak{g})$$

for some $r \geq 0$, then $\sigma = \sigma_3\sigma_2$ in the cases of (B_ℓ, α_3) with $\ell \geq 4$ and (D_ℓ, α_3) with $\ell \geq 5$. In this case, $\Phi_{\sigma_3\sigma_2} = \{\alpha_3, \alpha_2 + \alpha_3\}$ from the description of $\Phi_{\sigma_i\sigma_j}$ in p.475 of [Ya]. The generator $x_{-\sigma_3\sigma_2(\theta)} \otimes (x_{\alpha_3} \wedge x_{\alpha_2+\alpha_3})$ is contained in $\mathfrak{g}_{-1} \otimes (\wedge^2 \mathfrak{g}^*_{-1}) \subset C^{0,2}$ because the coefficient of $\sigma_3\sigma_2(\theta)$ in α_3 is 1. Therefore $H^{r,2}(\mathfrak{m}, \mathfrak{g}) = 0$ for all $r > 0$ and $H^{0,2}(\mathfrak{m}, \mathfrak{g})$ is contained in $\mathfrak{g}_{-1} \otimes (\wedge^2 \mathfrak{g}^*_{-1})$. Q.E.D.

Combining 2.3 and 2.4, we have the following.

Proposition 2.5. *Let $(\mathfrak{g}, \alpha)$ be a complex simple Lie algebra with a choice of a long simple root which is neither of symmetric type nor of contact type and $\mathfrak{m}$ be the corresponding nilpotent graded algebra. Then a regular differential system of type $\mathfrak{m}$ is locally isomorphic to the standard one on G/G', except when $(\mathfrak{g}, \alpha)$ is either (B_ℓ, α_3) with $\ell \geq 4$ or (D_ℓ, α_3) with $\ell \geq 5$.*

§3. Review on the theory of the variety of minimal rational tangents

Here we will collect some general facts about the variety of minimal rational tangents on Fano manifolds of Picard number 1. See [Hw01] and [Mo] for details.

We will start with recalling some definitions in projective differential geometry. Let V be a complex vector space and $\mathbb{P}V$ be its projectivization. Given a (not necessarily closed) complex submanifold $Z \subset \mathbb{P}V$ and a point $z \in Z$, its affine tangent space $\hat{T}_z$ is the subspace of V whose projectivization is the projective tangent space of Z in $\mathbb{P}V$ at z. The tangent space of Z at z is naturally isomorphic to

$$T_z(Z) \cong Hom(\hat{z}, \hat{T}_z/\hat{z})$$

where $\hat{z}$ denotes the 1-dimensional subspace of V corresponding to z. The second fundamental form II_z is a homomorphism

$$\mathrm{II}_z : S^2T_z(Z) \to T_z(\mathbb{P}V)/T_z(Z)$$

defined as the derivative of the Gauss map. The image of II_z is called the *first normal space* of Z at z and is denoted by $N_z^1(Z)$. The third fundamental form III_z is a homomorphism

$$\mathrm{III}_z : S^3T_z(Z) \to T_z(\mathbb{P}V)/(T_z(Z) + N_z^1(Z))$$

defined as the derivative of the second fundamental form. The fourth fundamental form is defined similarly.

Now let $\mathcal{V}$ be a vector bundle on a complex manifold X and $\pi : \mathbb{P}\mathcal{V} \to X$ be its projectivization. For a complex analytic subvariety $\mathcal{Z} \subset \mathbb{P}\mathcal{V}$ and a smooth curve $R \subset \mathcal{Z}$ we say that $\mathcal{Z}$ is *relatively immersed along* R if for each point $x \in \pi(R)$, $Z_x := \mathcal{Z} \cap \pi^{-1}(x)$ is immersed at $R \cap \pi^{-1}(x)$. In this case, the collection of the affine tangent space $\hat{T}_z$ as z varies on R give a vector bundle on R, called *the bundle of the relative affine tangent spaces* of $\mathcal{Z}$ along R. Similarly, *the relative tangent bundle, the relative fundamental forms, the relative first normal spaces* of $\mathcal{Z}$ along R can be defined.

Let X be a Fano manifold of dimension n and Picard number 1. Fix a minimal dominating rational component $\mathcal{K}$ and its variety of minimal rational tangents $\mathcal{C} \subset \mathbb{P}T(X)$. If a member C of $\mathcal{K}$ is immersed, its tangent directions form a smooth rational curve in $\mathbb{P}T(X)$, which we denote by $C^\sharp$ and call the *tangential lift* of C. By definition, $C^\sharp$ lies on the variety of minimal rational tangents $\mathcal{C} \subset \mathbb{P}T(X)$. It is well-known that a general member C of $\mathcal{K}$ is standard (e.g. Theorem 1.2 in [Hw01]).

This means that C is immersed and the pull-back of $T(X)$ to $C^\sharp$ splits as

$$\mathcal{O}(2) \oplus \mathcal{O}(1)^p \oplus \mathcal{O}^{n-1-p}$$

where $p + 2 = C \cdot K_C^{-1}$. When C is standard, $\mathcal{C}$ is relatively immersed along $C^\sharp$ in the sense of the previous paragraph (e.g. Proposition 1.4 in [Hw01]). Thus we can consider the relative tangent bundle, the bundle of the relative affine tangent spaces, the relative fundamental forms and the relative first normal spaces of $\mathcal{C}$ along $C^\sharp$. It is well-known that the bundle of the relative affine tangent spaces of $\mathcal{C}$ along $C^\sharp$, which is a vector bundle on $C^\sharp$, corresponds to the $\mathcal{O}(2) \oplus \mathcal{O}(1)^p$-part of the splitting of $T(X)$ pulled back to $C^\sharp$. The next result is essentially contained in the proof of Proposition 2.2 and Proposition 3.1 in [Mo]. Since it was stated there only for the special cases considered in [Mo], we will give a general proof here.

Proposition 3.1. *Let C be a standard rational curve belonging to $\mathcal{K}$ and $C^\sharp$ be its tangential lift. Then the relative tangent spaces of $\mathcal{C}$ along $C^\sharp$ is a vector bundle T^π on $C^\sharp$ of the splitting type $\mathcal{O}(-1)^p$ and the relative first normal spaces of $\mathcal{C}$ along $C^\sharp$ is a vector bundle N^π on $C^\sharp$ of splitting type $\mathcal{O}(-2)^m$ for some integer $m > 0$. In particular, the relative second fundamental forms of $\mathcal{C}$ along $C^\sharp$, forms a constant section of $Hom(S^2T^\pi, N^\pi)$.*

Proof. The relative affine tangent bundle of $\mathcal{C}$ along $C^\sharp$ has splitting type $\mathcal{O}(2) \oplus \mathcal{O}(1)^p$. In this splitting, the $\mathcal{O}(2)$ part is the relative tautological bundle of $\mathbb{P}T(X)$ along $C^\sharp$. Thus the relative tangent bundle of $\mathcal{C}$ along $C^\sharp$ has the splitting type $Hom(\mathcal{O}(2), \mathcal{O}(1)^p) = \mathcal{O}(-1)^p$. The relative second fundamental forms define a homomorphism

$$\mathrm{II} : S^2(\mathcal{O}(-1)^p) \to Hom(\mathcal{O}(2), \mathcal{O}^{n-1-p}) = \mathcal{O}(-2)^{n-1-p}.$$

Thus the image must be of the form $\mathcal{O}(-2)^m$ and II must be a constant section. Q.E.D.

Now let us assume that $\mathcal{C}_x$ is irreducible for a general $x \in X$. Let $D_x \subset T_x(X)$ be the linear span of $\mathcal{C}_x$, namely, $\mathbb{P}D_x$ is the smallest linear subspace of $\mathbb{P}T_x(X)$ containing $\mathcal{C}_x$. The collection of D_x as x varies over general points of X defines a subsheaf of $T(X)$. By taking double dual of D, we can assume that there exists a subvariety $\mathrm{Exc}(D) \subset X$ of codimension ≥ 2 such that $D|_{X\setminus \mathrm{Exc}(D)}$ is a subbundle of $T(X)|_{X\setminus \mathrm{Exc(D)}}$. We will call D the *linear differential system spanned by* $\mathcal{C}$. If the rank of D is smaller than $n = \dim X$, it is well-known that D is not integrable, namely, the Frobenius bracket at a general point $x \in X$,

$$[,] : \wedge^2 D_x \to T_x(X)/D_x$$

is not zero. In fact, if it is integrable, the leaves define a holomorphic fibration outside a set of codimension ≥ 2, which is a contradiction to the assumption that X has Picard number 1 (cf. Proposition 2.2 in [Hw01]). Given any subvariety of codimension ≥ 2, we can choose a member of $\mathcal{K}$ disjoint from it (e.g. Lemma 2.1 in [Hw01]). Thus we can assume that a general member of $\mathcal{K}$ is disjoint from $\mathrm{Exc}(D)$.

Proposition 3.2. *Let D be the linear differential system spanned by $\mathcal{C}$. Assume that the rank of D is smaller than n. Then for a general member $C \subset X \setminus \mathrm{Exc}(D)$ of $\mathcal{K}$, the pull-back of D to $C^\sharp$ is of the form*

$$\mathcal{O}(2) \oplus \mathcal{O}(1)^p \oplus \mathcal{O}^q \oplus \bigoplus_{i=1}^{r} \mathcal{O}(a_i)$$

where $q \geq m$, m being the rank of the relative first normal spaces of $\mathcal{C}$ along $C^\sharp$, $r > 0$ and $a_i < 0$.

Proof. Since the pull-back of D to $C^\sharp$ is a vector subbundle of $\mathcal{O}(2) \oplus \mathcal{O}(1)^p \oplus \mathcal{O}^{n-1-p}$ and $\mathcal{C}_x \subset \mathbb{P}D_x$ for each $x \in C$, it must be of the form

$$\mathcal{O}(2) \oplus \mathcal{O}(1)^p \oplus \bigoplus_{j=1}^{d-1-p} \mathcal{O}(b_j)$$

where $d = \mathrm{rank}(D)$ and $b_j \leq 0$. Moreover, from 3.1, at least m of b_j's must be 0. Thus it suffices to exclude the possibility that $b_j = 0$ for all j. Suppose this is the case. Then the pull-back of $T(X)/D$ to $C^\sharp$ must have degree 0. Since it is the quotient of $\mathcal{O}(2) \oplus \mathcal{O}(1)^p \oplus O^{n-1-p}$, we see that $T(X)/D$ has splitting type $\mathcal{O}^{n-d}$ where d is the rank of D. Then the Frobenius bracket $[,] : \wedge^2 D_x \to T_x(X)/D_x$ must satisfy $[\alpha, D_x] = 0$ for α corresponding to the tangent direction $\mathcal{O}(2)$ of C at x. Since this is true for general $[\alpha] \in \mathcal{C}_x$ while $\mathcal{C}_x$ spans D_x, we conclude that D is integrable, a contradiction. Q.E.D.

§4. Variety of minimal rational tangents on the rational homogeneous space associated to a long simple root

Let $S = G/G'$ be a rational homogeneous space of type $(\mathfrak{g}, \alpha)$ as explained in Section 2. There is a unique minimal dominating rational component. For a base point $s \in S$, let $\mathcal{C}_s \subset \mathbb{P}T_s(S)$ be the variety of minimal rational tangents. When α is a long simple root, or equivalently, when $\mathcal{C}_s$ is homogeneous, the following fact was proved in Proposition 1 and Proposition 7 of [HM02].

Proposition 4.1. *Let S be a rational homogeneous space of type $(\mathfrak{g}, \alpha)$ with α a long root. Let $\mathfrak{m}$ be the associated graded nilpotent Lie algebra. Let $\hat{\mathcal{C}}_s \subset E_s$ be the homogeneous cone of the variety of minimal rational tangents $\mathcal{C}_s$ at a base point $s \in S$. Define a graded Lie algebra $\mathfrak{n} = \mathfrak{n}_{-1} + \mathfrak{n}_{-2} + \cdots$ as the quotient of the graded free Lie algebra generated by $\mathfrak{n}_{-1} = E_s$ modulo the relations given by $[v_1, v_2] = 0$ for $v_1, v_2 \in \mathfrak{n}_{-1}$ if $v_1 \in \hat{\mathcal{C}}_s$ and $< v_1, v_2 >$ is tangent to $\hat{\mathcal{C}}_s$. Then $\mathfrak{n}$ is isomorphic to $\mathfrak{m}$ as graded nilpotent Lie algebras.*

This has several consequences.

Proposition 4.2. *In the setting of 4.1, suppose $(\mathfrak{g}, \alpha)$ is neither of symmetric type nor of contact type. Then the group of the graded Lie algebra automorphisms of $\mathfrak{n}$ acts irreducibly on $\mathfrak{n}_{-1} = E_s$ and $\mathcal{C}_s$ is the highest weight orbit of this action.*

In Proposition 1 of [HM02], it was proved that the variety of minimal rational tangents $\mathcal{C}_s$ is the highest weight orbit of the isotropy representation of G' on E_s. This does not imply 4.2 immediately. It is a result about the differential system E, while 4.2 is a result about the symbol algebra of E.

Proof. The Lie algebra $\mathrm{aut}(\mathfrak{m})$ of the automorphism group of $\mathfrak{m}$ is the 0-th prolongation of $\mathfrak{m}$ in the sense of p.430 of [Ya], which is exactly $\mathfrak{g}_0$ by Theorem 5.2 of [Ya]. By 4.1, the Lie subalgebra $\mathrm{aut}(\hat{\mathcal{C}}_s) \subset \mathbf{gl}(T_s(S))$ of infinitesimal linear automorphisms of the cone is contained in $\mathrm{aut}(\mathfrak{m})$. From the list of $\mathcal{C}_s$ in p. 176 of [HM02], $\mathrm{aut}(\hat{\mathcal{C}}_s)$ is isomorphic to $\mathfrak{g}_0$ and $\mathcal{C}_s$ is the highest weight orbit of the irreducible representation of $\mathfrak{g}_0$ on $\mathfrak{g}_{-1}$. It follows that

$$\mathrm{aut}(\mathfrak{m}) = \mathrm{aut}(\hat{\mathcal{C}}_s) = \mathfrak{g}_0$$

and $\mathcal{C}_s$ is the highest weight orbit of the $\mathrm{aut}(\mathfrak{m})$-representation on $T_s(S)$. Q.E.D.

Proposition 4.3. *Let S be a rational homogeneous space of type $(\mathfrak{g}, \alpha)$, with α a long root, neither of symmetric type nor of contact type. Let X be a Fano manifold of Picard number 1 with a choice of a minimal dominating rational component $\mathcal{K}$ such that the variety of minimal rational tangents at a general point $x \in X$ is isomorphic to that of S. Let D be the differential system spanned by the variety of minimal rational tangents. Then at a general point x of X,*

(1) the symbol algebra $\mathbf{sym}_x(D)$ is isomorphic to $\mathfrak{m}$, the symbol algebra of the standard differential system on S, and

(2) the variety of minimal rational tangents $\mathcal{C}_x$ is the highest weight orbit of the representation of the automorphisms of the symbol algebra $\mathbf{sym}_x(D)$ *on* $\mathbb{P}D_x$.

Proof. (1) is just Proposition 5 of [HM02]. (2) follows from 4.2. Q.E.D.

Proposition 4.4. *In the setting of 4.3, suppose the linear differential system D is equivalent to the standard system E on S in an analytic neighborhood of x. Then X is biholomorphic to S.*

Proof. By 4.3 (2), there exists a neighborhood U of x and a biholomorphic map $\varphi : U \to U'$ into an open subset U' of S such that $\varphi(\mathcal{C}|_U)$ agrees with the variety of minimal rational tangents of S restricted to U'. By the main theorem in p. 564 of [HM01], this implies that X is biholomorphic to S. Q.E.D.

Corollary 4.5. *Unless S is of type (B_ℓ, α_3) with $\ell \geq 4$ or (D_ℓ, α_3) with $\ell \geq 5$, Main Theorem follows from 2.5.*

Thus *from now on till the end of the paper, we will assume that S is of type (B_ℓ, α_3) with $\ell \geq 4$ or (D_ℓ, α_3) with $\ell \geq 5$.*

The homogeneous space S is the quadric Grassmannian of 3-dimensional isotropic subspaces in an orthogonal vector space. Let $p + 2 > 0$ be the degree of minimal rational curves of S with respect to K_S^{-1}.

Proposition 4.6. *Let S and p be as defined above. Let C be a minimal rational curve on S. Then*

(1) $\dim S = 3p + 3$ *and* $\operatorname{rank}(E) = 3p$,

(2) $E|_C = \mathcal{O}(2) \oplus \mathcal{O}(1)^p \oplus \mathcal{O}^{2p-3} \oplus \mathcal{O}(-1)^2$,

(3) $(T(S)/E)|_C = \mathcal{O}(1)^2 \oplus \mathcal{O}$ *and*

(4) $\dim[\mathfrak{g}_{\alpha_3}, \mathfrak{g}_1] = 2$.

Proof. Let $\alpha_1, \ldots, \alpha_\ell$ be a set of simple roots of Lie algebras of type B_ℓ or D_ℓ. For B_ℓ, we define

$$\beta_i := \sum_{k=i}^{\ell} \alpha_k \text{ for } 1 \leq i \leq \ell \text{ and } \gamma_{ij} := \sum_{k=i}^{j-1} \alpha_k \text{ for } 1 \leq i < j \leq \ell.$$

For D_ℓ, we define

$$\beta_i := \sum_{k=i}^{\ell-2} \alpha_k + \alpha_{\ell-1},\ \beta_i' := \sum_{k=i}^{\ell-2} \alpha_k + \alpha_\ell \text{ for } 1 \leq i \leq \ell - 2$$

$$\text{and } \gamma_{ij} := \sum_{k=i}^{j-1} \alpha_k \text{ for } 1 \leq i < j \leq \ell - 1.$$

Now from p.30 and p.35 of [Ti], we can explicitly write down the subsets of roots of $\mathfrak{g}$ defined in Section 2 as follows. For (B_ℓ, α_3),

$$\Phi_1 = \{\beta_1, \beta_2, \beta_3, \gamma_{ij}, \beta_i + \beta_j \mid 1 \leq i \leq 3 < j \leq \ell\}$$

$$\Phi_2 = \{\beta_1 + \beta_2, \beta_1 + \beta_3, \beta_2 + \beta_3\}.$$

For (D_ℓ, α_3),

$$\Phi_1 = \{\beta_1, \beta_2, \beta_3, \beta_1', \beta_2', \beta_3', \gamma_{ij}, \beta_i + \beta_j' \mid 1 \leq i \leq 3 < j \leq \ell - 1\}$$

$$\Phi_2 = \{\beta_1 + \beta_2', \beta_1 + \beta_3', \beta_2 + \beta_3'\}.$$

By the argument in the proof of Proposition 1 of [HM02], the splitting type of E on C is $\bigoplus_{\beta \in \Phi_1} \mathcal{O}(\beta(H_{\alpha_3}))$ where H_{α_3} denotes the coroot of α_3. This can be computed easily from the above list for Φ_1 to prove (2). In the same way, the splitting type of $T(S)/E$ on C is $\bigoplus_{\beta \in \Phi_2} \mathcal{O}(\beta(H_{\alpha_3}))$ and (3) can be checked from the above list for Φ_2. (1) is immediate from (2) and (3). (4) can be checked directly from the above list of roots. Q.E.D.

Proposition 4.7. *The variety of minimal rational tangents at a base point $s \in S$ satisfies the following.*

(1) $\mathcal{C}_s \subset \mathbb{P}E_s$ is isomorphic to the Segre embedding of $\mathbb{P}_2 \times \mathbf{Q}_{p-2}$ where $\mathbf{Q}_{p-2}$ denotes the hyperquadric of dimension $p-2$ in $\mathbb{P}_{p-1}$.

(2) The first normal space at each point of $\mathcal{C}_s$ has rank $2p-3$ and the fourth fundamental form of $\mathcal{C}_s$ is zero.

(3) Regard a tangential line to $\mathcal{C}_s$ as a one-dimensional subspace of $\wedge^2 E_s$. Then the kernel of the bracket $\wedge^2 E_s \to T_s(S)/E_s$, which is isomorphic to the Lie bracket $\wedge^2 \mathfrak{g}_{-1} \to \mathfrak{g}_{-2}$, is spanned by tangential lines to $\mathcal{C}_s$.

Proof. (1) follows from the description of the highest weight orbit of G_0 on $\mathbb{P}E_s$, as given by p.176 of [HM02]. From (1), there is no hyperplane section of $\mathcal{C}_s$ with multiplicity ≥ 4 and hyperplane sections with multiplicity 3 forms a pencil. It follows that the fourth fundamental form is 0 and the sum of the tangent space and the first normal space has codimension 2 in $\mathbb{P}E_s$. This implies (2). Noting that $\mu = 2$ in this case, (3) is a direct consequence of 4.1. Q.E.D.

Proposition 4.8. *Let $s \in S$ be a base point and $\mathcal{C}_s \subset \mathbb{P}E_s$ be the variety of minimal rational tangent. There exists a vector space W of dimension 3 with $\wedge^2 W \cong T_s(S)/E_s$ and a vector space F of dimension p equipped with a non-degenerate quadratic from $\omega : S^2 F \to \mathbb{C}$ such that*

$E_s \cong W \otimes F$. Furthermore, under the decomposition $S^2F = \mathbb{C} \oplus \omega^\perp$, the Lie bracket $\wedge^2 E_s \to T_s(S)/E$ is just the projection to $\wedge^2 W$ in

$$\begin{aligned}\wedge^2(W \otimes F) &= (\wedge^2 W \otimes S^2 F) \oplus (S^2 W \otimes \wedge^2 F)\\ &= \wedge^2 W \oplus (\wedge^2 W \otimes \omega^\perp) \oplus (S^2 W \otimes \wedge^2 F).\end{aligned}$$

Proof. This is the consequence of the following which can be checked easily. The semi-simple part of $\mathfrak{g}'$ under the Levi-decomposition is isomorphic to $\mathbf{sl}_3 \oplus \mathbf{so}_p$. Its representation on $\mathfrak{g}_{-1}$ is just the tensor product of the fundamental representations of $\mathbf{sl}_3$ and $\mathbf{so}_p$. Its representation on $\mathfrak{g}_{-2}$ is just the tensor product of the dual of the fundamental representation of $\mathbf{sl}_3$ and the trivial representation of $\mathbf{so}_p$. Q.E.D.

§5. Extension of the structure to a neighborhood of a standard rational curve

From now on we assume that X is a Fano manifold of Picard number 1 with a minimal dominating rational component $\mathcal{K}$ such that, for a general point $x \in X$, the projective subvariety $\mathcal{C}_x \subset \mathbb{P}T_x(X)$ is isomorphic to $\mathcal{C}_s \subset \mathbb{P}T_s(S)$.

Let $D \subset T(X)$ be the linear differential system on X spanned by $\mathcal{C}$ at a general point. As in Section 3, D is a subbundle of $T(X)$ on $X \setminus \mathrm{Exc}(D)$ and a general member of $\mathcal{K}$ is disjoint from $\mathrm{Exc}(D)$. By 4.3, the symbol algebra $\mathbf{sym}_x(D)$ of D at a general point x is isomorphic to the nilpotent algebra $\mathfrak{m} = \mathfrak{g}_{-1} + \mathfrak{g}_{-2}$.

Proposition 5.1. *In the above setting, let C be a general member of $\mathcal{K}$ disjoint from* Exc(D) *and let $C^\sharp \subset \mathcal{C}$ be the tangential lift of C. Then the pull-back of D to $C^\sharp$ splits as*

$$\mathcal{O}(2) \oplus \mathcal{O}(1)^p \oplus \mathcal{O}^{2p-3} \oplus \mathcal{O}(-1)^2$$

and the pull-back of $T(X)/D$ to $C^\sharp$ splits as $\mathcal{O}(1)^2 \oplus \mathcal{O}$. Moreover, the relative second fundamental forms and the relative third fundamental forms of $\mathcal{C}$ along $C^\sharp$ are all isomorphic to those of $\mathcal{C}_s \subset \mathbb{P}E_s$.

Proof. From 3.2 and 4.7 (2), the splitting type of D on $C^\sharp$ is of the form

$$\mathcal{O}(2) \oplus O(1)^p \oplus \mathcal{O}^{2p-3} \oplus \mathcal{O}(b_1) \oplus \mathcal{O}(b_2)$$

for some $0 \geq b_1 \geq b_2, b_2 < 0$.

The relative third fundamental forms of $\mathcal{C}$ along $C^\sharp$ defines a homomorphism

$$\mathrm{III} : S^3(\mathcal{O}(-1)^p) \to Hom(\mathcal{O}(2), \mathcal{O}(b_1) \oplus \mathcal{O}(b_2)) = \mathcal{O}(b_1 - 2) \oplus \mathcal{O}(b_2 - 2).$$

By 4.7 (2), III must be surjective over a general point $x \in C$. Thus $b_2 = -1$.

Suppose that $b_1 = 0$. Then by Chern number consideration, the splitting type of $T(X)/D$ on $C^\sharp$ must be $\mathcal{O}(1) \oplus \mathcal{O}^2$. Then the Frobenius bracket $[,] : \wedge^2 D \to T(X)/D$ must satisfy

$$[\mathcal{O}(2), D] = [\mathcal{O}(2), \mathcal{O}(2) \oplus \mathcal{O}(1)^p \oplus \mathcal{O}^{2p-2} \oplus \mathcal{O}(-1)] \subset \mathcal{O}(1).$$

Thus $\dim[\alpha, D_x] = 1$. But by 4.3, we know that $\mathbf{sym}_x(D)$ is isomorphic to $\mathfrak{m}$. This is a contradiction to 4.6 (4).

We conclude that $b_1 = b_2 = -1$ and III must be constant along $C^\sharp$. Moreover, $\dim[\alpha, D_x] = 2$. This implies that $T(X)/D$ has the splitting type of $\mathcal{O}(1)^2 \oplus \mathcal{O}$. Q.E.D.

Proposition 5.2. *Let $\mathcal{V}$ be a vector bundle on the unit disc $\Delta := \{t \in \mathbb{C}, |t| < 1\}$ and $\pi : \mathbb{P}\mathcal{V} \to \Delta$ be its projectivization. Let $\mathcal{Z}$ be a subvariety of $\mathbb{P}\mathcal{V}$ flat over Δ such that for each $t \neq 0$, $Z_t := \pi^{-1}(t) \cap \mathcal{Z}$ is isomorphic to the Segre embedding of $\mathbb{P}_m \times \mathbf{Q}_k$ for some positive integers m and k. Assume that there exists a section $R \subset \mathcal{Z}$ of π such that the relative second fundamental forms and the relative third fundamental forms of $\mathcal{Z}$ along R are constant. Then $Z_0 := \pi^{-1}(0) \cap \mathcal{Z}$ is also isomorphic to the Segre embedding of $\mathbb{P}_m \times \mathbf{Q}_k$.*

Proof. This is proved in the proof of Proposition 3.2 Case (E) in [Mo] when $m = 1$. The proof there works verbatim for arbitrary m. Q.E.D.

Proposition 5.3. *Let C be a general standard rational curve through x. Then for each $y \in C$, $\mathcal{C}_y \subset \mathbb{P}T_y(X)$ is isomorphic to $\mathcal{C}_s \subset \mathbb{P}T_s(S)$ and the symbol algebra $\mathbf{sym}_y(D)$ is isomorphic to $\mathfrak{m}$.*

Proof. The first statement is an immediate consequence of 5.1 and 5.2. From the first statement, as in 4.8, there exist a line bundle L, a vector bundle W' of rank 3 and a vector bundle F' of rank p equipped with a non-degenerate quadratic form $\omega' : S^2F' \to L$ in a neighborhood of C such that $D \cong W' \otimes F'$ and, at the point x, the Frobenius bracket annihilates $(\wedge^2 W'_x \otimes (\omega')^\perp_x) \oplus (S^2 W'_x \otimes \wedge^2 F'_x)$-part of the decomposition

$$\wedge^2 D \cong \wedge^2(W' \otimes F') = (\wedge^2 W' \otimes L) \oplus (\wedge^2 W' \otimes (\omega')^\perp) \oplus (S^2 W' \otimes \wedge^2 F').$$

From 5.1, we may assume that the splitting type of W' along $C^\sharp$ is $\mathcal{O}(1) \oplus \mathcal{O}^2$, the splitting type of F' along $C^\sharp$ is $\mathcal{O}(1) \oplus \mathcal{O}^{p-2} \oplus \mathcal{O}(-1)$ and L is the trivial line bundle along $C^\sharp$. It follows that $\wedge^2 W' \otimes L$ and $T(X)/D$ have the same splitting type along $C^\sharp$. Since the homomorphism $\wedge^2 W' \otimes$

$L \to T(X)/D$ induced by the bracket of D in a neighborhood of C is an isomorphism at a general point of C by 4.3, it must be an isomorphism at every point of C from the splitting type. This implies that $\mathbf{sym}_y(D)$ is isomorphic to $\mathfrak{m}$ at every $y \in C$. □

§6. Vanishing of the curvature

We keep the notation from the previous section. To prove Main Theorem for X, it suffices to show, by 2.5 and 4.4, the vanishing of the curvature function of the associated normal Cartan connection at a general point of X. For this, we need to interpret the curvature function as a holomorphic section of a vector bundle as follows. This is an analog of Proposition 2.5 in [Ho].

Proposition 6.1. *Let (M, D) be a regular differential system of type $\mathfrak{m}$. If there is no nonzero holomorphic section of the bundle $D \otimes \wedge^2 D^*$, then (M, D) is locally isomorphic to the standard differential system E on S.*

Proof. Let (P, ω) be a normal Cartan connection given by Proposition 2.2. Then the vector bundle $D \otimes \wedge^2 D^*$ is the associated vector bundle of the principal G'-bundle P with fiber $\mathfrak{g}_{-1} \otimes \wedge^2 \mathfrak{g}^*_{-1}$ on which G' acts as follows: for $c \in \mathfrak{g}_{-1} \otimes \wedge^2 \mathfrak{g}^*_{-1}$, $a \in G'$ and $v_1, v_2 \in \mathfrak{g}_{-1}$,

$$a \cdot c(X, Y) = Ad(b^{-1})c(\rho(b)v_1, \rho(b)v_2)$$

where $a = b \exp(A_1) \exp(A_2), b \in G_0$, $A_i \in \mathfrak{g}_i$ in the decomposition $G = G_0 \cdot \exp(\mathfrak{g}_1) \exp(\mathfrak{g}_2)$ given in Lemma 1.7 of [Ta] and $\rho : G' \longrightarrow GL(\mathfrak{m})$ is the isotropy representation.

Let K' be the $\mathfrak{g}_{-1} \otimes \wedge^2 \mathfrak{g}^*_{-1}$-component of K. The behavior of K under the G'-action is described in 2.1 (2). Here, we need to study the behavior of K'. We will use the following equivariance property of K from Lemma 2.4 of [Ta].

Lemma *Let m be any integer. If $K^p = 0$ for all $p < m$, then for any $z \in P$, $a \in G'$ and $v_1, v_2 \in \mathfrak{g}_{-1}$,*

$$K^m(za)(v_1, v_2) = Ad(b^{-1})(K^m(z)(\rho(a)v_1, \rho(a)v_2))$$

where b is the G_0-component of a in the decomposition

$$G = G_0 \cdot \exp(\mathfrak{g}_1) \exp(\mathfrak{g}_2)$$

given in Lemma 1.7 of [Ta].

Noting that $C^{0,2} = \mathfrak{g}_{-1} \otimes \wedge^2 \mathfrak{g}_{-1}^* + \mathfrak{g}_{-2} \otimes (\mathfrak{g}_{-1} \wedge \mathfrak{g}_{-2})^*$ and applying Lemma to K^0, we have

$$K'(za)(v_1, v_2) = Ad(b^{-1})K'(z)(\rho(a)v_1, \rho(a)v_2)$$

for $z \in P$, $a \in G'$ and $v_1, v_2 \in \mathfrak{g}_{-1}$. Since $\exp(\mathfrak{g}_1 + \mathfrak{g}_2)$ acts trivially on $\mathfrak{g}_{-1}$, K' is an equivariant function on P under the action of G'. So K' can be thought of as a section of $D \otimes \wedge^2 D^*$ on M.

If there is no nonzero section of $D \otimes \wedge^2 D^*$, then K' vanishes and thus the harmonic part $H(K)$ of the curvature function K of the Cartan connection (P, ω) is zero by 2.4. By Proposition 2.3 the whole curvature K vanishes and thus the differential system (M, D) is locally isomorphic to (S, E). Q.E.D.

In the setting of Main Theorem with S of type $(B_\ell, \{\alpha_3\})$ $(\ell \geq 4)$, or $(D_\ell, \{\alpha_3\})$ $(\ell \geq 5)$, 5.3 implies that in a neighborhood M of a general member C of $\mathcal{K}$, we are given a regular differential system of type $\mathfrak{m}$. By 4.4 and 6.1, to finish the proof of Main Theorem, it suffices to prove the following.

Proposition 6.2. *Let S, X and $\mathcal{K}$ be as before. Let D be the linear differential system of type $\mathfrak{m}$ in a neighborhood M of a general member of $\mathcal{K}$ given by 5.3. Then there is no nonzero holomorphic section of the vector bundle $D \otimes \wedge^2 D^*$ in that neighborhood.*

Proof. A holomorphic section of $D \otimes \wedge^2 D^*$ gives a homomorphism $\varphi : \wedge^2 D \to D$. We claim that φ factors through the Lie bracket $\wedge^2 D \to T(M)/D$ to a homomorphism $\varphi' : T(M)/D \to D$. By 4.7 (3), it suffices to show that for a general point $x \in M$, a general vector $v \in D_x$ with $[v] \in \mathcal{C}_x$ and a vector $w \in D_x$ such that $< v, w >$ is tangent to $\mathcal{C}_x$ at $[v]$, $\varphi(v \wedge w) = 0$. Let C be a member of $\mathcal{K}$ passing through x with $[v] = [T_x(C)]$. Since the bundle of the relative affine tangent spaces of $\mathcal{C}$ along $C^\sharp$ corresponds to the $(\mathcal{O}(2) \oplus \mathcal{O}(1)^p)$-part of the splitting of the pull-back of D to $C^\sharp$ given in 5.1, we have a holomorphic section $\tilde{v}$ (resp. $\tilde{w}$) of D pulled back to $C^\sharp$ such that $\tilde{v}(x) = v$ (resp. $\tilde{w}(x) = w$) and $\tilde{v}$ (resp. $\tilde{w}$) has two (resp. one) zeros. Then $\varphi(\tilde{v} \wedge \tilde{w})$ is a section of the pull-back of D to $C^\sharp$ with three zeroes. From the splitting type of D, we conclude that $\varphi(\tilde{v} \wedge \tilde{w}) = 0$, which implies that $\varphi(v \wedge w) = 0$. This proves the claim.

Now it remains to show that the induced homomorphism φ' from $T(M)/D$ to D is zero. Let us choose C through x with the tangent direction $[v] \in \mathcal{C}_x$ as above. Since the pull-back of $T(M)/D$ to $C^\sharp$ splits as $\mathcal{O}(1)^2 \oplus \mathcal{O}$, the image $\mathrm{Im}(\varphi')$ is contained in the $(\mathcal{O}(2) \oplus \mathcal{O}(1)^p \oplus \mathcal{O}^{2p-3})$-part of the splitting of the pull-back of D to $C^\sharp$. From 4.7 (2), at the point x, $(\mathcal{O}(2) \oplus \mathcal{O}(1)^p \oplus \mathcal{O}^{2p-3})_x$ is precisely the span of the tangent space and the first normal space of $\mathcal{C}_x$ at $[v]$. It follows that $[\mathrm{Im}(\varphi')_x] \subset \mathbb{P}D_x$ is contained in the span of the tangent space and the first normal space of $\mathcal{C}_x$ at $[v]$ for each general choice of $[v] \in \mathcal{C}_x$. Thus if $[\mathrm{Im}(\varphi')_x]$ is non-empty, it is contained in the common intersection, say, $Z \subset \mathbb{P}D_x$, of the spans of the tangent space and the first normal space of $\mathcal{C}_x$ at $[v]$ as $[v]$ varies over $\mathcal{C}_x$. But such Z must be a G_0-invariant subvariety of $\mathbb{P}D_x$ under the natural representation of G_0 on D_x. This is a contradiction because Z is linearly degenerate by its definition, while the representation of G_0 on D_x is irreducible. It follows that $\varphi' = 0$ at a general point x, and consequently $\varphi' = 0$. Q.E.D.

References

[Ho] J. Hong, Fano manifolds with geometric structures modeled after homogeneous contact manifolds, Internat. J. Math., **11** (2000), 1203–1230.

[Hw01] J.-M. Hwang, Geometry of minimal rational curves on Fano manifolds. School on Vanishing Theorems and Effective Results in Algebraic Geometry, Trieste, 2000, ICTP Lect. Notes, **6**, Abdus Salam Int. Cent. Theoret. Phys., Trieste, 2001, pp. 335–393.

[Hw06] J.-M. Hwang, Rigidity of rational homogeneous spaces, Proceedings of ICM 2006 Madrid, II, Eur. Math. Soc., 2006, pp. 613–626.

[HM97] J.-M. Hwang and N. Mok, Uniruled projective manifolds with irreducible reductive G-structures, J. Reine Angew. Math., **490** (1997), 55–64.

[HM01] J.-M. Hwang and N. Mok, Cartan-Fubini type extension of holomorphic maps for Fano manifolds of Picard number 1, J. Math. Pures Appl. (9), **80** (2001), 563–575.

[HM02] J.-M. Hwang and N. Mok, Deformation rigidity of the rational homogeneous space associated to a long simple root, Ann. Sci. Ecole Norm. Sup. (4), **35** (2002), 173–184.

[Ko] B. Kostant, Lie algebra cohomology and generalized Borel-Weil theorem, Ann. of Math. (2), **74** (1961), 329–387.

[Mo] N. Mok, Recognizing certain rational homogeneous manifolds of Picard number 1 from their varieties of minimal rational tangents, IMR preprint, 2005, no.9, http://www.hku.hk/math/imr/.

[Sh] R. W. Sharpe, Differential geometry: Cartan's generalization of Klein's Erlangen program, GTM 166, Springer.

[Ta] N. Tanaka, On the equivalence problems associated with simple graded Lie algebras, Hokkaido Math. J., **8** (1979), 23–84.

[Ti] J. Tits, Tabellen zu den einfachen Lie Gruppen und ihren Darstellungen, Lect. Notes Math., **40** (1967).

[Ya] K. Yamaguchi, Differential systems associated with simple grade Lie algebras, Adv. Stud. Pure Math., **22** (1993), 413–494.

Jaehyun Hong
Department of Mathematical Sciences
Seoul National University
San 56-1 Shinrim-dong, Kwanak-gu
Seoul 151-747 KOREA
E-mail address: jhhong@math.snu.ac.kr

Jun-Muk Hwang
Korea Institute for Advanced Study
207-43 Cheongnyangni 2-dong, Dongdaemun-gu
Seoul 130-722 KOREA
E-mail address: jmhwang@kias.re.kr

Advanced Studies in Pure Mathematics 50, 2008
Algebraic Geometry in East Asia — Hanoi 2005
pp. 237–249

Maximal divisorial sets in arc spaces

Shihoko Ishii

Abstract.

In this paper we introduce a maximal divisorial set in the arc space of a variety. The generalized Nash problem is reduced to a translation problem of the inclusion of two maximal divisorial sets. We study this problem and show a counter example to the most natural expectation even for a non-singular variety.

§1. Introduction

In [8], Nash posed a problem: if the set of the families of arcs through the singularities on a variety (these families are called the Nash components) corresponds bijectively to the set of essential divisors of resolutions of the singularities. This problem is affirmatively answered for some 2-dimensional singularities by A. Reguera, M. Lejeune-Jalabert, C. Plénat and P. Popescu-Pampu [7], [9], [10], [11], [12] and toric singularities of arbitrary dimension by S. Ishii and J. Kollár [6]. For non-normal toric variety of arbitrary dimension the answer is also affirmative ([5]). On the other hand this problem is negatively answered in general. The paper [6] gives a counter example of dimension greater than or equal to 4. Therefore the Nash problem should be changed to the problem to determine the divisors corresponding to the Nash components.

We can generalize this problem into the characterization problem for valuations corresponding to the irreducible components of contact loci $\mathrm{Cont}^{\geq m}(\mathfrak{a})$ which are introduced by L. Ein, R. Lazarsfeld and M. Mustaţă ([2]). In this paper we introduce the maximal divisorial set

Received April 3, 2006.
Revised June 14, 2006.
2000 *Mathematics Subject Classification.* Primary 14J17; Secondary 14M99.
Key words and phrases. arc space, valuation, Nash problem.
Partially supported by Grant-In-Aid of Ministry of Science and Education in Japan.

$C_X(v)$ in the arc space of a variety X corresponding to a divisorial valuation v. A fat irreducible component of a contact locus is a maximal divisorial set. In order to characterize the valuations corresponding to the irreducible components of a contact locus, it is essential to translate the inclusion relation between two maximal divisorial sets to a relation between the corresponding divisorial valuations. The most natural candidate as this relation is the value-inequality relation, i.e., $v(f) < v'(f)$ for every regular function f on the affine variety X. Actually, if the variety X and the valuations v, v' are toric, we have the equivalence: $v(f) \leq v'(f)$ for every regular function f on $X \Leftrightarrow C_X(v) \supset C_X(v')$. For non-toric valuations, we show that this equivalence does not hold even on $\mathbb{C}^2$.

This paper is organized as follows: In the second section we introduce the maximal divisorial set corresponding to a divisorial valuation and show the basic properties. In the third section we show some basic properties of a contact locus and formulate a generalized Nash problem. In the forth section we show an example of divisorial valuations v, v' over $\mathbb{C}^2$ such that the value-inequality relation does not imply the inclusion of the maximal divisorial sets corresponding to v and v'.

The author is grateful to the members of Singularity Seminar at Nihon University for useful suggestions and encouragement.

In this paper, a variety is always an irreducible reduced separated scheme of finite type over $\mathbb{C}$.

§2. Maximal divisorial sets in the arc space

Definition 2.1. Let X be a scheme of finite type over $\mathbb{C}$ and $K \supset \mathbb{C}$ a field extension. A morphism $\alpha : \operatorname{Spec} K[[t]] \longrightarrow X$ is called an *arc* of X. We denote the closed point of $\operatorname{Spec} K[[t]]$ by 0 and the generic point by η.

For $m \in \mathbb{N}$, a morphism $\beta : \operatorname{Spec} K[t]/(t^{m+1}) \longrightarrow X$ is called an *m-jet* of X. Denote the space of arcs of X by X_∞ and the space of m-jets of X by X_m.

The arc space X_∞ and m-jet space X_m are characterized by the following property:

Proposition 2.2. *Let X be a scheme of finite type over $\mathbb{C}$. Then, for an arbitrary $\mathbb{C}$-scheme Y,*

$$\operatorname{Hom}_{\mathbb{C}}(Y, X_\infty) \simeq \operatorname{Hom}_{\mathbb{C}}(Y \widehat{\times}_{\operatorname{Spec} \mathbb{C}} \operatorname{Spec} \mathbb{C}[[t]], X)$$

where $Y \widehat{\times}_{\operatorname{Spec} \mathbb{C}} \operatorname{Spec} \mathbb{C}[[t]]$ means the formal completion of $Y \times_{\operatorname{Spec} \mathbb{C}} \operatorname{Spec} \mathbb{C}[[t]]$ along the subscheme $Y \times_{\operatorname{Spec} \mathbb{C}} \{0\}$, and for $m \in \mathbb{N}$

$$\mathrm{Hom}_{\mathbb{C}}(Y, X_m) \simeq \mathrm{Hom}_{\mathbb{C}}(Y \times_{\mathrm{Spec}\,\mathbb{C}} \mathrm{Spec}\, K[t]/(t^{m+1}), X).$$

2.3. By thinking of the case $Y = \mathrm{Spec}\, K$ for an extension field K of $\mathbb{C}$, we see that K-valued points of X_∞ correspond to arcs $\alpha : \mathrm{Spec}\, K[[t]] \longrightarrow X$ bijectively. Based on this, we denote the K-valued point corresponding to an arc $\alpha : \mathrm{Spec}\, K[[t]] \longrightarrow X$ by the same symbol α. The canonical projection $X_\infty \longrightarrow X$, $\alpha \mapsto \alpha(0)$ is denoted by π_X. If there is no risk of confusion, we write just π.

A morphism $\varphi : Y \longrightarrow X$ of varieties induces a canonical morphism $\varphi_\infty : Y_\infty \longrightarrow X_\infty$, $\alpha \mapsto \varphi \circ \alpha$.

The concept "thin" in the following is first introduced in [2] and a "fat arc" is introduced and studied in [5].

Definition 2.4. Let X be a variety over $\mathbb{C}$. We say an arc $\alpha : \mathrm{Spec}\, K[[t]] \longrightarrow X$ is *thin* if α factors through a proper closed subset of X. An arc which is not thin is called a *fat arc.*

An irreducible closed subset C in X_∞ is called a *thin set* if the generic point of C is thin. An irreducible closed subset in X_∞ which is not thin is called a *fat set.*

Definition 2.5. Let $\alpha : \mathrm{Spec}\, K[[t]] \longrightarrow X$ be a fat arc of a variety X and $\alpha^* : \mathcal{O}_{X,\alpha(0)} \longrightarrow K[[t]]$ the local homomorphism induced from α. By Proposition 2.5, (i) in [5], α^* is extended to the injective homomorphism of fields $\alpha^* : K(X) \longrightarrow K((t))$, where $K(X)$ is the rational function field of X. Define a function $v_\alpha : K(X) \setminus \{0\} \longrightarrow \mathbb{Z}$ by

$$v_\alpha(f) = \mathrm{ord}_t\, \alpha^*(f).$$

Then, v_α is a valuation of $K(X)$. We call it the *valuation corresponding to* α.

Definition 2.6. Let X be a variety, $g : X_1 \longrightarrow X$ a proper birational morphism from a normal variety X_1 and $E \subset X_1$ an irreducible divisor. Let $f : X_2 \longrightarrow X$ be another proper birational morphism from a normal variety X_2. The birational map $f^{-1} \circ g : X_1 \dashrightarrow X_2$ is defined on a (nonempty) open subset E^0 of E. The closure of $(f^{-1} \circ g)(E^0)$ is well defined. It is called the *center* of E on X_2.

We say that E appears in f (or in X_2), if the center of E on X_2 is also a divisor. In this case the birational map $f^{-1} \circ g : X_1 \dashrightarrow X_2$ is a local isomorphism at the generic point of E and we denote the birational transform of E on X_2 again by E. For our purposes $E \subset X_1$ is identified with $E \subset X_2$. (Strictly speaking, we should be talking about the corresponding *divisorial valuation* instead.) Such an equivalence class is called a *divisor over* X.

Definition 2.7. A valuation v on the rational function field $K(X)$ of a variety X is called a *divisorial valuation over* X if $v = q\,\mathrm{val}_E$ for some $q \in \mathbb{N}$ and a divisor E over X. The divisor E is called *the center of* the valuation $v = q\,\mathrm{val}_E$. A fat arc α of X is called a *divisorial arc* if v_α is a divisorial valuation over X. A fat set is called a *divisorial set* if the generic point is a divisorial arc.

Definition 2.8. For a divisorial valuation v over a variety X, define the maximal divisorial set corresponding to v as follows:

$$C_X(v) := \overline{\{\alpha \in X_\infty \mid v_\alpha = v\}},$$

where $\overline{\{\ \}}$ is the Zariski closure in X_∞.

Proposition 2.9. *Let $\varphi : Y \longrightarrow X$ be a proper birational morphism of varieties and v a divisorial valuation over X. Then,*

(i) *v is also a divisorial valuation over Y and $C_X(v) = \overline{\varphi_\infty(C_Y(v))}$.*

(ii) *If an open subset $U \subset Y$ intersects the center of the valuation v on Y, Then $C_X(v) = \overline{\varphi_\infty(C_U(v))}$.*

Proof. The first assertion of (i) follows from constructing a suitable proper birational morphism dominating Y on which the corresponding divisor appears. The inclusion $\supset$ is obvious. For the opposite inclusion, take a fat arc $\alpha \in X_\infty$ such that $v_\alpha = v$. As the image $\alpha(\eta)$ is the generic point of X, it is in the open subset on which φ is isomorphic. Hence, $\alpha(\eta)$ is lifted on Y. Then, by the valuative criterion of properness, α is uniquely lifted to $\tilde{\alpha} \in Y_\infty$. We obtain that $\alpha = \varphi_\infty(\tilde{\alpha})$ and $v_{\tilde{\alpha}} = v$, which imply $\alpha \in \varphi_\infty(C_Y(v))$. For the statement (ii), let $\alpha : \operatorname{Spec} K[[t]] \longrightarrow Y$ be a fat arc of Y such that $v_\alpha = v$. Then, $\alpha(\eta)$ is the generic point of Y and $\alpha(0)$ is the generic point of the center of v on Y. One can see that both are on U, therefore $\alpha \in U_\infty$. Q.E.D.

In the following section, we will see that a maximal divisorial set is irreducible.

Example 2.10. For toric varieties, we use the terminologies in [3]. Let X be an affine toric variety defined by a cone σ in $N \simeq \mathbb{Z}^n$. Let M be the dual of N and $\langle\ ,\ \rangle : N \times M \longrightarrow \mathbb{Z}$ the canonical pairing. A $\mathbb{C}$-algebra $\mathbb{C}[M]$ consists of linear combinations over $\mathbb{C}$ of monomials x^u for $u \in M$. A point $v \in \sigma \cap N$ gives a toric valuation v by $v(x^u) = \langle v, u\rangle$. An irreducible locally closed subset $T_\infty(v)$ is defined in [4] as follows:

$$T_\infty(v) = \{\alpha \in X_\infty \mid \alpha(\eta) \in T,\ \mathrm{ord}_t\, \alpha^*(x^u) = \langle v, u\rangle \text{ for } u \in M\},$$

where T denotes the open orbit and also the torus acting on X and $\alpha^* : \mathbb{C}[\sigma^\vee \cap M] \longrightarrow K[[t]]$ is the ring homomorphism corresponding to

α. Then, $C_X(v) = \overline{T_\infty(v)}$. This is proved as follows: First, take any point $\alpha \in C_X(v)$ such that $v_\alpha = v$. Then, $\operatorname{ord}_t \alpha^*(x^u) = v(x^u) = \langle v, u \rangle$. Therefore, $\alpha \in T_\infty(v)$, which yields $C_X(v) \subset \overline{T_\infty(v)}$. Conversely, let $\beta \in \overline{T_\infty(v)}$ be the generic point. Then, by the upper semi-continuity (see for example [5, Proposition 2.7]) $v_\beta(f) \leq v_\alpha(f) = v(f)$ for every $f \in \mathbb{C}[\sigma^\vee \cap M]$. But, since v is toric, it is the minimal valuation satisfying $v_\beta(x^u) = v(x^u)$ for every $u \in \sigma^\vee \cap M$. Therefore, $v_\beta = v$, which implies $\beta \in C_X(v)$.

§3. Contact loci

Definition 3.1. Let $\psi_m : X_\infty \longrightarrow X_m$ be the canonical projection to the space of m-jets X_m. A subset $C \subset X_\infty$ is called a *cylinder* if there is a constructible set $S \subset X_m$ for some $m \in \mathbb{N}$ such that

$$C = \psi_m^{-1}(S).$$

For an arc $\alpha : \operatorname{Spec} K[[t]] \longrightarrow X$ of an affine variety $X = \operatorname{Spec} A$, we always denote by α^* the ring homomorphism $A \longrightarrow K[[t]]$ corresponding to α.

Let X be a variety and $\mathfrak{a}$ an ideal sheaf on X. For an arc $\alpha : \operatorname{Spec} K[[t]] \longrightarrow X$, there is an open affine subset $U \subset X$ such that α factors through U. We define $\operatorname{ord}_t \alpha^*(\mathfrak{a})$ as follows:

$$\operatorname{ord}_t \alpha^*(\mathfrak{a}) = \operatorname{ord}_t \alpha^*(\Gamma(U, \mathfrak{a})).$$

Definition 3.2 ([2]). For an ideal sheaf $\mathfrak{a}$ on a variety X, we define

$$\operatorname{Cont}^m(\mathfrak{a}) = \{\alpha \in X_\infty \mid \operatorname{ord}_t \alpha^*(\mathfrak{a}) = m\}$$

and

$$\operatorname{Cont}^{\geq m}(\mathfrak{a}) = \{\alpha \in X_\infty \mid \operatorname{ord}_t \alpha^*(\mathfrak{a}) \geq m\}.$$

These subsets are called *contact loci* of an ideal $\mathfrak{a}$. The subset $\operatorname{Cont}^{\geq m}(\mathfrak{a})$ is closed and $\operatorname{Cont}^m(\mathfrak{a})$ is locally closed. Both are cylinders.

Definition 3.3 ([2]). For a simple normal crossing divisor $E = \bigcup_{i=1}^s E_i$ on a non-singular variety X, we introduce the multi-contact loci for a multi-index $\nu \in \mathbb{Z}_{\geq 0}^s$:

$$\operatorname{Cont}^\nu(E) = \{\alpha \in X_\infty \mid \operatorname{ord}_t \alpha^*(I_{E_i}) = \nu_i\},$$

where I_{E_i} is the defining ideal of E_i. The multi-contact locus $\operatorname{Cont}^\nu(E)$ is irreducible if it is not empty.

Proposition 3.4. *Let $v = q\,\mathrm{val}_{E_0}$ be a divisorial valuation over a variety X. Let $\varphi : Y \longrightarrow X$ be a resolution of the singularities of X such that the irreducible divisor E_0 appears on Y. Then,*

$$C_X(v) = \overline{\varphi_\infty(\mathrm{Cont}^q(E_0))}.$$

In particular, $C_X(v)$ is irreducible.

Proof. By Proposition 2.9, it is sufficient to prove that

$$C_Y(v) = \overline{\mathrm{Cont}^q(E_0)}.$$

Let $\alpha \in \mathrm{Cont}^q(E_0)$ be the generic point, then $\alpha(0) = e$ is the generic point of E_0. Therefore, we obtain a local homomorphism

$$\alpha^* : \mathcal{O}_{Y,e} \longrightarrow K[[t]]$$

such that $\mathrm{ord}_t\, \alpha^*(\tau) = q$, where τ is a generator of the maximal ideal of $\mathcal{O}_{Y,e}$. Then, for every $f \in \mathcal{O}_{Y,e}$ written as $f = u\tau^n$, where u is a unit in $\mathcal{O}_{Y,e}$, it follows that $\mathrm{ord}_t\, \alpha^*(f) = qn = q\,\mathrm{val}_{E_0}(f) = v(f)$. This implies $\alpha \in C_Y(v)$.

Conversely, if $\beta \in C_Y(v)$ is an arc such that $v_\beta = v$, then $\mathrm{ord}_t\, \beta^*(\tau) = q$, which yields $\beta \in \mathrm{Cont}^q(E_0)$. Q.E.D.

The following is the characterization of ideals which have the same contact loci.

Proposition 3.5. *Let $\mathfrak{a}, \mathfrak{b} \subset A$ be ideals on an affine variety $X =$ Spec A. Then, the following are equivalent:*

(i) *$\overline{\mathfrak{a}} = \overline{\mathfrak{b}}$, where $\overline{\mathfrak{a}}, \overline{\mathfrak{b}}$ are the integral closures of $\mathfrak{a}$ and $\mathfrak{b}$, respectively;*
(ii) *$\mathrm{Cont}^m(\mathfrak{a}) = \mathrm{Cont}^m(\mathfrak{b})$ for every $m \in \mathbb{N}$;*
(iii) *$\mathrm{Cont}^{\geq m}(\mathfrak{a}) = \mathrm{Cont}^{\geq m}(\mathfrak{b})$ for every $m \in \mathbb{N}$.*

Proof. By $\mathrm{Cont}^m(\mathfrak{a}) = \mathrm{Cont}^{\geq m}(\mathfrak{a}) \setminus \mathrm{Cont}^{\geq m+1}(\mathfrak{a})$, the equivalence: *(ii)* $\Leftrightarrow$ *(iii)* is clear.

For *(i)* $\Rightarrow$ *(iii)*, it is sufficient to prove that $\mathrm{Cont}^{\geq m}(\mathfrak{a}) = \mathrm{Cont}^{\geq m}(\mathfrak{a}')$ $(m \in \mathbb{N})$ for an ideal $\mathfrak{a}'$ integral over $\mathfrak{a}$. As $\mathfrak{a}' \supset \mathfrak{a}$, it follows $\mathrm{Cont}^{\geq m}(\mathfrak{a}') \subset \mathrm{Cont}^{\geq m}(\mathfrak{a})$. To prove the opposite inclusion, take an arc $\alpha \in \mathrm{Cont}^{\geq m}(\mathfrak{a})$. It is sufficient to prove that $\mathrm{ord}_t\, \alpha^*(x) \geq m$ for every $x \in \mathfrak{a}'$. Assume $\mathrm{ord}_t\, \alpha^*(x) = d < m$. Since x is integral over $\mathfrak{a}$, there is a relation

$$(3.5.1) \qquad x^n + a_1 x^{n-1} + a_2 x^{n-2} + \ldots + a_n = 0$$

with $a_i \in \mathfrak{a}^i$. Here, noting that $\mathrm{ord}_t\, \alpha^*(a_i) \geq mi$, we obtain $\mathrm{ord}_t\, \alpha^*(a_i x^{n-i}) > nd$ for $i \geq 1$ and $\mathrm{ord}_t\, \alpha^*(x^n) = nd$. Therefore the order of the left hand side of (3.5.1) is nd, which is a contradiction to the equality.

For $(ii) \Rightarrow (i)$, take an element $x \in \mathfrak{b}$. Given a discrete valuation ring $R_v \supset A$, we obtain a ring homomorphism

$$A \subset R_v \longrightarrow \widehat{R_v} = K[[t]],$$

where K is the residue field of R_v and t is a generator of the maximal ideal of R_v. Let α be the arc corresponding to this ring homomorphism $A \longrightarrow K[[t]]$. Let $m = \operatorname{ord}_t \alpha^*(\mathfrak{a})$. Then $\alpha \in \operatorname{Cont}^m(\mathfrak{a}) = \operatorname{Cont}^m(\mathfrak{b})$, and therefore $\operatorname{ord}_t \alpha^*(\mathfrak{b}) = m$. Hence, $v(x) = \operatorname{ord}_t \alpha^*(x) \geq m = v(\mathfrak{a}R_v)$. This shows that x is integral over $\mathfrak{a}$, i.e., $\mathfrak{b} \subset \overline{\mathfrak{a}}$ by the valuative characterization of integrity. Similarly, we obtain $\mathfrak{a} \subset \overline{\mathfrak{b}}$. Now we have $\overline{\mathfrak{a}} = \overline{\mathfrak{b}}$ as required. Q.E.D.

It is proved in [2, Corollary 2.6] that a fat component of a contact locus is a divisorial set in case X is non-singular. The following gives more precise information also for singular X.

Proposition 3.6 ([1]). *Let X be an affine variety* $\operatorname{Spec} A$ *and* $\mathfrak{a} \subset A$ *a non-zero ideal. For $m \in \mathbb{N}$, a fat irreducible component of* $\operatorname{Cont}^{\geq m}(\mathfrak{a})$ *is a maximal divisorial set and the number of such components is finite.*

Here, we note that, for a divisorial valuation v over an affine variety $X = \operatorname{Spec} A$ and an ideal $\mathfrak{a} \subset A$, $C_X(v) \subset \operatorname{Cont}^{\geq m}(\mathfrak{a})$ if and only if $v(\mathfrak{a}) \geq m$.

Problem 3.7. A generalized Nash problem. *(Embedded version of Nash's problem in [2]) Let X be an affine variety* $\operatorname{Spec} A$ *and* $\mathfrak{a} \subset A$ *an ideal. Determine the set*

$$\mathcal{V}^m(\mathfrak{a}) := \{v \mid C_X(v) \textit{ is an irreducible component of } \operatorname{Cont}^{\geq m}(\mathfrak{a})\}.$$

Problem 3.8. The Nash problem. *Let $I_{\operatorname{Sing} X} \subset A$ be the defining ideal of the singular locus* $\operatorname{Sing} X$ *of X. Determine the set* $\mathcal{V}^1(I_{\operatorname{Sing} X})$.

3.9. Nash posed his problem in a different way ([8], see also [6]), but his problem is translated into the above problem. As one sees, the Nash problem is a special case of the generalized Nash problem. Nash predicted that the set $\mathcal{V}^1(I_{\operatorname{Sing} X})$ coincides with the set of the valuations of essential divisors ([8]). But, it is not true for four or higher dimensional case ([6]). So, Nash's prediction is still open for 2 and 3 dimensional cases.

3.10. The generalized Nash problem is to determine maximal $C_X(v)$'s contained in $\operatorname{Cont}^{\geq m}(\mathfrak{a})$. Therefore it is essential to determine the relation of valuations v, v' when $C_X(v) \supset C_X(v')$. The most natural candidate for the relation of v and v' is that $v(f) \leq v'(f)$ for every $f \in A$, which is denoted by:

$$v|_A \leq v'|_A.$$

If X is a toric variety and v, v' are toric divisorial valuation, then we obtain in [4]:

$$v|_A \leq v'|_A \quad \Leftrightarrow \quad C_X(v) \supset C_X(v').$$

Lemma 3.11. *Let v, v' be divisorial valuations over X.*

(i) *If $C_X(v) \supset C_X(v')$, then $v|_A \leq v'|_A$.*

(ii) *Assume that v is a toric valuation, then the converse also holds.*

Proof. Let α and α' be the generic points of $C_X(v)$ and $C_X(v')$, respectively. Then, $C_X(v) \supset C_X(v')$ yields $\mathrm{ord}_t\, \alpha^*(f) \leq \mathrm{ord}_t\, \alpha'^*(f)$ for every $f \in A$ by the upper semi-continuity. This yields $v|_A \leq v'|_A$.

Assume that v is toric. Let v'_0 be the toric valuation determined by $v'_0(x) = v'(x)$ for every monomial x. Then, by the definition of $T_\infty(v'_0)$, it follows $\alpha' \in T_\infty(v'_0)$. Hence, $C_X(v'_0) \supset C_X(v')$. On the other hand, v and v'_0 are both toric, it follows that $v|_A \leq v'_0|_A$ and therefore $C_X(v) \supset C_X(v'_0)$ by [4, Proposition 4.8]. Q.E.D.

So we expect that, at least for a simple variety like $\mathbb{C}^n$, the above equivalence holds for arbitrary divisorial valuations v, v'. In the next section we consider this problem and will give a negative answer.

§4. An example over $\mathbb{C}^2$

Proposition 4.1. *For a divisorial valuations v, v' over $X = \mathrm{Spec}\, A$, the following hold:*

(i) $C_X(v) = \overline{\bigcap_{f \in A\setminus\{0\}} \mathrm{Cont}^{v(f)}(f)}$;

(ii) *$C_X(v)$ is an irreducible component of $\bigcap_{f \in A\setminus\{0\}} \mathrm{Cont}^{\geq v(f)}(f)$;*

(iii) *the relation $v|_A \leq v'|_A$ holds if and only if the inclusion $C_X(v') \subset \bigcap_{f \in A\setminus\{0\}} \mathrm{Cont}^{\geq v(f)}(f)$ holds.*

Proof. For the proof of (i), take the generic point α of $C_X(v)$, then for every $f \in A \setminus \{0\}$ it follows that $\mathrm{ord}_t\, \alpha^*(f) = v(f)$ which means $\alpha \in \mathrm{Cont}^{v(f)}(f)$. On the other hand, let $\beta \in \bigcap_{f \in A\setminus\{0\}} \mathrm{Cont}^{v(f)}(f)$, then it follows $v_\beta = v$, which yields $\beta \in C_X(v)$. For the proof of (ii) note first that $C_X(v) \subset \bigcap_{f \in A\setminus\{0\}} \mathrm{Cont}^{\geq v(f)}(f)$. Let C' be an irreducible component of $\bigcap_{f \in A\setminus\{0\}} \mathrm{Cont}^{\geq v(f)}(f)$ containing $C_X(v)$. Let α and α' be the generic points of $C_X(v)$ and C', respectively. As $\alpha' \in \bigcap_{f \in A\setminus\{0\}} \mathrm{Cont}^{\geq v(f)}(f)$, we have $\mathrm{ord}_t\, \alpha'^*(f) \geq v(f)$ for every $f \in A\setminus\{0\}$. On the other hand, the inclusion $\alpha \in C'$ and the upper-semicontinuity yield $\operatorname{ord}_t \alpha'^*(f) \leq \operatorname{ord}_t \alpha^*(f) = v(f)$ for every $f \in A \setminus \{0\}$. Therefore we obtain that $v_{\alpha'} = v$ which implies $\alpha' \in C_X(v)$. For the proof of (iii), let β be the generic point of $C_X(v')$. Then, $v|_A \leq v'|_A$ if and only if $\operatorname{ord}_t \beta^*(f) = v'(f) \geq v(f)$ for every $f \in A \setminus \{0\}$ and this is equivalent to the inclusion $\beta \in \bigcap_{f\in A\setminus\{0\}} \operatorname{Cont}^{\geq v(f)}(f)$. Q.E.D.

The proposition suggests that $v|_A \leq v'|_A$ does not imply $C_X(v) \supset C_X(v')$, since there is no guarantee of the equality $\overline{\bigcap_{f\in A\setminus\{0\}} \operatorname{Cont}^{v(f)}(f)}$ $= \bigcap_{f\in A\setminus\{0\}} \operatorname{Cont}^{\geq v(f)}(f)$. In the following we show an example for which the equality actually does not hold.

In the rest of this section, $X = \mathbb{C}^2 = \operatorname{Spec} A$ and $A = \mathbb{C}[x, y]$. We construct an example of v, v' such that $v|_A \leq v'|_A$ but $C_X(v) \not\supset C_X(v')$.

4.2 (Construction). Let $X_\infty = \operatorname{Spec} \mathbb{C}[a_i, b_i \mid i \geq 0]$. Then

$$\operatorname{Cont}^2(x) \cap \operatorname{Cont}^3(y) = \operatorname{Spec} \mathbb{C}[a_2^{\pm 1}, a_3, .., b_3^{\pm 1}, b_4, ..].$$

Let $D \subset \operatorname{Cont}^2(x) \cap \operatorname{Cont}^3(y)$ be the closed subset defined by the ideal $(a_2^3 - b_3^2, 3a_2^2 a_3 - 2b_3 b_4)$ in $\operatorname{Cont}^2(x) \cap \operatorname{Cont}^3(y)$. Then, it is easy to check that D is irreducible. Let $\alpha \in D$ be the generic point and v be the divisorial valuation v_α corresponding to α. Let v' be the toric valuation with $v'(x) = 3, v'(y) = 4$

Theorem 4.3. *Let v, v' be as above. Then, the following hold:*

(i) *$\alpha^*(x^3 - y^2) = c_8 t^8 + c_9 t^9 + \ldots$, where c_8 is transcendental over $\mathbb{C}(a_2, a_3, b_3, b_4)$ which is a subfield of K, where $\alpha^* : A \longrightarrow K[[t]]$ is a ring homomorphism corresponding to α;*

(ii) *Let $Z \subset X$ be the closed subset defined by $x^3 - y^2 = 0$ and $\psi_m : X_\infty \longrightarrow X_m$ the canonical projection for $m \in \mathbb{N}$. Then,*

$$\begin{aligned} D &= \operatorname{Cont}^2(x) \cap \operatorname{Cont}^3(y) \cap \psi_7^{-1}(Z_7) \\ &= \operatorname{Cont}^2(x) \cap \operatorname{Cont}^3(y) \cap \operatorname{Cont}^{\geq 8}(x^3 - y^2); \end{aligned}$$

(iii) *$C_X(v) = \overline{D}$;*

(iv) *$v|_A \leq v'|_A$;*

(v) *$C_X(v) \not\supset C_X(v')$.*

Proof. (i) Note that K is the quotient field of the ring $\mathbb{C}[a_2^{\pm 1}, a_3, \ldots, b_3^{\pm 1}, b_4, ..]/(a_2^3 - b_3^2, 3a_2^2 a_3 - 2b_3 b_4)$. The image $\alpha^*(x^3 - y^2)$ is

$$(a_2^3 - b_3^2)t^6 + (3a_2^2 a_3 - 2b_3 b_4)t^7 + (3a_2 a_3^2 + 3a_2^2 a_4 - b_4^2 - 2b_3 b_5)t^8 + \ldots,$$

where the coefficients of t^6 and t^7 are zero in K and the coefficient of t^8 contains a_4 and b_5 which are algebraically independent over the subfield $\mathbb{C}(a_2, a_3, b_3, b_4)$ in K. Therefore c_8 is transcendental over $\mathbb{C}(a_2, a_3, b_3, b_4)$.

(ii) Let f_j be the coefficient of t^j in

$$(a_0 + a_1 t + ..)^3 - (b_0 + b_1 t + \ldots)^2.$$

Then, the closed subset $\psi_7{}^{-1}(Z_7)$ is defined by the equations $f_0(a_i, b_i) = \ldots = f_7(a_i, b_i) = 0$ in X_∞. Noting that $a_0 = a_1 = b_0 = b_1 = b_2 = 0$ in $\mathrm{Cont}^2(x) \cap \mathrm{Cont}^3(y)$, we obtain that $f_j = 0$ automatically holds for $j = 0, .., 5$ in $\mathrm{Cont}^2(x) \cap \mathrm{Cont}^3(y)$. On the other hand f_6 and f_7 coincide with $a_2^3 - b_3^2$ and $3a_2^2 a_3 - 2b_3 b_4$, respectively in $\mathrm{Cont}^2(x) \cap \mathrm{Cont}^3(y)$. Therefore, $D = \mathrm{Cont}^2(x) \cap \mathrm{Cont}^3(y) \cap \psi_7^{-1}(Z_7)$.

(iii) As $v_\alpha = v$, it is clear that $\overline{D} \subset C_X(v)$. On the other hand, by Proposition 4.1,

$$C_X(v) = \overline{\bigcap_{f \in A} \mathrm{Cont}^{v(f)}(f)}$$

and the right hand side is contained in $\mathrm{Cont}^2(x) \cap \mathrm{Cont}^3(y) \cap \mathrm{Cont}^{\geq 8}(x^3 - y^2)$.

For (iv) and (v) we need the following lemmas.

Lemma 4.4. *Let v_0 be the toric valuation with $v_0(x) = 2$ and $v_0(y) = 3$. Let $g \in A$ be a homogeneous element with respect to v_0, i.e., every monomial in g has the same value of v_0. Then, $g = (x^3 - y^2)^m g'$, where $m \geq 0$ and $v(g') = v_0(g')$. Here, g satisfies $v(g) > v_0(g)$ if and only if $m > 0$.*

Proof. First, if $v(g) = v_0(g)$, then g is written as in the lemma with $m = 0$. Now assume that $v(g) > v_0(g)$. Note that $\alpha^*(x) = a_2 t^2 + a_3 t^3 + \ldots$, and $\alpha^*(y) = b_3 t^3 + b_4 t^4 + \ldots$, where $\alpha^* : A \longrightarrow K[[t]]$ is the ring homomorphism corresponding to α. Here, K is the quotient field of $\mathbb{C}[a_2^{\pm 1}, a_3, .., b_3^{\pm 1}, b_4, ..]/(a_2^3 - b_3^2, 3a_2^2 a_3 - 2b_3 b_4)$. As g is homogeneous with respect to v_0, we obtain that

$$\alpha^*(g) = g(a_2, b_3) t^{v_0(g)} + (\text{higher order terms}).$$

Since $\mathrm{ord}_t\, \alpha^*(g) = v(g) > v_0(g)$, the leading coefficient $g(a_2, b_3) = 0$ in K. Therefore, $g(a_2, b_3) = 0$ in $\mathbb{C}[a_2, b_3]/(a_2^3 - b_3^2) \subset K$, which yields that $(x^3 - y^2) \mid g$. Now write $g = (x^3 - y^2) g'$. If g' satisfies still $v(g') > v_0(g')$, then, by the above discussion, it follows that $(x^3 - y^2) \mid g'$. We have $g = (x^3 - y^2)^2 g''$ and $v_0(g'') < v_0(g') < v_0(g)$. By this procedure, we obtain finally the statement of the lemma. Q.E.D.

Lemma 4.5. *Assume $v(g) > v_0(g)$ for a polynomial $g \in A$. Let $g = g_1 + g_2 + \ldots + g_r$ be the decomposition into homogeneous parts with $v_0(g_i) < v_0(g_{i+1})$. Then,*

$$v(g) = \min\{v(g_i) \mid i = 1, \ldots, r\}.$$

Proof. By Lemma 4.4 we obtain for each $i = 1, .., r$

$$g_i = (x^3 - y^2)^{n_i} g_i',$$

where $n_i \geq 0$ and g_i' satisfies $v(g_i') = v_0(g_i')$. As $v(x^3 - y^2) = 8$, it follows that $v(g_i) = 8n_i + v_0(g_i')$. Let $d = \min\{v(g_i) \mid i = 1, .., r\}$ and $I = \{i \mid v(g_i) = d\}$. Then, for two distinct element $i, j \in I$, it follows that $n_i \neq n_j$. Indeed, if $n_i = n_j$, then, $8n_i + v_0(g_i') = d = 8n_j + v_0(g_j')$, and therefore $v_0(g_i') = v_0(g_j')$. Hence, $v_0(g_i) = 6n_i + v_0(g_i') = 6n_j + v_0(g_j') = v_0(g_j)$, which is a contradiction to the assumption. Now, we have

$$\alpha^*(g) = \sum_{i \in I} \alpha^* \left((x^3 - y^2)^{n_i} g_i'\right) + \sum_{i \notin I} \alpha^* \left((x^3 - y^2)^{n_i} g_i'\right).$$

The right hand side is

$$\left(\sum_{i \in I} c_8^{n_i} g_i'(a_2, b_3)\right) t^d + \text{(higher order term)}.$$

If $v(g) = \mathrm{ord}_t\, \alpha^*(g) > d$, then $\#I \geq 2$ and we have a non-trivial algebraic relation

$$\sum_{i \in I} c_8^{n_i} g_i'(a_2, b_3) = 0$$

of c_8 over $\mathbb{C}(a_2, b_3)$. But this is a contradiction to (i). Therefore, $v(g) = d$. Q.E.D.

Proof of (iv) and (v) of Theorem 4.3. Take an arbitrary element $g \in A$. Let $g = g_1 + .. + g_r$ be the homogeneous decomposition as in Lemma 4.5. Let $g_i = (x^3 - y^2)^{n_i} g_i'$, where $v(g_i') = v_0(g_i')$ as in Lemma 4.4. Then $v'(g_i) \geq 8n_i + v_0(g_i') = v(g_i)$. By Lemma 4.5, $v(g) = \min\{v(g_i) \mid i = 1, .., r\} \leq \min\{v'(g_i) \mid i = 1, .., r\} \leq v'(g)$. This proves (iv).

Let $\beta \in \mathrm{Cont}^3(x) \cap \mathrm{Cont}^4(y)$ be the generic point. Then $\beta^*(x) = a_3 t^3 + a_4 t^4 + \ldots$ and $\beta^*(y) = b_4 t^4 + b_5 t^5 + \ldots$ Then

$$\beta \in D' = \mathrm{Spec}\, \mathbb{C}[a_2, a_3, .., b_3, b_4 \ldots]/(a_2^3 - b_3^2, 3a_2^2 a_3 - 2b_3 b_4) \subset X_\infty.$$

Here, D' has two component. One is $\overline{D}$ and we denote the other by D''. The component D'' is defined by $a_2 = b_3 = 0$ in D'. It is easy to see that β is the generic point of D''. Therefore, $C_X(v) = \overline{D}$ and $C_X(v') = D''$ have no inclusion relation. Q.E.D.

4.6. This theorem and 3.11, (i) shows that the relation $C_X(v) \supset C_X(v')$ is strictly stronger than the relation $v|_A \leq v'|_A$. Now we consider another relation of divisorial valuation. For $X = \mathbb{C}^2$ we have an order $\prec$ of divisors E and E' over X: $E \prec E'$ if there is a successive blowing-ups $\longrightarrow X_j \longrightarrow \cdots \longrightarrow X_i \longrightarrow X_{i-1} \longrightarrow \cdots X_1 \longrightarrow X$ such that E appears on X_i and E' appears on X_j with $i < j$, and the center of E' on X_i is contained in E.

Proposition 4.7. *Let* $v = \mathrm{val}_E$ *and* $v' = \mathrm{val}_{E'}$. *If* $E \prec E'$, *then* $C_X(v) \supset C_X(v')$. *But* $C_X(v) \supset C_X(v')$ *does not imply* $E \prec E'$.

Proof. Let $p \in E \subset X_i$ be the center of E' on X_i. Take a suitable affine neighborhood U of p, then $U = \operatorname{Spec} B \simeq \mathbb{C}^2$, p is the origin and E is an invariant divisor, where we put a suitable toric structure on U. Then we have $v|_B \leq v'|_B$. As v is toric on U, this inequality implies $C_U(v) \supset C_U(v')$ by Lemma 3.11 and therefore $C_X(v) \supset C_X(v')$ by Proposition 2.9.

For the second assertion, let $v = (2, 1)$ and $v' = (2, 3)$ in $\sigma \cap N$. We also denote the toric valuations corresponding to v and v' by the same symbol v and v', respectively. Then $v|_A \leq v'|_A$, where $A = \mathbb{C}[\sigma^\vee \cap M]$. Since v and v' are toric, it follows that $C_X(v) \supset C_X(v')$. But the divisors E and E' corresponding to v and v', respectively, does not satisfy $E \prec E'$. Indeed, let $\varphi_1 : X_1 \longrightarrow X$ be the blow-up at 0 and E_1 the exceptional divisor. There are two closed orbits p_1 and p_2 on E_1. The divisor E is the exceptional divisor of the blow-up $\varphi : X_2 \longrightarrow X_1$ at one of the closed orbits on E_1, let it be p_1. Then, the center of E' on X_2 is p_2. Therefore $E \not\prec E'$. Q.E.D.

By this, the relation of v and v' for $C_X(v) \supset C_X(v')$ is something between "$v \prec v'$" and "$v|_A \leq v'|_A$".

References

[1] T. De Fernex, L. Ein and S. Ishii, Divisorial valuations via arcs, preprint, arXive:math.AG/0701867, to appear in Publication RIMS, **44** (2) (2008).

[2] L. Ein, R. Lazarsfeld and M. Mustaţă, Contact loci in arc spaces, Comp. Math., **140** (2004), 1229–1244.

[3] W. Fulton, Introduction to Toric Varieties, Ann. of Math. Stud., **131** (1993), Princeton Univ. Press.

[4] S. Ishii, The arc space of a toric variety, J. Algebra, **278** (2004), 666–683.

[5] S. Ishii, Arcs, valuations and the Nash map, J. Reine Angew. Math., **588** (2005), 71–92.

[6] S. Ishii and J. Kollár, The Nash problem on arc families of singularities, Duke Math. J., **120** (2003), 601–620.

[7] M. Lejeune-Jalabert and A. J. Reguera-Lopez, Arcs and wedges on sandwiched surface singularities, Amer. J. Math., **121** (1999), 1191–1213.
[8] J. F. Nash, Arc structure of singularities, Duke Math. J., **81** (1995), 31–38.
[9] C. Plénat, Résolution du problème des arcs de Nash pour les points doubles rationnels $D_n (n \geq 4)$, Thèse, Univ. Paul Sabatier, Toulouse, September 2004.
[10] C. Plénat and P. Popescu-Pampu, A class of non-rational surface singularities for which the Nash map is bijective, preprint, arXive:math.AG/0410145.
[11] A. J. Reguera-Lopez, Families of arcs on rational surface singularities, Manuscripta Math., **88** (1995), 321–333.
[12] A. J. Reguera, Image of Nash map in terms of wedges, C. R. Acad. Sci. Ser. I, **338** (2004), 385–390.

Department of Mathematics
Tokyo Institute of Technology
Oh-Okayama, Meguro
Tokyo JAPAN
E-mail address: shihoko@math.titech.ac.jp

Advanced Studies in Pure Mathematics 50, 2008
Algebraic Geometry in East Asia — Hanoi 2005
pp. 251–267

Two non-conjugate embeddings of $S_3 \times \mathbb{Z}_2$ into the Cremona group II

Vasilii Alekseevich Iskovskikh

Abstract.

We give a detailed proof of the following main result of [Isk1]. There are two non-conjugate embeddings of $S_3 \times \mathbb{Z}_2$ into the Cremona group given by the linear action of this group on the plane and the action on the two-dimensional torus respectively.

§1. Introduction and the main result

1.1.

Let us first recall our main result which is inspired by [LPR]. Consider the following two actions of the group $G \simeq S_3 \times \mathbb{Z}_2$ on the rational surfaces P and T (where S_3 denotes the symmetric group and $\mathbb{Z}_2$ denotes the cyclic group of order two).

(I) P **- the plane** $x + y + z = 0$**:** S_3 acts by the permutations on the coordinates and $\mathbb{Z}_2$ is given by $(x, y, z) \to (-x, -y, -z)$;

(II) T **- the two-dimensional torus** $xyz = 1$**:** S_3 acts by the permutations on the coordinates and $\mathbb{Z}_2$ is given by $(x, y, z) \to (x^{-1}, y^{-1}, z^{-1})$.

In both cases we obtain two embeddings of this group into the two-dimensional Cremona group $Cr_2(\mathbb{C})$. A natural question arising here is whether the images of G under these embeddings are conjugate in $Cr_2(\mathbb{C})$ or not?

Our answer is the following proposition.

1.1.1. **Proposition** *The images of G under the above embeddings are not conjugate in $Cr_2(\mathbb{C})$. In the other words, there exists no G-equivariant birational map between G-surfaces P and T.*

Received April 4, 2006.
Revised September 27, 2006.
The work was partially supported by RFFI grant 08-01-00395-a, RFBR 06-01-72017-MNTI-a and CRDF grant RUM1-2692-MO-05, and grant of Nsh-1987.2008.1.

1.2. Commentary

Together with [LPR] this result means that the simple algebraic group of type $\mathbb{G}_2$ is not a Cayley group in the following sense. Let G as in [LPR] be a connected algebraic group (we assume it only in this commentary). Let $\mathfrak{g}$ be its Lie algebra. Consider G as an algebraic variety X with the action of G on itself by the conjugation gxg^{-1}, $g \in G$, $x \in X$, and the Lie algebra $\mathfrak{g}$ as an algebraic variety Y with adjoint action $Ad_G g(y) = gyg^{-1}$, $g \in G$, $y \in Y$. The group G is called *a Cayley group* if there exist G-equivariant birational map $\lambda \colon X \dashrightarrow Y$, *i. e.* $\lambda(gxg^{-1}) = Ad_G g(\lambda(x))$. The question about the Cayley property is reduced to some property of character lattice $\hat{T}$ of the maximal torus $T \in G$ with the natural action of the Weyl group W on $\hat{T}$. Many canonical linear algebraic groups are proved to be non-Cayley ones. However, there are also a lot of exceptions.

The maximal torus of the group of type $\mathbb{G}_2$ is two-dimensional, and its Weyl group is $W \simeq S_3 \times \mathbb{Z}_2$. The question about the Cayley property is exactly problem 1.1. Our result says that $\mathbb{G}_2$ is not Cayley. Surprisingly it was proved in [LPR] that the group $\mathbb{G}_2 \times G_m^2$ (where G_m is the multiplicative group of the ground field with the trivial action of the Weyl group) is a Cayley group. Thus, $\mathbb{G}_2$ is stable Cayley in this sense. It would be interesting to ask whether the group $\mathbb{G}_2 \times G_m$ is Cayley or not?

In our work [Isk1] we gave a sketchy proof of Proposition 1.1.1 based on the general method of factorization of G-equivariant birational maps between rational G-surfaces (see [Ma1], [Isk3]) with G denoting a finite group. In this paper we shall give a detailed proof of it since after the appearance of [Isk1] a great interest surrounded this fact.

1.3.

As in [Isk1] we shall use the concept of a rational G-surface in our proof. Here $G \subset Cr_2(\mathbb{C})$ denotes a finite subgroup of the Cremona group (see, for instance, [Ma1], [Isk2]). In our case $G \simeq S_3 \times \mathbb{Z}_2$ and it acts on the compactifications of Y and X of surfaces P and T corresponding respectively to the cases (I) and (II) above. Let us describe these actions more precisely.

The case (I): $Y = \mathbb{P}^2$. Let (u_0, u_1, u_2) be the homogeneous coordinates on $\mathbb{P}^2$ and $x = \frac{u_1}{u_0}$, $y = \frac{u_2}{u_0}$ and $z = -\frac{u_1+u_2}{u_0}$. Then the action of G on Y is given in the matrix form as follows (we use the homogeneous coordinates).

$$\text{The involution } \sigma_{xy} = (xy) = \begin{pmatrix} 1 & 0 & 0 \\ 0 & 0 & 1 \\ 0 & 1 & 0 \end{pmatrix};$$

$$\text{the 3-cycle } \sigma_{xyz} = (xyz) = \begin{pmatrix} 1 & 0 & 0 \\ 0 & 0 & 1 \\ 0 & -1 & -1 \end{pmatrix};$$

$$\text{the involution } \tau = \begin{pmatrix} 1 & 0 & 0 \\ 0 & -1 & 0 \\ 0 & 0 & -1 \end{pmatrix} \text{ is the generator of } \mathbb{Z}_2.$$

Fixed elements: the point $(1,0,0)$.

Invariant elements: the line $L_0 = (u_0 = 0)$. The group $\mathbb{Z}_2$ acts on L_0 trivially and S_3 acts as the standard irreducible two-dimensional linear representation. A unique G-invariant 0-orbit of length 2 is $\{(0,1,\lambda),(0,\lambda,1)\}$, where $\lambda = e^{\frac{2\pi i}{3}}$ is the cubic root of 1. There are also two G-invariant 0-orbits of length 3 given by $\{(0,0,1),(0,1,-1),(0,1,0)\}$ and $\{(0,1,1),(0,1,-2),(0,-1,1)\}$.

There is a G-equivariant pencil of lines with G-fixed point $(1,0,0)$ and G-invariant section (the line L_0). The group S_3 acts on the base of this pencil and on the section L_0. The group $\mathbb{Z}_2$ acts on the fibers by $t \to -t$.

The case (II): X is the most natural smooth compactification of the two-dimensional torus T: $(xyz = 1)$. Consider the threefold $\mathbb{P}^1 \times \mathbb{P}^1 \times \mathbb{P}^1$ with homogeneous coordinates $(x_1, x_0) \times (y_1, y_0) \times (z_1, z_0)$ and coordinates $(x = \frac{x_1}{x_0}, y = \frac{y_1}{y_0}, z = \frac{z_1}{z_0})$. This compactification is given by

$$x_1 y_1 z_1 = x_0 y_0 z_0. \tag{1}$$

Under the Segre map $\mathbb{P}^1 \times \mathbb{P}^1 \times \mathbb{P}^1 \hookrightarrow \mathbb{P}^7$, the equation (1) is the G-invariant equation of the hyperplane $\mathbb{P}^6 \supset X$. The surface X is given there by the quadratic equations that are restrictions on $\mathbb{P}^6$ of the equations which give the Segre map. Under this embedding X is a smooth projective del Pezzo surface of degree 6 in $\mathbb{P}^6$.

G acts in (1) as follows. S_3 acts by the coordinate permutations induced by the permutations of factors of $\mathbb{P}^1 \times \mathbb{P}^1 \times \mathbb{P}^1$ (for example $\sigma_{xy}\colon (x_1, x_0) \longleftrightarrow (y_1, y_0)$). $\mathbb{Z}_2$ acts as follows.

$$(x_1, y_1, z_1) \longleftrightarrow (x_0, y_0, z_0).$$

These actions are linear under the Segre map.

According to the classification of G-minimal rational G-surfaces (see [Isk2], [Mo]), $X \subset \mathbb{P}^6$ is the minimal del Pezzo G-surface of degree 6 with G-invariant Picard group $\operatorname{Pic}^G(X) = \mathbb{Z}(-K_X)$, where K_X denotes the canonical class. The configuration of (-1)-curves on X is a unique G-orbit and has the form of a regular hexagon. Its equation is the infinite hyperplane section

$$x_0 y_0 z_0 = 0. \tag{2}$$

These lines are one-dimensional fibers of three birational projections $X \to \mathbb{P}^1 \times \mathbb{P}^1$, two one-dimensional fibers in each of them.

Further, we need other G-birational equivalent to X models X_0, X_1, X_2 and the classification of some 0-dimensional (of small length) and 1-dimensional orbits on them.

1.4.

Start from the surface X given by (1). $\operatorname{Pic} X$ is generated by (-1)-curves and all (-1)-curves form a unique G-orbit, which is equivalent to $-K_X$, so $\operatorname{Pic}^G_X = \mathbb{Z}(-K_X)$. We may consider an affine model $T\colon (xyz = 1)$ for studying of G-orbits, because the orbit at infinity is known (it is all (-1)-curves) and there is no 0-orbit of length less than 6.

1.4.1. **Lemma** *Let $\mathcal{A} \subset T$ be a G-equivariant 0-orbit of length $d < 6$. Then $\mathcal{A}$ is one of the following.*

- $d = 1$: *a unique fixed point $P = (1,1,1)$;*
- $d = 2$: *a unique orbit $\{P_1 = (\lambda,\lambda,\lambda), P_{-1} = (\lambda^{-1},\lambda^{-1},\lambda^{-1})\}$, where $\lambda = e^{\frac{2\pi i}{3}}$ is a cubic root of 1 (notice that $\lambda^{-1} = \lambda^2$);*
- $d = 3$: *a unique orbit $\{Q_1 = (1,-1,-1), Q_2 = (-1,1,-1), Q_3 = (-1,-1,1)\}$;*
- $d = 4$: *there are no such orbits;*
- $d = 5$: *there are no such orbits, because $5 \nmid 12 = |G|$.*

Indeed, one can see this from the equation $xyz = 1$ and the action of G on T. In the case $d = 4$ the stabilizer of the point is $\langle \sigma_{xyz} \rangle = \mathbb{Z}_3$ acting by $(x,y,z) \to (y,z,x)$, which means that $x = y = z$.

1.4.2. We now find some G-orbits consisting of rational curves.

$\mathcal{T} = \Gamma_x + \Gamma_y + \Gamma_z \sim -K_X$. This G-orbit consists of the curves of genus 0 which contain the fixed point $P = (1,1,1)$ and is given by $x = 1$, $y = 1$, $z = 1$. In the projective form these equations on $\mathbb{P}^1 \times \mathbb{P}^1$ are as follows.

$$\Gamma_x\colon y_1 z_1 - y_0 z_0 = 0,$$

$$\Gamma_y\colon x_1 z_1 - x_0 z_0 = 0, \tag{3}$$

$$\Gamma_z\colon x_1y_1 - x_0y_0 = 0.$$

Each of these curves Γ is smooth of genus 0 with $-K_X\Gamma = 2$, $\Gamma^2 = 0$, and any two of them intersect by a unique point P. Blow up this point. Then after a G-equivariant contraction of these curves we have three G-conjugate points on the image of the exceptional (-1)-curve of the blowing-up on the model $X_2 \subset \mathbb{P}^3$. This is an example of G-equivariant link $\Phi_{6,1}$, *cf.* Section 2 below.

This birational G-map $\gamma\colon X \dashrightarrow X_2$ is a projection from the tangent plane to the fixed point P with the embedding $X \subset \mathbb{P}^6$ described above. Under this projection the point P is blown up, and three conics passing through P are contracted.

$\mathcal{D} = \Delta_x + \Delta_y + \Delta_z \sim -2K_X$. This is a triple of smooth curves of genus 0 with equations $y = z$, $z = x$, and $x = y$ respectively. In the projective form we have

$$\Delta_x\colon x_1y_1^2 - x_0y_0^2 = 0,$$

$$\Delta_y\colon y_1z_1^2 - y_0z_0^2 = 0, \tag{4}$$

$$\Delta_z\colon z_1x_1^2 - z_0x_0^2 = 0.$$

All these curves Δ intersect in three points P, P_1 and P_{-1} with $-K_X\Delta = 4$ and $\Delta^2 = 2$.

1.4.3. Curves Γ and Δ are components of the fibers of the G-invariant pencil of rational curves $\Pi = |-K_X - 2P - P_1 - P_{-1}|$ (*i. e.* the pencil of curves from the linear system $|-K_X|$ which contain points P_1, P_{-1} and twice point P). Moreover $\Gamma_x + \Delta_x \sim \Gamma_y + \Delta_y \sim \Gamma_z + \Delta_z$ are the fibers of this pencil. Equations (3) and (4) show that $\Gamma_x \cap \Delta_x = \{P, Q_1\}$, $\Gamma_y \cap \Delta_y = \{P, Q_2\}$, $\Gamma_z \cap \Delta_z = \{P, Q_3\}$.

$\mathcal{E} = E_x + E_y + E_z \sim -K_X$. This triple of smooth curves of genus 0 is given by $x = -1$, $y = -1$, $z = -1$ accordingly. One can see from these equations that $E_x \ni Q_2, Q_3$, $E_y \ni Q_1, Q_3$, $E_z \ni Q_1, Q_2$, and Q_i are the only intersection points for the components of $\mathcal{E}$. As above, $-K_XE = 2$, $E^2 = 0$, so after blowing up Q_1, Q_2 and Q_3 strict transforms E'_x, E'_y, E'_z form a G-conjugate triple of (-2)-curves which do not intersect each other. They may be G-equivariant contracted to three ordinary double points on a cubic surface in $\mathbb{P}^3$.

1.5.

Consider now a G-birational model X_0 of G-surface X, namely, the standard projectivisation of torus $T \subset \mathbb{A}^3 \subset \mathbb{P}^3$

$$X_0\colon xyz = w^3, \tag{5}$$

where (x, y, z, w) are homogeneous coordinates on $\mathbb{P}^3$. This model differs from X only at infinity $w = 0$. Three lines given by $x = w = 0$, $y = w = 0$ and $z = w = 0$ are on X_0 unlike on X there are 6 lines. It has also 3 ordinary double points $(0, 0, 1, 0)$, $(0, 1, 0, 0)$, and $(1, 0, 0, 0)$. G acts on X_0 not linearly (as on $X \subset \mathbb{P}^6$ it does). More precisely S_3 acts linearly by permutations of x, y, z and $\mathbb{Z}_2$ acts via the birational involution $(x, y, z, w) \rightarrowtail (x^{-1}, y^{-1}, z^{-1}, w^{-1})$.

However there is another cubic model $X_1 \subset \mathbb{P}^3$ such that G acts linearly on it. As X_0 it has three ordinary double points. This model is given by G-equivariant birational projection

$$p_1 \colon X \subset \mathbb{P}^6 \dashrightarrow \mathbb{P}^3 \supset X_1$$

from the plane $\langle Q_1, Q_2, Q_3 \rangle$, which is a linear span of $Q_1, Q_2, Q_3 \in X$ in $\mathbb{P}^6$. As noted in 1.4 conics E_x, E_y, E_z are contracted to the singular points.

Choose homogeneous coordinates (x, y, z, w) in $\mathbb{P}^3$ such that the image of P is the point $(0, 0, 0, 1)$ and $\mathbb{Z}_2$ acts via the involution $(x, y, z, w) \rightarrowtail (-x, -y, -z, w)$. Then S_3 acts by coordinate permutations as on X_0. Under the projection $p_1 \colon X \dashrightarrow X_1$ three singular points (the images of E_x, E_y, and E_z) lie on the plane at infinity $w = 0$. The images of conics Γ_x, Γ_y, and Γ_z are three lines passing through $P_0 = p_1(P)$. Thus P_0 is the Eckardt point whose tangent plane is G-invariant, *i. e.* it is given by the linear equation $x + y + z = 0$.

So the equation of $X_1 \subset \mathbb{P}^3$ is given by

$$X_1 \colon xyz - aw^2(x + y + z) = 0, \quad a \in \mathbb{C}. \tag{6}$$

For convenience one may assume $a = \frac{1}{3}$. Then the singular points are $(1, 0, 0, 0)$, $(0, 1, 0, 0)$, $(0, 0, 1, 0)$, and the images of P_1 and P_{-1} are $(1, 1, 1, 1)$ and $(-1, -1, -1, -1)$ respectively. Three lines $x = 0$, $y = 0$, $z = 0$ lie on the tangent plane $x + y + z = 0$.

1.6. Remark

As on each cubic surface, there is Geiser's birational involution concerned with P_0, *i. e.* the projection from this point to the plane composed with the automorphism of the double covering. Obviously this birational involution is G-equivariant and in our case it is biregular because P_0 is the Eckardt point. G-equivariant birational Bertini involution is concerned with two G-conjugate points $P_1, P_{-1} \in X$. Its action on $\operatorname{Pic} X_1$ differs from the standard one because X_1 is singular and the points P_1, P_{-1} are not in the general position. This means that the third intersection point of the line $\langle P_1, P_{-1} \rangle \subset \mathbb{P}^3$ and X_1 (i.e. P_0) lies on the (-1)-curves.

1.7.

Another G-equivariant birational model of X is the quadric $X_2 \subset \mathbb{P}^3$ of type

$$X_2\colon xy + yz + zx = 3w^2. \tag{7}$$

It can be obtained from X_1 by G-equivariant birational transform $(x, y, z, w) \mapsto (x^{-1}, y^{-1}, z^{-1}, w^{-1})$. Indeed this coordinate change with the multiplication to the common factor transform the equation (6) with $a = \frac{1}{3}$ to the equation (7). The action of G on (x, y, z, w) is still linear. Moreover X_2 is nothing but the image of $X \subset \mathbb{P}^6$ under the linear projection

$$p_2\colon X \subset \mathbb{P}^6 \dashrightarrow \mathbb{P}^3 \supset X_2$$

from the tangent plane to X at the G-fixed point $P \in X$. The projection p_2 blows up P to the conic $C_0\colon (w = 0) \subset X_2$ and contracts three conics $\Gamma_x, \Gamma_y, \Gamma_z \subset X$ to the G-invariant 0-orbit $\mathcal{A} = \{(1,0,0,0), (0,1,0,0), (0, 0,1,0)\} \subset C_0$.

The curve C_0 is the image of the point $P_0 \in X_1$ and the 0-orbit $\mathcal{A}$ is the image of the lines $x = 0, y = 0, z = 0$ under our map $X_1 \dashrightarrow X_2$.

1.8. Lemma

(i): *The surface X_2 is smooth and there are no G-fixed points on it;*

(ii): *The conics $C_0 = (w = 0)$ and $C_1 = (x + y + z = 0)$ are G-invariant and on C_0 acts only S_3 by a unique two-dimensional irreducible representation and G acts effective on C_1;*

(iii): *The G-invariant 0-orbits of length $d < 6$ on X_2 are:*

$d = 2$: $\{P_1 = (1,1,1,1), P_{-1} = (-1,-1,-1,1)\}$, *the images on X_2 of P_1 and P_{-1} on X (or X_1); $\{R_1 = (1, \lambda, \lambda^{-1}), R_2 = (\lambda, 1, \lambda^{-1})\} = C_0 \cap C_1$ - the base points of the conic pencil on X_2 generated by C_0 and C_1, $\lambda = e^{\frac{2\pi i}{3}}$*

$d = 3$: $\mathcal{A} = \{(1,0,0,0), (0,1,0,0), (0,0,1,0)\}$,
$\mathcal{B} = \{(-1,2,2,0), (2,-1,2,0), (2,2,-1,0)\}$, $\mathcal{A}, \mathcal{B} \subset C_0$;

$d = 4$: *there are no such orbits;*

$d = 5$: *there are no such orbits;*

(iv): *Let*

$$\Pi_0 = (t_0(x + y + z) + t_1 w = 0), \quad (t_0, t_1) \in \mathbb{P}^1,$$

be the pencil of conics generated by C_0 and C_1; then on the base of this pencil $\mathbb{P}^1$ acts only $\mathbb{Z}_2 = G/S_3\colon (t_0, t_1) \mapsto (-t_0, t_1)$ with two fixed points $(1,0)$ and $(0,1)$; there are two reducible *fibers* F_1 *and* F_{-1} *on* Π_0 *with intersection points* P_1 *and* P_{-1} *of the components with coordinates* $(1,-3)$ *and* $(1,3)$ *accordingly; on the irreducible fibers* G *acts without fixed points;*

(v): *there is another pencil of conics*

$$\Pi_1 = \{u_0(x-y)+u_1(y-z)+u_2(z-x) = 0;\ u_0+u_1+u_2 = 0,\ (u_0,u_1,u_2) \in \mathbb{P}^2\}$$

passing through $P_1, P_{-1} \in X_2$*; it is an image of* Π *from 1.4.3 under the projection* $p_2\colon X \dashrightarrow X_2$*; on the base of this pencil* $\mathbb{P}^1 = (u_0 + u_1 + u_2 = 0)$ *as usual acts* S_3 *by a two-dimensional irreducible representation,* $\mathbb{Z}_2$ *acts on each fiber with fixed points* $\Pi_1 \cap C_0$ *and invariant* 2*-section* $\Pi_1 \cap C_1$*.*

There are two reducible fibers G_1 *and* G_2 *on* Π_1 *with intersection points of the components* R_1 *and* R_2 *with coordinates* $(\lambda^{-1}, 1, \lambda)$ *and* $(\lambda^{-1}, \lambda, 1)$ *on the base accordingly.*

All these statements are straightforward by analyzing the equations and the action of G.

§2. Proof of the main result

2.1.

The idea of the proof of Proposition 1.1.1 is the same as in [Isk1]. We have to show that there is no G-equivariant birational map between the G-surfaces X and $Y = \mathbb{P}^2$ defined in section 1.3. According to the general theory (see [Isk3]), each G-equivariant birational map can be decomposed into a sequence of G-equivariant elementary birational maps—links, via the G-equivariant Sarkisov program. The classification of such links was obtained in the algebraic case for G a finite Galois group acting on smooth k-minimal rational surfaces over a perfect field k (see [Isk3]).

However according to the general concept of rational G-surfaces (see [Ma1], [Isk2]) this classification may be applied to the geometrical case, *i. e.* to the case of action of a finite group G on smooth projective minimal rational G-surfaces. Their classification is also known (see [Isk2] and G-equivariant 2-dimensional Mori theory [Mo]).

Some differences between geometrical and algebraical cases are concerned with the description of G-orbits. For example, in the geometrical case (with non-trivial G-action) G-orbits are closed proper subset. In the algebraical case for del Pezzo surfaces, say, of large degrees, the existence of one k-point (orbit of degree 1) implies the density everywhere of such k-points in the Zariski topology.

The decomposition theorem for a birational G-map to a sequence of links is known (see [Isk3], 2.5) and all links are determined by their centers, *i. e.* G-orbits of dimension 0, all points of which are in the general position on the corresponding G-surface. So for the proof of our main result we need the following

a) Choose from the general classification (see [Isk3], 2.6) the links which are concerned with our case. Indicate minimal G-surfaces on which these links act (as G-equivariant birational maps) starting from X.

b) Classify all zero-dimensional G-orbits of length $d < \deg X_i = K_{X_i}^2$ on all of such surfaces X_i, and choose those whose points are in the general position (this is necessary condition for the existence of links).

Recall that the points $x_1, x_2, \ldots, x_d \in X$ are in the general position, if

1) X is del Pezzo surface, $\sigma \colon X' \to X$ is the simultaneous blow up of all $x_1, \ldots, x_d$ and X' is also del Pezzo surface, or

2) $\pi \colon X \to \mathbb{P}^1$ is a conic bundle, then these points do not lie on the degenerated fibers and any two of $x_1, \ldots, x_d$ do not lie on the same fiber; this is necessary and sufficient condition for the existence of link with center in $x_1, \ldots, x_d$, *i. e.* the birational transform

$$\begin{array}{ccc} X & \dashrightarrow & X', \\ & \searrow \quad \swarrow & \\ & \mathbb{P}^1 & \end{array}$$

which is a blow up of $x_1, \ldots, x_d$ and a contraction of strict transforms of the fibers on which they lie.

2.2.

Suppose that there exists a G-equivariant birational map $\chi \colon X \dashrightarrow Y$. Then χ can be decomposed into a composition of links. According to the particular algorithm of decomposition and considering all possibilities for links we will show that there is no link whose image is $\mathbb{P}^2$, *i. e.* a contradiction with the decomposition theorem. Thus such a χ can not exist. This means that two embeddings of G to thc Cremona group, defined in (I) and (II), section 1.1, are not conjugate in $Cr_2(\mathbb{C})$.

By the mentioned algorithm, choose a very ample G-invariant linear system $\mathcal{H}' = \mathcal{H}_Y$ on Y, for example, $\mathcal{H}' = |-K_Y|$. Let $\mathcal{H} = \mathcal{H}_X = \chi_*^{-1}\mathcal{H}'$ be its strict transform on X. By the definition of strict transform, the linear system $\mathcal{H}$ has no fixed components and $\dim \mathcal{H} = \dim \mathcal{H}'$ ($\mathcal{H}$ has base points if χ is not morphism). $\mathrm{Pic}^G(X) = \mathbb{Z}(-K_X)$, so $\mathcal{H} \sim -aK_X$, $a \in \mathbb{Z}_{>0}$.

The map $\chi\colon X \to Y$ is not isomorphism. So by the G-equivariant Noether inequality (see [Isk3], 2.4) the linear system $\mathcal{H}$ has a maximal singularity which is a zero-dimensional G-orbit $x \subset X$ of length $d = d(x)$ and multiplicity $mult_x\mathcal{H} = r = r(x) > a$ at all points $x_i \in x$. We have $\mathcal{H}^2 = a^2K_X^2 > r^2d$ ($\mathcal{H}$ is a linear system without fixed components passing through all points of G-orbits x with multiplicity r and determining a birational map) so $d < K_X^2$. In our case $d < K_X^2 = 6$. The length of the orbit is divided by the order of the group. So the maximal singularities giving rise to our links in the decomposition can be only G-orbits of length $d = 1$, 2, 3 or 4. Such orbits may be found in Lemma 1.4.1.

2.3.

Now for every such G-orbit we have to verify whether its points are in the general position or not and if yes, then choose from the classification ([Isk3], 2.6) appropriate link $\Phi\colon X \dashrightarrow X_1$ (do not confuse with X_1 from paragraph 1.5). The decomposition algorithm is going as follows. The map $\chi_1\colon X_1 \dashrightarrow Y$ in the composition $\chi = \chi_1 \circ \Phi$ is given by the linear system $\mathcal{H}_1 = \Phi_*(\mathcal{H}) \sim -a_1K_{X_1}$ with $a_1 < a$. Since $a \in \mathbb{Z}_{>0}$ the decomposition process must be terminated on an isomorphism after a finite number of steps. We begin to find links. There is no G-orbit of length 4 on X (see 1.4.1). So let's start from $d = 3$.

The case $d = 3$. There is one such orbit (Q_1, Q_2, Q_3). It can be shown that in this case there is no such an "untwisting" link Φ. Indeed the points Q_1, Q_2, Q_3 on X are not in the general position. On the blow up of these three points $\sigma\colon X' \to X$ strict transforms E_x', E_y', E_z' of conics E_x, E_y, E_z (see 1.4.2) are G-invariant triple of (-2)-curves, so $-K_X$ is not ample $(-K_X \cdot E_x' = 0)$.

(As we have seen in paragraph 1.5, the projection $p_1\colon X \dashrightarrow X_1$ from the plane $\langle Q_1, Q_2, Q_3\rangle \subset \mathbb{P}^6 \dashrightarrow \mathbb{P}^3$ blows up these three points and contracts E_x, E_y, E_z to the ordinary double points on X_1.)

Notice that in [Isk1] there was an inaccuracy claiming that there is such a link $\Phi_{6,3}$. The reason was that we did not check the condition of generality (although this does not affect to the final result).

2.3.1. **Remark** The fact that the link $\Phi_{6,3}$ does not exists can be seen easily. For the existence of a link with center in the maximal singularity $x \subset X$ it is necessary and sufficient that under the blow up $\sigma\colon X' \to X$ of the cycle x on the surface X' there are two extremal rays, one of which is an exceptional divisor. Since $\mathrm{Pic}^G(X') = \mathbb{Z} \oplus \mathbb{Z}$, the Mori cone is generated by these extremal rays and $-K_{X'}$ is ample by

Kleiman's criterion. The link Φ is nothing but the blow up σ and the extremal contraction of the second ray.

In the classical terms, the non-general position condition for Q_1, Q_2 and Q_3 means that this G-orbit could not be a maximal singularity in the linear system $\mathcal{H}$, because if $r > a$, then E_x, E_y and E_z are fixed components of $\mathcal{H}$. Indeed $\mathcal{H} \subset |-aK_X - rQ_1 - rQ_2 - rQ_3|$, the cycles $E_x - Q_2 - Q_3$, $E_y - Q_1 - Q_3$, $E_z - Q_1 - Q_2$ are effective and must have non-negative intersection with $\mathcal{H}$. But $\mathcal{H} \cdot (E_x - Q_2 - Q_3) = (-aK_x - rQ_1 - rQ_2 - rQ_3) \cdot (E_x - Q_2 - Q_3) = 2a - 2r < 0$. So we would get a contradiction.

2.3.2. **The case $d = 2$.** There is such a unique orbit $x = \{P_1, P_{-1}\}$ (see 1.4.1). The pair P_1, P_{-1} is in the general position on X. Indeed, each of these points does not lie on the (-1)-curves (which lie on the infinity $x_0y_0z_0 = 0$), and there is no curve of genus 0 and degree 2 (with respect to $-K_X$) on which both of these points lie.

There is corresponding link $\Phi = \Phi_{6,2}\colon X \dashrightarrow X_1$ (see [Isk3], 2.6). Find out what is X_1. Let $\sigma\colon X' \to X$ be the blow up of $x = \{P_1, P_{-1}\}$. Then $-K_{X'}$ is ample and $K_{X'}^2 = 4$. The linear system $|-K_{X'}|$ gives a G-equivariant embedding $X' \subset \mathbb{P}^4$ into the del Pezzo surface of degree 4 (*i. e.* into an intersection of two quadrics). The image of the exceptional (-1)-curves are two skew lines whose linear span is a hyperplane $\mathbb{P}^3$. This image is G-invariant and intersects $X' \subset \mathbb{P}^4$ by another G-invariant pair of skew lines which is reducible curve of genus 1 together with the pair that we blow up.

Contracting this residual pair of lines we get a del Pezzo surface $X_1 \subset \mathbb{P}^6$ of degree 6.

In the general algebraic case the surface X_1 may be not isomorphic to X, *i. e.* $\Phi_{6,2}$ is not always a birational involution (but I do not know such an example). However, in our particular case $\Phi_{6,2}$ is a birational involution of Bertini type. Indeed, if $\sigma\colon X'' \to X' \to X$ is a blow up of G-orbit $\{Q_1, Q_2, Q_3\}$ together with $\{P_1, P_{-1}\}$, then $K_{X''}^2 = 1$. But $-K_{X''}$ is not ample (because X'' is a degenerated del Pezzo surface of degree 1 with three (-2)-curves, or, after their contraction, three ordinary double points). Nevertheless, the linear system $|-2K_{X''}|$ gives a contraction of (-2)-curves composed with double covering $X'' \to Q^* \subset \mathbb{P}^3$ of the quadratic cone branched over a fixed point and a curve on Q^* with three ordinary double points.

The involution of this double covering $X'' \to X$ induces the birational involution $\Phi_{6,2}$. It differs from the standard Bertini involution given by the Picard group, because 5 blown up points are not in the general position. The point P and ordinary double points are still fixed under this involution. It also induces the involution on the cubic surface X_1 considered in Remark 1.6.

The birational involution $\Phi_{6,2}$ maps the pencil of rational curves $\Pi = |-K_X - 2P - P_1 - P_{-1}|$ onto itself and interchanges the components of its three fibers: $\Gamma_x \leftrightarrow \Delta_x$, $\Gamma_y \leftrightarrow \Delta_y$, $\Gamma_z \leftrightarrow \Delta_z$ (see 1.4.3).

Apparently one can write the particular equation for the residual G-invariant pair of the curves of genus 0 on X (although we will not need this in the sequel). This 3-linear 3-homogeneous equation must be as follows.

$$H\colon \sum_{g\in G} g((x_1-\lambda x_0)(y_1-\lambda y_0)(z_1-\lambda^2 z_0)+(x-\lambda x_0)(y_1-\lambda^2 y_0)(z_1-\lambda^2 z_0)) = 0.$$

The curve $H \sim -K_X$ is G-invariant. It passes through P_1, P_{-1}, and has them as ordinary double singularities. This means that the curve is reducible.

The link $\Phi_{6,2}\colon X \dashrightarrow X$ maps the linear system $\mathcal{H} \sim -aK_X$ with the maximal singularity $x = \{P_1, P_{-1}\}$ of multiplicity $r = r(x)$ to the linear system $\mathcal{H}_1 \sim -a_1K_X$ with the base cycle $x_1 = x = \{P_1, P_{-1}\}$ with the multiplicity $r_1 < a_1$ (this is not maximal singularity for $\mathcal{H}_1$). The formulas for the coefficients are given in [Isk3], Theorem 2.6, case $K_X^2 = 6$, case d):

$$(8) \qquad a_1 = 2a - r(x) < a, \qquad r_1 = 3a - 2r(x) < a_1.$$

So, $x = \{P_1, P_{-1}\}$ already is not a maximal singularity for $\mathcal{H}_1$. The map $\chi_1 = \chi \circ \Phi^{-1}\colon X \dashrightarrow Y$ defined by $\mathcal{H}_1$ is not isomorphism, so, by Lemma – Noether's inequality, $\mathcal{H}_1$ must have a maximal singularity. Now it can be only the point P of multiplicity $r_1 = r(P) > a_1$.

2.3.3. **The case** $d = 1$. The only fixed point $x_1 = P$. It does not lie on the (-1)-curves, so it is in a general position. The corresponding link $\Phi_{6,1}\colon X \dashrightarrow X_2$, where $X_2 \subset \mathbb{P}^3$ is a smooth quadric with $\operatorname{Pic}^G(X_2) = \frac{1}{2}\mathbb{Z}(-K_X)$ considered in section 1.7, is the birational projection $p_2\colon X \dashrightarrow X_2 \subset \mathbb{P}^3$ from the tangent plane in $\mathbb{P}^6$ to X at P. This link blows up the point P and contracts three conics Γ_x, Γ_y, Γ_z to the G-orbit $\mathcal{A}$ of length 3 on the conic $C_0 \subset X_2$ which is the image of blowing up P. The stabilizator of $P \in X$ is the whole G which acts on the projectivisation of tangent plane to C_0 as S_3 by the standard irreducible 2-dimensional linear representation. This situation was completely investigated in the Lemma 1.8.

There is a G-invariant curve $C_0 \sim -\frac{1}{2}K_{X_2}$, which generates $\operatorname{Pic}^G(X_2)$ on X_2. In the notations from 1.8 one sees that the formulas for the action of $\Phi_{6,1}$ can be given as follows.

$$-K_X \rightarrowtail -\frac{3}{2}K_{X_2} - 2\mathcal{A} \sim 3C_0 - 2\mathcal{A},$$

$$P \rightarrowtail -\frac{1}{2}K_{X_2} - \mathcal{A} \sim C_0 - \mathcal{A}, \tag{9}$$

$$\mathcal{H}_1 \subset |-a_1K_X - r_1P| \rightarrowtail \mathcal{H}_2 \subset |-a_2K_{X_2} - r_2\mathcal{A}|, \quad \text{where}$$

$$a_2 = \frac{3}{2}a_1 - \frac{1}{2}r_1, \qquad r_2 = 2a_1 - r_1.$$

Under our maximality hypothesis of the singularity at P we have $r_0 = mult_P\mathcal{H}_1 > a_1$, so $a_2 < a_1$ and $r_2 < a_2$. This means that the G-orbit $\mathcal{A}$ is not a maximal singularity for $\mathcal{H}_2$.

2.4.

Now we are on the quadric $X_2 \subset \mathbb{P}^3$ with linear system $\mathcal{H}_2 \sim -a_2K_{X_2}$. Denote this system by $\mathcal{H}$ and put $a = a_2$ for simplicity. The map $\chi_2 \colon X_2 \dashrightarrow Y = \mathbb{P}^2$ is not isomorphism, so, by Noether inequality, there is a maximal singularity, G-orbit x of multiplicity $r = r(x) > a$ and of length $d < K_{X_2}^2 = 8$. $d|12 = |G|$ so the only possibilities for d are $d = 1, 2, 3, 4, 6$. We can exclude the cases $d = 1$ and $d = 4$ by Lemma 1.8. It remains to consider the cases $d = 2, 3, 6$.

2.4.1. **The case** $d = 2$. There are two G-orbits $\{R_1, R_2\}$ and $\{P_1, P_{-1}\}$ (see 1.8). First we start from $x = \{R_1, R_2\} = C_0 \cap C_1$ of multiplicity $r = r(x) > a$. Such R_1, R_2 do not lie on one line on the quadric X_2, because each line intersects C_0 at a unique point (C_0 is the hyperplane section). So R_1 and R_2 are in the general position. There exists a link $\Phi_{8,2}$ (see [Isk3], 2.6, the case $K_X^2 = 8$, a), $d = 2$), which is a blow up $X_2 \dashrightarrow X_1$ of the points $\{R_1, R_2\}$. There is a G-invariant structure of conic bundle $\pi \colon X_1 \to \mathbb{P}^1$ on X_1, which is given by the pencil Π_0 from 1.8 (iv). Let $\mathcal{H}_1$ be the strict transform on X_1 of linear system $\mathcal{H}$. Then $\mathcal{H}_1 \sim -a_1K_{X_1} + b_1f_1$, where $f_1 \cong C_0'$ is a G-invariant fiber. We have

$$K_{X_1}^2 = 6, \qquad -K_{X_1} = E + C_0' + C_1' \sim E + 2f_1,$$

where E is an exceptional divisor on $X_1 \to X_2$, *i. e.* a G-invariant pair of (-1)-curves, and C_0', C_1' are strict transforms of C_0, C_1. By [Isk3], 2.6

$$a_1 = 2a - r(x) < a, \tag{10}$$

$$b_1 = 2(r(x) - a).$$

From (10) one can see that $a_1 < a$ and $b_1 > 0$. By the Noether inequality (because $X_1 \not\cong Y$) there is a maximal singularity in $\mathcal{H}_1$.

2.4.2. **Lemma** *There is no zero-dimensional G-orbit x_1 of length d_1 on X_1, whose points are in the general position in the sense of* 2.1 2), *i. e. whose points do not lie on the degenerated fibers and no two points among them lie on the same fiber of morphism $\pi\colon X_1 \to \mathbb{P}^1$ (this is a condition of existence of "untwisting" link for $\mathcal{H}_1$ on the conic bundle with $b_1 \geq 0$).*

Proof. By Lemma 1.8 (iv), a cyclic group of order 2 acts on $\mathbb{P}^1$, so for the generality condition the length d_1 must be less or equal than 2. There are no G-fixed points on X_1 (as on X_2). The orbit $\{P_1, P_{-1}\}$ of length 2 by Lemma 1.8 lies on the degenerated fibers. Also on the non-degenerated fibers S_3 acts without fixed points. Q.E.D.

So the decomposition algorithm in this situation is impossible. This means that $\{R_1, R_2\}$ is not maximal, and we need to come back to the quadric X_2. Consider the other possibilities.

2.4.3. **The case** $d = 2$ with maximal singularity $x = \{P_1, P_{-1}\}$, $r(x) > a$. As in 2.4.1 the pair $\{P_1, P_{-1}\}$ is in the general position and there exists a link $\Phi_{8,2}$ (see [Isk3], Theorem 2.6, the case $K_X^2 = 8$, c), $d = 2$), which is the blow up $X \dashrightarrow X_1$ of P_1, P_{-1}. There is a structure of conic bundle $\pi\colon X_1 \to \mathbb{P}^1$ on X_1 that is given by the pencil Π_1 from 1.8 (v).

The difference from the previous case is the following. The action of $G = S_3 \times \mathbb{Z}_2$ on $\pi\colon X_1 \to \mathbb{P}^1$ is given by two commuting actions. S_3 acts on the base $\mathbb{P}^1$ without fixed points and this action can be shifted to X_1 by changing the fibers. $\mathbb{Z}_2$ acts on $X_1 \to \mathbb{P}^1$ only on the fibers. The fixed points in this case are the intersection points of the fibers with the curve $C_0' \subset X_1$, which is the pre-image of the curve $C_0 \subset X_2$. So $\mathbb{Z}_2$ acts as the classical De Jonquieres involution. The fixed curve C_0' determines $\pi\colon X_1 \to \mathbb{P}^1$. Indeed the factor by this involution is ruled surface $\mathbb{F}_1 \to \mathbb{P}^1$ branched over C_0'. The pair of the (-1)-curves E covers the exceptional section $S_1 \subset \mathbb{F}_1$.

Notice that there are no fixed points on the base $\mathbb{P}^1$ (so there is no G-invariant fiber as in the previous case). But the class of the fiber $[f_1]$ in $\operatorname{Pic}^G(X)$ is G-invariant, so the action of this link is the same as in (10).

By the Noether inequality there must be a maximal singularity in the linear system $\mathcal{H} \sim -a_1K_{X_1} + b_1f_1$ which is G-orbit $x_1 \subset X_1$ of multiplicity $r_1 = r(x_1) > a_1$, whose points are in the general position in the sense of 2.4.2.

There are no G-fixed points on X_1 and the only length 2 G-invariant orbit $\{R_1, R_2\}$ lies in the degenerated fibers (see 1.8 (v)).

Two G-orbits of length 3, $\mathcal{A}$ and $\mathcal{B}$, satisfy the generality condition. Both of them lie on C'_0 and they are interchanged under the involution of double covering $\pi|_{C'_0}: C'_0 \to \mathbb{P}^1$ (do not confuse with the involution of the action of $\mathbb{Z}_2$).

There are also G-orbits of length 6. By the generality condition they lie on C'_0, one point on the fiber. The other intersection points of C'_0 with fibers are complementary G-orbit of length 6.

There is an "untwisting" link concerned with every maximal singularity $x_1 \subset X_1$ of length 3 or 6. That is, the elementary transform

(11)
$$\begin{array}{ccc} \Phi\colon X_1 & \overset{\mathcal{E}_{x_1}}{\dashrightarrow} & X'_1. \\ & \searrow \quad \swarrow & \\ & \mathbb{P}^1 & \end{array}$$

Recall that the elementary transform blows up the points of orbit $x_1 \subset X_1$ and contracts the strict transforms of the fibers on which these points lie. In the general case an elementary transform is not a birational involution. But in our case it is so. Indeed the transform $\mathcal{E}_{X_1}$ induces the isomorphism on the curve C'_0 and, as noted above, C'_0 completely determines $\pi: X_1 \to \mathbb{P}^1$. So $X'_1 = X_1$ in (11).

The birational involution $\Phi = \mathcal{E}_x$ acts by the following formulas from [Isk3], 2.6.

$$\begin{aligned} -K_{X_1} &\mapsto -K_{X_1} + d_1 f_1 - 2x_1, & a'_1 &= a_1, \\ f_1 &\mapsto f_1, & b'_1 &= b_1 + d_1(a_1 - r_1), \\ x_1 &\mapsto d_1 f_1 - x_1, & r'_1 &= 2(r_1 - a_1), \end{aligned}$$

where d_1 is the length of the orbit x_1. As the last formula shows, the multiplicities of maximal singularities decrease. After a finite number of such transforms we can "untwist" all maximal singularities (which means that the multiplicities of the base points in the modified linear system will be not greater then a). By the Noether inequality the coefficient at b_1 at the fiber will be less than 0.

In this situation the link $\Phi_{8,2}^{-1}: X_1 \to X_2$ (the inverse to $\Phi_{8,2}$) contracts G-invariant pair of (-1)-curves E (the sections of the pencil $\pi: X_1 \to \mathbb{P}^1$) to G-orbit of length two $x = \{P_1, P_{-1}\}$. It acts by the following formulas.

(12)
$$-K_{X_1} \mapsto -K_{X_2} - x, \qquad a = a_1 + \frac{2}{3}b_1,$$

$$2f_1 \rightarrowtail -K_{X_2} - 2x, \qquad r(x) = a_1 + b_1.$$

These formulas show that $a < a_1$ and $r(x) < a$, because $b_1 < 0$. It means that the link is untwisted and $x = \{P_1, P_{-1}\}$ already is not a maximal singularity.

2.5.

The last cases are those whose singularities of $\mathcal{H}$ are G-orbits of length 3 or 6. There are only two orbits of length 3: $\mathcal{A}$ and $\mathcal{B}$ (see 1.8). Let us first consider

2.5.1. **The case** $d = 6$. Suppose that there is a maximal singularity, G-orbit x of length 6, all whose points are in the general position and the multiplicity $r = r(x) > a$, where $\mathcal{H} \sim -aK_{X_2}$.

Then there is an untwisting link $\Phi_{8,6}$, the Geiser's involution (see [Isk3], 2.6). Indeed, blowing up x we get a del Pezzo surface of degree 2, on which the classical Geiser's involution acts biregularly. Now apply this involution and contract the blown up divisor. The link $\Phi_{8,6}$ acts by the following formulas.

$$a_1 = 7a - 6r(x) < a,$$

$$r(x_1) = 8a - 7r(x) < a_1.$$

2.5.2. **The case** $d = 3$. If the maximal singularity is $x = \mathcal{A}$, then the link $\Phi_{8,3}$ with center in this singularity is the inverse map $\Phi_{6,3}^{-1}\colon X_2 \dashrightarrow X$ and we are in the situation from which we started. If the maximal singularity is $x = \mathcal{B}$ (we missed this case in [Isk1]), then the link $\Phi'_{8,3}\colon X_2 \dashrightarrow X'$ acts from X_2 to del Pezzo surface of degree 6 X'. But $X' \cong X$, because the birational involution on X with the center in $\{P_1, P_{-1}\}$ considered in 2.3.2 interchanges $\{\Gamma_2, \Gamma_y, \Gamma_z\} \leftrightarrow \{\Delta_x, \Delta_y, \Delta_z\}$. Thus it interchanges the cycles $\mathcal{A}$ and $\mathcal{B}$ on C_0 that correspond to these directions in the tangent plane to $P \in X$.

So we completed our consideration of all possibilities but have not found any G-equivariant G-map $X \dashrightarrow Y = \mathbb{P}^2$. So there is no such a map, which prove the Proposition 1.1.1.

2.6. Commentary

In the proof we determined that the smooth relative minimal G-models on which there exists G-equivariant map from X are only the quadric X_2 or the conic bundle with two different actions of G. It is interesting to find by this method all G-surfaces on which $Y = \mathbb{P}^2$ may be G-equivariant mapped.

The following fact is also of interest. If we consider only S_3 instead of G with the similar actions on X and Y, then these surfaces are S_3-birational equivalent. Indeed in this case there are three fixed points

P, P_1 and P_{-1} on X. Therefore there are two fixed points P_1, P_{-1} on the quadric X_2. The stereographic projection from one of them is the required G-equivariant birational isomorphism, so both of these embeddings of S_3 in $Cr_2(\mathbb{C})$ are conjugate. On the other hand, the involution $(x, y) \to (x^{-1}, y^{-1})$ of course is conjugate to the linear one.

Acknowledgement. I would like to thank the organizers of the International Symposium "Algebraic Geometry in East Asia. II" held at Hanoi October 10-14, 2005 for the kind invitation and support.

References

[Isk1] V. A. Iskovskikh, Two non-conjugate embeddings of $S_3 \times \mathbb{Z}_2$ into the Cremona group, Proc. Steklov Inst. of Math., **241** (2003), 93–97.

[Isk2] V. A. Iskovskikh, Minimal models of rational surfaces over arbitrary fields, Math. USSR–Izv., **14** (1980), 17–39.

[Isk3] V. A. Iskovskikh, Factorization of birational maps of rational surfaces from the viewpoint of Mori theory, Russian Math. Surveys, **51** (1996), 585–652.

[LPR] N. Lemire, V. L. Popov and Z. Reichstein, Caley groups, J. Amer. Math. Soc., **19** (2006), 921–967.

[Ma1] Yu. I. Manin, Rational surfaces over perfect fields II, Math. USSR-Sb., **1** (1968), 141–168.

[Ma2] Yu. I. Manin, Cubic forms: Algebra, geometry, arithmetic, North-Holland Math. Library, **4**, Amsterdam, 1974, 292 p.

[Mo] S. Mori, Threefolds whose canonical bundles are not numerically effective, Ann. of Math. (2), **116** (1982), 133–176.

Steklov Institute of Mathematics
8 Gubkin street, Moscow 119991
RUSSIA
E-mail address: iskovsk@mi.ras.ru

Advanced Studies in Pure Mathematics 50, 2008
Algebraic Geometry in East Asia — Hanoi 2005
pp. 269–287

On the Castelnuovo-Severi inequality for a double covering

Changho Keem[1] **and Akira Ohbuchi**[2]

§0. Introduction, motivation and the results

Let C be a smooth projective irreducible complex algebraic curve of genus $g \geq 2$. We denote g^1_d by a 1-dimensional possibly incomplete linear system of degree d on C. For any $d \geq g+1$, every curve C of genus g has a base point free g^1_d which may be taken as a general pencil of a general element in $W^{d-g}_d(C) = J(C)$. If C is a hyperelliptic curve with the hyperelliptic pencil g^1_2, it is well-known that any base point free pencil of degree $d \leq g$ is a subsystem of the complete rg^1_2 where $r = \frac{d}{2}$; cf. [1, p.109]. In particular, the only base point free and complete pencil on a hyperelliptic curve is the g^1_2. On the other hand, a non-hyperelliptic curve C has a base point free and complete pencil of degree g, by taking off $g-2$ general points from the very ample canonical linear system $|K_C|$.

Furthermore, a theorem of Harris asserts that any non-hyperelliptic curve of genus g has a base point free and complete pencil of degree $g-1$; cf. [1, p.372]. However, this seemingly simple fact requires a proof which is somewhat involved. Especially, in case C is a bi-elliptic curve, one needs to show that the variety $W^1_{g-1}(C)$ consisting of special pencils of degree $g-1$ is reducible by using enumerative methods; see also [3, Proposition 3.3],[6, Proposition 2.5] for the other proofs concerning the existence of a base point free and complete pencil g^1_{g-1} on a bi-elliptic curve. At this point, it is worthwhile to recall the following classical result known as Castelnuovo-Severi inequality.

Received August 17, 2006.
Revised January 12, 2007.
[1] Supported by Korea Research Foundation (Grant # 2001-DS0003).
[2] Supported by Brain Pool Program of KOFST and partially supported by Grant-in-Aid for Scientific Research (15540035), JSPS.

Proposition 0.1 (Castelnuovo-Severi inequality, [1, p.366]**).** *Let C, B_1, B_2 be curves of respective genera g, g_1, g_2. Assume that*

$$\pi_i : C \to B_i,\ \ i = 1, 2,$$

is a d_i-sheeted mapping such that

$$(\pi_1, \pi_2) : C \to B_1 \times B_2$$

is birational to its image. Then

$$g \leq (d_1 - 1)(d_2 - 1) + d_1 g_1 + d_2 g_2.$$

As an easy application of Proposition 0.1, we make a note of the following remarks.

Remark 0.2. (i) *Let C be a hyperelliptic curve with the 2-sheeted covering $\pi_1 : C \to \mathbb{P}^1$ induced by the unique hyperelliptic pencil g^1_2. Let g^1_d be a base point free pencil not composed with the g^1_2. In other words, g^1_d induces a covering $\pi_2 : C \to \mathbb{P}^1$ of degree d such that $(\pi_1, \pi_2) : C \to \mathbb{P}^1 \times \mathbb{P}^1$ is birational to its image. By the Castelnuovo-Severi inequality, we have $g \leq d - 1$. This recovers the fact that any base point free pencil of degree $d \leq g$ is a subsystem of a multiple of the hyperelliptic pencil, which was mentioned earlier.*
(ii) *More generally, let $\pi : C \to E$ be a double covering of a smooth curve E of genus h. Let g^1_d be a base point free pencil of degree d not composed with the involution determined by π (composed with π for short). Again by the Castelnuovo-Severi inequality, we have*

$$d \geq g - 2h + 1. \tag{0.1}$$

Therefore it follows that any base point free pencil of degree $d \leq g - 2h$ is of the the form $\pi^* g^1_e$ for some g^1_e on E.

In case $h = 1$, the theorem of Harris quoted earlier indicates that the inequality (0.1) is indeed sharp on a bi-elliptic curve. For the case $h = 2$, it only has been known that there exists a base point free and complete pencil of degree $g - 2$ not composed with the double covering under somewhat unsatisfactory genus assumption $g \geq 11$, whereas the existence of a base point free and complete g^1_{g-3} has remained open; cf. [2, Proposition 2.6]. Therefore, we would like to raise the following questions regarding the sharpness of the inequality (0.1).

Question 0.3. (i) *Let $\pi : C \to E$ be double covering of a smooth curve E of genus h. Does there exist a base point free pencil of degree $g-2h+1$ not composed with π ?*
(ii) *Let $\pi : C \to E$ be double covering of a smooth curve E of genus h. Does there exists a base point free pencil of degree d not composed with π for every $d \geq g-2h+1$?*
(iii) *What is the optimal range for the genus g of the double covering with respect to the genus h of the base curve E ensuring affirmative answers to the questions above ? Or, find examples of double coverings for which questions* (i) *or* (ii) *fail.*

We may even pose a more naive question: Given a smooth curve E of genus h, does there exist a smooth double covering $C \xrightarrow{\pi} E$ of genus g possessing a base point free pencil of degree $g-2h+1$ not composed with π ? However this turns out to be relatively easy to answer.

Example 0.4. *Given a smooth curve E of genus $h \geq 0$ and an integer $g \geq 4h$, let $C \subset \mathbb{P}^1 \times E$ be a general divisor linearly equivalent to $D := 2p \times E + \mathbb{P}^1 \times N$ with* $\deg N = g-2h+1$ *and $p \in \mathbb{P}^1$. By the condition $g \geq 4h$, D is very ample and hence C is a smooth curve of genus g by the adjunction formula. Furthermore, the two projection maps of $E \times \mathbb{P}^1$ to E and $\mathbb{P}^1$ restricted to C correspond to a degree two morphism $C \xrightarrow{\pi} E$ and a base point free and complete pencil g^1_{g-2h+1} not composed with π.*

Motivated by Example 0.4, the main result of this paper is the following theorem which provides an affirmative answer to the Question 0.3 (i).

Theorem A. *Let C be a curve of genus g which admits a double covering $\pi : C \longrightarrow E$ with $g(E) = h \geq 0$ and $g \geq 4h$. Then C has a base point free and complete g^1_{g-2h+1} not composed with π.*

By using Theorem A, we are also able to answer the Question 0.3 (ii) in the affirmative.

Theorem B. *Let C be a curve of genus g which admits a double covering $\pi : C \longrightarrow E$ with $g(E) = h$ and $g \geq 8h-4$. Then there exists a base point free pencil of degree d not composed with π for any degree d with $d \geq g-2h+1$.*

The organization of this paper is as follows. In §1, after giving a general theory between a double covering $\pi : C \to E$ and an embedding of C into a ruled surface (see Proposition 1.5), we prove necessary and sufficient conditions for the existence of base point free and complete g^1_{g-2h+1} (see Theorem 1.1). This can be done by observing the relationship between the above associated embedding of C into a ruled surface and the embedding $(\pi, g^1_{g-2h+1}) : C \hookrightarrow E \times \mathbb{P}^1$ by using elementary transformations. In §2, we prove that the necessary and sufficient condition in §1 for the existence of such g^1_{g-2h+1} holds for any smooth double covering under the numerical assumption $g \geq 4h$ by using Theorem 2.1, thereby proving Theorem A. This will be carried out by using an elementary theory of determinantal varieties. We then proceed to prove Theorem B by the excess linear series argument. In §3, we mainly deal with the Question 0.3 (iii). Specifically, we show that the numerical assumption $g \geq 4h$ in Theorem A is the best possible one by constructing an example of a double covering of $g = 4h - 1$ without a base point free and complete g^1_{g-2h+1}. We also exhibit an example of a double covering with a base point free and complete g^1_{g-2h+1} under the same numerical condition $g = 4h - 1$. Throughout we use the same notations and conventions as in [1].

§1. Curves on ruled surfaces

In this section we study double coverings on a ruled surface. In particular we collect and develop some methods realizing a double covering with a base point free pencil of particular degree as a smooth divisor on a ruled surface. The goal of this section is to prove the following result:

Theorem 1.1. *Let $C = \mathbf{Spec}(\mathcal{O}_E \oplus \mathcal{O}_E(-N)) \xrightarrow{\pi} E$ be a smooth double covering and let $\iota : C \to \mathbb{P}(\mathcal{O}_E \oplus \mathcal{O}_E(-N))$ be an embedding associated with π such that $\rho_N \iota = \pi$. Then the following four conditions are equivalent:*

1) *C has a base point free and complete g^1_{g-2h+1} which is not composed with π, and $\pi_* D \in |N|$ for $D \in g^1_{g-2h+1}$.*

2) *There is a section $H \in |T_N + \rho_N^*(N)|$ such that $H|_{\iota(C)} = D_1 + D_2$ with $\pi_* D_1, \pi_* D_2 \in |N|$ and $D_1 \sim \sigma^* D_2$.*

3) *There is a divisor $H \in |T_N + \rho_N^* N|$ such that $H \cap T_N = \emptyset$ satisfying $H|_{\iota(C)} = D_1 + D_2$ with $\pi_* D_1, \pi_* D_2 \in |N|$.*

4) *There is a divisor $A \in |\pi^* N| \setminus \{\pi^* L \mid L \in |N|\}$ such that $\pi_* A = N_1 + N_2$ and $N_1, N_2 \in |N|$.*

Let M be an effective divisor on a smooth projective curve E of genus h and let $\mathcal{O}_E(M)$ be the line bundle associated with M. Throughout this paper, we denote the structure morphism of the ruled surface $\mathbb{P}(\mathcal{O}_E \oplus \mathcal{O}_E(-M))$ by

$$\rho_M : \mathbb{P}(\mathcal{O}_E \oplus \mathcal{O}_E(-M)) \to E$$

and its minimal section by T_M; by the minimal section, we always mean the section of minimal degree on a normalized ruled surface. For $P \in \mathbb{P}(\mathcal{O}_E \oplus \mathcal{O}_E(-M))$, let F be the fibre over $p = \rho_M(P)$. In the blowing-up

$$\eta : S_P \to \mathbb{P}(\mathcal{O}_E \oplus \mathcal{O}_E(-M))$$

of the ruled surface $\mathbb{P}(\mathcal{O}_E \oplus \mathcal{O}_E(-M))$ at P, let e be the exceptional divisor of η, f the proper transform of F and

$$\tau : S_P \to S'$$

the contraction of f. We put $P' = \tau(f) \in S'$. Since S' is an elementary transformation of $\mathbb{P}(\mathcal{O}_E \oplus \mathcal{O}_E(-M))$ with center P, S' is a ruled surface over E; cf. [4, p.416]. We define ρ' as its ruling $S' \to E$.

We choose a section $H_M \in |T_M + \rho_M^* M|$ and hence $H_M \cap T_M = \emptyset$. Let $\widetilde{T_M}$ and $\widetilde{H_M}$ be the proper transforms of T_M and H_M on S_P respectively, and set $T' = \tau(\widetilde{T_M})$, $H' = \tau(\widetilde{H_M})$. Since $H_M \cap T_M = \emptyset$, we have $H' \cap T' = \emptyset$ for $P \in T_M \cup H_M$ which implies

$$S' \cong \mathbb{P}(\mathcal{O}_E \oplus \mathcal{O}_E(-M'))$$

for some $M' \in \mathrm{Div}(E)$; cf. [4, p.383]. Let C_0 be an irreducible curve on $\mathbb{P}(\mathcal{O}_E \oplus \mathcal{O}_E(-M))$ with $C_0 \sim 2T_M + \rho_M^*(Z)$ for some $Z \in \mathrm{Div}(E)$, let $\phi : C \to C_0$ be its normalization, let $\widetilde{C_0}$ be the proper transform of C_0 on S_P, let $C_0' = \tau(\widetilde{C_0})$ and let $\phi' : C \to C_0'$ be its normalization. Let $\pi = \rho_M \phi$. Note that $\pi = \rho' \phi'$ and $\pi : C \to E$ is a double covering and we denote the associated involution by σ.

From now, we assume that $P \in T_M \cup H_M$. First, we consider the case, the point $P \in T_M \cup H_M$ is a smooth point of C_0. By $(\widetilde{C_0} + e.f + e) = (C_0.F) = 2$ and $(\widetilde{C_0}.e) = 1$, we have $(\widetilde{C_0}.f) = 1$. Hence C_0' is non-singular at P'. Therefore $C_0 \cong \widetilde{C_0} \cong C_0'$, when C is non-singular.

Lemma 1.2. (i) *T' is a minimal section $T_{M'}$ on $\mathbb{P}(\mathcal{O}_E \oplus \mathcal{O}_E(-M'))$.*
(ii) *In case $P \in H_M$ and* $\deg(M - p) \geq 0$, *we have*

$$\mathbb{P}(\mathcal{O}_E \oplus \mathcal{O}_E(-M')) \cong \mathbb{P}(\mathcal{O}_E \oplus \mathcal{O}_E(-(M - p))),$$

$H' \sim T' + {\rho_{M-p}}^*(M-p), C_0' \sim 2T' + {\rho_{M-p}}^*(Z-p)$ and $\phi'^* H' = \phi^* H_M - P$.

(iii) *In case $P \in T_M$, we have*

$$\mathbb{P}(\mathcal{O}_E \oplus \mathcal{O}_E(-M')) \cong \mathbb{P}(\mathcal{O}_E \oplus \mathcal{O}_E(-(M + p))),$$

$H' \sim T' + {\rho_{M+p}}^*(M + p)$, $C_0' \sim 2T' + {\rho_{M+p}}^*(Z + p)$ and $\phi'^* H' = \phi^* H_M + \sigma^* P$.

Proof. We only give a proof for the case $P \in H_M$ and $\deg(M-p) \geq 0$; the case $P \in T_M$ is similar. Since $H_M|_{C_0} = \widetilde{H_M} + e|_{\widetilde{C_0}}$ and $(\widetilde{C_0}.e) = 1$, we have

$$H'|_{C_0'} = \widetilde{H_M}|_{\widetilde{C_0}},\ e|_{\widetilde{C_0}} = P,\ H'|_{C_0'} = H_M|_{C_0} - P.$$

Now we show that T' is a minimal section. Since $P \notin T_M$,

$$\eta^* T_M = \widetilde{T_M} \cong T_M \cong T'.$$

Since T_M is a (minimal) section, $T_M \cong E$ and $\mathcal{O}_{T_M}(T_M) \cong \mathcal{O}_E(-M)$. Therefore we have

$$\mathcal{O}_E(-M) \cong \mathcal{O}_{T_M}(T_M) \cong \mathcal{O}_{\widetilde{T_M}}(\eta^* T_M) = \mathcal{O}_{\widetilde{T_M}}(\widetilde{T_M}).$$

Since $\mathcal{O}_{T'}(T') \cong \mathcal{O}_{\widetilde{T_M}}(\tau^* T') = \mathcal{O}_{\widetilde{T_M}}(\widetilde{T_M} + f)$,

$$\mathcal{O}_{T'}(T') \cong \mathcal{O}_E(-(M-p)). \tag{1.2.1}$$

To see $\mathbb{P}(\mathcal{O} \oplus \mathcal{O}(-M')) \cong \mathbb{P}(\mathcal{O} \oplus \mathcal{O}(-(M-p)))$, we argue as follows. If $(T'.T_{M'}) < 0$, $T' = T_{M'}$ which implies $M' \sim -T'|_{T'} \sim M-p$ and we are done for this case. Therefore we may assume $(T'.T_{M'}) \geq 0$. Let $T' \sim aT_{M'} + \rho_{M'}^* B$ with $\deg B = b$ and let $(T_{M'}^2) = -n'$. By the assumption $\deg(M-p) \geq 0$, we have $(T'^2) = a(2b - an') \leq 0$. Since T' is a section, $a = 1$ and $b \geq an'$ by $(T'.T_{M'}) \geq 0$, which implies $b = 0$. On the other hand, since T' is effective

$$\{0\} \neq \Gamma(S', \mathcal{O}(T')) \cong \Gamma(E, \mathcal{O}_E(B) \oplus \mathcal{O}_E(B - M')) \tag{1.2.2}$$

by projection formula. When $M' > 0$, $\deg(B - M') = \deg(-M') = -n' < 0$ implying $B \sim 0$ and hence $T' = T_{M'}$. Therefore it follows that $M' \sim -T'|_{T'} \sim M - p$ by (1.2.1). When $M' = 0$, we have either $B \sim 0$ or $M' \sim B$ by (1.2.2). Since $M' = \deg M' = 0$, $T_{M'} + \rho_{M'}^* M'$ is linearly equivalent to a minimal section $T_{-M'} \subset \mathbb{P}(\mathcal{O}_E \oplus \mathcal{O}_E(M')) \cong \mathbb{P}(\mathcal{O}_E \oplus \mathcal{O}_E(-M'))$. Therefore we have either $T' = T_{M'}$ when $B \sim 0$ or $T' = T_{-M'}$ when $M' \sim B$. In either cases, T' is a minimal section satisfying (1.2.1). Therefore

$$\mathbb{P}(\mathcal{O}_E \oplus \mathcal{O}_E(-M')) \cong \mathbb{P}(\mathcal{O}_E \oplus \mathcal{O}_E(-(M-p))) \text{ and } T' = T_{M-p}.$$

Now we prove $H' \sim T' + \rho_{M-p}{}^*(M-p)$. Let $H' \sim T' + \rho_{M-p}{}^*(G)$ for some $G \in \mathrm{Div}E$. Since $H' \cap T' = \emptyset$, $H'|_{T'} \sim 0$ and hence $T'|_{T'} + G \sim 0$ which implies $G \sim -T'|_{T'} \sim M - p$. Hence $H' \sim T' + \rho_{M-p}{}^*(M-p)$.

Finally we prove $C'_0 \sim 2T' + \rho_{M-p}{}^*(Z-p)$. Since C'_0 is smooth and $\tau^* C'_0 \sim \widetilde{C_0} + f$,

$$\tau^* C'_0 \sim \eta^* C_0 - e + f. \tag{1.2.3}$$

Since $\eta^* C_0 \sim 2\widetilde{T_M} + \eta^* \rho_M^*(Z - p + p) \sim 2\widetilde{T_M} + \eta^* \rho_M^*(Z-p) + (e+f)$,

$$\begin{aligned} \tau^* C'_0 &\sim 2(\widetilde{T_M} + f) + \tau^* \rho_{M-p}{}^*(Z-p) \\ &= 2\tau^* T' + \tau^* \rho_{M-p}^*(Z-p) = \tau^*(2T' + \rho_{M-p}^*(Z-p)) \end{aligned}$$

by (1.2.3). Hence $C'_0 \sim 2T' + \rho_{M-p}{}^*(Z-p)$.

Q.E.D.

Next, we consider the case, the point $P \in T_M \cup H_M$ is a singular point of C_0. Let $F = \rho_M^* p$ where $p = \rho_M(P)$. Since $\rho_M \phi = \pi : C \to E$ is a double covering, $\phi^* F = \pi^* p$ which means P is a double point or a cusp.

Lemma 1.3. (i) *In case* $P \in H_M$ *and* $\deg(M-p) \geq 0$, *we have*

$$S' \cong \mathbb{P}(\mathcal{O}_E \oplus \mathcal{O}_E(-(M-p))),$$

$H' \sim T' + \rho_{M-p}{}^*(M-p), C'_0 \sim 2T'' + \rho_{M-p}{}^*(Z)$ and $\phi'^* H' = \phi^* H_M - \pi^* p$.

(ii) *In case* $P \in T_M$, *we have*

$$S' \cong \mathbb{P}(\mathcal{O}_E \oplus \mathcal{O}_E(-(M+p))),$$

$H' \sim T' + \rho_{M+p}{}^*(M+p), C'_0 \sim 2T' + \rho_{M+p}{}^*(Z)$ and $\phi^* H' = \phi' * H_M + \pi^* p$.

Proof. We only give a proof for the case $P \in H_M$ and $\deg(M-2p) \geq 0$; the case $P \in T_M$ is similar. By Lemma 1.2, $S'' \cong \mathbb{P}(\mathcal{O}_E \oplus \mathcal{O}_E(-(M-2p)))$, $H'' \sim T'' + \rho_{M-2p}{}^*(M-2p)$. We now prove $C''_0 \sim 2T'' + \rho_{M-2p}{}^*(Z-2p)$. Since $P \in C_0$ is a double point, $\eta_1^* C_0 \sim \widetilde{C_0} + 2e_1$. Therefore $2 = (C_0.F) = (\widetilde{C_0} + 2e_1.e_1 + f_1)$ which implies $(\widetilde{C_0}.e_1) = 2$. Hence $(\widetilde{C_0}.f_1) = 0$, so we have $\widetilde{C_0} = \tau_1^* C'_0$ because τ_1 is a contraction of f_1. Since $1 = (T_M.F) = (\widetilde{T_M} + e_1.e_1 + f_1)$ and $(\widetilde{T_M}.e_1) = 1$, $(\widetilde{T_M}.f_1) = 0$. Therefore $\tau_1^* T' = \widetilde{T_M}$ which implies

$$\tau_1^* C'_0 \sim \eta_1^*(2T_M + \rho_M^* Z) - 2e_1 \sim 2\tau_1^* T' + \eta_1^* \rho_M^* Z.$$

Since $\eta_1^* \rho_M^* Z \sim \tau_1^* \rho_1^* Z$, $\tau_1^* C'_0 \sim \tau_1^*(2T' + \rho_1^* Z)$, i.e.

$$C'_0 \sim 2T' + \rho_1^* Z.$$

Q.E.D.

Finally, we consider the case, the point $P \in T_M \cup H_M$ does not lie on C_0.

Lemma 1.4. (i) *In case* $P \in H_M$ *and* $\deg(M-p) \geq 0$, *we have*

$$S' \cong \mathbb{P}(\mathcal{O}_E \oplus \mathcal{O}_E(-(M-p))),$$

$H' \sim T' + {\rho_{M-p}}^*(M-p), C_0' \sim 2T'' + {\rho_{M-p}}^*(Z-2p)$ and $\phi_1^* H' = \phi^* H_M$.

(ii) *In case* $P \in T_M$, *we have*

$$S' \cong \mathbb{P}(\mathcal{O}_E \oplus \mathcal{O}_E(-(M+p))),$$

$H' \sim T' + {\rho_{M+p}}^*(M+p), C_0' \sim 2T' + {\rho_{M+p}}^*(Z+2p)$ and $\phi^* H' = \phi'^* H_M$.

Proof. We only give a proof for the case $P \in H_M$ and $\deg(M-2p) \geq 0$; the case $P \in T_M$ is similar. By Lemma 1.2, $S'' \cong \mathbb{P}(\mathcal{O}_E \oplus \mathcal{O}_E(-(M-2p)))$, $H'' \sim T'' + {\rho_{M-2p}}^*(M-2p)$. We now prove $C_0'' \sim 2T'' + {\rho_{M-2p}}^*(Z-2p)$. Since $P \in C_0$ is a double point, $\eta_1^* C_0 \sim \widetilde{C_0} + 2e_1$. Therefore $2 = (C_0.F) = (\widetilde{C_0} + 2e_1.e_1 + f_1)$ which implies $(\widetilde{C_0}.e_1) = 2$. Hence $(\widetilde{C_0}.f_1) = 0$, so we have $\widetilde{C_0} = \tau_1^* C_0'$ because τ_1 is a contraction of f_1. Since $1 = (T_M.F) = (\widetilde{T_M} + e_1.e_1 + f_1)$ and $(\widetilde{T_M}.e_1) = 1$, $(\widetilde{T_M}.f_1) = 0$. Therefore $\tau_1^* T' = \widetilde{T_M}$ which implies

$$\tau_1^* C_0' \sim \eta_1^*(2T_M + \rho_M^* Z) - 2e_1 \sim 2\tau_1^* T' + \eta_1^* \rho_M^* Z.$$

Since $\eta_1^* \rho_M^* Z \sim \tau_1^* \rho_1^* Z$, $\tau_1^* C_0' \sim \tau_1^*(2T' + \rho_1^* Z)$, i.e.

$$C_0' \sim 2T' + \rho_1^* Z.$$

Q.E.D.

We recall some basics of a double covering of a curve E of genus h; see [5] for a full treatment. For $N \in E_{g-2h+1}$, let R be an effective divisor on E with $\mathcal{O}_E(R) \cong \mathcal{O}_E(2N)$. Given an isomorphism

$$\phi : \mathcal{O}_E(-N)^{\otimes 2} \xrightarrow{\sim} \mathcal{O}_E(-R) \subset \mathcal{O}_E,$$

one defines an $\mathcal{O}_E$-algebra structure on $\mathcal{O}_E \oplus \mathcal{O}_E(-N)$ by

$$(a, b) \cdot (c, d) = (ac + \phi(bd), ad + bc).$$

One then has a double covering $\pi : C = \mathbf{Spec}(\mathcal{O}_E \oplus \mathcal{O}_E(-N)) \to E$ with $\pi_* \mathcal{O}_C \cong \mathcal{O}_E$. The virtual genus of C is g, i.e. $\dim H^1(C, \mathcal{O}_C) = g$. Note that $(a,b) \mapsto (a,-b)$ is an $\mathcal{O}_E$-algebra isomorphism of order 2 which induces an involution $\sigma : C \to C$ over E. Conversely, every double covering over E is of this form. We also recall that a double covering $\pi : C = \mathbf{Spec}(\mathcal{O}_E \oplus \mathcal{O}_E(-N)) \to E$ is an irreducible reduced non-singular curve if and only if R is reduced. Let $\lambda : \pi^*\mathcal{E} \to \mathcal{O}_C$ be the restriction of a natural map $\lambda : \pi^*\pi_*(\mathcal{O}_C) \to \mathcal{O}_C$ to $\pi^*\mathcal{E}$. Since λ is surjective, we have a morphism

$$\iota : C \to \mathbb{P}(\mathcal{E}) = \mathbb{P}(\mathcal{O}_E \oplus \mathcal{O}_E(-N))$$

with $\rho_N \iota = \pi$.

Proposition 1.5. *ι is embedding and $\iota(C) \sim 2(T_N + \rho_N^*(N))$ on $\mathbb{P}(\mathcal{E}) = \mathbb{P}(\mathcal{O}_E \oplus \mathcal{O}_E(-N))$.*

Proof. Let $U = \mathrm{Spec}(A) \subset E$ be an affine open subset and $\mathrm{Spec}B = \pi^{-1}U$. Let $\mathcal{E}|_U = \pi_*(\mathcal{O}_C)|_U = \mathcal{O}_U e_1 \oplus \mathcal{O}_U e_2$ where $e_1 = 1_A$ and $\pi^* e_1 = (e_1, 0)$ is the identity of $B = \Gamma(U, \pi_*\mathcal{O}_C)$. Let $(\pi^{-1}U)_{\pi^* e_1} = \{P \in \pi^{-1}U \mid \pi^* e_1(P) \neq 0\}$. Note that $(\pi^{-1}U)_{\pi^* e_1} = \pi^{-1}U = \mathrm{Spec}(B)$ is affine and the homomorphism $A[e_1, e_2] \to \Gamma(\pi^{-1}(U), \mathcal{O}_C) = B$ defined by $e_i \mapsto \dfrac{\pi^* e_i}{\pi^* e_1} = \pi^* e_i$ $(i = 1, 2)$ is surjective. By [4, p.151 Proposition 7.2], $\iota|_{\pi^{-1}U}$ is embedding and hence ι is embedding. Since $\rho_{N*}\mathcal{O}(T_N) \cong \pi_*\mathcal{O}_C$,

$$\pi^*\rho_{N*}\mathcal{O}(T_N + \rho_N^* N) \cong \pi^*(\rho_{N*}\mathcal{O}(T_N) \otimes \mathcal{O}_E(N)) \cong \pi^*\pi_*(\mathcal{O}_C) \otimes \mathcal{O}_C(\pi^*(N))$$

by the projection formula. Therefore $\lambda \otimes \mathcal{O}_C(\pi^*(N)) : \pi^*\rho_{N*}\mathcal{O}(T_N + \rho_N^*(N)) \to \mathcal{O}_C(\pi^*(N))$ again defines ι. This means $\phi_{|\pi^*(N)|} = \phi_{|T_N + \rho_N^*(N)|}\iota$ where $\phi_{|\pi^*(N)|} : C \to \mathbb{P}(\Gamma(C, \mathcal{O}_C(\pi^*(N))))$ is a morphism defined by $|\pi^*(N)|$ and $\phi_{|T_N + \rho_N^*(N)|} : \mathbb{P}(\mathcal{O}_E \oplus \mathcal{O}_E(-N)) \to \mathbb{P}(\Gamma(\mathcal{O}(T_N + \rho_N^*(N))))$ is a morphism defined by $|T_N + \rho_N^*(N)|$. Since ι is an embedding and $\phi_{|T_N + \rho_N^*(N)|}$ is an birational morphism only contracting T_N, $\phi_{|\pi^*(N)|}$ is a birational morphism onto its image. Hence

$$(\iota(C).T_N + \rho_N^*(N)) = \deg\phi_{|\pi^*(N)|}(C) = 2(g - 2h + 1).$$

Since ι is an embedding and $\rho_N \iota = \pi$, $(\iota(C).\rho_N^*(N)) = \deg\pi^*(N) = 2(g-2h+1)$ which implies $(\iota(C).T_N) = 0$, i.e. $\iota(C) \cap T_N = \emptyset$ because $\iota(C)$ and T_N are irreducible. Therefore $\iota(C)|_{T_N} \sim 0$. Let $\iota(C) \sim 2T_N + \rho_N^* B$. Then

$$0 \sim \iota(C)|_{T_N} \sim 2T_N|_{T_N} + B.$$

Since $T_N|_{T_N} \sim -(N)$, $\iota(C) \sim 2(T_N + \rho_N^*(N))$.

Q.E.D.

Corollary 1.6. *Let $\pi : C \to E$ be a smooth double covering. Then C is isomorphic to $\mathbf{Spec}(\mathcal{O}_E \oplus \mathcal{O}_E(-N))$ over E if and only if it has an embedding $\iota : C \hookrightarrow \mathbb{P}(\mathcal{O}_E \oplus \mathcal{O}_E(-N))$ with $\rho_N \iota = \pi$ and $\iota(C) \sim 2(T_N + \rho_N^* N)$.*

Proof. Assume that there is an embedding $\iota : C \hookrightarrow \mathbb{P}(\mathcal{O}_E \oplus \mathcal{O}_E(-N))$ with $\rho_N \iota = \pi$ and $\iota(C) \sim 2(T_N + \rho_N^* N)$. Let R' be the branch locus of the double covering π. Then there is a divisor N' on E with an isomorphism $\phi' : \mathcal{O}_E(N')^{\otimes 2} \xrightarrow{\sim} \mathcal{O}_E(R')$ such that:

$\mathcal{O}_E \oplus \mathcal{O}_E(-N')$ is an $\mathcal{O}_E$-algebra by $(a,b)\cdot(c,d) = (ac + \phi'(bd), ad + bc)$
$C \cong \mathbf{Spec}(\mathcal{O}_E \oplus \mathcal{O}_E(-N'))$ over E.

We need to show $N \sim N'$. By the Hurwitz relation, $K_C \sim \pi^*(K_E + N')$. On the other hand, we have

$$K_{\mathbb{P}(\mathcal{O}_E \oplus \mathcal{O}_E(-N))} + \iota(C)|_{\iota(C)} \sim \rho_N^*(K_E + N)|_{\iota(C)} \sim \pi^*(K_E + N)$$

by the adjunction formula and the assumption $\iota(C) \sim 2(T_N + \rho_N^* N)$. Therefore we have $K_E + N' \sim K_E + N$ and hence $N \sim N'$. For the converse part, we denote $\iota = \iota_0$. Then the result is clear by Proposition 1.5.

Q.E.D.

Remark 1.7. *Let $\iota : C \to \mathbb{P}(\mathcal{O}_E \oplus \mathcal{O}_E(-N))$ with $\iota(C) \sim 2(T_N + \rho_N^*(N))$ be an embedding associated with a double covering $C \cong$ $\mathbf{Spec}(\mathcal{O}_E \oplus \mathcal{O}_E(-N)) \to E$. Since*

$$\Gamma(\mathbb{P}(\mathcal{O}_E \oplus \mathcal{O}_E(-N)), \mathcal{O}(T_N + \rho_N^*(N))) \cong \Gamma(E, \mathcal{O}_E(N) \oplus \mathcal{O}_E),$$

we have

$$|\pi^*(N)| = \{H|_{\iota(C)} \mid H \in |T_N + \rho_N^*(N)|\}. \qquad (1.7.1)$$

Since $T_N|_{\iota(C)} \sim 0$, we have

$$\{\pi^* L \mid L \in |N|\} = \{(T_N + \rho_N^* L)|_{\iota(C)} \mid L \in |N|\}. \qquad (1.7.2)$$

Now we prove Theorem 1.1.

Proof of Theorem 1.1: 1) $\Rightarrow$ 2): Since C has a base point free g^1_{g-2h+1} not composed with π, the morphism $\phi = (g^1_{g-2h+1}, \pi) : C \to \mathbb{P}^1 \times E = \mathbb{P}(\mathcal{O}_E \oplus \mathcal{O}_E)$ is a birational morphism. Note that $\phi(C) \sim 2T_0 + \rho_0^*(J)$ where $J = \pi_* L$ and $L \in g^1_{g-2h+1}$. Take $a \neq b \in \mathbb{P}^1$ and put $T_0 = \{a\} \times E$, $H_0 = \{b\} \times E$ which implies $T_0 \cap H_0 = \emptyset$. Let $D_1' = \phi^* T_0$, $D_2' = \phi^* H_0$ and note that $D'_1, D'_2 \in g^1_{g-2h+1}$. Applying Lemma 1.2-(iii) and Lemma 1.3-(ii) to every $P \leq D'_1$, we get a ruled surface $\rho_M : \mathbb{P}(\mathcal{O}_E \oplus \mathcal{O}_E(-M)) \to E$ and a non-singular curve C' on $\mathbb{P}(\mathcal{O}_E \oplus \mathcal{O}_E(-M))$ such that $C' \cong C$ and $C' \sim 2(T_M + \rho_M^* M)$. We put $D_1 = \sigma^* D'_1$ and $D_2 = D'_2$ which implies $D_1 \sim \sigma^* D_2$. Note that $\pi_* D'_1 \sim \pi_* D'_2$ since $D'_1 \sim D'_2$. By Theorem 1.6, $M \sim N$ and we finally have $\pi_* D_1, \pi_* D_2 \in |N|$, $H' \sim T_N + \rho_N^*(N)$, $H'|_{C_2} = D_1 + D_2$.

2) $\Rightarrow$ 4): We take a section $H \in |T_N + \rho_N^* N|$, D_1, $D_2 \in \mathrm{Div}(C)$ satisfying the condition 2). Since $H \cap T_N = \emptyset$, $H \notin \{T_N + \rho^* N_0 \mid N_0 \in |N|\}$. We put $A = H|_{\iota(C)}$, $N_1 = \pi_* D_1$ and $N_2 = \pi_* D_2$. Then we have $A \in |\pi^* N| \setminus \{\pi^* N_0 \mid N_0 \in |N|\}$ by Remark 1.7 and $\pi_* A = N_1 + N_2$, $N_1, N_2 \in |N|$.

4) $\Rightarrow$ 3): We take a divisor $A \in |\pi^* N| \setminus \{\pi^* N_0 \mid N_0 \in |N|\}$ such that $\pi_* A = N_1 + N_2$ and $N_1, N_2 \in |N|$. By (1.7.1) there is a divisor $H \in |T_N + \rho_N^* N|$ such that $H|_{\iota(C)} = A$. Since $\pi_* A = N_1 + N_2$, there exist two effective divisors D_1, $D_2 \in \mathrm{Div}(C)$ such that $H|_{\iota(C)} = D_1 + D_2$ and $\pi_* D_1, \pi_* D_2 \in |N|$. By (1.7.2), $H \cap T_N$ is finite and hence $H \cap T_N = \emptyset$ by $(H.T_N) = 0$.

3) $\Rightarrow$ 1): Take a divisor $H \in |T_N + \rho_N^* N|$ such that $H \cap T_N = \emptyset$ satisfying $H|_{\iota(C)} = D_1 + D_2$ with $\pi_* D_1, \pi_* D_2 \in |N|$. We prove that H is a section. Since $(H.\rho_N^* p) = 1$, there exists an irreducible divisor $\widehat{H}$ and a divisor B such that $H = \widehat{H} + B$ with $(\widehat{H}.\rho_N^* p) = 1$ and $(B.\rho_N^* p) = 0$. Therefore $B = \rho^*(p_1 + \cdots + p_s)$ for some $p_1, \cdots p_s \in E$. By $(H.T_N) = 0$, $(\widehat{H}.T_N) + s = 0$. If $s > 0$, then $(\widehat{H}.T_N) < 0$ implying $\widehat{H} = T_N$ and hence $H \cap T_N \neq \emptyset$, a contradiction. Therefore $s = 0$ and $H = \widehat{H}$, i.e. H is a section. Applying Lemma 1.2-(ii) and Lemma 1.3-(i) to every $P \leq D_1$, we get a ruled surface $\rho_0 : \mathbb{P}(\mathcal{O}_E \oplus \mathcal{O}_E) = \mathbb{P}^1 \times E \to E$, a non-singular curve C' such that $C' \cong C$, and $C' \sim 2T_0 + \rho_0^*(N) = 2\{\mathrm{pt}\} \times E + \mathbb{P}^1 \times N$ with $\deg N = g - 2h + 1$. Therefore the second projection $\mathbb{P}(\mathcal{O}_E \oplus \mathcal{O}_E) \cong \mathbb{P}^1 \times E \to \mathbb{P}^1$ restricted on C' induces a base point free g^1_{g-2h+1} not composed with π.

Q.E.D.

§2. Proof of Theorem A

Let

$$\phi : \Gamma(E, \mathcal{O}_E(N)) \otimes \Gamma(E, \mathcal{O}_E(N)) \to \Gamma(E, \mathcal{O}_E(2N))$$

be the natural cup product map. Our eventual goal is to prove Theorem A, but for most of this paper, we prove the following theorem.

Theorem 2.1. *Let* $C = \mathbf{Spec}(\mathcal{O}_E \oplus \mathcal{O}_E(-N)) \xrightarrow{\pi} E$ *be a double covering of genus* g *over a curve* E *of genus* h *with* $g \geq 4h-2$. *Choose* $r \in \Gamma(E, \mathcal{O}_E(2N))$ *whose zero is the branch locus of* π. *Then* C *has a base point free and complete* g^1_{g-2h+1} *not composed with* π *if and only if* $r \in \mathrm{im}[\phi : \Gamma(E, \mathcal{O}_E(N)) \otimes \Gamma(E, \mathcal{O}_E(N)) \to \Gamma(E, \mathcal{O}_E(2N))]$.

We put $V = \Gamma(E, \mathcal{O}_E(N))$. We assume that $g \geq 4h-2$. Since $\deg(\mathcal{O}_E(N)) = g-2h+1 \geq 2h-1$, $\mathcal{O}_E(N)$ is non-special and hence $\dim V = g-3h+2$ by the Riemann-Roch Theorem. Let $\mathrm{M}(m)$ be the variety of $m \times m$ complex matrices. Let

$$M_k(g-3h+2) \subset M(g-3h+2)$$

be the kth determinantal variety, i.e. the subvariety of $M(g-3h+2)$ defined by the ideal generated by $(k+1)\times(k+1)$-minors of (a_{ij}). The codimension of $M_k(g-3h+2)$ is

$$\mathrm{codim} M_k(g-3h+2) = (g-3h+2-k)^2 \tag{2.1}$$

by [1, p.67 Proposition]. Let $e_1, \cdots, e_{g-3h+2}$ be a basis of V and let

$$\chi : V \otimes V \to \mathrm{M}(g-3h+2)$$

be the natural isomorphism defined by $\displaystyle\sum_{i,j=1,\cdots,g-3h+2} a_{ij} e_i \otimes e_j \mapsto (a_{ij})$.

Lemma 2.2. $\chi^{-1}(M_1(g-3h+2)) = \{u \otimes v \mid u, v \in V\}$.

Proof. Since

$$M_1(g-3h+2) = \{(a_{ij}) \mid a_{ij} = u_i v_j,\ i,j = 1, \cdots, g-3h+2\},$$

we have $\chi^{-1}(M_1(g-3h+2)) = \{u \otimes v \mid u = \sum u_i e_i, v = \sum v_j e_j\}$.

Q.E.D.

We put

$$M_1 = \chi^{-1}(M_1(g-3h+2)) \text{ and } M_0 = \{a \otimes a \mid a \in V\},$$

which are affine cones, i.e. if $c \in M_i$ and $\lambda \in \mathbb{C}$, then $\lambda c \in M_i$ for $i = 0, 1$. For an affine cone A, we denote $\mathbb{P}(A)$ by $A/\mathbb{C}^*$. For an element

$z \in A$, we denote $[z]$ by $\mathbb{C}z/\mathbb{C}^* \in \mathbb{P}(A)$. Let S^2V be the subspace of $V \otimes V$ generated by $\{a \otimes b + b \otimes a \mid a, b \in V\}$. Then S^2V is indeed the second symmetric product of V containing M_0.

Lemma 2.3. *$\mathbb{P}(M_0) \subset \mathbb{P}(\mathrm{S}^2V)$ is the image of $\mathbb{P}(V)$ under the Veronese embedding.*

Proof. Let $a = \sum_{i=1}^{g-3h+2} a_i e_i \in V$. Then

$$a \otimes a = \sum_{i=1}^{g-3h+2} a_i^2 e_i \otimes e_i + \sum_{i<j} a_i a_j (e_i \otimes e_j + e_j \otimes e_i)$$

which gives a coordinate of the Veronese embedding $\mathbb{P}(V) \hookrightarrow \mathbb{P}(\mathrm{S}^2V)$.

Q.E.D.

By (2.1) and Lemma 2.3, we have the following:

Corollary 2.4. $\dim\mathbb{P}(M_1) = 2g - 6h + 2$ *and* $\dim\mathbb{P}(M_0) = g - 3h + 1$.

Let $\widetilde{V}$ be a vector space. For affine cones $S, T \subset \widetilde{V}$, we put

$$S * T = \{\lambda\widetilde{x} + \mu\widetilde{y} \mid \widetilde{x} \in S, \widetilde{y} \in T, \lambda, \mu \in \mathbb{C}\}.$$

Note that $S * T$ is again an affine cone and we may consider $\mathbb{P}(S * T) \subset \mathbb{P}(\widetilde{V})$.

Lemma 2.5. *Let $M^* = M_0 * M_1 \subset V \otimes V$. Then* $\dim\mathbb{P}(M^*) = 3g - 9h + 4$.

Proof. We define a morphism

$$\theta : V \oplus V \oplus V \to M^*$$

by $\theta(x, y, u) = x \otimes y + u \otimes u$, which is surjective. Let $x, y, u \in V$ be general elements and let $x', y', u' \in V$ be arbitrary elements. We may assume that x, y, u are linearly independent. We put

$$x = \sum_{i=1}^{g-3h+2} x_i e_i, \ y = \sum_{i=1}^{g-3h+2} y_i e_i, \ u = \sum_{i=1}^{g-3h+2} u_i e_i$$

and

$$x' = \sum_{i=1}^{g-3h+2} x'_i e_i, \ y' = \sum_{i=1}^{g-3h+2} y'_i e_i, \ u' = \sum_{i=1}^{g-3h+2} u'_i e_i.$$

Assume that $\theta(x,y,u)=\theta(x',y',u')$. Then

$$x\otimes y+u\otimes u=x'\otimes y'+u'\otimes u'. \tag{2.5.1}$$

Since $V\otimes V$ can be decomposed as $(V\otimes e_1)\oplus\cdots\oplus(V\otimes e_{g-3h+2})$,

$$y_ix+u_iu=y_i'x'+u_i'u' \ (i=1,\cdots,g-3h+2)$$

by (2.5.1). Since $x,y,u\in V$ are general elements, we may assume that $\det\begin{pmatrix} y_i & u_i \\ y_j & u_j \end{pmatrix}\neq 0$ for any $i,j=1,\cdots g-3h+2$ with $i\neq j$, and hence x,u are linear combinations of x',u'. Note that x,y are linearly independent, we have

$$x'=\alpha x+\beta u \text{ and } u'=\gamma x+\delta u$$

for some $\alpha,\beta,\gamma,\delta\in\mathbb{C}$. Since $V\otimes V=(e_1\otimes V)\oplus\cdots\oplus(e_{g-3h+2}\otimes V)$ and $x,y,u\in V$ are general elements, we again have $x_iy+u_iu=x_i'y'+u_i'u'$ for $i=1,\cdots,g-3h+2$ and hence

$$y'=\xi y+\eta u \text{ and } u'=\lambda y+\mu u$$

for some $\xi,\eta,\lambda,\mu\in\mathbb{C}$. Especially

$$u'=\gamma x+\delta u=\lambda y+\mu u.$$

Since x,y,u are linearly independent, $\gamma=\lambda=0$ and $\delta=\mu$. Therefore $x'=\alpha x+\beta u$, $y'=\xi y+\eta u$ and $u'=\delta u$. By (2.5.1), $\alpha\xi=1,\alpha\eta=0,\beta\xi=0,\beta\eta+\delta^2=1$. Therefore $\beta=\eta=0,\alpha\xi=1,\delta^2=1$, i.e.

$$x'=\alpha x, y'=\frac{1}{\alpha}y \text{ and } u'=\pm u.$$

Therefore $\theta^{-1}(x\otimes y+u\otimes u)$ is 1-dimensional for general elements $x,y,u\in V$. Hence $\dim M^*=3\dim V-1=3g-9h+5$, i.e. $\dim\mathbb{P}(M^*)=3g-9h+4$.
Q.E.D.

We take $0\neq\kappa\in\mathrm{im}\phi\subset\Gamma(E,\mathcal{O}_E(2N))$, $\widetilde{\kappa}\in\phi^{-1}(\kappa)$ and consider a linear subspace

$$L_\kappa=\{\lambda\widetilde{\kappa}+x\mid\lambda\in\mathbb{C},x\in\ker\phi\}=\mathbb{C}\widetilde{\kappa}+\ker\phi\subset V\otimes V.$$

Note that $\mathbb{C}\widetilde{\kappa}\cap\ker\phi=\{0\}$ and hence $\dim L_\kappa=\dim\ker\phi+1\geq\dim V\otimes V-\dim\Gamma(E,\mathcal{O}_E(2N))+1=\dim V\otimes V-(2g-5h+3)+1$, therefore

$$\dim\mathbb{P}(L_\kappa)\geq\dim\mathbb{P}(V\otimes V)-(2g-5h+2). \tag{2.2}$$

We now prove Theorem 2.1.

Proof of Theorem 2.1:Let $s \in \Gamma(C, \mathcal{O}_C(\pi^*N))$ and let $(s)_0 = A$. Since $s\sigma^*s = \pi^*\lambda$ for some $\lambda \in \Gamma(E, \mathcal{O}_E(2N))$, we put $\mathrm{Nm}_{C/E}(s) = \lambda$ and call it the Norm of s for the Galois covering $\pi : C \to E$. Since $\pi_*\mathcal{O}_C(\pi^*N) \cong \mathcal{O}_E(N) \oplus \mathcal{O}_E$, there is an isomorphism

$$\Gamma(C, \mathcal{O}_E(\pi^*N)) \cong \Gamma(E, \mathcal{O}_E(N)) \oplus \Gamma(E, \mathcal{O}_E).$$

Therefore s can be written as $s = (\alpha, \beta)$ for some $\alpha \in \Gamma(E, \mathcal{O}_E(N))$, $\beta \in \Gamma(E, \mathcal{O}_E)$ and

$$A = (s)_0 \in \{\pi^*N_0 \mid N_0 \in |N|\} \text{ if and only if } \beta = 0.$$

By the $\mathcal{O}_E$-algebra structure on $\mathcal{O}_E \oplus \mathcal{O}_E(-N)$, we have

$$s\sigma^*s = (\alpha^2 - r\beta^2, 0) \in \Gamma(E, \mathcal{O}_E(2N)) \oplus \Gamma(E, \mathcal{O}_E(N)).$$

Therefore $\mathrm{Nm}_{C/E}(s) = \alpha^2 - r\beta^2$. Since π_*A is defined by $\mathrm{Nm}_{C/E}(s)$, $\pi_*A = (\alpha^2 - r\beta^2)_0$. Let

$$W = \{\mathrm{Nm}_{C/E}(s) \mid s \in \Gamma(C, \mathcal{O}_C(\pi^*N))\} \subset \Gamma(E, \mathcal{O}_E(2N)).$$

Since $M_0 = \{a \otimes a \mid a \in \Gamma(E, \mathcal{O}_E(N))\}$ and $\mathrm{Nm}_{C/E}(\pi^*a) = a^2$ for any $a \in \Gamma(E, \mathcal{O}_E(N))$,

$$\phi(M_0) = \{\mathrm{Nm}_{C/E}(\pi^*a) \mid a \in \Gamma(E, \mathcal{O}_E(N))\} \subset \Gamma(E, \mathcal{O}_E(2N)).$$

Then $W = \{\alpha^2 - r\beta^2 \mid \alpha \in \Gamma(E, \mathcal{O}_E(N)), \beta \in \Gamma(E, \mathcal{O}_E)\}$ which implies $\mathbb{P}(W) = \mathbb{P}(\phi(M_0) * \mathbb{C}r)$.

We now assume C has a base point free and complete g^1_{g-2h+1} not composed with π. By Theorem 1.1, there exist $l, m \in \Gamma(E, \mathcal{O}_E(N))$ such that $lm \in W \setminus \phi(M_0)$. Then there exists $a_0 \in \Gamma(E, \mathcal{O}_E(N))$ such that $r \in \mathbb{P}(\mathbb{C}a_0^2 + \mathbb{C}lm)$. Hence $r = \alpha a_0^2 + \beta lm = \phi(\alpha a_0 \otimes a_0 + \beta l \otimes m)$ for some $\alpha, \beta \in \mathbb{C}$ which implies $r \in \mathrm{im}\phi$.

Next we assume $r \in \mathrm{im}(\phi)$. Since $\mathbb{P}(L_r)$ is a linear subspace of $\mathbb{P}(V \otimes V)$,

$$\dim\mathbb{P}(M^*) \cap \mathbb{P}(L_r) \geq \dim\mathbb{P}(M^*) - (2g - 5h + 2) = g - 4h + 2 \geq 0$$

by (2.2) and Lemma 2.5. Hence there exists $\widetilde{x} \in M^*$ such that $\phi(\widetilde{x}) = r$. Since $\widetilde{x} = l \otimes m + a \otimes a$ for some $l, m, a \in V$, we have

$$r = lm + a^2.$$

Assume $[l] \neq [m]$. Then $lm \in W \setminus \phi(M_0)$. We now put $\alpha = \sqrt{-1}a$, $\beta = \sqrt{-1}(\neq 0)$ and $s = (\alpha, \beta)$. Let $A = (s)_0 \in |\pi^*N|$. Since $r = lm + a^2$,

$\alpha^2 - r\beta^2 = lm$. Therefore $\pi_* A = N_1 + N_2$, $N_1, N_2 \in |N|$ and $A \in |\pi^* N| \setminus \{\pi^* N_0 \mid N_0 \in |N|\}$. By Theorem 1.1, there exists a base point free g^1_{g-2h+1} not composed with π. Assume $[l] = [m]$. We may assume that $l = m$. Then $r = (l + \sqrt{-1}a)(l - \sqrt{-1}a)$. When $(l + \sqrt{-1}a)_0 = (l - \sqrt{-1}a)_0$, we have $(l)_0 = (a)_0$ which implies the branch locus $(r)_0$ is not reduced. This is a contradiction, since C is non-singular. Therefore $(l + \sqrt{-1}a)_0 \neq (l - \sqrt{-1}a)_0$. We put $s = (0, 1)$ and let $A = (s)_0 \in |\pi^* N|$. Then $\pi_* A = (r)_0$. Since $r = (l + \sqrt{-1}a)(l - \sqrt{-1}a)$, we again have $\pi_* A = N_1 + N_2$, $N_1, N_2 \in |N|$ and $A \in |\pi^* N| \setminus \{\pi^* N_0 \mid N_0 \in |N|\}$, which implies that there exists a base point free g^1_{g-2h+1} not composed with π by Theorem 1.1.

Q.E.D.

Finally we prove Theorem A:

Proof of Theorem A: Since $g \geq 4h$, $\deg(\mathcal{O}_E(N)) = g - 2h + 1 \geq 2h + 1$. Therefore $\mathcal{O}_E(N)$ is normally generated and hence ϕ is automatically surjective. Hence we have Theorem A by Theorem 2.1.

Q.E.D.

We are now ready to prove Theorem B as a corollary to Theorem A.

Proof of Theorem B

Claim. Fix an integer $e \geq 1$. Let C be a smooth curve of genus $g \geq 4e - 4$, not necessarily a double covering. Let Σ^1_d be the union of those components of $W^1_d(C)$ whose general element is base point free and complete. If $\Sigma^1_{g-e+1} \neq \emptyset$ then $\dim \Sigma^1_{g-e+1}$ has the expected dimension and $\Sigma^1_{g-e+2} \neq \emptyset$.

Proof of the Claim. Since it is assumed that $\Sigma^1_{g-e+1} \neq \emptyset$, any component of Σ^1_{g-e+1} has dimension at least $\rho(g - e + 1, g, 1) = g - 2e$. Suppose there exists a component $\Sigma \subset \Sigma^1_{g-e+1}$ such that $\dim \Sigma = n \geq g - 2e + 1$ and take a general $L \in \Sigma$. By the base point free pencil trick and the description of the tangent space to the scheme $W^r_d(C)$ in general, we have

$$\begin{aligned} h^0(C, L^2) &= 2(g - e + 1) - g + 1 + h^1(C, L^2) \\ &= 2(g - e + 1) - g + 1 + \ker \mu_0 \\ &\geq g - 2e + 3 + n - \rho(g - e + 1, g, 1) = n + 3, \end{aligned}$$

where $\mu_0 : H^0(C, L) \otimes H^0(C, KL^{-1}) \to H^0(C, K)$ is the natural map given by multiplication of sections; cf. [1, p 189]. Therefore it follows that

$$g - 2e + 1 \leq n \leq \dim W^{n+2}_{2g-2e+2}(C) \leq \dim W^{g-2e+3}_{2g-2e+2}(C) = 2e - 4,$$

contrary to the assumption $g \geq 4e - 4$. And this completes the proof of the first assertion of the Claim. Suppose now that $\Sigma^1_{g-e+2} = \emptyset$. Then, we have

$$W^1_{g-e+2}(C) = [\Sigma^1_{g-e+1} + W_1(C)] \cup [W^1_{g-e}(C) + W_2(C)].$$

Since

$$\dim[\Sigma^1_{g-e+1} + W_1(C)] = \rho(g - e + 1, g, 1) + 1 < \rho(g - e + 2, g, 1),$$

it follows that the closed locus $\Sigma^1_{g-e+1} + W_1(C)$ is contained in $W^1_{g-e}(C) + W_2(C)$. Note that a general element in the locus $\Sigma^1_{g-e+1} + W_1(C)$ is a complete pencil with only one base point, whereas a complete pencil in $W^1_{g-e}(C) + W_2(C)$ has at least two base points, which is an absurdity. This completes the proof of the Claim.

We now take $e = 2h$ in the Claim. By Theorem A, we have $\Sigma^1_{g-2h+1} \neq \emptyset$ and hence $\Sigma^1_{g-2h+2} \neq \emptyset$ by the Claim. By taking $e' = 2h - 1$ in the Claim, we again have $\Sigma^1_{g-e'+2} = \Sigma^1_{g-2h+3} \neq \emptyset$; note that $g \geq 8h - 4 > 4e' - 4$. We may continue this process by taking smaller $e's$ and we are done.

Q.E.D.

§3. Examples

In this final section, we exhibit two examples which show that the genus assumption $g \geq 4h$ in Theorem A is the best possible one. We first give an example of a double covering $C \xrightarrow{\pi} E$ of genus $g = 4h - 1$ without a base point free and complete g^1_{g-2h+1} not composed with π. We also give another example of a double covering $C \xrightarrow{\pi} E$ possessing a base point free and complete g^1_{g-2h+1} not composed with π under the same genus assumption $g = 4h - 1$. In these examples we shall make use of the following well-known fact regarding the normal generation of line bundles on a hyperelliptic curve.

Remark 3.1. *Let E be a hyperelliptic curve of genus h. A very ample line bundle on E of degree $2h$ is not normally generated; cf.* [1, p.221 C-3].

Example 3.2. *There exists a double covering $C \xrightarrow{\pi} E$ of genus $g = 4h-1$ which does not have a base point free and complete g^1_{g-2h+1} not composed with π.*

Proof. Let E be a hyperelliptic curve of genus $h \geq 2$. Given $N \in \mathrm{Div}(E)$, we consider the natural cup product map

$$\phi : \Gamma(E, \mathcal{O}_E(N)) \otimes \Gamma(E, \mathcal{O}_E(N)) \to \Gamma(E, \mathcal{O}_E(2N)).$$

For $h = 2$, let $\mathcal{O}_E(N)$ be a base point free line bundle of degree $2h = 4$. Note that $\mathcal{O}_E(N)$ is not very ample. Assume that ϕ is surjective. By using [1, p.222 C-4] inductively, we easily see that

$$\Gamma(E, \mathcal{O}_E(N))^{\otimes k} \to \Gamma(E, \mathcal{O}(kN))$$

is surjective for every $k \geq 1$, i.e. $\mathcal{O}_E(N)$ is normally generated. Hence $\mathcal{O}_E(N)$ is very ample which is a contradiction. Therefore we may choose $r \notin \mathrm{im}\phi$ such that $(r)_0 = R$ is reduced. For $h \geq 3$, by a well-known theorem of Halphen, we may take a very ample line bundle $\mathcal{O}_E(N)$ of degree $2h = g - 2h + 1$. By Remark 3.1, $\mathcal{O}_E(N)$ is not normally generated and hence ϕ is not surjective by [1, p.222]. Therefore we may again choose $r \notin \mathrm{im}\phi$ such that $(r)_0 = R$ is reduced.

Let $C \xrightarrow{\pi} E$ be a double covering of genus $g = 4h - 1$ with the branch locus R. By Theorem 2.1, C does not have a base point free and complete g^1_{g-2h+1} not composed with π.

Q.E.D.

For an example of a double covering of genus $g = 4h - 1$ with a base point free and complete g^1_{g-2h+1}, we have implictly exhibited such one for $h \geq 3$ in the Example 0.4. We simply note that, in the Example 0.4, it is possible to take a very ample $N \in E_{g-2h+1}$ even in the range $3h + 2 \leq g \leq 4h - 1$ for any curve E of genus $h \geq 3$. One may also construct such an example by a similar method as in Example 3.2.

Example 3.3. *There is double covering $C \xrightarrow{\pi} E$ of gneus $g = 4h - 1$ which has a base point free and complete g^1_{g-2h+1}.*

Proof. Let $\mathcal{O}_E(N)$ be a line bundle of degree $2h = g - 2h + 1$. Since $|N|$ is base point free , we may take $r \in \mathrm{im}\phi$ whose zero $R = (r)_0$ is reduced and let $C \xrightarrow{\pi} E$ be a double covering of genus genus $g = 4h - 1$ with the branch locus R. Then C has a base point free and complete g^1_{g-2h+1} not composed with π by Theorem 2.1.

Q.E.D.

References

[1] E. Arbarello, M. Cornalba, P. A. Griffiths and J. Harris, Geometry of Algebraic Curves I, Springer-Verlag, 1985.

[2] E. Ballico and C. Keem, Variety of linear systems on double covering curves, J. Pure Appl. Algebra, **128** (1998), 213–224.

[3] C. Ciliberto and E. Sernesi, Singularities of the theta divisor and congruences of planes, J. Algebraic Geom., **1** (1992), 231–250.

[4] R. Hartshorne, Algebraic Geometry, Springer-Verlag, 1977.

[5] D. Mumford, Prym varieties I, Contribution to Analysis, Acad. Press, 1974, 325–355.

[6] V. V. Shokurov, Distinguishing Prymians from Jacobians, Invent. Math., **65** (1981), 209–219.

Changho Keem
Department of Mathematics
Seoul National University
Seoul 151-742 SOUTH KOREA
E-mail address: ckeem@math.snu.ac.kr

Akira Ohbuchi
Department of Mathematics
Faculty of Integrated Arts and Sciences
Tokushima University
Tokushima 770-8502 JAPAN
E-mail address: ohbuchi@ias.tokushima-u.ac.jp

Advanced Studies in Pure Mathematics 50, 2008
Algebraic Geometry in East Asia — Hanoi 2005
pp. 289–297

Poincaré polynomial of a class of signed complete graphic arrangements

Guangfeng Jiang,[1] Jianming Yu[2] and Jianghua Zhang

Abstract.

We compute the Poincaré polynomial of hyperplane arrangements associated with a class of signed complete graphs. We also make a factorization of the Poincaré polynomial over the integers.

§1. Introduction

Let $\mathbf{V}$ be an n-dimensional vector space over a field $\mathbb{K}$. Let S be the symmetric algebra over the dual space $\mathbf{V}^* := \mathrm{Hom}_{\mathbb{K}}(\mathbf{V}, \mathbb{K})$. If $x_1, \ldots, x_n$ is a basis of $\mathbf{V}^*$, then there are identifications $S = \mathbb{K}[x_1, \ldots, x_n]$ and $\mathbf{V} = \mathbb{K}^n$. A hyperplane H in $\mathbb{K}^n$ is by definition the zero set of a degree one polynomial α_H in the variables $x_1, \cdots, x_n$. An arrangement of hyperplanes $\mathcal{A}$ in $\mathbb{K}^n$ is a finite collection of hyperplanes.

Let $L(\mathcal{A})$ be the collection of all non-empty intersections of hyperplanes from $\mathcal{A}$, which is a partial ordered set with the order defined by the inverse inclusion. The rank of an element $X \in L(\mathcal{A})$ is defined by $r(X) = \mathrm{codim}(X)$. Let μ be the Möbius function of $L(\mathcal{A})$, and denote $\mu(X) = \mu(\mathbf{V}, X)$. The Poincaré polynomial of L is defined by $\pi(L,t) = \sum_{X \in L} \mu(X)(-t)^{\mathrm{r}(X)}$.

If $\mathbb{K}$ is the field $\mathbb{C}$ of complex numbers, the complement M of $\mathcal{A}$ is of interest from topological point of view. One of the central topics in studying hyperplane arrangements is to describe the topology of M by

---

Received March 17, 2006.

Revised June 30, 2006.

2000 *Mathematics Subject Classification.* Primary 52C35; Secondary 06C05, 05C22.

Key words and phrases. signed graph, hyperplane arrangement, free arrangement.

[1]Supported by NSF of China under grant 10671009, and SRF for ROCS, SEM..

[2]Supported by NSF of China under grant 10671009.

the data from $L(\mathcal{A})$. For example, it is well known that the cohomology $H^*(M;\mathbb{C})$ of M is isomorphic to the Orlik-Solomon algebra $\mathcal{OS}(\mathcal{A})$ [7] and the Poincaré polynomial of $\mathcal{OS}(\mathcal{A})$ equals to $\pi(L,t)$.

Graphic arrangements have been studied by a number of authors [5, 4, 3, 12], since they are related to the arrangements associated with classical groups [1, 2, 9, 10]. We study a special class of arrangements associated with signed complete graphs [13] and prove the following factorization formula.

Theorem 1. *If a signed complete graph Σ_n with n vertices is switching equivalent to the signed complete graph $\Sigma_n^{(3)}$ with negative part a triangle, the Poincaré polynomial of the corresponding arrangement $\mathcal{A}(\Sigma_n)$ is*

$$\pi(\mathcal{A}(\Sigma_n),t) = (t+1)(2t+1)\cdots[(n-3)t+1]Q(t), \tag{1}$$

where

$$Q(t) := [3(n-3)(n-2)+1]\,t^3 + (n^2-6)t^2 + (2n-3)t + 1.$$

§2. Preliminaries

2.1. Intersection lattice

An arrangement $\mathcal{A}$ is called central if the intersection of all the hyperplanes in $\mathcal{A}$ is not empty. If $\mathcal{A}$ is central, $L = L(\mathcal{A})$ is a geometric lattice. The minimal element of L is denoted by $\hat{0}$ and the maximal element is denoted by $\hat{1}$.

The meet of $X, Y \in L$ is defined by $X \wedge Y = \cap\{Z \in L \mid Z \supseteq X \cup Y\}$, and if $X \cap Y \neq \emptyset$, their join is defined by $X \vee Y = X \cap Y$. A pair $(X,Y) \in L \times L$ is called a modular pair if for all $Z \leq Y$ one has $Z \vee (X \wedge Y) = (Z \vee X) \wedge Y$. A pair $(X,Y) \in L \times L$ is modular if and only if $r(X) + r(Y) = r(X \vee Y) + r(X \wedge Y)$. An element X is called a modular element if it forms a modular pair with each $Y \in L$.

An element in a geometric lattice L is called an atom if it covers the minimal element of L. If L is the intersection lattice of an arrangement, each hyperplane is an atom of L. Let $A(L)$ be the collection of the atoms in L. For any $X \in L$, let $A(L)_X = \{Y \in A(L) \mid Y \leq X\}$.

Lemma 2. *For $X, Y \in L$,*

$$X \wedge Y = \hat{0} \Longleftrightarrow A(L)_X \cap A(L)_Y = \emptyset.$$

Proof. Obviously, the following formula holds.

$$A(L)_X \cap A(L)_Y = A(L)_{X \wedge Y}.$$

The lemma follows from this formula. ◇

2.2. Stanley Theorem

Let L be a geometric lattice, $\mathcal{OS}(L)$ the Orlik-Solomon algebra generated by the atoms of L. By [7], $\mathcal{OS}(L(\mathcal{A}))$ is the same as $\mathcal{OS}(\mathcal{A})$ for a hyperplane arrangement $\mathcal{A}$. Note that, for $Y \in L$, $L_Y = \{Z \in L \mid Z \leq Y\}$ is also a geometric lattice.

Stanley Theorem [8, 11]. Let L be a geometric lattice, and $X \in L$ be a modular element. Then

$$\pi(\mathcal{OS}(L), t) = \pi(\mathcal{OS}(L_X), t) \sum_{Z \in L, Z \wedge X = \hat{0}} \mu(Z)(-t)^{r(Z)}, \tag{2}$$

where μ is the Möbius function of L.

2.3. Signed complete graph

A signed complete graph $\Sigma_n = (K_n, \sigma)$ consists of an ordinary complete graph K_n with n vertices, and an arc labelling mapping $\sigma : E \longrightarrow \{\pm\}$, where E is the edge set of K_n. Let $E_+ = \sigma^{-1}(+)$ and $E_- = \sigma^{-1}(-)$ denote the sets of the positive and negative edges respectively. An edge $\{ij\} \in E_+$ is denoted by $\{ij\}^+$ and is pictured as a line segment connecting the vertices i and j. An edge $\{ij\} \in E_-$ is denoted by $\{ij\}^-$ and is pictured as a dashed line segment connecting the vertices i and j. For general study of singed graphs we refer the reader to Zaslavsky [13].

Given a signed complete graph $\Sigma_n = (K_n, \sigma)$, define an arrangement $\mathcal{A}(\Sigma_n)$ in $\mathbb{K}^n$ as follows:

$$\{x_i - x_j = 0\} \in \mathcal{A}(\Sigma_n) \text{ if } \{ij\} \in E_+$$

and

$$\{x_i + x_j = 0\} \in \mathcal{A}(\Sigma_n) \text{ if } \{ij\} \in E_-$$

2.4. Switching equivalence

For a signed complete graph $\Sigma_n = (K_n, \sigma)$, we consider the following operations on Σ_n.

1) A permutation of the labels on the vertices of Σ_n;
2) Switching a vertex $i_0 \in [n] = \{1, 2, \ldots, n\}$ is to switch the sign of the edge $\{i_0 i\}$ for each $i \in N_{i_0} := \{i \in [n] \mid \{i_0 i\} \in E\}$, the neighborhood of i_0. We call this *the vertex switching*, or, *switching* i_0.

The first operation is essentially a permutation on the coordinates, which allows one to consider unlabeled graphs. For a vertex v, switching v corresponds to switching the sign of the coordinate x_v. Obviously,

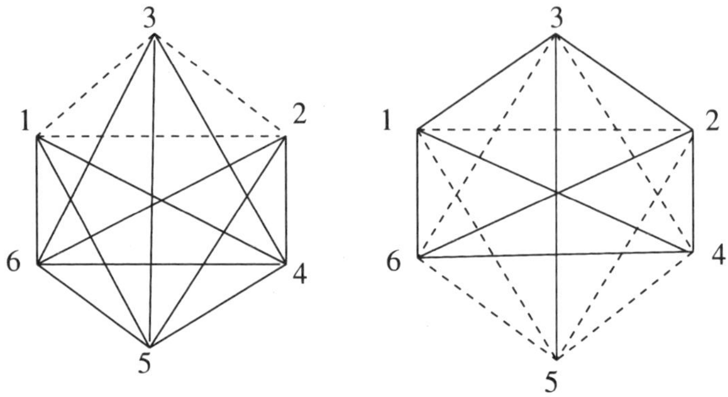

Fig. 1

the vertex switching operations form a group, which acts on the set of signed graphs with fixed number of vertices. Since a coordinate transformation on the vector space $\mathbf{V}$ does not affect the intersection lattice of an arrangement, the first two operations preserve the modular elements.

Two signed graphs Σ'_n and Σ''_n are *switching equivalent* if there exists a series of vertex switchings such that Σ'_n can be transformed into Σ''_n up to a permutation of the labels on the vertices. In this case, we also say that the corresponding arrangements are *switching equivalent.* For example, the two signed complete graphs with 6 vertices in figure 1 are switching equivalent by first switching the vertex 5, then the vertex 3 of the graph on the right hand side, one gets the graph on the left hand side.

§3. Modular elements

Signed complete graphs with at most 6 vertices were classified under the switching equivalence in [6]. One class of signed complete graphs with n vertices is denoted by $\Sigma_n^{(3)}$ in which the negative part is a triangle. We denote the vertices of the triangle by 1,2,3 and label the other vertices by 4, …, n. The arrangement associated with $\Sigma_n^{(3)}$ consists of the following hyperplanes.

$$\begin{array}{lll} H_{12}: \ x_1 + x_2 = 0, & H_{13}: \ x_1 + x_3 = 0, & H_{23}: \ x_2 + x_3 = 0, \\ H_{1m}: \ x_1 - x_m = 0, & H_{2m}: \ x_2 - x_m = 0, & 4 \le m \le n, \\ H_{ij}: \ x_i - x_j = 0, & 3 \le i < j \le n. & \end{array}$$

For the case of $n = 6$ see figure 1.

To simplify notation, in the following we set $L = L(\mathcal{A}(\Sigma_n^{(3)}))$.

Lemma 3. *The element* $X = \bigcap_{3\le i<j\le n} H_{ij} = \{x \in \mathbb{K}^n \mid x_3 = \cdots = x_n\} \in L$ *is modular.*

Proof. It is enough to prove that for each $Y \in L$ with $X \wedge Y = \hat{0}$, (X, Y) is modular pair. This is equivalent to $r(X) + r(Y) = r(X \vee Y)$ since $r(X \wedge Y) = r(\hat{0}) = 0$. By lemma 2 and the definition of rank function, it is sufficient to prove that for each $Y \in L$ with $A(L)_X \cap A(L)_Y = \emptyset$, the equation

$$\dim Y - \dim(X \cap Y) = n - 3 \tag{3}$$

holds.

Let

$$A_1 = \{H_{1j} \mid j = 4, \ldots n\}, A_2 = \{H_{2j} \mid j = 4, \ldots n\}, A_3 = \{H_{12}, H_{13}, H_{23}\}.$$

Then

$$A(L)_X \cap A(L)_Y = \emptyset \implies A(L)_Y \subset A_1 \sqcup A_2 \sqcup A_3$$

with

$$|A(L)_Y \cap A_m| \le 1, m = 1, 2, \quad |A(L)_Y \cap A_3| \le 3.$$

Hence, there are 16 possibilities for $(|A(L)_Y \cap A_1|, |A(L)_Y \cap A_2|, |A(L)_Y \cap A_3|)$:

$$\begin{matrix} (0,0,0) & (0,1,0) & (1,0,0) & (0,0,1) \\ (1,1,0) & (0,1,1) & (1,0,1) & (0,0,2) \\ (1,1,1) & (0,1,2) & (1,0,2) & (0,0,3) \\ (1,1,2) & (0,1,3) & (1,0,3) & (1,1,3) \end{matrix} \tag{4}$$

The case $(0,0,0)$ can not appear. If $(|A(L)_Y \cap A_1|, |A(L)_Y \cap A_2|, |A(L)_Y \cap A_3|) = (0,0,1)$, $A(Y)$ consists of one of the hyperplanes from A_3. Then $\dim Y = n - 1$, $\dim(X \cap Y) = 2$, and equation (3) holds.

If $(|A(L)_Y \cap A_1|, |A(L)_Y \cap A_2|, |A(L)_Y \cap A_3|) = (0,0,2)$, $A(Y)$ consists of two of the hyperplanes from A_3. Then $\dim Y = n - 2$, $\dim(X \cap Y) = 1$, and equation (3) holds.

If $(|A(L)_Y \cap A_1|, |A(L)_Y \cap A_2|, |A(L)_Y \cap A_3|) = (0,0,3)$, $A(Y) = A_3$. Then $\dim Y = n - 3$, $\dim(X \cap Y) = 0$, and equation (3) holds. Similar treatment will prove the cases $(0,1,0), (1,0,0), (0,1,1), (1,0,1)$ and $(1,1,0)$.

For the case $(0,1,2)$, if $A(Y) \cap A_3 = \{H_{12}, H_{13}\}$ and $A(Y) \cap A_2 = \{H_{2j}\}$ for some $j \ge 4$, then Y is the intersection of $x_1 + x_2 = 0, x_1 +$

$x_3 = 0$, and $x_2 - x_j = 0$. This implies that Y is contained in the hyperplane $H_{3j} : x_3 - x_j = 0$, which is impossible since $X \subset H_{3j}$. If $A(Y) \cap A_3 = \{H_{12}, H_{23}\}$, and $A(Y) \cap A_2 = \{H_{2j}\}$ for some $j \geq 4$, then $\dim Y = n - 3$, $\dim(X \cap Y) = 0$, and equation (3) holds. Similar treatment works for $A(Y) \cap A_3 = \{H_{13}, H_{23}\}$.

One can treat the case $(1, 0, 2)$ in a similar way.

The cases $(0, 1, 3)$ and $(1, 0, 3)$ do not appear, since otherwise there would be $Y \subset H_{3k}$ for some $k > 3$.

If $(|A(L)_Y \cap A_1|, |A(L)_Y \cap A_2|, |A(L)_Y \cap A_3|) = (1, 1, 1)$, $A(Y)$ consists of $H_{1k} : x_1 - x_k = 0$, $H_{2j} : x_2 - x_j = 0$ and one of the hyperplanes from A_3. It is easy to see that $\dim Y = n - 3$ and $\dim(X \cap Y) = 0$.

Let $(|A(L)_Y \cap A_1|, |A(L)_Y \cap A_2|, |A(L)_Y \cap A_3|) = (1, 1, 2)$, there are four cases for $A(Y)$:

1) $A(Y) = \{H_{1k}, H_{2j}, H_{12}, H_{13}\}$ which implies that $Y \subset H_{3j}$, a contradiction;

2) $A(Y) = \{H_{1k}, H_{2j}, H_{12}, H_{23}\}$ which implies that $Y \subset H_{3k}$, a contradiction;

3) $A(Y) = \{H_{1k}, H_{2j}, H_{13}, H_{23}\}$ with $k \neq j$ which implies that $Y \subset H_{kj}$, a contradiction;

4) $A(Y) = \{H_{1k}, H_{2k}, H_{13}, H_{23}\}$ with $\dim Y = n - 3, X \cap Y = 0$, hence equation (3) holds.

The case $(|A(L)_Y \cap A_1|, |A(L)_Y \cap A_2|, |A(L)_Y \cap A_3|) = (1, 1, 3)$ does not appear since otherwise we would have $Y \subset H_{jk}$. ◇

Let

$$\mathcal{A}_0 = \emptyset,\ \mathcal{A}_1 = \{H_{34}\},\ \mathcal{A}_2 = \{H_{ij} \mid 3 \leq i < j \leq 5\},\ \ldots,\ \mathcal{A}_{n-3} = \{H_{ij} \mid 3 \leq i < j \leq n\}.$$

and

$$X_k = \bigcap_{H \in \mathcal{A}_k} H, \quad k = 0, 1, 2, \ldots n - 3.$$

Note that $X_0 = \hat{0}$ and $X_{n-3} = X$. By [4], we have the following

Lemma 4. *There is a chain of modular elements*

$$\hat{0} < X_1 < X_2 < \cdots < X_{n-4} < X_{n-3} = X.$$

◇

For X defined in lemma 3, it follows from lemma 4 that $L_X = L(\mathcal{A}(\Sigma_n^{(3)}))_X$ is a modular lattice. By [4], we have

$$\pi(\mathcal{OS}(L_X), t) = (t + 1)(2t + 1) \cdots ((n - 3)t + 1). \tag{5}$$

§4. Proof of the main result.

It remains to calculate

$$Q = \sum_{Z \in L, Z \wedge X = \hat{0}} \mu(Z)(-t)^{r(Z)}.$$

Note that in our case X is modular with $r(X) = n - 3$. For $Z \in L, Z \wedge X = \hat{0}$, we have

$$\begin{aligned} n \geq r(X \vee Z) &= r(X \vee Z) + r(X \wedge Z) = r(X) + r(Z) \\ &= n - 3 + r(Z) \Rightarrow r(Z) \leq 3. \end{aligned}$$

Hence

(6) $$Q = \alpha(-t)^3 + \beta t^2 + \gamma(-t) + 1,$$

where

$$\alpha = \sum_{\substack{Z \in L, Z \wedge X = \hat{0} \\ r(Z)=3}} \mu(Z), \quad \beta = \sum_{\substack{Z \in L, Z \wedge X = \hat{0} \\ r(Z)=2}} \mu(Z), \quad \gamma = \sum_{\substack{Z \in L, Z \wedge X = \hat{0} \\ r(Z)=1}} \mu(Z).$$

By lemma 2,

(7) $$\gamma = -|A(L) \setminus A(L)_X| = -\left[\binom{n}{2} - \binom{n-2}{2}\right] = -(2n-3).$$

Next we compute β. Since $A(L)_Z \cap A(L)_X = \emptyset$ and $r(Z) = 2$, for each $k = 1, 2$, $A(L)_Z$ may contain H_{ki} or H_{kj} $(i \neq j)$, but does not contain both at the same time for, otherwise, there would be $H_{ki} \cap H_{kj} \supset X$. Hence, $A(L)_Z$ contains two hyperplanes, and $\mu(Z) = 1$.

There are three cases to be considered.

1) one of the two hyperplanes in $A(L)_Z$ comes from $\{H_{1i} \mid i = 4, \ldots, n\}$, and the other one comes from $\{H_{2i} \mid i = 4, \ldots, n\}$. Hence there are as many as $(n-3)^2$ possibilities;

2) the two hyperplanes in $A(L)_Z$ comes from H_{12}, H_{23}, H_{13}, there are 3 possibilities;

3) one of the two hyperplanes in $A(L)_Z$ comes from $\{H_{ki} \mid k = 1, 2, i = 4, \ldots, n\}$ and the other one comes from A_3. There are $3 \times 2(n-3)$ possibilities.

Hence

(8) $$\beta = (n-3)^2 + 3 + 6(n-3) = n^2 - 6.$$

Now we compute α. The point is that in some cases, $A(L)_Z$ contains more atoms than the the minimal possible.

Since $r(Z)=3$, $(|A(L)_Z\cap A_1|, |A(L)_Z\cap A_2|, |A(L)_Z\cap A_3|)$ has only the following possibilities.

$$(0,0,3),(0,1,2),(1,0,2),(1,1,1),(1,1,2),(1,0,3),(0,1,3),(1,1,3).$$

The cases $(1,0,3)$, $(0,1,3)$, $(1,1,3)$ are excluded by the condition $Z\wedge X=\hat{0}$.

For the case $(0,0,3)$, $Z=H_{12}\cap H_{13}\cap H_{23}$ and $\mu(Z)=-1$, This contributes -1 to α.

In case $(0,1,2)$, for $4\le k\le n$, we have

$$Z_1^{(k)}=H_{12}\cap H_{23}\cap H_{2k},\ \ Z_2^{(k)}=H_{12}\cap H_{13}\cap H_{2k},\ \ Z_3^{(k)}=H_{13}\cap H_{23}\cap H_{2k}.$$

Since $Z_2^{(k)}\subset H_{3k}$, $Z_2^{(k)}$ should be excluded. It is obvious that

$$A(L)_{Z_1^{(k)}}=\{H_{12},H_{23},H_{2k}\},\quad A(L)_{Z_3^{(k)}}=\{H_{13},H_{23},H_{1k},H_{2k}\}.$$

So $Z_3^{(k)}$ should belong to the case $(1,1,2)$, which will be considered later. Since $\mu(Z_1^{(k)})=-1$, this case contributes $-(n-3)$ to α.

The case $(1,0,2)$ is similar to the case $(0,1,2)$. For $4\le k\le n$, we have

$$\tilde{Z}_1^{(k)}=H_{12}\cap H_{23}\cap H_{1k},\ \ \tilde{Z}_2^{(k)}=H_{12}\cap H_{13}\cap H_{1k},\ \ \tilde{Z}_3^{(k)}=H_{13}\cap H_{23}\cap H_{1k}.$$

Since $\tilde{Z}_1^{(k)}\subset H_{3k}$, $\tilde{Z}_1^{(k)}$ should be excluded. It is obvious that $\tilde{Z}_3^{(k)}=Z_3^{(k)}$. Since $\mu(\tilde{Z}_2^{(k)})=-1$. This case contributes $-(n-3)$ to α.

In case $(1,1,1)$, there are $3(n-3)^2$ possibilities all together. Since $H_{13}\cap H_{1k}\cap H_{2k}=H_{23}\cap H_{1k}\cap H_{2k}=Z_3^{(k)}$ which should be in the case $(1,1,2)$, this case contributes $-(3(n-3)^2-2(n-3))$ to α.

We consider the case $(1,1,2)$. For $4\le j,k\le n$, we have

$$Z_4^{(jk)}=H_{12}\cap H_{13}\cap H_{1j}\cap H_{2k},\quad Z_5^{(jk)}=H_{12}\cap H_{23}\cap H_{1j}\cap H_{2k},$$
$$Z_6^{(jk)}=H_{13}\cap H_{23}\cap H_{1j}\cap H_{2k}.$$

It is obvious that $Z_4^{(jk)}\subset H_{3k}$, $Z_5^{(jk)}\subset H_{3j}$, and for $j\neq k$, $Z_6^{(jk)}\subset H_{jk}$, which are excluded by the condition $Z\wedge X=\hat{0}$. For $j=k$, we have $Z_6^{(jk)}=Z_6^{(kk)}=Z_3^{(k)}$, and $\mu(Z_3^{(k)})=-3$. Hence, this case contributes $-3(n-3)$ to α.

Hence

$$\alpha=-(1+3(n-2)(n-3)). \tag{9}$$

Combine formulae (2), (7), (8), and (9), we obtain the formula (1) in Theorem 1.

References

[1] V. I. Arnold, Wave front evolution and equivariant Morse lemma, Comm. Pure Appl. Math., **29** (1976), 557–582.

[2] V.I. Arnold, Indices of singular points of 1-forms on a manifold with boundary, convolution of invariants of reflection groups, and singular projections of smooth surfaces, Russian Math. Surveys, **34** (1979), 1–42.

[3] G. D. Bailey, Inductively factored signed-graphic arrangements of hyperplanes, preprint.

[4] P. H. Edelman and V. Reiner, Free hyperplane arrangements between A_{n-1} and B_n, Math. Z., **215** (1994), 347–365.

[5] T. Józefiak and B. E. Sagan, Basic derivations for subarrangements of Coxeter Arrangements, J. Algebraic Combin., **2** (1993), 291–320.

[6] G. Jiang and J. Yu, Supersolvability of Complementary Signed-Graphic hyperplane arrangements, Australas. J. Combin., **30** (2004), 261–275.

[7] P. Orlik and H.Terao, Arrangements of Hyperplanes, Springer-Verlag, 1992.

[8] R. P. Stanley, Modular elements of geometric lattices, Algebra Universalis, **1** (1971), 214–217.

[9] K. Saito, Theory of logarithmic differential forms and logarithmic vector fields, J. Fac. Sci. Univ. Tokyo Sect. IA Math., **27** (1981), 265–291.

[10] H. Terao, Free arrangements of hyperplanes and unitary reflection groups, Proc. Japan Acad. Ser. A Math. Sci., **56** (1980), 389–392.

[11] H. Terao, Modular elements of lattices and topological fibration, Adv. in Math., **62** (1986), 135–154.

[12] T. Zaslavsky, The geometry of root system and signed graphes, Amer. Math. Monthly, **88** (1981), 88–105.

[13] T. Zaslavsky, Signed graphs, Discrete Appl. Math., **4** (1982), 47–74; Erratum, Discrete Appl. Math., **5** (1983), 248 p.

Guangfeng Jiang
Department of Mathematics and Information Science
Beijing University of Chemical Technology
Beijing P. R. CHINA

Jianming Yu
Institute of Mathematics
Academy of Mathematics and System Sciences
Chinese Academy of Sciences
Beijing P. R. CHINA

Jianghua Zhang
Department of Fundamental Courses
Beijing Institute of Clothing Technology
Beijing P. R. CHINA

Advanced Studies in Pure Mathematics 50, 2008
Algebraic Geometry in East Asia — Hanoi 2005
pp. 299–331

Singularities of dual varieties in characteristic 2

Ichiro Shimada

Abstract.

We investigate unibranched singularities of dual varieties of even-dimensional smooth projective varieties in characteristic 2.

§1. Introduction

We work over an algebraically closed field k.

Let $X \subset \mathbb{P}^m$ be a smooth projective variety of dimension $n > 0$. We assume that X is not contained in any hyperplane of $\mathbb{P}^m$. Then we can consider the dual projective space

$$\mathbf{P} := (\mathbb{P}^m)^\vee$$

of $\mathbb{P}^m$ as the parameter space of the linear system $|M|$ of hyperplane sections of X, where M is a linear subspace of $H^0(X, \mathcal{O}_X(1))$. We use the same letter to denote a point $H \in \mathbf{P}$ and the corresponding hyperplane $H \subset \mathbb{P}^m$. Let $\mathcal{D} \subset X \times \mathbf{P}$ be the universal family of the hyperplane sections. Then $\mathcal{D}$ is a smooth scheme of dimension $n + m - 1$. The support of $\mathcal{D}$ is equal to the set

$$\{ (p, H) \in X \times \mathbf{P} \mid p \in H \}.$$

Let $\mathcal{C} \subset \mathcal{D}$ be the critical subscheme of the second projection $\mathcal{D} \to \mathbf{P}$. (See Notation and Terminology below for the definition of critical subschemes.) Then $\mathcal{C}$ is smooth, irreducible and of dimension $m - 1$. If $\mathcal{N}$ is the conormal sheaf of $X \subset \mathbb{P}^m$, then $\mathcal{C}$ is isomorphic to $\mathbb{P}^*(\mathcal{N})$ ([12, Remarque 3.1.5]). The support of $\mathcal{C}$ is equal to the set

$$\{ (p, H) \in \mathcal{D} \mid \text{the divisor } H \cap X \text{ of } X \text{ is singular at } p \}.$$

Received April 3, 2006.
Revised May 30, 2006.
2000 *Mathematics Subject Classification.* Primary 14B05; Secondary 14C20.

We denote the projections by

$$\pi_1 : \mathcal{C} \to X \quad \text{and} \quad \pi_2 : \mathcal{C} \to \mathbf{P}.$$

The image of π_2 is called the *dual variety* of $X \subset \mathbb{P}^m$. Let $\mathcal{E} \subset \mathcal{C}$ be the critical subscheme of π_2. By [12, Proposition 3.3], $\mathcal{E}$ is set-theoretically equal to

$$\{ (p, H) \in \mathcal{C} \mid \text{the Hessian at } p \in \operatorname{Sing}(H \cap X) \text{ is degenerate} \}.$$

We will study the singularities of the dual variety by investigating the morphism π_2 at points of $\mathcal{E}$.

Since the paper of Wallace [22], properties of dual varieties peculiar to positive characteristics have been studied mainly from the point of view of the (failure of the) reflexivity. See Kleiman's paper [14] for the definition and a detailed account of the reflexivity. Many studies have been done for the analysis of the situation in which the reflexivity does not hold. See [5, 6, 7, 8, 9, 10, 11, 16], for example.

A well-known example of the situations in which the reflexivity fails is as follows. Suppose that

$$\operatorname{char} k = 2 \quad \text{and} \quad \dim X \equiv 1 \bmod 2. \tag{1.1}$$

Then the critical subscheme $\mathcal{E}$ of $\pi_2 : \mathcal{C} \to \mathbf{P}$ coincides with $\mathcal{C}$ ([12, Section 1.2], [14, Corollary (18)]), and hence either the dual variety is not a hypersurface in $\mathbf{P}$, or X is not reflexive by the generalized Monge-Segre-Wallace criterion ([13, Theorem (4.4)], [14, Theorem (4)]).

Except for the case (1.1), however, the reflexivity is recovered when the linear system $|M|$ of the hyperplane sections are sufficiently ample. We have the following theorem ([12, Théorème 2.5], [7, Theorem (5.4)]):

Theorem 1.1. *Suppose that* $\operatorname{char} k \neq 2$ *or* $\dim X$ *is even. If* X *is embedded in* $\mathbb{P}^m$ *by a complete linear system of the form* $|\mathcal{A}^{\otimes d}|$ *with* $\mathcal{A}$ *being a very ample line bundle and* $d \geq 2$*, then the dual variety of* $X \subset \mathbb{P}^m$ *is a hypersurface of* $\mathbf{P}$*, and* $X \subset \mathbb{P}^m$ *is reflexive.*

See also [19, Proposition 4.9] or Proposition 6.12 of this paper for other sufficient conditions for the dual variety to be a hypersurface, and for the reflexivity to hold.

In our previous paper [19], we have discovered that, even when the reflexivity holds, the singularities of dual varieties in characteristic 3 still possess a peculiar feature. We assume that the linear system $|M|$ is sufficiently ample. In particular, the dual variety is a hypersurface in $\mathbf{P}$

and X is reflexive. We consider the projective plane curve obtained by cutting the dual variety by a general plane in $\mathbf{P}$. If k is of characteristic > 3 or 0, then the plane curve has only ordinary cusps as its unibranched singularities. In characteristic 3, the plane curve has singular points of E_6-type instead of ordinary cusps.

In this paper, we investigate the singularities of the dual variety in the case where $\operatorname{char} k = 2$ and $\dim X$ is even.

The dual variety is the *discriminant variety* associated with the linear system of hyperplane sections. See [2] for the definition of discriminant varieties. In fact, our results are proved not only for dual varieties but for discriminant varieties associated with (not necessarily very ample) linear systems. Here in Introduction, however, we present our results for dual varieties.

Let k be of characteristic 2, and let $\dim X$ be even. For simplicity, we assume that $|M|$ is sufficiently ample so that the evaluation homomorphism

$$v_p^{[4]} : M \to \mathcal{L}_p/\mathfrak{m}_p^5\mathcal{L}_p$$

is surjective at every point p of X, where $\mathfrak{m}_p$ is the maximal ideal of $\mathcal{O}_{X,p}$, and $\mathcal{L}_p$ is the $\mathcal{O}_{X,p}$-module $\mathcal{L}\otimes\mathcal{O}_{X,p}$. Under this assumption, the critical subscheme $\mathcal{E}$ is an irreducible divisor of $\mathcal{C}$ and hence the dual variety is a hypersurface of $\mathbf{P}$.

We can define the *universal Hessian*

$$\mathcal{H} : \pi_1^*T(X)\otimes\pi_1^*T(X) \to \pi_1^*\mathcal{L}\otimes\pi_2^*\mathcal{O}_{\mathbf{P}}(1)$$

on $\mathcal{C}$, where $T(X)$ is the tangent bundle of X. As was proved in [19, Proposition 3.14], the subscheme $\mathcal{E}$ coincides with the degeneracy subscheme of the homomorphism

$$\pi_1^*T(X) \to \pi_1^*\mathcal{L}\otimes\pi_2^*\mathcal{O}_{\mathbf{P}}(1)\otimes(\pi_1^*T(X))^\vee$$

induced from $\mathcal{H}$, where $(\pi_1^*T(X))^\vee$ is the dual vector bundle of $\pi_1^*T(X)$. (See Notation and Terminology below for the definition of degeneracy subschemes.) The peculiarity of the geometry of the dual variety in characteristic 2 stems from the fact that the universal Hessian is not only symmetric but also *anti*-symmetric; that is, $\mathcal{H}(x\otimes x) = 0$ and $\mathcal{H}(x\otimes y)+\mathcal{H}(y\otimes x) = 0$ hold for any local sections x and y of $\pi_1^*T(X)$. From this fact, it follows that the irreducible divisor $\mathcal{E}$ is written as $2\mathcal{R}$, where $\mathcal{R}$ is a reduced divisor of $\mathcal{C}$. We denote by $\varpi_2 : \mathcal{R} \to \mathbf{P}$ the projection.

Definition 1.2. We put $R := k[[t_1, \dots, t_{m-4}]]$. It turns out that the equations

$$
\begin{aligned}
(1.2)\quad u_1y^2 + v_1 &= u_2y^2 + v_2 = wv_1 + u_1v_2y = \\
&= wv_2 + u_2v_1y = v_1v_2 + wy = w^2 + u_1u_2y^2 = 0
\end{aligned}
$$

define an $(m-1)$-dimensional singular scheme in the $(m+2)$-dimensional smooth scheme $\operatorname{Spec} R[[u_1, v_1, u_2, v_2, w, y]]$. We will denote this $(m-1)$-dimensional singular scheme by Γ_{m-1}.

The scheme Γ_{m-1} is singular along the $(m-2)$-dimensional locus defined by

$$v_1 = v_2 = w = y = 0.$$

See §4 for the geometric meaning of Γ_{m-1}.

Let $P = (p, H) \in X \times \mathbf{P}$ be a general point of the reduced irreducible divisor $\mathcal{R}$ of $\mathcal{C}$. Then we have the following:

(I) There exist isomorphisms of local rings

$$\mathcal{O}_{\pi_2^{-1}(H),P} \cong k[[s,t]]/(s^2,t^2) \quad \text{and} \quad \mathcal{O}_{\varpi_2^{-1}(H),P} \cong k[[s,t]]/(s^2,t^2)$$

over k. In particular, the projection $\varpi_2 : \mathcal{R} \to \mathbf{P}$ is inseparable of degree 4 over its image.

(II) The formal completion

$$(\pi_2)^{\wedge}_P : \operatorname{Spec}(\mathcal{O}_{\mathcal{C},P})^{\wedge} \to \operatorname{Spec}(\mathcal{O}_{\mathbf{P},H})^{\wedge}$$

of the projection $\pi_2 : \mathcal{C} \to \mathbf{P}$ at P factors through a singular scheme isomorphic to Γ_{m-1}.

(III) Let Λ and L be general linear subspaces of $\mathbf{P}$ of dimension 2 and 3, respectively, such that $H \in \Lambda \subset L$. We put $C_\Lambda := \pi_2^{-1}(\Lambda)$ and $S_L := \pi_2^{-1}(L)$. Then S_L is smooth of dimension 2 at P, and C_Λ is a curve on S_L that has an ordinary cusp at P.

(IV) Let $\nu : \widetilde{C}_\Lambda \to C_\Lambda$ be the normalization of C_Λ at P, and let z be a formal parameter of $\widetilde{C}_\Lambda$ at the inverse image $P' \in \widetilde{C}_\Lambda$ of P. Then the formal completion at P' of the composite of $\nu : \widetilde{C}_\Lambda \to C_\Lambda$ and $\pi_2 \,|\, C_\Lambda : C_\Lambda \to \Lambda$ is written as

$$
\begin{aligned}
((\pi_2 \,|\, C_\Lambda) \circ \nu)^* x &= a\, z^4 + (\text{terms of degree} \geq 6) \quad \text{and} \\
((\pi_2 \,|\, C_\Lambda) \circ \nu)^* y &= b\, z^4 + (\text{terms of degree} \geq 6)
\end{aligned}
$$

for some $a, b \in k$, where (x, y) is a formal parameter system of Λ at H.

This paper is organized as follows. First we define several notions in §2. The conditions on $P = (p, H) \in \mathcal{R}$ for the facts (I)-(IV) above to hold will be stated more precisely in terms of the singularity of the divisor $H \cap X$ of X at p. For this purpose, we define some classes of hypersurface singularities in §3. In §4, the $(m-1)$-dimensional singular scheme Γ_{m-1} is introduced. After recalling the definitions and results given in [19, Section 3] in §5, we prove the main results above in more refined forms in §6. In §7, we give a remark about the degree of $\mathcal{R}$ with respect to $\mathcal{O}_{\mathbf{P}}(1)$, and derive a nontrivial divisibility relation among Chern numbers of X from the fact (I) above.

The author would like to thank Professor Hajime Kaji for many valuable comments and suggestions.

Notation and Terminology.

(1) We work over an algebraically closed field k. By a *variety*, we mean a reduced irreducible quasi-projective scheme over k. A *point* of a variety means a closed point. Let X be a variety, and $P \in X$ a point. We denote by $(X, P)^{\wedge}$ the scheme $\mathrm{Spec}(\mathcal{O}_{X,P})^{\wedge}$, where $(\mathcal{O}_{X,P})^{\wedge}$ is the formal completion of the local ring $\mathcal{O}_{X,P}$ of X at P.

(2) Let X be a smooth variety. We denote by $T_P(X)$ the Zariski tangent space to X at a point $P \in X$, and by $T(X)$ the tangent bundle of X.

(3) Let E and F be vector bundles on a variety X with rank e and f, respectively, and let $\sigma : E \to F$ be a bundle homomorphism. We put $r := \min(e, f)$. The *degeneracy subscheme* of σ is the closed subscheme of X defined locally on X by all the r-minors of the $f \times e$-matrix expressing σ.

(4) Let $f : X \to Y$ be a morphism from a smooth variety X to a smooth variety Y. The *critical subscheme* of f is the degeneracy subscheme of the homomorphism $df : T(X) \to f^*T(Y)$.

(5) For a formal power series F with coefficients in k, we denote by $F^{[d]}$ the homogeneous part of degree d in F.

§2. Preliminaries

2.1. The quotient morphism by a tangent subbundle

Let X be a smooth variety defined over an algebraically closed field of characteristic $p > 0$.

Definition 2.1. A subbundle $\mathcal{N}$ of $T(X)$ is called *integrable* if $\mathcal{N}$ is closed under the pth power operation and the bracket product of derivations.

Proposition 2.2 ([18] Théorème 2). *Let $\mathcal{N}$ be an integrable subbundle of $T(X)$. Then there exists a unique morphism $q : X \to X^{\mathcal{N}}$ with the following properties;*

(i) *q induces a homeomorphism on the underlying topological spaces,*
(ii) *q is a radical covering of height 1, and*
(iii) *the kernel of $dq : T(X) \to q^* T(X^{\mathcal{N}})$ coincides with $\mathcal{N}$.*

Moreover, the variety $X^{\mathcal{N}}$ is smooth, and the morphism q is finite of degree p^r, where r is the rank of $\mathcal{N}$.

Definition 2.3. For an integrable subbundle $\mathcal{N}$ of $T(X)$, the morphism $q : X \to X^{\mathcal{N}}$ is called the *quotient morphism* by $\mathcal{N}$.

From the construction of the quotient morphism given in the proof of [18, Théorème 2], we obtain the following:

Proposition 2.4. *Let $f : X \to Y$ be a morphism from a smooth variety X to a smooth variety Y such that the kernel $\mathcal{K}$ of $df : T(X) \to f^* T(Y)$ is a subbundle of $T(X)$. Then $\mathcal{K}$ is integrable, and the morphism f factors through the quotient morphism $q : X \to X^{\mathcal{K}}$ by $\mathcal{K}$.*

2.2. Anti-symmetric forms

Definition 2.5. Let V be a finite dimensional vector space over a field K. A bilinear form $\varphi : V \times V \to K$ is called *anti-symmetric* if the following conditions are satisfied:

(i) $\varphi(v, v) = 0$ for any $v \in V$, and
(ii) $\varphi(v, w) + \varphi(w, v) = 0$ for any $v, w \in V$.

Note that, in characteristic 2, the condition (ii) does not imply the condition (i).

Definition 2.6. An $n \times n$-matrix $A = (a_{ij})$ with components in a commutative ring is called *anti-symmetric* if the following conditions are satisfied:

(i) $a_{ii} = 0$ for $i = 1, \dots, n$, and
(ii) $a_{ij} + a_{ji} = 0$ for $i, j = 1, \dots, n$.

An anti-symmetric matrix in characteristic 2 is just a symmetric matrix with zero diagonal components.

The following results are well-known. We put

$$J_2 := \begin{bmatrix} 0 & 1 \\ -1 & 0 \end{bmatrix},$$

and let $J_{2r,n}$ be the $n \times n$ matrix

$$J_{2r,n} := \begin{bmatrix} J_2 & & & 0 \\ & \ddots & & \\ & & J_2 & \\ & 0 & & 0 \end{bmatrix}$$

with r copies of J_2 along the diagonal.

Lemma 2.7. *Let A be an anti-symmetric matrix of type $n \times n$ with components in a field K. Then the rank of A is even, and there exists $T \in GL(n, K)$ such that ${}^tTAT$ is equal to $J_{2r,n}$, where $2r :=$ rank(A).*

For a positive integer m, we define a homogeneous polynomial f_{2m} of degree m in variables x_{ij} $(1 \le i < j \le 2m)$ by

$$f_{2m} := \sum \operatorname{sign} \begin{pmatrix} 1 & 2 & \dots & & 2m \\ i_1 & j_1 & \dots & i_m & j_m \end{pmatrix} x_{i_1 j_1} x_{i_2 j_2} \cdots x_{i_m j_m},$$

where the summation ranges over all the lists $[[i_1, j_1], \dots, [i_m, j_m]]$ satisfying the conditions $i_\nu < j_\nu$ $(\nu = 1, \dots, m)$, $i_1 < i_2 < \cdots < i_m$, and $\{i_1, j_1, \dots, i_m, j_m\} = \{1, 2, \dots, 2m\}$.

Lemma 2.8. *Let*

$$A = \begin{bmatrix} 0 & x_{12} & x_{13} & \dots \\ -x_{12} & 0 & x_{23} & \dots \\ -x_{13} & -x_{23} & 0 & \dots \\ & \dots & & \ddots \end{bmatrix} \tag{2.1}$$

be an anti-symmetric matrix of type $2m \times 2m$ with components being variables x_{ij} $(i < j)$. Then $\det A$ is equal to f_{2m}^2.

For a positive integer i, we put

$$\tau(i) := \begin{cases} i-1 & \text{if } i \text{ is even}, \\ i+1 & \text{if } i \text{ is odd}. \end{cases}$$

Corollary 2.9. *Let A be an anti-symmetric matrix as in Lemma 2.8. We put*

$$g_A := \sqrt{\det(J_{2r,2m} + A)}\,,$$

which is a polynomial of x_{ij} by Lemma 2.8.

(1) *If $2r \le 2m-4$, then g_A has no terms of degree ≤ 1.*

(2) *Suppose that $2r = 2m-2$. Then g_A is equal, up to sign, to*

$$z + \sum_{j=1}^{m-1} \varepsilon_j\, x_{2j-1,2j}\, z + \sum_{i=1}^{2m-2} \varepsilon'_i\, x_{i,2m-1}\, x_{\tau(i),2m} + (\text{terms of degree} \ge 3),$$

where $z := x_{2m-1,2m}$, and ε_j and ε'_i are ± 1.

Using (2.1), we consider the affine space

$$\mathbb{A}^{m(2m-1)} = \operatorname{Spec} k[\ldots, x_{ij}, \ldots] \quad (1 \le i < j \le 2m)$$

as the space of anti-symmetric matrices of type $2m \times 2m$.

Lemma 2.10. *The hypersurface in $\mathbb{A}^{m(2m-1)}$ defined by $f_{2m} = 0$ is irreducible.*

Proof. By Lemma 2.7, the hypersurface defined by $f_{2m} = 0$ is the closure of the locus

$$\{\, {}^tT J_{2m-2,2m} T \mid T \in GL(2m,k) \,\},$$

and hence is irreducible. Q.E.D.

2.3. The formula of Harris-Tu-Pragacz

Let X be a smooth variety, E a vector bundle on X, and L a line bundle on X.

Definition 2.11. A bundle homomorphism

$$\sigma : E \otimes_{\mathcal{O}_X} E \to L$$

is called *anti-symmetric* if the following conditions are satisfied:

(i) $\sigma(x \otimes x) = 0$ for any local section x of E, and

(ii) $\sigma(x \otimes y) + \sigma(y \otimes x) = 0$ for any local sections x and y of E.

Definition 2.12. For an anti-symmetric homomorphism $\sigma : E \otimes E \to L$, we define the *degeneracy subscheme* $\mathbf{R}(\sigma)$ as follows. Let U be an arbitrary Zariski open subset of X over which E and L are trivialized. Let A_U be the anti-symmetric matrix with components in $\Gamma(U, \mathcal{O}_X)$ expressing σ over U. By Lemma 2.8, there exists $f_U \in \Gamma(U, \mathcal{O}_X)$ such that $\det A_U = f_U^2$. Then the closed subschemes $\{f_U = 0\}$ of Zariski open subsets U patch together to form a closed subscheme $\mathbf{R}(\sigma)$ of X, which is the degeneracy subscheme of σ.

Definition 2.13. For a closed subscheme W of X, we denote by $[W]$ the class of W in the Chow group of X.

The following was proved by Harris-Tu [4] in characteristic 0 and by Pragacz [17] in any characteristics.

Theorem 2.14. *Suppose that the rank e of E is even, and that $\mathbf{R}(\sigma)$ is of codimension 1 in X. Then we have*

$$[\mathbf{R}(\sigma)] = \left(\frac{e}{2}c_1(L) - c_1(E)\right) \cap [X]$$

in the Chow group of X.

Remark 2.15. Suppose that $\mathbf{R}(\sigma)$ is of codimension 1 in X. Let

$$\sigma^{\sim} : E \to L \otimes E^{\vee}$$

be the homomorphism induced from σ, where $E^{\vee}$ is the dual vector bundle of E. Then the degeneracy subscheme of $\sigma^{\sim}$ is also a divisor of X, and is equal to $2\,\mathbf{R}(\sigma)$ as a divisor of X.

2.4. Ordinary cusps

Let P be a point of a smooth surface S, and let (s,t) be a formal parameter system of S at P. For $\phi \in (\mathcal{O}_{S,P})^{\wedge} = k[[s,t]]$, we denote by $\phi^{[d]}$ the homogeneous part of degree d in (s,t). Let C be a divisor of S that contains P and is singular at P.

Definition 2.16. We say that $\phi \in k[[s,t]]$ is *degenerate* if $\phi^{[0]} = \phi^{[1]} = 0$ and $\phi^{[2]} = l(s,t)^2$ hold for some linear form $l(s,t)$ of (s,t). It is obvious that this definition of degeneracy does not depend on the choice of the formal parameter system (s,t). We say that C has a *degenerate singularity* at P if a (and hence any) formal power series defining C at P is degenerate.

The equivalence of the following conditions on the singularity of the curve C at P, which is well-known in characteristic 0, holds also in characteristic 2.

(i) For an arbitrary formal parameter system (s,t) of S at P, any formal power series ϕ defining C at P is degenerate and the linear form $\sqrt{\phi^{[2]}}$ is not zero and does not divide $\phi^{[3]}$.

(ii) There exist a formal parameter system (s, t) of S at P and a defining formal power series ϕ of C at P such that ϕ is degenerate and that the linear form $\sqrt{\phi^{[2]}}$ is not zero and does not divide $\phi^{[3]}$.

(iii) There exists a formal parameter system (s, t) of S at P such that C is defined by $s^2 + t^3 = 0$ locally at P.

Definition 2.17. We say that C has an *ordinary cusp* at P if the conditions (i)-(iii) above are satisfied.

§3. Hypersurface singularities in characteristic 2

From now on until the end of the paper, we assume that the base field k is of characteristic 2.

Let X be a smooth variety of dimension n. Let p be a point of X, and let D be a hypersurface of X that passes through p and is singular at p. Let $(x_1, \dots, x_n)$ be a formal parameter system of X at p. Suppose that D is defined by $\phi = 0$ locally at p, where ϕ is a formal power series of $(x_1, \dots, x_n)$. The $n \times n$ matrix

$$H_{\phi,p} := \left(\frac{\partial^2 \phi}{\partial x_i \partial x_j}(p) \right)$$

defining the Hessian

$$T_p(X) \times T_p(X) \quad \to \quad k$$

of D at p is *anti*-symmetric because we are in characteristic 2.

Definition 3.1. A formal parameter system $(x_1, \dots, x_n)$ of X at p is said to be *admissible with respect to* ϕ if the matrix $H_{\phi,p}$ is of the form $J_{2r,n}$, where $2r$ is the rank of the Hessian of D at p.

Remark 3.2. By Lemma 2.7, a formal parameter system admissible with respect to ϕ is obtained from an arbitrarily given formal parameter system by a *linear* transformation of parameters.

From now on to the end of this section, we assume that n is even.

We put

$$\rho := f_n\left(\dots, \frac{\partial^2 \phi}{\partial x_i \partial x_j}, \dots\right) = \sqrt{\det\left(\frac{\partial^2 \phi}{\partial x_i \partial x_j}\right)},$$

where f_n is the polynomial defined in §2.2.

Definition 3.3. We put $(X,p)^\wedge := \mathrm{Spec}(\mathcal{O}_{X,p})^\wedge$. Let $C(D,p)$ be the subscheme of $(X,p)^\wedge$ defined by

$$\phi = \frac{\partial\phi}{\partial x_1} = \cdots = \frac{\partial\phi}{\partial x_n} = 0,$$

and let $R(D,p)$ be the subscheme of $(X,p)^\wedge$ defined by

$$\phi = \frac{\partial\phi}{\partial x_1} = \cdots = \frac{\partial\phi}{\partial x_n} = \rho = 0.$$

It is obvious that each of $C(D,p)$ and $R(D,p)$ depends only on D and p, and not on the choice of ϕ and $(x_1,\ldots,x_n)$.

Using the admissible formal parameter system with respect to ϕ, or the subschemes $C(D,p)$ and $R(D,p)$ of $(X,p)^\wedge$, we can define various classes of hypersurface singularities in characteristic 2.

Suppose that the rank of the Hessian matrix $H_{\phi,p}$ is $n-2$. Let

$$(x,\xi) = (x_1,\ldots,x_{n-2},\xi_1,\xi_2)$$

be an admissible formal parameter system with respect to ϕ. We put

$$G := \{\, M \in GL(n-2,k) \mid {}^tMJ_{n-2,n-2}M = J_{n-2,n-2} \,\}.$$

The following lemma is trivial:

Lemma 3.4. *Let*

$$(x',\xi') = (x'_1,\ldots,x'_{n-2},\xi'_1,\xi'_2)$$

be another formal parameter system of X at p. Then (x',ξ') is admissible with respect to ϕ if and only if (x,ξ) and (x',ξ') are related by the transformation of the form

$$\left[\begin{array}{c} x \\ \hline \xi \end{array}\right] = \left[\begin{array}{c|c} A & 0 \\ \hline * & B \end{array}\right] \left[\begin{array}{c} x' \\ \hline \xi' \end{array}\right] + (\text{terms of degree} \geq 2)$$

such that $A \in G$ and $B \in GL(2,k)$.

We denote the coefficients in the formal power series expansion of ϕ in the parameters (x,ξ) as indicated in Table 3.1. The following Proposition follows immediately from Lemma 3.4.

$$
\begin{array}{llllllll}
 & x_1 x_2 + \ldots + x_{n-3} x_{n-2} + & & & & & & \\
+ & a_1 x_1^2 & + & \cdots & + & a_{n-2} x_{n-2}{}^2 & + & \alpha_1 \xi_1^2 + \alpha_2 \xi_2^2 + \\
+ & b_1 x_1 \xi_1^2 & + & \cdots & + & b_{n-2} x_{n-2} \xi_1^2 & + & \beta_1 \xi_1^3 + \beta_2 \xi_1^2 \xi_2 + \\
+ & c_1 x_1 \xi_2^2 & + & \cdots & + & c_{n-2} x_{n-2} \xi_2^2 & + & \gamma_1 \xi_1 \xi_2^2 + \gamma_2 \xi_2^3 + \\
+ & d_1 x_1 \xi_1 \xi_2 & + & \cdots & + & d_{n-2} x_{n-2} \xi_1 \xi_2 & + & \\
+ & \multicolumn{7}{l}{\text{(terms of degree 3 in } (x,\xi) \text{ and degree} \geq 2 \text{ in } x)} \\
+ & \multicolumn{7}{l}{f_{40}\, \xi_1^4 + f_{31}\, \xi_1^3 \xi_2 + f_{22}\, \xi_1^2 \xi_2^2 + f_{13}\, \xi_1 \xi_2^3 + f_{04}\, \xi_2^4 +} \\
+ & \multicolumn{7}{l}{\text{(terms of degree 4 in } (x,\xi) \text{ and degree} \geq 1 \text{ in } x)} \\
+ & \multicolumn{7}{l}{\text{(terms of degree} \geq 5 \text{ in } (x,\xi))}
\end{array}
$$

Table 3.1. The coefficients of a formal power series in (x, ξ)

Proposition 3.5. *Suppose that the rank of $H_{\phi,p}$ is $n-2$. Then the following condition is independent of the choice of ϕ and (x, ξ): at least one of the coefficients $\alpha_1, \alpha_2, \beta_1, \beta_2, \gamma_1, \gamma_2$ is not zero.*

Definition 3.6. We say that the singularity of D at p is *of type* $(\widetilde{\mathrm{A}})$ if $H_{\phi,p}$ is of rank $n-2$ and at least one of the coefficients α_1, α_2, β_1, β_2, γ_1 or γ_2 is not zero.

Proposition-Definition 3.7. (1) *The following three conditions are equivalent:*

(i) $\dim_k \mathcal{O}_{C(D,p),p} \leq 4$,
(ii) $\mathcal{O}_{C(D,p),p} \cong k[[s,t]]/(s^2, t^2)$, *and*
(iii) *the rank of the Hessian $H_{\phi,p}$ is $n-2$, and*

$$\operatorname{rank} \begin{bmatrix} \alpha_1 & \beta_1 & \beta_2 \\ \alpha_2 & \gamma_1 & \gamma_2 \end{bmatrix} = 2. \tag{3.1}$$

We say that the singularity of D at p is of type (C) *if these conditions are satisfied.*

(2) *We put*

$$T_1 := \sum_{i=1}^{n-2} d_i b_{\tau(i)} + f_{31} \quad \textit{and} \quad T_2 := \sum_{i=1}^{n-2} d_i c_{\tau(i)} + f_{13}.$$

The following three conditions are equivalent:

(i) $\dim_k \mathcal{O}_{R(D,p),p} \leq 4$,
(ii) $\mathcal{O}_{R(D,p),p} \cong k[[s,t]]/(s^2,t^2)$, *and*
(iii) *the rank of the Hessian* $H_{\phi,p}$ *is* $n-2$, *and*

$$\text{rank}\begin{bmatrix} \alpha_1 & \beta_1 & \beta_2 & T_1 \\ \alpha_2 & \gamma_1 & \gamma_2 & T_2 \end{bmatrix} = 2. \tag{3.2}$$

We say that the singularity of D *at* p *is* of type (R) *if these conditions are satisfied.*

Remark 3.8. As a corollary of Proposition-Definition 3.7, we see that the conditions (3.1) and (3.2) do not depend on the choice of the admissible formal parameter system (x,ξ). This fact can also be proved directly by means of Lemma 3.4.

For the proof of Proposition-Definition 3.7, we use the following easy lemma:

Lemma 3.9. *Let* $g_1,\ldots,g_l$ *be degenerate formal power series in* $k[[s,t]]$, *and let* J *be the ideal of* $k[[s,t]]$ *generated by* $g_1,\ldots,g_l$. *We denote by* a_i *and* b_i *the coefficients of* s^2 *and* t^2 *in* g_i, *respectively;*

$$g_i = a_i s^2 + b_i t^2 + (\text{terms of degree} \geq 3).$$

Then the following equivalence holds:

$$\dim_k k[[s,t]]/J \leq 4 \iff J=(s^2,t^2) \iff \text{rank}\begin{bmatrix} a_1 & \ldots & a_l \\ b_1 & \ldots & b_l \end{bmatrix} = 2.$$

Proof of Proposition-Definition 3.7. Since the proof of the assertion (1) is similar to and simpler than that of (2), we prove only the assertion (2). For simplicity, we put

$$D_i\phi := \frac{\partial \phi}{\partial x_i},$$

and let I denote the ideal of $(\mathcal{O}_{X,P})^\wedge$ generated by $\phi, D_1\phi,\ldots,D_n\phi$ and ρ. Suppose that rank $H_{\phi,p} = n-2l$, and let $(x_1,\ldots,x_n)$ be an admissible formal parameter system with respect to ϕ. Then the linear parts of ϕ and $D_i\phi$ are given as follows;

$$\phi^{[1]} = 0, \quad (D_i\phi)^{[1]} = \begin{cases} x_{\tau(i)} & \text{if } i \leq n-2l \\ 0 & \text{if } i > n-2l. \end{cases}$$

By Corollary 2.9, we also have

$$\rho^{[1]} = \begin{cases} 0 & \text{if } n-2l \leq n-4 \\ d_1x_1+\cdots+d_{n-2}x_{n-2} & \text{if } n-2l = n-2. \end{cases}$$

Therefore the classes of $1 \in k$ and $x_{n-2l+1}, \dots, x_n$ span a $(2l+1)$-dimensional linear subspace in $(\mathcal{O}_{R(D,p),p})^\wedge = k[[x_1, \dots, x_n]]/I$. Consequently the condition that $\dim_k \mathcal{O}_{R(D,p),p} \le 4$ implies that $\operatorname{rank} H_{\phi,p}$ is $n-2$. Hence it suffices to show the equivalence of the conditions (i), (ii) and (iii) under the assumption that $\operatorname{rank} H_{\phi,p} = n-2$.

We use the admissible formal parameter system (x, ξ) and Table 3.1. Let I' be the ideal of $k[[x, \xi]]$ generated by $D_1\phi, \dots, D_{n-2}\phi$, and put

$$s := \xi_1 \bmod I', \quad t := \xi_2 \bmod I'.$$

For $i = 1, \dots, n-2$, we have

$$D_i\phi = x_{\tau(i)} + b_i\xi_1^2 + c_i\xi_2^2 + d_i\xi_1\xi_2 + \\ + \begin{pmatrix} \text{terms of degree} \ge 2 \text{ in } (x,\xi) \text{ and degree} \\ \ge 1 \text{ in } x, \text{ or degree} \ge 3 \text{ in } (x,\xi) \end{pmatrix}.$$

Hence we obtain

$$x_i \bmod I' = b_{\tau(i)}s^2 + c_{\tau(i)}t^2 + d_{\tau(i)}st + (\text{terms of degree} \ge 3 \text{ in } (s,t))$$

for $i = 1, \dots, n-2$. In particular, we have $k[[x,\xi]]/I' = k[[s,t]]$. We put

$$g_1 := \phi \bmod I', \quad g_2 := (D_{n-1}\phi) \bmod I', \quad g_3 := (D_n\phi) \bmod I'.$$

Then we have

$$\begin{aligned} g_1 &= \alpha_1 s^2 + \alpha_2 t^2 + (\text{terms of degree} \ge 3 \text{ in } (s,t)), \\ g_2 &= \beta_1 s^2 + \gamma_1 t^2 + (\text{terms of degree} \ge 3 \text{ in } (s,t)), \\ g_3 &= \beta_2 s^2 + \gamma_2 t^2 + (\text{terms of degree} \ge 3 \text{ in } (s,t)). \end{aligned}$$

On the other hand, we see that

$$\frac{\partial^2\phi}{\partial\xi_1\partial\xi_2} \bmod I', \quad \frac{\partial^2\phi}{\partial x_i\partial\xi_1} \bmod I', \quad \text{and} \quad \frac{\partial^2\phi}{\partial x_i\partial\xi_2} \bmod I'$$

are equal to

$$f_{31}\, s^2 + f_{13}\, t^2 + \sum_{i=1}^{n-2} d_i\,(b_{\tau(i)}s^2 + c_{\tau(i)}t^2 + d_{\tau(i)}st) + \\ + (\text{terms of degree} \ge 3 \text{ in } (s,t)),$$

$$d_i\, t + (\text{terms of degree} \ge 2 \text{ in } (s,t)) \quad \text{and}$$
$$d_i\, s + (\text{terms of degree} \ge 2 \text{ in } (s,t)),$$

respectively. Moreover, we have

$$\frac{\partial^2 \phi}{\partial x_i \partial x_j} \bmod I' = \begin{cases} 1 + (\text{terms of degree} \geq 1 \text{ in } (s,t)) & \text{if } i = \tau(j), \\ (\text{terms of degree} \geq 1 \text{ in } (s,t)) & \text{if } i \neq \tau(j), \end{cases}$$

for $i, j = 1, \dots, n-2$. By Corollary 2.9 and using $\sum_{i=1}^{n-2} d_i d_{\tau(i)} = 0$, we obtain

$$g_4 := \rho \bmod I' = T_1 s^2 + T_2 t^2 + (\text{terms of degree} \geq 3 \text{ in } (s,t)).$$

Since

$$(\mathcal{O}_{R(D,p),p})^\wedge = k[[x,\xi]]/I = k[[s,t]]/(g_1, g_2, g_3, g_4),$$

we obtain the equivalence of the conditions (i)-(iii) from Lemma 3.9. Q.E.D.

The following remark is crucial in the proof of Proposition 6.12.

Remark 3.10. Each of the conditions that the singularity of D at p is of type $(\widetilde{\mathrm{A}})$, (C) and (R) is an *open* condition on the coefficients of the formal power series ϕ with $\operatorname{rank} H_{\phi,p} \leq n-2$ in the following sense. We denote by $\mathfrak{m}$ the maximal ideal of $k[[x_1, \dots, x_n]]$. For a positive integer $d \geq 2$, we put

$$V_d := \mathfrak{m}^2 / \mathfrak{m}^{d+1}.$$

For $\phi \in \mathfrak{m}^2$, let $\bar{\phi} \in V_d$ denote the class of ϕ modulo $\mathfrak{m}^{d+1}$. By Lemma 2.10, the locus

$$\Delta := \{ \bar{\phi} \in V_2 \mid \operatorname{rank} H_{\phi,p} \leq n-2 \}$$

is a closed irreducible hypersurface of V_2. For $d \geq 3$, let Δ_d denote the pull-back of Δ by the natural linear homomorphism $V_d \to V_2$. Note that the types $(\widetilde{\mathrm{A}})$, (C) and (R) of the hypersurface singularities are defined by using the coefficients of terms of degree ≤ 3, ≤ 3 and ≤ 4 of the formal power series defining the hypersurface, respectively. Hence the loci

$$\begin{aligned} \Delta(\widetilde{\mathrm{A}}) &:= \{ \bar{\phi} \in V_3 \mid \text{the singularity of } \phi = 0 \text{ at } p \text{ is of type } (\widetilde{\mathrm{A}}) \}, \\ \Delta(\mathrm{C}) &:= \{ \bar{\phi} \in V_3 \mid \text{the singularity of } \phi = 0 \text{ at } p \text{ is of type (C)} \}, \\ \Delta(\mathrm{R}) &:= \{ \bar{\phi} \in V_4 \mid \text{the singularity of } \phi = 0 \text{ at } p \text{ is of type (R)} \}, \end{aligned}$$

are well defined. The loci $\Delta(\widetilde{\mathrm{A}})$ and $\Delta(\mathrm{C})$ are Zariski open dense subsets of Δ_3, and the locus $\Delta(\mathrm{R})$ is a Zariski open dense subset of Δ_4.

§4. The singular scheme Γ_{m-1}

We define a homomorphism

$$\gamma^* : k[[u_1, v_1, u_2, v_2, w, y]] \to k[[x_1, x_2, z]]$$

over k by

$$\gamma^* u_1 = x_1^2, \quad \gamma^* v_1 = x_1 z, \quad \gamma^* u_2 = x_2^2, \\ \gamma^* v_2 = x_2 z, \quad \gamma^* w = x_1 x_2 z, \quad \gamma^* y = z,$$

and denote by

$$\Gamma := \operatorname{Spec}(k[[u_1, v_1, u_2, v_2, w, y]] \,/\, \operatorname{Ker} \gamma^*)$$

the scheme-theoretic image of the morphism

$$\gamma : \operatorname{Spec} k[[x_1, x_2, z]] \to \operatorname{Spec} k[[u_1, v_1, u_2, v_2, w, y]]$$

induced from γ^*. The critical subscheme of γ is defined by $z^2 = 0$. Calculating the Gröbner basis (see [1]) of the ideal

$$\langle\, u_1 - x_1^2,\ v_1 - x_1 z,\ u_2 - x_2^2,\ v_2 - x_2 z,\ w - x_1 x_2 z,\ y - z \,\rangle$$

in the polynomial ring $k[x_1, x_2, z, u_1, v_1, u_2, v_2, w, y]$ under the pure lexicographic order

$$x_1 > x_2 > z > w > v_2 > v_1 > y > u_2 > u_1,$$

we see that Γ is a 3-dimensional singular scheme defined by the equations (1.2) given in Introduction. The singular locus of Γ is equal to the 2-dimensional plane

$$v_1 = v_2 = w = y = 0.$$

For a k-algebra R, we denote by

$$\gamma^R : \operatorname{Spec} R[[x_1, x_2, z]] \to \operatorname{Spec} R[[u_1, v_1, u_2, v_2, w, y]]$$

the morphism obtained from γ by $k \hookrightarrow R$, and by

$$\Gamma^R := \operatorname{Spec}(R[[u_1, v_1, u_2, v_2, w, y]] \,/\, \operatorname{Ker}(\gamma^R)^*),$$

the scheme-theoretic image of γ^R.

Theorem 4.1. *Let V be a smooth variety of dimension $l \geq 3$, and let $\phi : V \to W$ be a morphism to a smooth variety W. Suppose that there exists a smooth hypersurface H of V such that, at every point P of H, the kernel of*

$$d_P\phi \; : \; T_P(V) \; \to \; T_{\phi(P)}(W)$$

is of dimension 2 and is contained in $T_P(H)$.

Let P be a point of H. We denote by R the ring of formal power series $k[[x_3, \dots, x_{l-1}]]$ of $l-3$ variables. Then there exists an isomorphism

$$\iota \; : \; (V, P)^\wedge \; \overset{\sim}{\to} \; \operatorname{Spec} R[[x_1, x_2, z]]$$

over k with the following properties:

(i) *the divisor $(H, P)^\wedge$ of $(V, P)^\wedge$ is mapped by the isomorphism ι to the divisor given by $z = 0$, and*

(ii) *there exists a unique morphism $\psi : \Gamma^R \to (W, \phi(P))^\wedge$ such that the formal completion $\phi_P^\wedge$ of ϕ at P factors as the following diagram:*

$$\begin{array}{ccc} (V, P)^\wedge & \overset{\phi_P^\wedge}{\longrightarrow} & (W, \phi(P))^\wedge \\ \iota \downarrow \wr & & \uparrow \psi \\ \operatorname{Spec} R[[x_1, x_2, z]] & \underset{\gamma^R}{\longrightarrow} & \Gamma^R. \end{array}$$

Proof. We denote by

$$\phi_H \; : \; H \; \to \; W$$

the restriction of ϕ to H. The kernel $\mathcal{K}$ of the homomorphism

$$d\phi_H \; : \; T(H) \; \to \; \phi_H^* T(W)$$

is an integrable subbundle of $T(H)$ with rank 2 by the assumption. Let P be an arbitrary point of H. By [18, Proposition 6], there exist a basis (D_1, D_2) of the $\mathcal{O}_{H,P}$-module $\mathcal{K} \otimes \mathcal{O}_{H,P}$ and a system of uniformizing variables $(y_1, \dots y_{l-1})$ of H at P such that

$$D_i(y_j) = \delta_{ij} \qquad (i = 1, 2, \;\; j = 1, \dots, l-1).$$

In other words, the submodule $\mathcal{K} \otimes \mathcal{O}_{H,P}$ of $T(H) \otimes \mathcal{O}_{H,P}$ is equal to

$$\mathcal{O}_{H,P} \frac{\partial}{\partial y_1} \; \oplus \; \mathcal{O}_{H,P} \frac{\partial}{\partial y_2}.$$

We can choose a formal parameter system $(x_1, \dots, x_{l-1}, z)$ of V at P such that H is defined by $z = 0$, and that the restriction $x_i \mid H$ of x_i to H is equal to y_i. Then, for every element $g \in (\mathcal{O}_{W,\phi(P)})^\wedge$, we have

$$\frac{\partial\,(\phi_H^* g)}{\partial y_1} = \frac{\partial\,(\phi^* g)}{\partial x_1}\Big|_H \equiv 0 \quad \text{and} \quad \frac{\partial\,(\phi_H^* g)}{\partial y_2} = \frac{\partial\,(\phi^* g)}{\partial x_2}\Big|_H \equiv 0;$$

that is, $\phi^* g$ is contained in the subring A of $(\mathcal{O}_{V,P})^\wedge = k[[x_1, \dots, x_{l-1}, z]]$ defined by

$$A := \left\{ f \in (\mathcal{O}_{V,P})^\wedge \;\middle|\; \frac{\partial f}{\partial x_1} \text{ and } \frac{\partial f}{\partial x_2} \text{ are contained in the ideal } (z) \right\}.$$

Putting $R := k[[x_3, \dots, x_{l-1}]]$, we have

$$A \;=\; R[[x_1^2, x_2^2, z]] \;+\; z\,R[[x_1, x_2, z]] \;\;\subset\;\; R[[x_1, x_2, z]] \;=\; (\mathcal{O}_{V,P})^\wedge.$$

Therefore, the subring A is the image of the homomorphism

$$(\gamma^R)^* \;:\; R[[u_1, v_1, u_2, v_2, w, y]] \;\to\; R[[x_1, x_2, z]].$$

We denote by

$$j \;:\; R[[u_1, v_1, u_2, v_2, w, y]]\,/\,(\mathrm{Ker}\,(\gamma^R)^*) \;\xrightarrow{\sim}\; A$$

the induced isomorphism. Let $(t_1, \dots, t_q)$ be a formal parameter system of W at $\phi(P)$, where $q = \dim W$. Since $\phi^* t_i \in A$, there exists a unique element

$$\psi_i \;\in\; R[[u_1, v_1, u_2, v_2, w, y]]\,/\,(\mathrm{Ker}\,(\gamma^R)^*)$$

such that $j(\psi_i) = \phi^* t_i$. Define a morphism

$$\psi \;:\; \Gamma^R \;\to\; (W, \phi(P))^\wedge$$

by $\psi^* t_i = \psi_i$. Then $\phi_P^\wedge$ factors as $\psi \circ \gamma^R \circ \iota$, where ι is the isomorphism given by the canonical isomorphism $R[[x_1, x_2, z]] \xrightarrow{\sim} (\mathcal{O}_{V,P})^\wedge$. Q.E.D.

Definition 4.2. When R is the ring of formal power series in $(m-4)$ variables with coefficients in k, the scheme-theoretic image Γ^R is an $(m-1)$-dimensional singular scheme, which we will denote by Γ_{m-1}.

Remark 4.3. A normal form theorem similar to Theorem 4.1 for a morphism from a smooth surface to a smooth curve was proved in [20].

§5. The discriminant variety

In this section, we recall the definitions and results in [19, Section 3], which are valid in any characteristics.

Let $\overline{X}$ be a projective variety of dimension $n > 0$, $\mathcal{L}$ a line bundle on $\overline{X}$, and M a linear subspace of $H^0(\overline{X}, \mathcal{L})$ with dimension $m+1 \geq 2$. We denote by

$$\mathbf{P} := \mathbb{P}_*(M)$$

the parameter space of the m-dimensional linear system $|M|$ of divisors on $\overline{X}$ corresponding to M, and put

$$X := \overline{X} \setminus (\mathrm{Sing}(\overline{X}) \cup \mathrm{Bs}(|M|)),$$

where $\mathrm{Sing}(\overline{X})$ is the singular locus of $\overline{X}$ and $\mathrm{Bs}(|M|)$ is the base locus of $|M|$. We denote by $\Psi : X \to \mathbf{P}^\vee$ the morphism induced by $\mathbf{P}$, and set

$$X^\circ := \left\{ p \in X \;\middle|\; \begin{array}{l} \text{the homomorphism } d_p\Psi : T_p(X) \to T_{\Psi(p)}(\mathbf{P}^\vee) \\ \text{is injective} \end{array} \right\}.$$

Note that $X^\circ = \overline{X}$ if $\overline{X}$ is smooth and $|M|$ is very ample.

Assumption 5.1. Throughout the paper, we assume that $m > n$, and that X° is dense in $\overline{X}$.

For a non-zero vector $f \in M$, we denote by $[f] \in \mathbf{P}$ the corresponding point of $\mathbf{P}$, by $\overline{D}_{[f]}$ the divisor of $\overline{X}$ defined by $f = 0$, and by $D_{[f]}$ the intersection $\overline{D}_{[f]} \cap X$. Let

$$\mathrm{pr}_1 : X \times \mathbf{P} \to X \quad \text{and} \quad \mathrm{pr}_2 : X \times \mathbf{P} \to \mathbf{P}$$

be the projections. We put

$$\widetilde{\mathcal{L}} := \mathrm{pr}_1^* \mathcal{L} \otimes \mathrm{pr}_2^* \mathcal{O}_{\mathbf{P}}(1).$$

There exists a canonical isomorphism $M^\vee \xrightarrow{\sim} H^0(\mathbf{P}, \mathcal{O}_{\mathbf{P}}(1))$ unique up to multiplicative constants. Combining this isomorphism with the inclusion $M \hookrightarrow H^0(X, \mathcal{L})$, we obtain a natural homomorphism

$$\mathrm{Hom}(M, M) = M \otimes M^\vee \;\to\; H^0(X \times \mathbf{P}, \widetilde{\mathcal{L}}).$$

We denote by

$$\sigma \in H^0(X \times \mathbf{P}, \widetilde{\mathcal{L}})$$

the global section of $\widetilde{\mathcal{L}}$ corresponding to the identity homomorphism of M. Note that σ is determined uniquely up to multiplicative constants. The zero locus of σ is equal to

$$\{\ (p, [f]) \in X \times \mathbf{P}\ \mid\ p \in D_{[f]}\ \}.$$

Let $\mathcal{D}$ be the subscheme of $X \times \mathbf{P}$ defined by $\sigma = 0$, and let

$$p_1 : \mathcal{D} \to X \quad \text{and} \quad p_2 : \mathcal{D} \to \mathbf{P}$$

be the projections. Then $\mathcal{D}$ is smooth of dimension $m + n - 1$, and $p_2 : \mathcal{D} \to \mathbf{P}$ is the universal family of the divisors $D_{[f]}$ ($[f] \in \mathbf{P}$). We have a natural homomorphism

$$d\sigma\ :\ T(X \times \mathbf{P}) \otimes \mathcal{O}_{\mathcal{D}}\ \to\ \widetilde{\mathcal{L}} \otimes \mathcal{O}_{\mathcal{D}}$$

defined by $\Theta \mapsto (\Theta\sigma) \,|\, \mathcal{D}^{\circ}$, where Θ is a local section of $T(X \times \mathbf{P})$ considered as a derivation of $\mathcal{O}_{X \times \mathbf{P}}$. We denote by

$$d\sigma_X\ :\ p_1^* T(X)\ \to\ \widetilde{\mathcal{L}} \otimes \mathcal{O}_{\mathcal{D}}$$

the restriction of $d\sigma$ to the direct factor $p_1^*T(X)$ of

$$T(X \times \mathbf{P}) \otimes \mathcal{O}_{\mathcal{D}} = p_1^*T(X) \oplus p_2^*T(\mathbf{P}).$$

Let $\mathcal{C}$ be the critical subscheme of $p_2 : \mathcal{D} \to \mathbf{P}$.

Proposition 5.2. (1) *The scheme $\mathcal{C}$ is equal to the degeneracy subscheme of $d\sigma_X$.*

(2) *The intersection $\mathcal{C}^{\circ}$ of $\mathcal{C}$ and $X^{\circ} \times \mathbf{P}$ is smooth, irreducible and of dimension $m - 1$.*

Corollary 5.3. (1) *The support of $\mathcal{C}$ is equal to the set*

$$\{\ (p, [f]) \in \mathcal{D}\ \mid\ D_{[f]} \text{ is singular at } p\ \}.$$

(2) *Let $P = (p, [f])$ be a point of $\mathcal{C}$, and let $\pi_2' : \mathcal{C} \to \mathbf{P}$ be the projection. Then, as a subscheme of $(\mathrm{pr}_2^{-1}([f]), P)^{\wedge} \cong (X, p)^{\wedge}$, the completion $(\pi_2'^{-1}([f]), P)^{\wedge}$ of the fiber of π_2' at P coincides with $C(D_{[f]}, p)$ defined in Definition 3.3.*

We denote by

$$\pi_1 : \mathcal{C}^{\circ} \to X^{\circ} \quad \text{and} \quad \pi_2 : \mathcal{C}^{\circ} \to \mathbf{P}$$

the projections from $\mathcal{C}^{\circ} = \mathcal{C} \cap (X^{\circ} \times \mathbf{P})$. We can define the *universal Hessian*

$$\mathcal{H}\ :\ \pi_1^*T(X^{\circ}) \otimes \pi_1^*T(X^{\circ})\ \to\ \widetilde{\mathcal{L}} \otimes \mathcal{O}_{\mathcal{C}^{\circ}}$$

on $\mathcal{C}^\circ$. Because we are in characteristic 2, the universal Hessian $\mathcal{H}$ on $\mathcal{C}^\circ$ is *anti*-symmetric. Let

$$\mathcal{H}^\sim : \pi_1^* T(X^\circ) \to \widetilde{\mathcal{L}} \otimes (\pi_1^* T(X^\circ))^\vee$$

be the homomorphism obtained from $\mathcal{H}$. Let $\mathcal{E}$ be the critical subscheme of the projection $\pi_2 : \mathcal{C}^\circ \to \mathbf{P}$.

Proposition 5.4. *The scheme $\mathcal{E}$ is equal to the degeneracy subscheme of $\mathcal{H}^\sim$.*

Corollary 5.5. *The support of $\mathcal{E}$ is equal to the set of all point $(p, [f]) \in \mathcal{C}^\circ$ such that the Hessian of the hypersurface singularity $p \in D_{[f]}$ is degenerate.*

Since the rank of an anti-symmetric form is always even, we obtain the following:

Corollary 5.6. *Suppose that n is odd. Then $\mathcal{E}$ coincides with $\mathcal{C}^\circ$.*

§6. Main results

From now on until the end of the paper, we assume that n is even.

Definition 6.1. Let $\mathcal{R} \subset \mathcal{C}^\circ$ be the degeneracy subscheme $\mathbf{R}(\mathcal{H})$ of the anti-symmetric form $\mathcal{H}$, and let

$$\varpi_1 : \mathcal{R} \to X^\circ \quad \text{and} \quad \varpi_2 : \mathcal{R} \to \mathbf{P}$$

be the projections.

Proposition 6.2. (1) *The subscheme $\mathcal{R}$ is of codimension ≤ 1 in $\mathcal{C}^\circ$. If $\mathcal{R}$ is of codimension* 1 *in $\mathcal{C}^\circ$, then $\mathcal{E}$ coincides with $2\mathcal{R}$ as a divisor of $\mathcal{C}^\circ$.*

(2) *Let $P = (p, [f])$ be a point of $\mathcal{R}$. Then, as a subscheme of $(\mathrm{pr}_2^{-1}([f]), P)^\wedge \cong (X, p)^\wedge$, the completion $(\varpi_2^{-1}([f]), P)^\wedge$ of the fiber of ϖ_2 at P coincides with $R(D_{[f]}, p)$ defined in Definition 3.3.*

Proof. The assertion (1) follows from Remark 2.15 and Proposition 5.4. The assertion (2) follows from the definition. Q.E.D.

It is possible that $\mathcal{R}$ is of codimension 0 in $\mathcal{C}^\circ$. The following is a classical example due to Wallace [22]. See also [21].

Example 6.3. Let X be the Fermat hypersurface of degree $2^\nu + 1$ in $\mathbb{P}^{n+1}$ with $\nu \ge 1$, and let $|M|$ be the complete linear system $|\mathcal{O}_X(1)|$. Then the rank of the Hessian of $D_{[f]}$ at p is zero at every point $(p, [f])$ of $\mathcal{C}^\circ = \mathcal{C}$. In particular, $\mathcal{R}$ coincides with $\mathcal{C}^\circ$.

Let $P = (p, [f])$ be a point of $\mathcal{R}$, so that $D_{[f]}$ is singular at p and the Hessian of $D_{[f]}$ at p is of rank $\leq n-2$. We investigate the formal completion of the morphism $\pi_2 : \mathcal{C}^\circ \to \mathbf{P}$ at P. For this purpose, we introduce a good formal parameter system of $X^\circ \times \mathbf{P}$ at P.

Because $p \in X^\circ$, we can choose a basis $\boldsymbol{b}_0, \dots, \boldsymbol{b}_m$ of the vector space M in such a way that the following conditions (i)-(iv) hold. Let $\tilde{\boldsymbol{b}}_i$ be the global section of $\mathcal{L}$ corresponding to $\boldsymbol{b}_i$, and let D_i be the divisor defined by $\tilde{\boldsymbol{b}}_i = 0$. Then

(i) $\boldsymbol{b}_0 = f$, so that $D_0 = D_{[f]}$ passes through p and is singular at p,
(ii) $p \in D_i$ for $i = 0, 1, \dots, m-1$, and $p \notin D_m$,
(iii) $D_1, \dots, D_n$ are smooth at p and intersect transversely at p, and
(iv) $D_{n+1}, \dots, D_{m-1}$ are singular at p.

For $i = 0, \dots, m-1$, we put

$$\phi_i := \tilde{\boldsymbol{b}}_i / \tilde{\boldsymbol{b}}_m,$$

which is a rational function on X regular at p. Then $(\phi_1, \dots, \phi_n)$ forms a local parameter system of X at p by (iii). We put

$$x_i := \phi_i \quad (i = 1, \dots, n).$$

We will regard $(x_1, \dots, x_n)$ as a formal parameter system of X at p, and consider ϕ_0 and $\phi_{n+1}, \dots, \phi_{m-1}$ as formal power series of $(x_1, \dots, x_n)$, which have no terms of degree ≤ 1 by (i) and (iv). Note that the rank of the anti-symmetric matrix $H_{\phi_0, p}$ is $\leq n-2$, because $P \in \mathcal{R}$. Using Remark 3.2, we can further assume the following by a linear transformation of $\boldsymbol{b}_1, \dots, \boldsymbol{b}_n$:

(v) the formal parameter system $(x_1, \dots, x_n)$ is admissible with respect to ϕ_0.

Let $(Y_0, \dots, Y_m)$ be the linear coordinates of M with respect to the basis $\boldsymbol{b}_0, \dots, \boldsymbol{b}_m$, and let $(y_1, \dots, y_m)$ be the affine coordinate system of $\mathbf{P}$ given by

$$y_i := Y_i / Y_0.$$

Then the point $[f] \in \mathbf{P}$ is the origin. We regard

$$(x_1, \dots, x_n, y_1, \dots, y_m)$$

as a formal parameter system of $X^\circ \times \mathbf{P}$ at $P = (p, [f])$.

We can regard the coordinates Y_i as global sections of $\mathcal{O}_{\mathbf{P}}(1)$. Recall that σ is the canonical section of $\widetilde{\mathcal{L}} := \mathrm{pr}_1^* \mathcal{L} \otimes \mathrm{pr}_2^* \mathcal{O}_{\mathbf{P}}(1)$. Multiplying

σ by a suitable non-zero constant, we have

$$\sigma = \sum_{i=0}^{m} \tilde{\boldsymbol{b}}_i \otimes Y_i.$$

We put

$$\Phi := \left(\sum_{i=0}^{m} \tilde{\boldsymbol{b}}_i \otimes Y_i \right) / (\tilde{\boldsymbol{b}}_m \otimes Y_0),$$

which is a rational function of $X^\circ \times \mathbf{P}$ regular at P. By definition, we have

$$\Phi = \phi_0 + y_1 x_1 + \cdots + y_n x_n + y_{n+1}\phi_{n+1} + \cdots + y_{m-1}\phi_{m-1} + y_m.$$

The scheme $\mathcal{D}$ is defined by $\Phi = 0$ locally at P. For simplicity, we put

$$D_i \Phi := \frac{\partial \Phi}{\partial x_i}.$$

By Proposition 5.2, the subscheme $\mathcal{C}^\circ$ of $X^\circ \times \mathbf{P}$ is defined by

$$\Phi = D_1\Phi = \cdots = D_n\Phi = 0$$

locally at P. We put

$$R := f_n\left(\ldots, \frac{\partial^2 \Phi}{\partial x_i \partial x_j}, \ldots \right) = \sqrt{\det\left(\frac{\partial^2 \Phi}{\partial x_i \partial x_j} \right)},$$

where f_n is the polynomial defined in §2.2. The subscheme $\mathcal{R}$ of $X^\circ \times \mathbf{P}$ is defined by

$$\Phi = D_1\Phi = \cdots = D_n\Phi = R = 0$$

locally at P.

Note that the divisor $D_{[f]}$ of X is defined by $\phi_0 = 0$ locally at p. Let $2r$ be the rank of the Hessian $H_{\phi_0,p}$ of $D_{[f]}$ at p. Since $(x_1, \ldots, x_n)$ is admissible with respect to ϕ_0, we see that the linear parts of the formal power series Φ and $D_i\Phi$ are as follows:

$$(6.1) \qquad \begin{array}{lcll} \Phi^{[1]} & = & y_m, & \\ (D_i\Phi)^{[1]} & = & x_{\tau(i)} + y_i & \text{for } i = 1, \ldots, 2r, \\ (D_i\Phi)^{[1]} & = & y_i & \text{for } i = 2r+1, \ldots, n. \end{array}$$

When $2r = n - 2$, we use the parameter system $(x_1, \ldots, x_{n-2}, \xi_1, \xi_2)$ admissible with respect to ϕ_0 as in §3, and adopt Table 3.1 to denote the coefficients of ϕ_0. We also denote by $e_j \in k$ the coefficients of the term $\xi_1\xi_2$ in ϕ_j for $j = n+1, \dots, m-1$. By Corollary 2.9, we see that the linear part of R is equal to

$$
\begin{aligned}
R^{[1]} &= \begin{cases} 0 & \text{if } 2r < n-2, \\ \left(\dfrac{\partial^2 \Phi}{\partial \xi_1 \partial \xi_2} \right)^{[1]} & \text{if } 2r = n-2, \end{cases} \\
&= \begin{cases} 0 & \text{if } 2r < n-2, \\ d_1 x_1 + \cdots + d_{n-2} x_{n-2} + \\ \quad + e_{n+1} y_{n+1} + \cdots + e_{m-1} y_{m-1} & \text{if } 2r = n-2. \end{cases}
\end{aligned}
\tag{6.2}
$$

6.1. A normal form theorem

We put

$$\mathcal{R}^{\mathrm{sm}} := \{\, P \in \mathcal{R} \mid \mathcal{R} \text{ is smooth of dimension } m-2 \text{ at } P \,\},$$

and denote by $\varpi_2^{\mathrm{sm}} : \mathcal{R}^{\mathrm{sm}} \to \mathbf{P}$ the projection.

Theorem 6.4. *Let $P = (p, [f])$ be a point of $\mathcal{R}$.*

(1) If $P \in \mathcal{R}^{\mathrm{sm}}$, then the rank of the Hessian $H_{\phi_0,p}$ of $D_{[f]}$ at p is $n-2$. Conversely, suppose that $H_{\phi_0,p}$ is of rank $n-2$. Then P is a point of $\mathcal{R}^{\mathrm{sm}}$ if and only if at least one of

$$d_1, \dots, d_{n-2}, e_{n+1}, \dots, e_{m-1}$$

is not zero.

(2) The kernel $\mathcal{K}$ of the homomorphism

$$d\varpi_2^{\mathrm{sm}} : T(\mathcal{R}^{\mathrm{sm}}) \to (\varpi_2^{\mathrm{sm}})^* T(\mathbf{P})$$

is a subbundle of $T(\mathcal{R}^{\mathrm{sm}})$ with rank 2.

Proof. The assertion (1) follows immediately from (6.1) and (6.2). Suppose that $P \in \mathcal{R}^{\mathrm{sm}}$. Then the kernel of $d_P\varpi_2^{\mathrm{sm}} : T_P(\mathcal{R}^{\mathrm{sm}}) \to T_{[f]}(\mathbf{P})$ is of dimension 2 and generated by the vectors

$$\left(\frac{\partial}{\partial x_{n-1}} \right)_P \quad \text{and} \quad \left(\frac{\partial}{\partial x_n} \right)_P$$

of $T_P(X^\circ \times \mathbf{P})$. Since this fact holds at every point P of $\mathcal{R}^{\mathrm{sm}}$, we see that $\mathcal{K}$ is a subbundle of $T(\mathcal{R}^{\mathrm{sm}})$ with rank 2. Q.E.D.

Corollary 6.5. *Suppose that $(n, m) = (2, 3)$. Then $\mathcal{R}^{\mathrm{sm}}$ is empty.*

By definition, the kernel $\mathcal{K}$ of $d\varpi_2^{\mathrm{sm}}$ is an integrable subbundle of $T(\mathcal{R}^{\mathrm{sm}})$. From Proposition 2.4, we obtain the following:

Corollary 6.6. *The morphism $\varpi_2^{\mathrm{sm}} : \mathcal{R}^{\mathrm{sm}} \to \mathbf{P}$ factors through the quotient morphism $q : \mathcal{R}^{\mathrm{sm}} \to (\mathcal{R}^{\mathrm{sm}})^{\mathcal{K}}$ by $\mathcal{K}$, which is finite of degree 4.*

From Theorems 4.1 and 6.4, we obtain the following normal form theorem for the morphism $\pi_2 : \mathcal{C}^\circ \to \mathbf{P}$ at a point $P = (p, [f])$ of $\mathcal{R}^{\mathrm{sm}}$.

Corollary 6.7. *Let $P = (p, [f])$ be a point of $\mathcal{R}^{\mathrm{sm}}$. Then there exist an isomorphism*

$$\iota : (\mathcal{C}^\circ, P)^\wedge \ \tilde{\to}\ \operatorname{Spec} k[[w_1, \dots, w_{m-1}]]$$

and a morphism

$$\psi : \Gamma_{m-1} \to (\mathbf{P}, [f])^\wedge$$

such that the formal completion

$$(\pi_2)^\wedge_P : (\mathcal{C}^\circ, P)^\wedge \to (\mathbf{P}, [f])^\wedge$$

of π_2 at P factors as $\psi \circ \gamma^R \circ \iota$, where

$$\gamma^R : \operatorname{Spec} k[[w_1, \dots, w_{m-1}]] \to \Gamma_{m-1}$$

is the morphism defined in §4.

6.2. General plane sections

Let $P = (p, [f])$ be a point of $\mathcal{R}$. We choose linear subspaces L and Λ of $\mathbf{P}$ with dimension 3 and 2, respectively, such that

$$[f] \in \Lambda \subset L.$$

We set

$$S_L := \pi_2^{-1}(L) \subset \mathcal{C}^\circ \quad \text{and} \quad C_\Lambda := \pi_2^{-1}(\Lambda) \subset \mathcal{C}^\circ,$$

and denote by

$$\pi_L : S_L \to L \quad \text{and} \quad \pi_\Lambda : C_\Lambda \to \Lambda$$

the restrictions of $\pi_2 : \mathcal{C}^\circ \to \mathbf{P}$ to S_L and to C_Λ, respectively.

Proposition 6.8. *Suppose that L is chosen generically over k. The scheme S_L is smooth of dimension 2 at P if and only if the rank of the Hessian $H_{\phi_0,p}$ of $D_{[f]}$ at p is $n-2$.*

Proof. We can assume that L is defined by

$$y_i = A_i\, y_{n-1} + B_i\, y_n + C_i\, y_m \qquad (i \neq n-1, n, m), \tag{6.3}$$

where A_i, B_i, C_i are generic over k. Then the assertion follows from the linear parts (6.1) of the defining equations of $\mathcal{C}^\circ$. Q.E.D.

Proposition 6.9. *Suppose that L and Λ are chosen generically over k, and that S_L is smooth of dimension 2 at P.*

(1) *If C_Λ is of codimension 1 in S_L, then C_Λ has a degenerate singularity at P.*

(2) *The following two conditions are equivalent:*

(i) *C_Λ is of codimension 1 in S_L, and the multiplicity of C_Λ at P is 2,*

(ii) *the singularity of $D_{[f]}$ at p is of type $(\widetilde{\mathrm{A}})$.*

(3) *Suppose that C_Λ is of codimension 1 in S_L, that the multiplicity of C_Λ at P is 2, and that P is a point of $\mathcal{R}^{\mathrm{sm}}$. Then P is an ordinary cusp of C_Λ if and only if the singularity of $D_{[f]}$ at p is of type* (R).

Proof. By Proposition 6.8 and the assumption, the rank of $H_{\phi_0,p}$ is $n-2$. We use the formal parameter system $(x_1, \dots, x_{n-2}, \xi_1, \xi_2)$ admissible with respect to ϕ_0 as in §3, and refer to Table 3.1 for the name of coefficients of ϕ_0. We also use the defining equations (6.3) of the linear subspace L. By (6.1) and (6.3), we see that

$$s := \xi_1 \,|\, S_L \quad \text{and} \quad t := \xi_2 \,|\, S_L$$

form a formal parameter system of S_L at P. We put

$$u := y_{n-1} \,|\, L, \quad v := y_n \,|\, L \quad \text{and} \quad w := y_m \,|\, L,$$

which form an affine coordinate system of L with the origin $[f]$. For a formal power series F of $(x_1, \dots, x_{n-2}, \xi_1, \xi_2, y_1, \dots, y_m)$, we denote by F_L the formal power series of

$$(x_1, \dots, x_{n-2}, s, t, u, v, w)$$

obtained from F by making the substitutions

$$\begin{aligned} &y_{n-1} = u, \quad y_n = v, \quad y_m = w, \\ &y_i = A_i u + B_i v + C_i w \quad (i \neq n-1, n, m), \quad \text{and} \\ &\xi_1 = s, \quad \xi_2 = t. \end{aligned}$$

Regarding

$$\Phi_L = (D_1\Phi)_L = \cdots = (D_n\Phi)_L = 0$$

as equations with indeterminates $x_1, \ldots, x_{n-2}, u, v, w$ and with coefficients in $(\mathcal{O}_{S_L,P})^\wedge = k[[s,t]]$, and solving them, we obtain the formal power series expansion

$$\begin{array}{lllll} x_i \mid S_L & = & X_i(s,t) & = & \sum X_{i,\mu\nu}\, s^\mu\, t^\nu \quad (i = 1, \ldots, n-2), \\ u \mid S_L & = & U(s,t) & = & \sum U_{\mu\nu}\, s^\mu\, t^\nu, \\ v \mid S_L & = & V(s,t) & = & \sum V_{\mu\nu}\, s^\mu\, t^\nu, \\ w \mid S_L & = & W(s,t) & = & \sum W_{\mu\nu}\, s^\mu\, t^\nu, \end{array}$$

of the functions $x_1|S_L, \ldots, x_{n-2}|S_L, u|S_L, v|S_L, w|S_L$ on S_L in (s,t). The formal power series U, V and W give the formal completion of $\pi_L : S_L \to L$ at P. We will calculate the homogeneous parts $U^{[d]}$, $V^{[d]}$, $W^{[d]}$ of degree d of them up to $d = 3$. From $\Phi_L = 0$ and $(D_{n-1}\Phi)_L = (D_n\Phi)_L = 0$, we obtain

$$U^{[1]} = V^{[1]} = W^{[1]} = 0.$$

From $(D_i\Phi)_L = 0$ for $i = 1, \ldots, n-2$, we obtain

$$X_i^{[1]} = 0 \qquad (i = 1, \ldots, n-2).$$

Looking at the homogeneous parts of degree 2 in (s,t) of Φ_L and $(D_i\Phi)_L$, we obtain the following equations:

$$\begin{array}{lcl} \alpha_1\, s^2 + \alpha_2\, t^2 + W^{[2]} & = & 0, \\ b_i\, s^2 + c_i\, t^2 + d_i\, st + X_{\tau(i)}^{[2]} + A_i\, U^{[2]} + B_i\, V^{[2]} + C_i\, W^{[2]} & = & 0 \\ \qquad\qquad (i \le n-2), & & \\ \beta_1\, s^2 + \gamma_1\, t^2 + U^{[2]} & = & 0, \\ \beta_2\, s^2 + \gamma_2\, t^2 + V^{[2]} & = & 0. \end{array}$$

Thus we obtain

$$\begin{array}{lllll} U^{[2]} & = & \beta_1\, s^2 & + & \gamma_1\, t^2, \\ V^{[2]} & = & \beta_2\, s^2 & + & \gamma_2\, t^2, \\ W^{[2]} & = & \alpha_1\, s^2 & + & \alpha_2\, t^2, \end{array}$$

and

$$\begin{array}{lll} X_i^{[2]} & = & b_{\tau(i)}\, s^2 + c_{\tau(i)} t^2 + d_{\tau(i)}\, st \\ & & + A_{\tau(i)} U^{[2]} + B_{\tau(i)} V^{[2]} + C_{\tau(i)} W^{[2]} \\ & = & (b_{\tau(i)} + A_{\tau(i)}\, \beta_1 + B_{\tau(i)}\, \beta_2 + C_{\tau(i)}\, \alpha_1)\, s^2 \\ & & + (c_{\tau(i)} + A_{\tau(i)}\, \gamma_1 + B_{\tau(i)}\, \gamma_2 + C_{\tau(i)}\, \alpha_2)\, t^2 \\ & & + d_{\tau(i)}\, st. \end{array}$$

Looking at the homogeneous part of degree 3 in (s,t) of Φ_L, we get the equation

$$\beta_1 s^3 + \beta_2 s^2 t + \gamma_1 st^2 + \gamma_2 t^3 + U^{[2]} s + V^{[2]} t + W^{[3]} = 0.$$

Thus we obtain

$$W^{[3]} = 0.$$

Looking at the homogeneous part of degree 3 in (s,t) of $(D_{n-1}\Phi)_L$ and $(D_n\Phi)_L$, we obtain the equations

$$\sum_{i=1}^{n-2} d_i X_i^{[2]} t + f_{31}s^2 t + f_{13}t^3 + U^{[3]} + \sum_{j=n+1}^{m-1} (A_j U^{[2]} + B_j V^{[2]} + C_j W^{[2]})\, e_j t = 0,$$

$$\sum_{i=1}^{n-2} d_i X_i^{[2]} s + f_{31}s^3 + f_{13}st^2 + V^{[3]} + \sum_{j=n+1}^{m-1} (A_j U^{[2]} + B_j V^{[2]} + C_j W^{[2]})\, e_j s = 0,$$

where e_j is the coefficient of $\xi_1\xi_2$ in ϕ_j. We put

$$\begin{aligned} Q(s,t) \;&:=\; \sum_{i=1}^{n-2} d_i X_i^{[2]} + f_{31}s^2 + f_{13}t^2 + \sum_{j=n+1}^{m-1} (A_j U^{[2]} + B_j V^{[2]} + C_j W^{[2]})\, e_j \\ &=\; Q_s\, s^2 + Q_t\, t^2, \end{aligned}$$

where Q_s and Q_t are given in Table 6.1. (Note that the coefficient of st in $Q(s,t)$ is $\sum_{i=1}^{n-2} d_i d_{\tau(i)} = 0$.) Then we obtain

$$U^{[3]} = t\,Q(s,t) \quad \text{and} \quad V^{[3]} = s\,Q(s,t).$$

Suppose that Λ is defined in L by

$$Du + Ev + Fw = 0,$$

where D, E, F are generic over k. From the assumption, A_i, B_i, C_i and D, E, F are algebraically independent over k. The scheme C_Λ is defined in S_L by $\Gamma = 0$ locally at P, where Γ is the formal power series of (s,t) defined by

$$\Gamma := DU + EV + FW.$$

$$
\begin{aligned}
Q_s \;:=\; & \sum_{i=1}^{n-2}(b_{\tau(i)} + A_{\tau(i)}\beta_1 + B_{\tau(i)}\beta_2 + C_{\tau(i)}\alpha_1)d_i + \\
& + f_{31} + \sum_{j=n+1}^{m-1}(A_j\beta_1 + B_j\beta_2 + C_j\alpha_1)e_j, \\
Q_t \;:=\; & \sum_{i=1}^{n-2}(c_{\tau(i)} + A_{\tau(i)}\gamma_1 + B_{\tau(i)}\gamma_2 + C_{\tau(i)}\alpha_2)d_i + \\
& + f_{13} + \sum_{j=n+1}^{m-1}(A_j\gamma_1 + B_j\gamma_2 + C_j\alpha_2)e_j.
\end{aligned}
$$

Table 6.1. The coefficients Q_s and Q_t

Obviously, we have $\Gamma^{[0]} = \Gamma^{[1]} = 0$. We also have

$$\Gamma^{[2]} = (D\beta_1 + E\beta_2 + F\alpha_1)\, s^2 + (D\gamma_1 + E\gamma_2 + F\alpha_2)\, t^2 = \ell^2,$$

where

$$\ell := \sqrt{D\beta_1 + E\beta_2 + F\alpha_1}\; s + \sqrt{D\gamma_1 + E\gamma_2 + F\alpha_2}\; t.$$

Hence, if $\Gamma \neq 0$, then the curve C_Λ on S_L has a degenerate singularity at P. Moreover, $\Gamma^{[2]}$ is not zero if and only if at least one of $\alpha_1, \alpha_2, \beta_1, \beta_2, \gamma_1, \gamma_2$ is not zero. Thus the assertions (1) and (2) are proved.

Suppose that $P \in \mathcal{R}^{\mathrm{sm}}$ and that $\Gamma^{[2]} = \ell^2 \neq 0$. The degenerate singular point P of C_Λ is an ordinary cusp if and only if the linear form ℓ does not divide

$$\Gamma^{[3]} = (D\,t + E\,s)\,(\sqrt{Q_s}\,s + \sqrt{Q_t}\,t)^2.$$

The two linear forms ℓ and $Dt + Es$ are proportional if and only if

$$D^2\,(D\beta_1 + E\beta_2 + F\alpha_1) + E^2\,(D\gamma_1 + E\gamma_2 + F\alpha_2) = 0 \tag{6.4}$$

holds. Since D, E, F are generic over k, and at least one of α_1, α_2, β_1, β_2, γ_1, γ_2 is not zero by $\Gamma^{[2]} \neq 0$, the equality (6.4) does not hold. The two linear forms ℓ and $\sqrt{Q_s}s + \sqrt{Q_t}t$ are proportional if and only if

$$Q_t\,(D\beta_1 + E\beta_2 + F\alpha_1) + Q_s\,(D\gamma_1 + E\gamma_2 + F\alpha_2) = 0 \tag{6.5}$$

holds. Since $P \in \mathcal{R}^{\rm sm}$, we see from Theorem 6.4 (1) that at least one of $d_1, \dots, d_{n-2}, e_{n+1}, \dots, e_{m-1}$ is not zero. Expanding the left hand side of (6.5) as a polynomial of A_i, B_i, C_i and D, E, F, we see that (6.5) holds for a generic choice of A_i, B_i, C_i, D, E, F if and only if the condition (3.2) does *not* hold. Therefore the assertion (3) is proved. Q.E.D.

Proposition 6.10. *Suppose that C_Λ is of codimension 1 in S_L at P, and is reduced and irreducible locally at P. Let $\nu : \widetilde{C}_\Lambda \to C_\Lambda$ be the normalization of C_Λ at P, and let z be a formal parameter of $\widetilde{C}_\Lambda$ at the point $P' \in \widetilde{C}_\Lambda$ such that $\nu(P') = P$. Then the image of*

$$(\pi_\Lambda \circ \nu)^{*\wedge}_{P'} : (\mathcal{O}_{\Lambda,[f]})^\wedge \to (\mathcal{O}_{\widetilde{C}_\Lambda,P'})^\wedge$$

is contained in the subring

$$T := \{\, f = \sum f_\mu z^\mu \mid f_1 = f_2 = f_3 = f_5 = 0 \,\}$$

of $(\mathcal{O}_{\widetilde{C}_\Lambda,P'})^\wedge = k[[z]]$.

Proof. We use the same notation as in the proof of Proposition 6.9. The formal completion at P' of the composite of $\nu : \widetilde{C}_\Lambda \to C_\Lambda$ and the inclusion $\iota : C_\Lambda \hookrightarrow S_L$ is given by

$$\begin{aligned} (\iota \circ \nu)^* s &= a\, z^2 + b\, z^3 + (\text{terms of degree} \geq 4) \quad \text{and} \\ (\iota \circ \nu)^* t &= c\, z^2 + d\, z^3 + (\text{terms of degree} \geq 4), \end{aligned}$$

where $a, b, c, d \in k$. Since Λ is general in L, the variables

$$u' := u \,|\, \Lambda \quad \text{and} \quad v' := v \,|\, \Lambda$$

form a formal parameter system of Λ at $[f]$. The formal completion at P' of the morphism $\pi_\Lambda \circ \nu : \widetilde{C}_\Lambda \to \Lambda$ is given by

$$\begin{aligned} (\pi_\Lambda \circ \nu)^* u' &= U(\, az^2 + bz^3 + \cdots ,\ cz^2 + dz^3 + \cdots) \quad \text{and} \\ (\pi_\Lambda \circ \nu)^* v' &= V(\, az^2 + bz^3 + \cdots ,\ cz^2 + dz^3 + \cdots). \end{aligned}$$

Since $U^{[1]}$ and $V^{[1]}$ are both zero, and the coefficients of st in $U^{[2]}$ and $V^{[2]}$ are also zero, we see that $(\pi_\Lambda \circ \nu)^* u'$ and $(\pi_\Lambda \circ \nu)^* v'$ are contained in the subring T of $k[[z]]$. Q.E.D.

6.3. The case where the linear system is sufficiently ample

Definition 6.11. For a point p of X and a positive integer d, we denote by

$$v_p^{[d]} : M \to \mathcal{L}_p/\mathfrak{m}_p^{d+1}\mathcal{L}_p$$

the evaluation homomorphism at p, where $\mathfrak{m}_p$ is the maximal ideal of $\mathcal{O}_{X,p}$, and $\mathcal{L}_p$ is the $\mathcal{O}_{X,p}$-module $\mathcal{L} \otimes \mathcal{O}_{X,p}$.

Proposition 6.12. (1) *Suppose that $v_p^{[2]}$ is surjective at every point p of X°. Then $\mathcal{R}$ is irreducible.*

(2) *Suppose that $v_p^{[3]}$ is surjective at every point p of X°. Then $\mathcal{R}^{\mathrm{sm}}$ is dense in $\mathcal{R}$. Moreover, if $P = (p, [f])$ is a general point of $\mathcal{R}$, then the singularity of $D_{[f]}$ at p is of type* ($\widetilde{\mathrm{A}}$) *and of type* (C).

(3) *Suppose that $v_p^{[4]}$ is surjective at every $p \in X^\circ$. If $P = (p, [f])$ is a general point of $\mathcal{R}$, then the singularity of $D_{[f]}$ at p is of type* (R).

Proof. Considering the first projection $\mathcal{R} \to X^\circ$, we deduce the statements of Proposition from Remark 3.10 and Theorem 6.4. Q.E.D.

Combining Proposition 6.12 with the results proved so far, we obtain the facts (I)-(IV) stated in Introduction.

§7. Divisibility of a Chern number

We assume that $\overline{X}$ is smooth and $\mathcal{L}$ is very ample, and put $|M| = |\mathcal{L}|$. Then we have $X^\circ = X = \overline{X}$ and $\mathcal{C} = \mathcal{C}^\circ$. We further assume that $\mathcal{R}^{\mathrm{sm}}$ is dense in $\mathcal{R}$. Under these assumptions, we calculate

$$\deg \mathcal{R} := \int_{X\times\mathbf{P}} c_1(\mathrm{pr}_2^*\,\mathcal{O}_{\mathbf{P}}(1))^{m-2} \cap [\mathcal{R}].$$

We put

$$\lambda := c_1(\mathrm{pr}_1^*\,\mathcal{L}) \quad \text{and} \quad h := c_1(\mathrm{pr}_2^*\,\mathcal{O}_{\mathbf{P}}(1)).$$

Then we have

$$c_1(\widetilde{\mathcal{L}}) = h + \lambda.$$

For simplicity, we write $c_i(X)$ for $\mathrm{pr}_1^*\,c_i(X)$. By Proposition 5.2 and Thom-Porteous formula [3, Chapter 14], we have

$$[\mathcal{C}] = \left((\lambda + h)\sum_{i=0}^{n}(-1)^i c_i(X)(\lambda+h)^{n-i}\right) \cap [X \times \mathbf{P}]$$

in the Chow group of $X \times \mathbf{P}$. By Theorem 2.14, we have

$$[\mathcal{R}] = \left(\frac{n}{2}(h+\lambda) - c_1(X)\right) \cap [\mathcal{C}].$$

We put

$$G := \left(\frac{n}{2}(h+\lambda) - c_1(X)\right)\left((\lambda+h)\sum_{i=0}^{n}(-1)^i c_i(X)(\lambda+h)^{n-i}\right).$$

Then we obtain

$$\deg \mathcal{R} = \int_X (\text{the coefficient of } h^2 \text{ in } G) \cap [X].$$

By Corollary 6.6, the integer $\deg \mathcal{R}$ is divisible by 4. If the coefficient of h^2 in G is regarded as a polynomial of λ, then the constant term is equal to $(-1)^n(n\, c_n(X)/2 + c_1(X)c_{n-1}(X))$. Putting $\mathcal{L} = \mathcal{A}^{\otimes 4}$ with $\mathcal{A}$ a sufficiently ample line bundle, we obtain the following:

Corollary 7.1. *Let X be a smooth projective variety in characteristic 2 of dimension n being even. Then the integer*

$$\int_X \left(\frac{n}{2} c_n(X) + c_1(X)c_{n-1}(X)\right) \cap [X]$$

is divisible by 4.

In fact, this divisibility relation can be deduced from the Hirzebruch-Riemann-Roch theorem by means of the argument of Libgober and Wood. See [15, Remark 2.4].

References

[1] D. Cox, J. Little and D. O'Shea, Ideals, varieties, and algorithms, second ed., An introduction to computational algebraic geometry and commutative algebra, Undergrad. Texts Math., Springer-Verlag, 1997.

[2] I. Dolgachev and A. Libgober, On the fundamental group of the complement to a discriminant variety, Algebraic geometry, Chicago, Ill., 1980, Lecture Notes in Math., **862**, Springer-Verlag, 1981, pp. 1–25.

[3] W. Fulton, Intersection theory, second ed., Ergeb. Math. Grenzgeb. (3), Folge. A Series of Modern Surveys in Mathematics [Results in Mathematics and Related Areas. (3), A Series of Modern Surveys in Mathematics], **2**, Springer-Verlag, 1998.

[4] J. Harris and L. W. Tu, On symmetric and skew-symmetric determinantal varieties, Topology, **23** (1984), 71–84.

[5] A. Hefez and A. Thorup, Reflexivity of Grassmannians and Segre varieties, Comm. Algebra, **15** (1987), 1095–1108.

[6] A. Hefez, Nonreflexive curves, Compositio Math., **69** (1989), 3–35.

[7] A. Hefez and S. L. Kleiman, Notes on the duality of projective varieties, Geometry today, Rome, 1984, Progr. Math., **60**, Birkhäuser Boston, 1985, pp. 143–183.

[8] M. Homma, On duals of smooth plane curves, Proc. Amer. Math. Soc., **118** (1993), 785–790.

[9] H. Kaji, On the Gauss maps of space curves in characteristic p, Compositio Math., **70** (1989), 177–197.

[10] ———, On the duals of Segre varieties, Geom. Dedicata, **99** (2003), 221–229.

[11] H. Kaji and A. Noma, On the generic injectivity of the Gauss map in positive characteristic, J. Reine Angew. Math., **482** (1997), 215–224.

[12] N. M. Katz, Pinceaux de Lefschetz: théorème d'existence, Groupes de monodromie en géométrie algébrique. II, Springer-Verlag, 1973, Séminaire de Géométrie Algébrique du Bois-Marie 1967–1969 (SGA 7 II), Dirigé par P. Deligne et N. Katz, Lecture Notes in Math., **340**, pp. 212–253.

[13] S. L. Kleiman, About the conormal scheme, Complete intersections, Acireale, 1983, Lecture Notes in Math., **1092**, Springer-Verlag, 1984, pp. 161–197.

[14] ———, Tangency and duality, Proceedings of the 1984 Vancouver conference in algebraic geometry CMS Conf. Proc., **6**, Amer. Math. Soc., Providence, RI, 1986, pp. 163–225.

[15] A. S. Libgober and J. W. Wood, Uniqueness of the complex structure on Kähler manifolds of certain homotopy types, J. Differential Geom., **32** (1990), 139–154.

[16] R. Pardini, Some remarks on plane curves over fields of finite characteristic, Compositio Math., **60** (1986), 3–17.

[17] P. Pragacz, Cycles of isotropic subspaces and formulas for symmetric degeneracy loci, Topics in algebra, Part 2, Warsaw, 1988, Banach Center Publ., **26**, PWN, Warsaw, 1990, pp. 189–199.

[18] C. S. Seshadri, L'opéation de Cartier. Applications, Séminaire C. Chevalley, 3ième année: 1958/59, Variétés de Picard, École Normale Supérieure, Paris, 1960, pp. 83–135.

[19] I. Shimada, Singularities of dual varieties in characteristic 3, Geom. Dedicata, **120** (2006), 141–177.

[20] ———, On supercuspidal families of curves on a surface in positive characteristic, Math. Ann., **292** (1992), 645–669.

[21] ———, Lattices of algebraic cycles on Fermat varieties in positive characteristics, Proc. London Math. Soc. (3), **82** (2001), 131–172.

[22] A. H. Wallace, Tangency and duality over arbitrary fields, Proc. London Math. Soc. (3), **6** (1956), 321–342.

Department of Mathematics
Faculty of Science
Hokkaido University
Sapporo 060-0810 JAPAN
E-mail address: shimada@math.sci.hokudai.ac.jp

Advanced Studies in Pure Mathematics 50, 2008
Algebraic Geometry in East Asia — Hanoi 2005
pp. 333–344

On the Castelnuovo-Weil lattices, I

Viet Nguyen-Khac and Tetsuji Shioda

Abstract.

The Castelnuovo-Weil lattice of a curve refers to the ring of correspondence of the curve or the endomorphism ring of its Jacobian variety, viewed as a lattice with respect to the natural trace form. The main subject is the minimal norm $\mu(C)$ of this lattice, and we discuss the question as to whether $\mu(C) = 2g$ holds true.

§1. Introduction

In the development of algebraic geometry, curves have always played an important role as the basic objects of dimension one. Yet some higher dimensional varieties have been indispensable for studying curves themselves, such as the Jacobian varieties, introduced by Riemann, and the correspondence theory of curves, started by Hurwitz and developped by the Italian geometers, especially by Castelnuovo and Severi.

Given two curves C and C', a correspondence between them is simply a divisor D on the product surface $C \times C'$. A correspondence is said to be equivalent to 0 (denoted $D \equiv 0$) if it is linearly equivalent to a divisor of the form $\mathfrak{a} \times C' + C \times \mathfrak{b}$ where $\mathfrak{a}$ or $\mathfrak{b}$ is a divisor on C or C'. Castelnuovo's theory gives a numerical criterion for equivalence; namely it attaches to every correspondence an integer $\gamma(D) \geq 0$, called the "equivalence defect", such that

$$\gamma(D) = 0 \Longleftrightarrow D \equiv 0.$$

Let D' be the correspondence between C' and C obtained by switching the two factors of the product $C \times C'$. The composition of correspondences can be defined in a natural way, and then $D' \circ D$ is a correspondence between C and itself. If Z is any such correspondence, the number

Received September 19, 2006.
Revised May 9, 2007.
2000 *Mathematics Subject Classification.* Primary 11G30, 11G10, 14H31.

of its fixed points is given by the intersection number $(Z.\Delta)$ of Z with the diagonal Δ on $C \times C$. Put

$$S(Z) = d(Z) + d'(Z) - (Z.\Delta) \tag{1.1}$$

where $d(Z), d'(Z)$ are the degrees, i.e. the intersection numbers of Z with $x \times C, C \times x'$. As is easily seen, $S(Z)$ depends only on the equivalence class ξ of Z so that one can write $S(Z) = \sigma(\xi)$. Then σ has the properties of a trace on the ring of correspondence classes. The equivalence defect can then be expressed as $\gamma(D) = S(D' \circ D)$, and the Castelnuovo's inequality asserts that

$$S(D' \circ D) \geq 0, \quad S(D' \circ D) = 0 \Leftrightarrow D \equiv 0. \tag{1.2}$$

In other words, we have

$$\sigma(\xi'\xi) \geq 0, \quad \sigma(\xi'\xi) = 0 \Leftrightarrow \xi = 0. \tag{1.3}$$

In this form, it is the fundamental theorem on correspondences, and Weil has given two methods of proof valid in any characteristic: the first one in terms of correspondence theory ([13]), and the second one in terms of endomorphisms of Jacobian variety ([14], cf. [15]). We recall some of the arguments in §2. [There is a third proof based on the Hodge's index theorem (cf. [4], [5]).]

The most remarkable application of this positivity theorem is Weil's proof of the Riemann hypothesis for curves over a finite field.

Now the ring $\mathfrak{C} = \mathfrak{C}(C)$ of correspondence classes on C can be regarded as a positive-definite even integral lattice for which the norm of an element $\xi \in \mathfrak{C}$ is given by $\sigma(\xi' \cdot \xi)$. We propose to call this lattice the *Castelnuovo-Weil lattice* of a curve C (CWL in short).

In this paper, we discuss what we can say about the minimal norm $\mu(C)$ of this lattice:

$$\mu(C) := \text{Min } \sigma(\xi' \cdot \xi) \quad (\xi \in \mathfrak{C}, \xi \neq 0). \tag{1.4}$$

§2. Problem setting

Let k be a field. By a curve, we mean a smooth, geometrically irreducible, projective curve defined over k. Let C be a curve of genus $g \geq 1$ and let J_C denote the Jacobian variety of C.

Given any correspondence D on $C \times C$, the map $cl(P - Q) \mapsto cl(D(P) - D(Q))$ defines an endomorphism α_D of J_C. Here $cl(\mathfrak{a})$ denotes the linear equivalence class of a divisor $\mathfrak{a}$ on C, and for any $P \in C$, $D(P)$ is the image of P under the correspondence D, i.e. a divisor on C such that $D.(P \times C) = P \times D(P)$. Then $D \mapsto \alpha_D$ induces a ring homomorphism from $\mathfrak{C}(C)$ to $\mathrm{End}(J_C)$, and this gives rise to an isomorphism:

$$\mathfrak{C}(C) \cong \mathrm{End}(J_C), \quad \xi = [D] \mapsto \alpha = \alpha_D \tag{2.1}$$

(see [14, VI, §43, Cor. 2]).

Further, by considering the representation on the Tate module $T_\ell(J_C)$, ℓ being a prime number $\neq \mathrm{char}(k)$, the trace $\sigma : \mathfrak{C}(C) \to \mathbb{Z}$ is transformed to the trace Tr on $\mathrm{End}(T_\ell(J_C))$ in the sense of linear algebra. So we can work equally in either $\mathrm{End}(J_C)$ or $\mathfrak{C}(C)$.

Now the class of the diagonal on $C \times C$, $\delta = [\Delta]$, is the identity of the ring $\mathfrak{C}(C)$, and we have

$$\sigma(\delta) = \mathrm{Tr}(1) = 2g. \tag{2.2}$$

This follows from (1.1) since we have $d = d' = 1$ for $Z = \Delta$ and $(\Delta^2) = 2 - 2g$ (the Euler number of C) in terms of correspondences, while it is obvious in the linear algebra setting.

Hence the minimal norm is always smaller or at most equal to $2g$:

$$\mu(C) \leq 2g. \tag{2.3}$$

More than fifteen years ago one of the authors has posed a naive question:

Question 2.1. *Does the equality $\mu(C) = 2g$ always hold?*

In the positive direction, this is true for small genus and also for the "generic" curve of any genus g. Namely we have:

Proposition 2.2. *Question 2.1 is true for curves of genus $g \leq 2$ and hyperelliptic curves of genus 3.*

The proof will be postponed till the end of this section.

Proposition 2.3. *For the generic curve of any genus g over an algebraically closed field k, we have $\mu(C) = 2g$.*

Proof This follows from Mori's result that one has $\mathrm{End}(J_C) \simeq \mathbb{Z}$ for a generic curve; see [7]. If we denote by $h \geq 1$ the rank of $\mathrm{End}(J_C)$, then the Picard number of the product surface $X = C \times C$ is equal to $\rho = h + 2$ by (2.1). We claim that $h = 1$, i.e. $\rho = 3$ for C generic. Indeed, considering the moduli space of genus g curves, if C specializes to C' and $X = C \times C$ to $X' = C' \times C'$, then we have $\rho \leq \rho'$ by the intersection theory. By Mori, we have $h' = 1$ and $\rho' = 3$ for some C'. This shows that $\rho = 3$ and hence $h = 1$ for C generic. *q.e.d.*

For the proof of Proposition 2.2, let us recall Weil's first proof ([13]) for the positivity (1.3):

Theorem 2.4 (Castelnuovo-Weil). *We have $\sigma(\alpha.\alpha') > 0$ for any non-zero α.*

For proving this, Weil ([13, II]) argues as follows. Assume that D is an effective p_1-flat correspondence (i.e. it has no vertical components). Letting $d = d(D)$, $e = d'(D)$, we write the value of D at P as $D(P) = \sum_{i=1}^{d} Q_i$. Because of p_1-flatness assumption, one sees that the 1-cycle $U := (C \times D).(D' \times C)$ in $C \times C \times C$ is well defined, so by the definition of the composition product, we have $D \circ D' = p_{13*}(U)$.

Let V be the cycle in $C \times C \times C$ obtained from U by interchanging two first factors. Then V is p_1-flat w.r.t. the first factor in $C \times C \times C$ and $D \circ D' = p_{23*}(V)$. Also $V(P) = D(P) \times D(P) = \sum_{i=1}^{d} Q_i \times Q_i + \sum_{1 \le i \ne j \le d} Q_i \times Q_j$. So V can be written in the form $V = V_0 + V_1$, where $V_0(P) = \sum_{i=1}^{d} Q_i \times Q_i$ and $V_1(P) = \sum_{1 \le i \ne j \le d} Q_i \times Q_j$. Furthermore it can be verified that $p_{23*}(V_0) = e\Delta$. Thus one obtains $D \circ D' = e\Delta + T$, where $T := p_{23*}(V_1)$. Obviously $d(T) = d'(T) = e(d-1)$.

In particular, if $d = 1$, then the part V_1 vanishes. Hence one concludes $D \circ D' = e\Delta$. It follows that, if $d = 1$,

$$\sigma(\alpha.\alpha') = e\sigma(\delta) = 2ge. \tag{2.4}$$

[This fact can be also shown by a direct argument as follows (cf. [13, II, §9, Prop.4]). Let $D'(Q) = \sum_{i=1}^{e} P_i$. Since $P_i \times Q$, $1 \le i \le e$ are lying in different fibre components, so $D(P_i) = Q$, $1 \le i \le e$. It follows that $D(D'(Q)) = eQ$. Therefore $D \circ D' = e\Delta$, because $D \circ D'$ is symmetric and by the definition of the composition product.]

Let $\mathfrak{d}$ be a divisor on C such that there are d linearly independent functions $\varphi_1, \ldots, \varphi_d \in \mathfrak{L}(\mathfrak{d}) := \{f \in \overline{k}(C) : \quad (f) + \mathfrak{d} \ge 0\} \cup \{0\}$. We assume also $Q_1, \dots, Q_d \notin \text{Supp}(\mathfrak{d})$ and

$$(*) \qquad \begin{aligned} &\varphi_i(P_j) \neq \infty, \text{ for all } i, j = 1, \dots, d, \\ &\det\left(\varphi_i(Q_j)\right)_{1 \leq i,j \leq d} \neq 0. \end{aligned}$$

A computation on the d-fold product $C \times \cdots \times C$ (provided $d \geq 2$) shows that $\deg(T.\Delta) \leq 2e \deg(\mathfrak{d})$. Consequently for $\alpha := [D]$ one obtains

$$(2.5) \qquad \sigma(\alpha.\alpha') \geq 2e\left[d + g - 1 - \deg(\mathfrak{d})\right],$$

Letting $\mathfrak{d} = \mathfrak{k}$ be the canonical divisor of C and taking $d = g$ ([13, II,§14]), we get

$$(2.6) \qquad \sigma(\alpha.\alpha') \geq 2e,$$

provided $g \geq 2$ (the case $g = 1$ is obvious). Since $e > 0$ for non-zero α, this proves Theorem 2.4.

Now we turn to the proof of Proposition 2.2. Keeping the notation above, note first that if $e \geq g$, then we have $\sigma(\alpha.\alpha') \geq 2g$. So it remains to treat the case (i) $e = 1, g \leq 3$ and (ii) $g = 3, e = 2$. Suppose that $e = 1$; then, after changing the symmetric role between d, e and by (2.4), we obtain $\sigma(\alpha.\alpha') = 2gd \geq 2g$. In general, the problem is reduced to the existence of special linear systems on C satisfying condition $(*)$. Next, in the case when C is a hyperelliptic curve of genus 3, we proceed as follows. Assuming $e = d = 2$ (otherwise inequality (2.5) proves Prop. 2.2), we take $|\mathfrak{d}| = g_2^1$ and Q_1, Q_2 not lying in the same fiber of the hyperelliptic morphism. It is easy to make the choice so that $(*)$ is satisfied, hence the proof of Proposition 2.2 follows.

As we see in §4, Question 2.1 will have a counterexample. Yet we do not know any counterexample in the hyperelliptic case:

Question 2.5. *Does $\mu(C) = 2g$ hold for any hyperelliptic curve C of genus g?*

§3. μ-problem for polarized abelian varieties

Let us consider the corresponding problem for polarized abelian varieties, leaving for a moment the case of Jacobian varieties.

Let (A, L) be a polarized abelian variety of dimension d over k. Recall that L is an ample line bundle. It corresponds to an isogeny $\varphi_L \colon A \to A^\vee = Pic^0(A)$, $\quad x \mapsto \tau_x^* L \otimes L^{-1}$ (τ_x is the translate to x). So for $\xi \in \mathrm{End}(A)$, $\xi^\vee : A^\vee \to A^\vee$ we know that the Rosati conjugate $\xi' \in \mathrm{End}^0(A)$ is defined as $\xi' = \varphi_L^{-1}.\xi^\vee.\varphi_L : A \to A$. In particular, if L is a principal polarization, i.e. $\deg(\varphi_L) = 1$, then one sees that $\xi' \in \mathrm{End}(A)$ (in the case $A = \mathrm{Jac}(C)$ this fact is clearly true via correspondence language above). For instance by using ℓ-adic representation on Tate's modules one can see in the form of matrices: $\xi \longleftrightarrow \xi_T$, $\varphi_L \longleftrightarrow m$, $\xi' \longleftrightarrow m^{-1}.{}^t\xi_T.m$ which is the conjugate of the transpose matrix.

The conclusion of the Castelnuovo-Weil theorem for (A, L) can be deduced from the ampleness of L and the following well-known formula ([6], V, §3, [8], §21, [1], V, 17.3) for $\mathrm{Tr}(\xi.\xi')$:

$$\mathrm{Tr}(\xi.\xi') = \frac{2d}{L^d}(L^{d-1}.\xi^*(L)).$$

In particular, if the curve C is equipped with a k-rational point, say P_0, then for the canonically polarized Jacobian (J_C, Θ) one gets

$$\mathrm{Tr}(\xi.\xi') = 2(\psi_{P_0}(C).\xi^*(\Theta)), \tag{3.1}$$

where $\psi_{P_0} : C \to J_C$, $P \mapsto cl(P - P_0)$ is the canonical embedding.

By analogy with J_C one defines

$$\mu(A) := \min \mathrm{Tr}(\xi.\xi'),$$

where ξ runs over all non-zero endomorphism of A.

The first thing to observe in this case is that $\mu(A) < 2d$ can happen even for principally polarized abelian varieties of dimension $d = 2$. Indeed, we have:

Proposition 3.1. *Let $A = E \times E$ be the selfproduct of an elliptic curve E with the decomposable principal polarization $\Theta = o \times E + E \times o$. Then we have $\mu(A) = 2 < 2d = 4$.*

Proof Immediate. N.B. This does not contradict the genus 2 case of Prop.2.2, because (A, Θ) cannot be isomorphic to a canonically polarized Jacobian of genus 2 curve by [16].

Nevertheless, for isogenies, we have the following proposition.

Proposition 3.2. *Let (A, L) be a polarized abelian variety of dimension d. Let $\xi \in \mathrm{End}(A)$ be an isogeny. Then $\mathrm{Tr}(\xi.\xi') \geq 2d$, where $\xi' \in \mathrm{End}^0(A)$ denotes the Rosati conjugate w.r.t. the polarization L.*

Proof. Since $\xi.\xi'$ is symmetric and positive, so by [6], V, §3, we know that $2d$ eigenvalues $\lambda_1, \ldots, \lambda_{2d}$ of $\xi.\xi'$ are real positive numbers. On the other hand ξ and ξ' (as in $\mathrm{End}^0(A)$) have the same characteristic polynomial. Hence

$$\deg(\xi) = \deg(\xi') = \det(\xi_T) = \det({}^t\xi_T),$$

so $\deg(\xi.\xi') = (\det(\xi_T))^2$ is a non-zero integer, as ξ is an isogeny. It remains to use the (Cauchy) arithmetic-geometric mean inequality

$$\frac{1}{2d}\sum_{i=1}^{2d} \lambda_i \geq \sqrt[2d]{\prod_{i=1}^{2d} \lambda_i} = \sqrt[2d]{\deg(\xi.\xi')} \geq 1$$

as required. *q.e.d.*

Corollary 3.3. *In the notations above,* $\mathrm{Tr}(\xi.\xi') = 2d$ *characterizes the automorphisms of* (A, L).

Corollary 3.4. *Let* (A, L) *be a simple polarized abelian variety of dimension* d. *Then* $\mu(A) = 2d$ *holds.*

§4. Refined Question

As is well-known, the existence of a covering from C to a curve of genus ≥ 1 gives rise to a decomposition of J_C, so it may also affect the existence of special linear systems on C. The aim of this section is to clarify this situation, which will lead to a counterexample to Question 2.1 and to the refined conjecture below. Let $f\colon C \to C_1$ be a finite covering of curves of genera g, g_1 respectively, and let $d = \deg(f)$. Then f induces two homomorphisms (the norm and conorm maps):

$$f_*\colon J_C \to J_{C_1},\ f_*(\mathcal{O}_C(\textstyle\sum n_i P_i)) := \mathcal{O}_{C_1}(\textstyle\sum n_i f(P_i)),$$

$$f^*\colon J_{C_1} \to J_C,\ f^*(\mathcal{O}_{C_1}(\textstyle\sum m_i Q_i)) := \mathcal{O}_C(\textstyle\sum m_i f^*(Q_i))$$

(sometimes f_* is denoted as N_f). We have the following two simple properties:

1) $f_* = f^{*\vee}$ (this justifies the term "norm" and "conorm").

2) $f_*.f^* = d_{J_{C_1}}$ the multiplication by d on J_{C_1}.

Let us mention also the following fact. We identify $J_C \simeq J_C^\vee$ and $J_{C_1} \simeq J_{C_1}^\vee$ via canonical polarizations φ_Θ and φ_{Θ_1} respectively. If $\imath :=$

$f^*\colon J_{C_1} \to J_C$ is injective, then one can consider the pull-back of Θ via $\imath$ which defines a polarization on J_{C_1}. We have the following standard commutative diagram

$$\begin{array}{ccc} J_C & \xrightarrow{\varphi_\Theta} & J_{C_1}^\vee \\ {\scriptstyle \imath}\uparrow & & \downarrow{\scriptstyle \imath^\vee} \\ J_{C_1} & \xrightarrow{\varphi_{\imath^*\Theta}} & J_{C_1}^\vee \end{array}$$

so $\varphi_{\imath^*\Theta} = f_*.f^* = d_{J_{C_1}}$. It means that $\imath^*\Theta = d\Theta_1$. In other words, if f^* is injective, then $\varphi_{\imath^*\Theta}$ (the polarization on J_{C_1} induced by f^*) is d times canonical one (i.e. φ_{Θ_1}).

Let $\xi_1 \in \mathrm{End}(J_{C_1})$. Then $\xi := f^*.\xi_1.f_* \in \mathrm{End}(J_C)$, or equivalently $d^2\xi_1 = f_*.\xi.f^*$. If C has a k-rational point, say P_0, $P_0' := f(P_0)$, then by using (3.1) and the standard commutative diagram above one has

$$\begin{aligned} \mathrm{Tr}(\xi.\xi') &= 2(\psi_{P_0}(C).\xi^*(\Theta)) \\ &= 2(f_*\psi_{P_0}(C).\xi_1^*.\imath^*(\Theta)) \\ &= 2(\psi_{P_0'}(f_*(C)).d\xi_1^*(\Theta_1)) = d^2\mathrm{Tr}(\xi_1.\xi_1'). \end{aligned}$$

Proposition 4.1. *Taking* $\xi_1 = \delta_{C_1}$ *one gets* $\mathrm{Tr}(\xi.\xi') = 2d^2g_1$.

Remark 4.2. It can be seen easily that in Proposition 4.1

$$\xi \longleftrightarrow \Gamma_f^* := (id \times f)^*(\Gamma_f) = (f \times f)^*(\Delta_{C_1}),$$

where $\Gamma_f \subset C \times C_1$ denotes the graph of f. So using bilinear form $\sigma(\cdot,\cdot)$ we obtain another proof of this statement.

Counterexample to Question 2.1 Take a bi-elliptic curve C of genus $g \geq 5$ (which clearly exists). Namely assume that C admits a double covering to an elliptic curve C_1. Then, with the notation above, we have $d = 2$ and

$$\mathrm{Tr}(\xi.\xi') = 2d^2 = 8 < 2g.$$

Thus the naive question 2.1 is false for any $g \geq 5$.

We refine this situation as follows.

Question 4.3. *In the notations of* §1 *we have* $\mu(C) = \mu_0$, *where* $\mu_0 := \mathrm{Min}\{2d^2g_1\}$, *the* Min *is taken over all coverings* $f\colon C \to C_1$ *with* $d = \deg f$ *and* $g_1 := g(C_1) \geq 1$.

Precisely Question 4.3 consists of two parts. Let L be the sublattice generated by all endomorphisms corresponding to the coverings $f\colon C \to C_1$ with $g_1 \geq 1$. Then the problem is, in fact, divided into two statements: $\mu(C) = \mu(L)$ and $\mu(L) = \mu_0$. As a heuristic consideration for the second part one might involve the following descending argument. Assume we have two endomorphisms γ_1, γ_2 corresponding to the coverings $f\colon C \to C_i$ of degree d_i and $g_i := g(C_i) \geq 1,\ i = 1, 2$. Let us denote by C_{12} the (normalized) image of the natural morphism $f_1 \times f_2\colon C \to C_1 \times C_2$ and consider the induced morphism $f_{12}\colon C \to C_{12}$, say of degree d_{12} and $g_{12} := g(C_{12})$. Then by using the Castelnuovo-Weil inequality (Theorem 2.4) one can see easily that either $d_{12}^2 g_{12} \leq \mathrm{Min}\{d_1^2 g_1, d_2^2 g_2\}$, or the angle between γ_1 and γ_2 is big (the angle between two endomorphisms can be defined in a natural way via the positive definite bilinear form $\sigma(\cdot,\cdot)$ defined in §2).

Proposition 4.4. *Question 4.3 implies Question 2.5.*

Proof. This is another application of Theorem 2.4. Indeed applying what we said above to the natural morphism $f_1 \times \iota\colon C \to C_1 \times \mathbb{P}^1$ where $f_1\colon C \to C_1$ is a morphism of degree d_1 and $\iota\colon C \to \mathbb{P}^1$ is the canonical hyperelliptic morphism, in view of Theorem 2.4, we have

$$g \leq (d_1 - 1) + d_1 g_1.$$

The right-hand side is clearly $< d_1^2 g_1$, provided $d_1 \geq 2$. *q.e.d.*

It should be noted that Question 4.3 has a close connection with the following important question in the theory of abelian varieties: which curves can appear in a polarized abelian variety? We shall treat this feature in a future publication.

§5. Some consequences

Let C be a curve of genus g over a finite field $\mathbb{F}_q$. The Hasse-Weil theorem says that the number $N_q(C)$ of $\mathbb{F}_q$-rational points on C can be at most $q + 1 + 2g\sqrt{q}$. If this bound attained such a curve C is called maximal.

Theorem 5.1. *If C is not maximal, and $\mu(C) = 2g$, then*

$$N_q(C) < q + 1 + 2g\sqrt{q - 3/4}.$$

Proof. Let us consider 2-dimensional lattice L_2 associated to

$$f(m,n) := 2gqm^2 + 2(q+1-N_q(C))mn + 2gn^2$$

whose discriminant $D(L_2)$ is $(q+1-N_q(C))^2 - 4g^2q$. Note that $f(m,n)$ is nothing but $\sigma(m\Gamma_\pi + n\Delta, m\Gamma_\pi + n\Delta)$ in the notation of §§2, 4, where π denotes the Frobenius morphism. So the positive definiteness of $\sigma(\cdot,\cdot)$ gives us the mentioned Hasse-Weil bound. Moreover if C is not maximal, then $\operatorname*{Min}_{(m,n)\neq(0,0)} f(m,n) \geq \mu(C)$. In the dimension 2 the Hermite constant

$$\gamma(L_2) := \operatorname*{Min}_{(m,n)\neq(0,0)} f(m,n)/\sqrt{-D(L_2)}$$

is known to be $\leq \sqrt{4/3}$ ([2], I). Moreover the equality holds for the form $2x^2 + 2xy + 2y^2$. Hence

$$N_q(C) < q + 1 + 2g\sqrt{q - \frac{3}{4}\left(\frac{\mu(C)}{2g}\right)^2}$$

and one obtains the statement of the theorem. *q.e.d.*

If $q = 2, 3$ one may take simply $m = n = 1$ in the argument above to get

$$N_q(C) \leq q + 1 + gq$$

that is Serre's bound ([9]) for this case. It should be noted that Serre's bound

$$N_q(C) \leq q + 1 + gm, \quad m := [2\sqrt{q}]$$

can be obtained in a similar vein as of §3 by dealing with $\xi_r := \pi + r\delta \in \mathrm{End}(J_C) \otimes \mathbb{R}$, where π denotes the Frobenius endomorphism and r is a (real) root of $x^2 - (m+1)x + q = 0$. To this end it suffices to use arguments of §3 taking into account

$$\xi_r \xi_r' = r\big(\pi + \pi' + (m+1)\delta\big)$$

and

$$\sqrt[g]{\det\big((\xi_r)_T\big)} \geq r.$$

In conclusion we make a remark concerning a further generalization. Taking $X = C_1 \times C_2$ in the Weil formalism of correspondences, with two different curves C_1, C_2 as in §1, one can define more generally the CWL of type $\mathfrak{C}(C_1, C_2)$ (cf. [12]). The case $C_2 = E-$ an elliptic curve leads to the well-known Mordell-Weil lattices treated in [3], [10] or [11].

For arbitrary C_1, C_2 the question about μ seems very difficult and still unknown.

Acknowledgement. The second-named author acknowledges the support from Grant-in-Aid for Scientific Research No.17540044. He thanks the organizers of the International Symposium "Algebraic Geometry in East Asia. II" held at Hanoi, October 10-14, 2005 for the kind invitation and hospitality where this work has been finished. We would like to thank the referee for the very constructive suggestions.

References

[1] G. Cornell and J. Silverman (ed.), Arithmetic Geometry, Springer-Verlag, 1986.

[2] J. Conway and N. Sloane, Sphere Packings, Lattices and Groups. 2nd ed., Springer-Verlag, 1993.

[3] N. Elkies, Mordell-Weil lattices in characteristic 2, I: Construction and first properties, Int. Math. Res. Notices, **8** (1994), 353–361. .

[4] A. Grothendieck, Sur une note de Mattuck-Tate, J. Reine Angew. Math., **200** (1958), 208–215.

[5] R. Hartshorne, Algebraic Geometry, Graduate Texts in Math., **52**, Springer-Verlag, 1977.

[6] S. Lang, Abelian Varities, Interscience, New York, 1959.

[7] S. Mori, The endomorphism rings of some abelian varities, I, Japan J. Math., **2** (1976), 109–130; II, Japan J. Math., **3** (1977), 105–109.

[8] D. Mumford, Abelian Varieties, Oxford Univ. Press, 1970.

[9] J.-P. Serre, Sur le nombre de points rationnels d'une courbe algébrique sur un corps fini, C. R. Acad. Sci. Paris Sér. I Math., **296** (1983), 397–402.

[10] T. Shioda, On the Mordell-Weil lattices, Comment. Math. Univ. St. Paul., **39** (1990), 211–240.

[11] T. Shioda, Mordell-Weil lattices and sphere packings, Amer. J. Math., **113** (1991), 931–948.

[12] T. Shioda, Correspondence of elliptic curves and Mordell-Weil lattices of certain elliptic $K3$ surfaces, In: Algebraic Cycles and Motives, Cambridge Univ. Press, to appear.

[13] A. Weil, Sur les Courbes Algébriques et les Variétés qui s'en déduisent, Hermann, Paris, 1948.

[14] A. Weil, Variétés Abéliennes et Courbes Algébriques, Hermann, Paris, 1948, the combined edition, 1971.

[15] A. Weil, Abstract versus classical algebraic geometry, Proc. Intern. Math. Congr. Amsterdam, 1954, III, 550–558; Collected Papers, II, Springer-Verlag, 1980, 180–188.

[16] A. Weil, Zum Beweis des Torellischen Satzes, Gött. Nachr., 1957, 33–53; Collected Papers, II, 1980, 307–327.

Viet Nguyen-Khac
Institute of Mathematics
18 Hoang Quoc Viet
10307 Hanoi VIETNAM

Current address
Vietnam Information Security Lab.
17 Hoang Sam str., Nghia Do - Cau Giay
Hanoi VIETNAM
E-mail address: nkviet@math.ac.vn

Tetsuji Shioda
Department of Mathematics
Rikkyo University
3-34-1 Nishi-Ikebukuro Toshima-ku
Tokyo 171-8501 JAPAN
E-mail address: shioda@rkmath.rikkyo.ac.jp

Advanced Studies in Pure Mathematics

A SERIES OF UP-TO-DATE GUIDES OF LASTING INTEREST TO ADVANCED MATHEMATICS

Volume 1 Algebraic Varieties and Analytic Varieties.
Edited by S. Iitaka. February, 1983

Volume 2 Galois Groups and their Representations..
Edited by Y. Ihara.. December, 1983.

Volume 3 Geometry of Geodesics and Related Topics.
Edited by K. Shiohama. June, 1984

Volume 4 Group Representations and Systems of Differential Equations.
Edited by K. Okamoto. March, 1985

Volume 5 Foliations.
Edited by I. Tamura. February, 1986

Volume 6 Algebraic Groups and Related Topics.
Edited by R. Hotta. March, 1985

Volume 7 Automorphic Forms and Number Theory.
Edited by I. Satake. February, 1986

Volume 8 Complex Analytic Singularities.
Edited by T. Suwa and P. Wagreich. February, 1987

Volume 9 Homotopy Theory and Related Topics.
Edited by H. Toda. February, 1987

Volume 10 Algebraic Geometry, Sendai, 1985.
Edited by T. Oda. July, 1987

Volume 11 Commutative Algebra and Combinatorics.
Edited by M. Nagata and H. Matsumura.
October, 1987

Volume 12 Galois Representations and Arithmetic Algebraic Geometry.
Edited by Y. Ihara. November, 1987

Volume 13 Investigations in Number Theory.
Edited by T. Kubota. March, 1988

Volume 14 Representations of Lie Groups, Kyoto, Hiroshima, 1986.
Edited by K. Okamoto and T. Oshima. February, 1989

Volume 15 Automorphic Forms and Geometry of Arithmetic Varieties.
Edited by K. Hashimoto and Y. Namikawa. July, 1989

Volume 16 Conformal Field Theory and Solvable Lattice Models.
Edited by M. Jimbo, T. Miwa and A. Tsuchiya.
June, 1988

Volume 17 Algebraic Number Theory—in honor of K. Iwasawa.
Edited by J. Coates, R. Greenberg, B. Mazur and I. Satake. August, 1989

Volume 18 I—Recent Topics in Differential and Analytic Geometry.
Edited by T. Ochiai. November, 1990
II—Kähler Metric and Moduli Spaces.
Edited by Edited by T. Ochiai. November, 1990

Volume 19 Integrable Systems in Quantum Field Theory and Statistical Mechanics.
Edited by M. Jimbo, T. Miwa and A. Tsuchiya.
November, 1989

Volume 20 Aspects of Low Dimensional Manifolds.
Edited by Y. Matsumoto and S. Morita.
December, 1992

Volume 21 Zeta Functions in Geometry.
Edited by N. Kurokawa and T. Sunada.
December, 1992

Volume 22 Progress in Differential Geometry.
Edited by K. Shiohama. September, 1993

Volume 23 Spectral and Scattering Theory and Applications.
Edited by K. Yajima. February, 1994

Volume 24 Progress in Algebraic Combinatorics.
Edited by E. Bannai and A. Munemasa. May, 1996

Volume 25 CR-Geometry and Overdetermined Systems.
Edited by T. Akahori, G. Komatsu, K. Miyajima, M. Namba and K. Yamaguchi. July, 1997

Volume 26 Analysis on Homogeneous Spaces and Representation Theory of Lie Groups, Okayama-Kyoto.
Edited by M. Kashiwara, T. Kobayashi, T. Matsuki, K. Nishiyama and T. Oshima. April, 2000

Volume 27 Arrangements – Tokyo 1998.
Edited by M. Falk and H. Terao. August, 2000

Volume 28 Combinatorial Methods in Representation Theory.
Edited by M. Kashiwara, K. Koike, S. Okada, I. Terada, and H.-F. Yamada. December, 2000

Volume 29	Singularities – Sapporo 1998. Edited by J.-P. Brasselet and T. Suwa. December, 2000
Volume 30	Class Field Theory – Its Centenary and Prospect. Edited by K. Miyake. April, 2001
Volume 31	Taniguchi Conference on Mathematics Nara '98. Edited by M. Maruyama and T. Sunada. May, 2001
Volume 32	Groups and Combinatorics — in memory of Michio Suzuki. Edited by E. Bannai, H. Suzuki, H. Yamaki and T. Yoshida. October, 2001
Volume 33	Computational Commutative Algebra and Combinatorics. Edited by T. Hibi. February, 2002
Volume 34	Minimal Surfaces, Geometric Analysis and Symplectic Geometry. Edited by K. Fukaya, S. Nishikawa and J. Spruck. June, 2002
Volume 35	Higher Dimensional Birational Geometry. Edited by S. Mori and Y. Miyaoka. July, 2002
Volume 36	Algebraic Geometry 2000, Azumino. Edited by S. Usui, M. Green, L. Illusie, K. Kato, E. Looijenga, S. Mukai and S. Saito. November, 2002
Volume 37	Lie Groups, Geometric Structures and Differential Equations — One Hundred Years after Sophus Lie —. Edited by T. Morimoto, H. Sato and K. Yamaguchi. December, 2002
Volume 38	Operator Algebras and Applications. Edited by H. Kosaki. January, 2004
Volume 39	Stochastic Analysis on Large Scale Interacting Systems. Edited by T. Funaki and H. Osada. March, 2004
Volume 40	Representation Theory of Algebraic Groups and Quantum Groups. Edited by T. Shoji, M. Kashiwara, N. Kawanaka, G. Lusztig and K. Shinoda. March, 2004
Volume 41	Stochastic Analysis and Related Topics in Kyoto *In honour of Kiyosi Itô.* Edited by H. Kunita, S. Watanabe and Y. Takahashi. March, 2004

Volume 42 Complex Analysis in Several Variables
– Memorial Conference of Kiyoshi Oka's Centennial Birthday, Kyoto/Nara 2001.
Edited by K. Miyajima, M. Furushima, H. Kazama, A. Kodama, J. Noguchi, T. Ohsawa, H. Tsuji and T. Ueda. June, 2004

Volume 43 Singularity Theory and Its Applications.
Edited by S. Izumiya, G. Ishikawa, H. Tokunaga, I. Shimada and T. Sano. January, 2007

Volume 44 Potential Theory in Matsue.
Edited by H. Aikawa, T. Kumagai, Y. Mizuta and N. Suzuki. August, 2006

Volume 45 Moduli Spaces and Arithmetic Geometry (Kyoto, 2004).
Edited by S. Mukai, Y. Miyaoka, S. Mori, A. Moriwaki and I. Nakamura. January, 2007

Volume 46 Singularities in Geometry and Topology 2004.
Edited by J.-P. Brassele and T. Suwa. February, 2007

Volume 47 Asymptotic Analysis and Singularities.
1—Hyperbolic and dispersive PDEs and fluid mechanics.
2—Elliptic and parabolic PDEs and related problems.
Edited by H. Kozono, T. Ogawa, K. Tanaka, Y. Tsutsumi and E. Yanagida. October, 2007

Volume 48 Finsler Geometry, Sapporo 2005
— In Memory of Makoto Matsumoto.
Edited by S. V. Sabau and H. Shimada.
November, 2007

Volume 49 Probability and Number Theory — Kanazawa 2005.
Edited by S. Akiyama, K. Matsumoto, L. Murata and H. Sugita. December, 2007

Volume 50 Algebraic Geometry in East Asia — Hanoi 2005.
Edited by K. Konno and V. Nguyen-Khac.
February, 2008

TO BE CONTINUED